COMBINATORICS '86

ANNALS OF DISCRETE MATHEMATICS 37

General Editor: Peter L. HAMMER
Rutgers University, New Brunswick, NJ, U.S.A.

Advisory Editors:
C. BERGE, Université de Paris, France
M. A. HARRISON, University of California, Berkeley, CA, U.S.A.
V. KLEE, University of Washington, Seattle, WA, U.S.A.
J.-H. VAN LINT, California Institute of Technology, Pasadena, CA, U.S.A.
G.-C. ROTA, Massachusetts Institute of Technology, Cambridge, MA, U.S.A.

NORTH-HOLLAND – AMSTERDAM • NEW YORK • OXFORD • TOKYO

COMBINATORICS '86

Proceedings of the International Conference on Incidence Geometries and Combinatorial Structures Passo della Mendola, Trento, Italy, 30 June – 5 July, 1986

Edited by

A. BARLOTTI
Università di Firenze, Firenze, Italy

M. MARCHI
Università Cattolica del S. Cuore, Brescia, Italy

G. TALLINI
Università 'La Sapienza', Roma, Italy

1988

NORTH-HOLLAND – AMSTERDAM • NEW YORK • OXFORD • TOKYO

ISBN: 0 444 70369 1

Publishers:

ELSEVIER SCIENCE PUBLISHERS B.V.
P.O. BOX 1991
1000 BZ AMSTERDAM
THE NETHERLANDS

Sole distributors for the U.S.A. and Canada:

ELSEVIER SCIENCE PUBLISHING COMPANY, INC.
52 VANDERBILT AVENUE
NEW YORK, N.Y. 10017
U.S.A.

LIBRARY OF CONGRESS
Library of Congress Cataloging-in-Publication Data

International Conference on Incidence Geometries and Combinatorial Structures (1986 : Trento, Italy)
Combinatorics '86 : proceedings of the International Conference on Incidence Geometries and Combinatorial Structures, held at Passo della Mendola, Trento, Italy, 30 June-5 July, 1986 / edited by A. Barlotti, M. Marchi, G. Tallini.
p. cm. -- (Annals of discrete mathematics ; 37)
Bibliography: p.
ISBN 0-444-70369-1 (U.S.)
1. Combinatorial geometry--Congresses. I. Barlotti, A. (Adriano), 1923- . II. Marchi, M. III. Tallini, G. (Giuseppe), 1930- . IV. Title. V. Series.
QA167.I59 1986
516'.12--dc19 87-32079
CIP

PRINTED IN THE NETHERLANDS

FOREWORD

This volume contains the proceedings of the Conference "COMBINATORICS '86" held at the Centro di cultura of the Università Cattolica of Milano, in Passo della Mendola (Trento - Italy) (30 June - 5 July 1986).

The Conference is the latest in a series of international meetings on the same subject held in Italy in the recent past: Roma, June 1981 (Conference in honour of B. Segre); Passo della Mendola, July 1982; Roma, May 1983; Bari, September 1984. The Proceedings of the first and the last of these meetings are published in this Series "Annals of Discrete Mathematics" by North Holland Publ. Co.

The participants in Combinatorics '86 numbered about 150 and represented more than a dozen nationalities.

The subjects covered by this volume concern recent developments in combinatorial and incidence geometry in all its different aspects, and its links with foundations of geometry, graph theory, algebraic structures, and applications to coding theory and computer sciences.

We are indebted to the Università Cattolica for the organization of the Conference and to the Italian Consiglio Nazionale delle Ricerche for financial support.

We are also profoundly grateful to the referees for their assistance.

A. BARLOTTI
M. MARCHI
G. TALLINI

OPENING WELCOME

Giuseppe Tallini

First of all I wish to heartily welcome the many participants, both from Italy and abroad, in "Combinatorics '86". This conference is another step in an already long series. I like to recall here the Rome conference in September '73, organized by the Accademia dei Lincei, again in Rome that in memory of B. Segre, in June 1981, the conferences held in July '82 at Passo della Mendola, in May '83 in Rome at the Istituto Nazionale di Alta Matematica "F. Severi", in September '84 in Bari. I hope that our present conference will be scientifically fruitful and at the same time enjoyable and satisfying. I also hope that this series will continue, say with a conference in Ravello (Naples) in 1988. Nowadays, these meetings and specialized seminars have the effective role of spreading ideas and promoting research, not only because of the information they present to the listeners during the formal talks, but mainly because of the opportunities they offer of discussions among the participants.

The vitality of combinatorics and of the Foundations of Geometry is asserted on the one hand by the many specialized journals and a vast literature, and on the other hand by the numerous conferences which are recurrently held all over the world. This vitality resides, in my opinion, in the fact that today combinatorics has moved to the forefront of applied mathematics because of its important interactions with computer science and statistics. Moreover, the theoretical interest of our discipline lies in the fascination of the discrete with which it deals. The thrill of combinatorial problems stems from the fact that the configurations to be studied in relation to a given problem, though finitely many, actually come in so huge numbers that they behave as infinitely many both for humans and computers. That is, we are daily confronted by a finiteness which for us is really infinite. For example, given a set of ten objects, if we wanted to decide if there exists a sharply 2-transitive set of permutations (which already accounts for 90 elements) and wished to employ our computer, then we would have to examine $\binom{10!}{90}$ cases and verify the required property for each of them. Such a check is impossible within the lifetime of mankind, even with the help of the most sophisticated computer. By the way, the number we were talking about considerably exceeds the number of atoms in the universe! For this reason, in order to solve combinatorial problems the geometric intuition of the researcher must intervene in a critical way, that is there is a quiddity man has while a computer does not (up to now).

The problems which arise are of a varied nature and the suitable techniques to deal with them are to be devised for each situation. Indeed, one of the special features of combinatorics is the often sporadic nature of solutions which comes from its links with number theory.

The branches of combinatorics are many and various. All of them are amply and appropriately represented in this conference whose scientific output, I hope, will reflect the fervour of the intense studies on these subjects and provide a new impulse to future developments.

CONTENTS

PARTICIPANTS

ABATANGELO L. Maria	Bari - Italy
ABATANGELO Vito	Bari - Italy
ANTONUCCI Salvatore	Napoli - Italy
ARTMANN Benno	Darmstadt - Germany
ARTZY Rafael	Haifa - Israel
BADER Laura	Roma - Italy
BAKER Catharine Anne	Sackville - Canada
BARLOTTI Adriano	Firenze - Italy
BASILE Alessandro	Perugia - Italy
BATTEN Lynn Margaret	Winnipeg, Manitoba - Canada
BÉNÉTEAU Lucien	Toulouse - France
BENZ Walter	Hamburg - Germany
BERARDI Luigia	L'Aquila - Italy
BERNARDI Marco Paolo	Pavia - Italy
BERTANI Laura	Parma - Italy
BETTEN Dieter	Kiel - Germany
BEUTELSPACHER Albrecht	München - Germany
BICHARA Alessandro	Roma - Italy
BILIOTTI Mauro	Lecce - Italy
BIONDI Paola	Napoli - Italy
BISCARINI Paola	Perugia - Italy
BONETTI Flavio	Ferrara - Italy
BONISOLI Arrigo	Modena - Italy
BONNEAU	Le Chesnay - France
BORZACCHINI Luigi	Bari - Italy
BRINI Andrea	Bologna - Italy
BROUWER Andries Evert	Amsterdam - Holland
BROWN Julia M. Nowlin	Downsview- Ontario - Canada
BRYLAWSKI Tom	Chapel Hill - N. Carolina - U.S.A.
CAGGEGI Andrea	Napoli - Italy
CAPODAGLIO Rita	Bologna - Italy
CAPURSI Mauro	Bari - Italy
CECCHERINI Pier Vittorio	Roma - Italy
CHEROWITZO William	Denver - Colorado - U.S.A.
CIVOLANI Nino	Potenza - Italy
CLAY James R.	Tucson - Arizona - U.S.A.
CORSINI Piergiulio	Udine - Italy
D'ANTONA Ottavio	Pavia - Italy
DE FINIS Massimo	Roma - Italy
de RESMINI Marialuisa J.	Roma - Italy
DE SALVO Mario	Messina - Italy
DE SOETE Marjke	Gent - Belgium
DE VITO Paola	Napoli - Italy
DI MARTINO Lino	Milano - Italy

DICUONZO Vincenzo	Roma - Italy
DODUNEKOV Stefan Manev	Sofia - Bulgaria
DUBIKAJTIS L.	Arcavacata di Rende - Cosenza - Italy
EBERT Gary L.	Newark - Delaware - U.S.A.
EUGENI Franco	L'Aquila - Italy
EVANS David Mark	Tübingen - Germany
FAINA Giorgio	Perugia - Italy
FERRERO Giovanni	Parma - Italy
FIORI Carla	Modena - Italy
FIORINI Stanley	Msida - Malta
FISHER J. Chris	Regina - Canada
FRANK Rolfdieter	Hamburg - Germany
FUNK Martin	Potenza - Italy
GARNER Cyril	Ottawa - Canada
GERBER P. Dean	
GEVAERT H.	Gent - Belgium
GHINELLI Dina	Roma - Italy
GIONFRIDDO Mario	Catania - Italy
GRUNDHÖFER Theo	Tübingen - Germany
HANSSENS Guy	Gent - Belgium
HARTMANN Peter	München - Germany
HEISE Werner	München - Germany
HERZER Armin	Mainz - Germany
HIRSCHFELD James	Brighton - Great Britain
IDEN Oddvar	Bergen - Norway
INNAMORATI Stefano	L'Aquila - Italy
JHA Vikram	Glasgow - Great Britain
JUNKERS Wilhelm	Dusseldorf - Germany
KALHOFF Franz Bernhard	Dortmund - Germany
KARZEL Helmut	München - Germany
KAYA Rustem	Eskisehir - Turkey
KEEDWELL Anthony Donald	Guildford - Great Britain
KEPPENS Dirk F.J.	Gent - Belgium
KERBY William	Hamburg - Germany
KEY Jennifer D.	Birmingham - Great Britain
KNARR Norbert	Kiel - Germany
KNÜPPEL Frieder	Kiel - Germany
KORCHMAROS Gabor	Potenza - Italy
KROLL Hans-Joachim	München - Germany
LANCELLOTTI Paola	Modena - Italy
LARATO Bambina	Bari - Italy
LAURI J.	Msida - Malta
LESTER June	Los Angeles - California - U.S.A.
LIZZIO Angelo	Catania - Italy
LO RE Pia Maria	Napoli - Italy
LORIMER J.W. Michael	Toronto - Canada
LÖWE Stefan	Braunschweig - Germany
LÖWEN Rainer	Tübingen - Germany
LUNARDON Guglielmo	Napoli - Italy
LÜNEBURG H.	Kaiserslautern - Germany
MANEV Nickolai Lazarov	Sofia - Bulgaria
MARCHI Mario	Brescia - Italy

MARINO Maria Corinna	Messina - Italy
MÄURER Helmut	Darmstadt - Germany
MAYER Erika	L'Aquila - Italy
MELONE Nicola	Napoli - Italy
MENICHETTI Giampaolo	Bologna - Italy
MEYER Rita	Hannover - Germany
MIGLIORATO Renato	Messina - Italy
MIGLIORI Grazia	Roma - Italy
MISFELD Jürgen	Hannover - Germany
NAGAMUNY Reddy	Tirupati - India
NOLTE Wolfgang	Darmstadt - Germany
OLANDA Domenico	Napoli - Italy
OTT Udo	Braunschweig - Germany
PASINI Antonio	Siena - Italy
PAYNE Stanley E.	Denver - Colorado - U.S.A.
PELLEGRINI Silvia	Brescia - Italy
PELLEGRINO Consolato	Modena - Italy
PERELLI CIPPO Claudio	Brescia - Italy
PERTICHINO Michele	Bari - Italy
PETIT Jean-Claude	Limoges - France
PFALZGRAF Jochem	Saarbrücken - Germany
PFLUGFELDER Hala	Philadelphia - U.S.A.
PIANTA Silvia	Brescia - Italy
PIEPER-SEIER Irene	Oldenburg - Germany
PIRILLO Giuseppe	Firenze - Italy
PLAUMANN Peter	Erlangen - Germany
POTT Alexander	Giessen - Germany
PRIESS-CRAMPE Sibylla	München - Germany
PROCESI CIAMPI Rita	Roma - Italy
PUCCIO Luigia	Messina - Italy
QUATTROCCHI Gaetano	Catania - Italy
QUATTROCCHI Pasquale	Modena - Italy
RAGUSO Grazia	Bari - Italy
RAJOLA Sandro	Roma - Italy
RAO Salvatore	Napoli - Italy
RELLA Luigia	Bari - Italy
REMUS Dieter	Hagen - Germany
ROSA Alexander	Hamilton - Ontario - Canada
ROSATI Luigi Antonio	Firenze - Italy
ROTA Rosaria	Roma - Italy
ROZERA Guglielmo	Roma - Italy
RUOFF Dieter	Regina - Canada
SAELI Donato	Potenza - Italy
SASSO-SANT Maic	Saarlouis - Germany
SCAFATI TALLINI Maria	Roma - Italy
SCAPELLATO Raffaele	Parma - Italy
SCHULZ Ralph H.	Berlin - Germany
SEIDEL Hans-Peter	Tübingen - Germany
SENATO Domenico	Napoli - Italy
SIEMON Helmut	Reichenberg - Germany
SIEMONS Johannes	Norwich - Great Britain
SIMONIS Juriaan	Delft - Holland

SPANICCIATI Renata	Roma - Italy
SPARTALIS Stefanos	Xanthi - Greece
STANGARONE Rosa	Bari - Italy
STEINKE Gunter	Kiel - Germany
STRAMBACH Karl	Erlangen - Germany
SYCHOWICZ Andrzej	Roma - Italy
SZAMBIEN Horst	Hannover - Germany
SZCZERBA Leslaw	Warsaw - Poland
TALLINI Giuseppe	Roma - Italy
THOMSEN Momme Johs	Hamburg - Germany
TONCHEV Vladimir D.	Sofia - Bulgaria
VAN MALDEGHEM Hendrik J.J.	Gent - Belgium
VEDDER Klaus	Giessen - Germany
VENEZIA Antonietta	Roma - Italy
VOUGIOUKLIS Thomas	Xanthi - Greece
WAGNER Ascher	Birmingham - Great Britain
WOLFF Karl Erich	Darmstadt - Germany
ZAGAGLIA Norma	Milano - Italy
ZAKS Joseph	Haifa - Israel
ZANELLA Corrado	Roma - Italy
ZEITLER Herbert	Bayreuth - Germany
ZIESCHANG Paul Hermann	Kiel - Germany
ZIZIOLI Elena	Brescia - Italy

Annals of Discrete Mathematics 37 (1988) 1–6

NET OF RATIONALITY IN A MINKOWSKI PLANE

Rafael ARTZY

Department of Mathematics
University of Haifa, 31999 Haifa, Israel

ABSTRACT

In one of the derived affine planes of a Minkowski plane π over an infinite KT-nearfield, an infinite net is constructed by means of joining points, intersecting lines, and drawing parallels to lines through points, and so on ad infinitum, starting out from two given nonparallel points and the generators through them. It is shown that the result is a plane (a "Möbius net") over a prime field. The construction is based on proving that the validity of certain special Desargues conditions in 4-webs is implied by the Rectangle Axiom in π. Algebraically this confirms the theorem of Kerby and Wefelscheid which states that every KT-nearfield has a prime subfield.

1. THE MINKOWSKI PLANE

DEFINITION. A *Minkowski plane* consists of a set Π of points and of three disjoint subsets of the power set of Π: the set of cycles, the set of (+)generators, and the set of (-)generators. Two points are called (+)parallel (notation $\|_+$) if some (+)generator contains both, (-)parallel (notation $\|_-$) if some (-)generator contains both. Parallel ($\|$) will mean "$\|_+$ or $\|_-$". Nonparallel means "neither $\|_+$ nor $\|_-$". Points lying on the same cycle are called concyclic. Points will be denoted by capital letters, cycles by lower case letters.

We postulate

I. For every point P there is a unique (+)generator and a unique (-)generator containing P.

II. Every generator intersects every cycle in exactly one point. Every (+)generator intersects every (-)generator in exactly one point.

III. For every three nonparallel points A, B, C there is a unique cycle containing them, denoted ABC.

IV. For each cycle c, each point P on c, and each Q in $\Pi \backslash c$, nonparallel to P, there is a unique cycle k containing P and Q such that $c \cap k = \{P\}$.

V. There is a cycle c with $|c| \geq 3$, and $\Pi \backslash c \neq \emptyset$.

DEFINITION. The *derived plane* with respect to the point P is the affine plane obtained from a Minkowski plane as follows: its points are $\Pi \backslash \{P\}$, and its lines are all generators of the Minkowski plane which do not contain P and all those cycles that contain P, with the provision that P be removed from all these cycles.

W. Benz's *Rectangle Axiom* Γ [cf.2] reads:

Axiom Γ. In a Minkowski plane, let each of the point quadruples P_i, Q_i, R_i (i=1,2,3,4) be concyclic, S_i another point quadruple, $P_i \parallel_+ Q_i \parallel_- R_i \parallel_+ S_i \parallel_- P_i$. Then the S_i are concyclic.

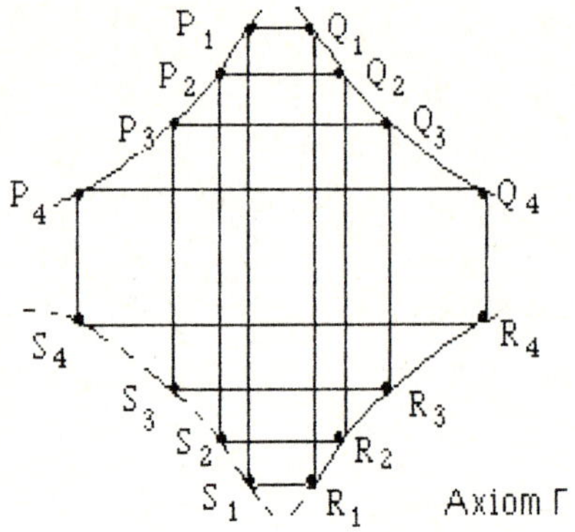

Axiom Γ

DEFINITION. A *KT-nearfield* is a planar nearfield in whose multiplicative structure there exists an involutory automorphism σ identically satisfying the functional equation $(x\sigma+1)\sigma + (x+1)\sigma = 1$.

1.1. RESULT (for instance, [4]). A Minkowski plane π can be coordinatized by a KT-nearfield if and only if Γ holds in π.

1.2. PROPOSITION (Fig.1). In a Minkowski plane with Γ, let $p := P_1P_2P_3$, $q := Q_1Q_2Q_3$, $r := R_1R_2R_3$, $s := S_1S_2S_3$, $A \in p \cap q$, $C \in r$, $A \parallel_- C$, and for i=1,2,3, $P_i \parallel_+ Q_i \parallel_- R_i \parallel_+ S_i \parallel_- P_i$. Then $C \in s$.

Proof. Suppose $C \notin s$. Let S be a point on s such that $S \neq C$, $S \parallel_+ C$, $P \in p$, $Q \in q$, $S \parallel_- P \parallel_+ Q$. In view of Γ, we have $Q \parallel_- C$. But $A \parallel_- C$ and $A \in q$, hence $A = Q$. Since $A \in p$, this implies $A = P$, and thus also $S = C$ and therefore $C \in s$.

1.3. PROPOSITION (Converse of 1.2). In Fig.1, let, for i = 1, 2, 3, $P_i \parallel_+ Q_i \parallel_- R_i \parallel_+ S_i \parallel_+ P_i$, $A \in p \cap q$, $C \in r \cap s$, $p = P_1P_2P_3$, $q = Q_1Q_2Q_3$, $r = R_1R_2R_3$, $s = S_1S_2C$. Then S_3 lies on s.

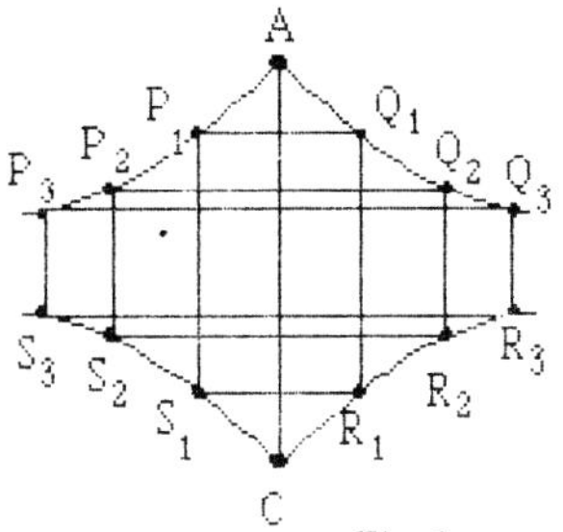

Fig.1

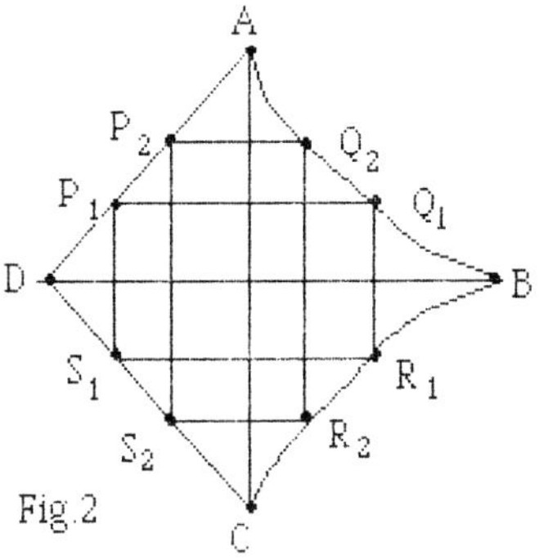

Fig.2

Proof. Suppose $S_3 \notin s$. Let $s' := S_1S_2S_3$. Then, by 1.1, $C \in s'$, and $s = s'$.

1.4. PROPOSITION (Fig.2). In a Minkowski plane with Γ, let $p = P_1P_2A$, $r = R_1R_2C$, $s = S_1S_2C$, $B \in q \cap r$, $B \parallel_+ D$, $D \in p$, $P_i \parallel_+ Q_i \parallel_- R_i \parallel_+ S_i \parallel_- P_i$, for $i=1,2$. Then D lies on s.

Proof. Suppose $D \notin s$. Then, by 1.3, there are S on s with $S \neq D$, $R \in r$, $Q \in q$, such that $D \parallel_- S \parallel_+ R \parallel_- Q \parallel_+ D$ and $B \in q$. Hence $B = Q$. Since $B \in r$, this implies $B = R$ and thus also $S = D$, $D \in s$.

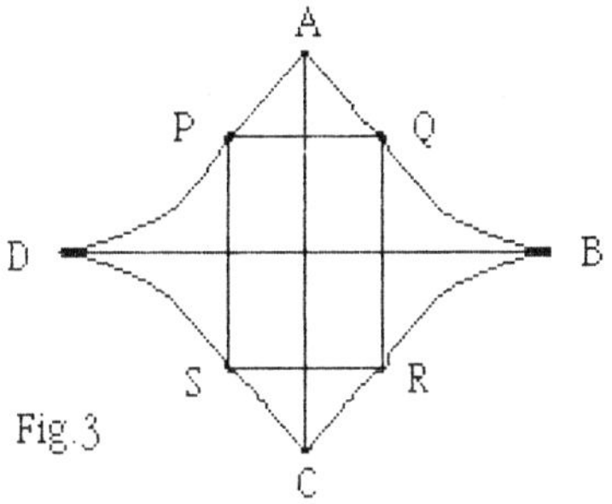

Fig.3

1.5. PROPOSITION (Fig.3). In a Minkowski plane with Γ, let $p = DPA$, $q = AQB$, $r = BRC$, $s = CSD$, $A \parallel_- C$. Then $s \cap p = \{D\}$ if and only if $q \cap r = \{B\}$, and $B \parallel_+ D$.

Proof. Let $s \cap p = \{D\}$. Suppose $r \cap q = \{B, B'\}$, B nonparallel to B'. Then there is a point S' on s such that $S' \parallel_+ B'$, a point P' on p such that $P' \parallel_- S'$, a point Q' on q such that $Q' \parallel_+ P'$, and by 1.4, $Q' \parallel_- B'$. But both Q' and B' lie on q, thus $B' = Q'$. This implies that P' and S' must coincide and lie on both p and s, that is, $P' = S' = p \cap s = D$, and $B' \parallel_+ D$. Hence $B = B'$. The converse is proved analogously.

1.6. PROPOSITION (Fig.4). With the same hypotheses as in 1.5, let $q \cap r = \{B\}$ and $B \in p$. Then $p \cap s = \{B\}$.

Proof. Suppose $B \notin p \cap s$. Then, by 1.4, $|p \cap s| = 1$, and $B \parallel_+ D$, $p \cap s = \{D\}$. But $B \in p$, hence $B = D$.

1.7. COROLLARY (Fig.4). In the derived affine plane with respect to B, the following special Desargues theorem Δ2 holds true: Let $P \parallel_+ Q \parallel_- R \parallel_+ S \parallel_- P$, $A \parallel_- C$, and the straight lines $CR \parallel AQ$ (usual line parallelism !). Then $PA \parallel CS$.

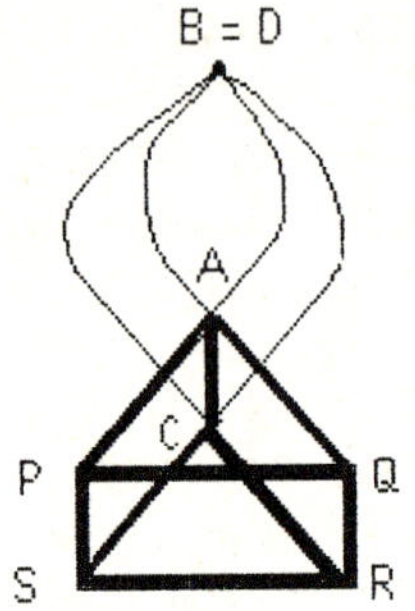

Fig.4

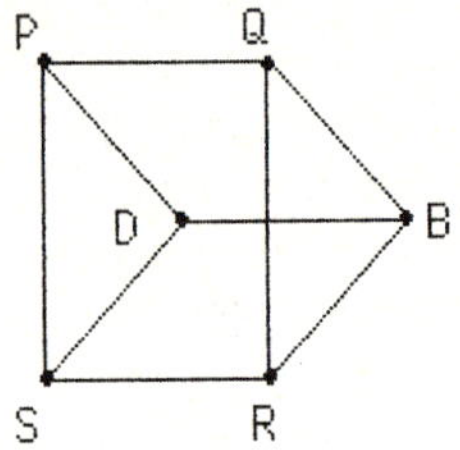

Fig.5

Proof. In the derived plane, all cycles through B become straight lines. Straight lines are parallel if, as cycles, their single point of intersection is B.

1.8. COROLLARY (Fig.5). The special Desargues theorem Δ1, obtained from Δ2 by interchanging $\parallel_+$ and $\parallel_-$, holds true: Let $P \parallel_+ Q \parallel_- R \parallel_+ S \parallel_- P$, $D \parallel_+ B$, and the lines $PD \parallel QB$. Then $SD \parallel RB$.

We summarize:
1.9. THEOREM. The special Desargues theorems Δ1 (Fig.4) and Δ2 (Fig.5) are valid in every derived affine plane of a Minkowski plane with axiom Γ.

2. THE NET

In an affine plane, let three noncollinear points be given. Starting from them, we join existing points, intersect existing lines, and draw parallels through existing points to existing lines. We continue these operations ad infinitum, without assuming the validity of any incidence theorems. The resulting plane is called an affine *minimal free extension plane*. The plane is called "minimal" because no such plane can be obtained if the number of starting points is less than 3.

If, however, we assume the validity of the Desargues theorem (and no other incidence condition) from the start, the resulting plane is called an affine *net of rationality* or an affine *Möbius net*. It is well known that a Möbius net is coordinatized by a prime field, that is, in the case of the absence of any finiteness conditions, the field $\mathbb{Q}$ of rational numbers. In other words, if the three starting points have, for instance, the coordinates (0,0), (1,0), (0,1), then all the points of the Möbius net have coordinates of the form (r,s) with $r, s \in \mathbb{Q}$. On the other hand, every point with such rational coordinates indeed exists in the Möbius net.

However, the affine Möbius net can already be obtained from the affine minimal free extension plane by imposing on it incidence theorems which are weaker than the full Desargues theorem. An analogous situation exists also in projective planes, where the projective Möbius net can be based on the threefold degeneration of the Desargues theorem ([5], p.275). In the following, we will describe such weaker conditions on which to base the construction of the affine Möbius net.

A *parallel n-web* is a partial affine plane in which each line is parallel to exactly one of n given mutually nonparallel lines. Thus, in a parallel n-web, there is not always a line connecting two given points, however Euclid's parallel axiom holds in every parallel n-web. The n line families will be numbered: we will speak of the 1-lines, 2-lines,..., n-lines of the n-web.

In one derived affine plane of a Minkowski plane π, consider the minimal affine free extension produced by three distinct points L, M, and N such that $L \parallel_+ M \parallel_- N$. Let the (+)generators be the 1-lines of a parallel web and the (-)generators its 2-lines. All the parallels to the line LN will be called 3-lines. We complete this parallel 3-web to a parallel 4-web W_1 by adding one further family of parallels to it. We repeat this procedure with every other existing fourth family of parallels and obtain thus 4-webs W_2, W_3, .. The union of the 4-webs W_1, W_2, . . . is a partial affine plane ψ.

2.1. RESULT [1]. A minimal affine free extension is constructed, starting out from the points L, M, and N. A 3-web is chosen so that LM is a 1-line, MN a 2-line, and LN a 3-line. The 3-web is completed to a 4-web by adding one further family of parallels to it. This is repeated for every existing family of parallels distinct from the families 1, 2, 3. If the special Desargues conditions $\Delta 1$ and $\Delta 2$ are made to hold in each of these 4-webs, then the result is a Möbius net.

2.2. THEOREM. If the Axiom Γ holds in the infinite Minkowski plane π, then the partial plane ψ is an affine Möbius net, and hence an affine plane over a prime field.

Proof: follows immediately from 1.9 and 2.1.

For the points of ψ we introduce coordinates (x,y), with rational x and y. Through every noncollinear triple of points which are mutually nonparallel in π, we draw the hyperbola with asymptotes x = const and y = const. These asymptotes then represent generators in π. The result is the classical hyperbola model of a Minkowski plane [2]. We could be tempted to call this plane a Minkowski Möbius net. However, it has a flaw: we cannot guarantee the existence of points of intersection for hyperbolas with lines or hyperbolas with hyperbolas. Intersections of hyperbolas with generators have rational coordinates, though, and we do have here a Minkowski plane over the rational field $\mathbb{Q}$. We have thus proved the following theorem, a special case of a well known result [3].

2.3. THEOREM. Every infinite KT-nearfield has a prime subfield F, such that $x\sigma = x^{-1}$ for every $x \neq 0$ in F.

REFERENCES

[1] Artzy, R., 4-webs and the Möbius net. Riveon Lematematika 7 (1954), 1-7.
[2] Benz, W., Vorlesungen über Geometrie der Algebren (Springer Verlag, Berlin-Heidelberg-New York, 1973).
[3] Kerby, W. and Wefelscheid, H., Über eine scharf 3-fach transitiven Gruppen zugeordnete algebraische Struktur. Abh. Math. Seminar Hamburg 37 (1972), 225-235.
[4] Percsy, N., A remark on the introduction of coordinates in Minkowski planes. J. of Geometry 12 (1979), 175-183.
[5] Pickert, G., Projektive Ebenen (Springer Verlag, Berlin- Heidelberg-New York, 1975).

Annals of Discrete Mathematics 37 (1988) 7–20

A NEW CLASS OF TRANSLATION PLANES

R. D. Baker and G. L. Ebert

Department of Mathematical Sciences
University of Delaware
Newark, Delaware 19716

We define a nest of reguli to be a collection P of reguli in a regular spread S of $PG(3,q)$ such that every line of S is contained in exactly 0 or 2 reguli of P. Let U denote the lines of S contained in the reguli of some nest. If V is a partial spread of $PG(3,q)$ covering the same points as U but having no lines in common with U, then V will be called a replacement set for U. Clearly, $(S-U) \cup V$ is a spread of $PG(3,q)$, yielding a (potentially new) translation plane of order q^2 which is 2-dimensional over its kernel.

Nests of size $(q+3)/2$ were first studied (under another name) by Bruen and later by many others. Whether such $(q+3)/2$-nests exist for $q > 13$ and whether such nests are necessarily reversible are still open questions. In this paper we consider nests of size q. We exhibit an infinite family of q-nests, one for each odd prime q, and show that each nest is reversible. The translation planes so obtained appear to be new, at least for $q \geq 11$.

1. INTRODUCTION

André [2] and Bruck and Bose [4] have shown that there is a one-to-one correspondence between spreads of odd-dimensional projective spaces and translation planes. Spreads of $PG(3,q)$ and the corresponding "two-dimensional" translation planes are those that have been most heavily studied. Since every spread of $PG(3,q)$ may be obtained by replacing a subset of lines in a regular spread, we construct new spreads by finding appropriate replaceable subsets of a regular spread.

In section 2 we define a nest of reguli in a regular spread, generalizing the notion of chains that was earlier introduced by Bruen [5]. In section 3 we show that nests of size q exist for any odd prime q, and in section 4 show that this leads to a replaceable subset of lines in a regular spread of $PG(3,q)$. The corresponding translation planes of order q^2 appear to be new, at least for $q \geq 11$. In section 5 we look at some geometric properties of the spreads obtained, discuss derivation of these spreads, and consider the orbit structure induced by their collineation groups.

2. PRELIMINARY RESULTS

Let $\sum = PG(3,q)$ denote projective 3-space over the finite field $GF(q)$. A *spread* of $\sum$ is any collection of $q^2 + 1$ skew lines, necessarily partitioning the points of $\sum$, and a *regulus* of $\sum$ is any set R of $q + 1$ skew lines such that any line transversal to 3 lines of R is transversal to all lines of R. Any three skew lines of $\sum$ uniquely determine a regulus, and a spread Ω is called *regular* if and only if the regulus determined by any three of its lines is contained in Ω. The translation plane obtained from a regular spread is desarguesian.

We define a *nest* of reguli to be a collection P of reguli in a regular spread Ω of $\sum$ such that every line of Ω is contained in exactly 0 or 2 reguli of P. Letting U denote the lines of Ω contained in the reguli of P, a set V will be called a *replacement set* for U if V is a partial spread of $\sum$ covering the same points as U but having no lines in common with U. Clearly, $(\Omega - U) \cup V$ then becomes a spread of $\sum$ yielding a (potentially new) translation plane of order q^2.

Nests of size $(q+3)/2$ were first studied (under the name "chains") by Bruen [5] and Bruen and Thas [6], and later by many others (see [1], [7], [11], [14], for instance). Whether such $(q+3)/2$-nests exist for $q > 13$ and whether such nests are necessarily reversible are still open questions. In this paper we study nests of size q, show that they always exist for odd primes q, and show that each such nest constructed is reversible. The translation planes so obtained appear to be new, at least for $q \geq 11$.

In [3] Bruck has shown there is a one-to-one correspondence between the lines and reguli of a regular spread Ω and the points and circles of a miquelian inversive plane $M(q)$ of order q. The reader is referred to [9] for the definition and relevant properties of a miquelian plane. Among the several models for $M(q)$, we will use the one given by W. F. Orr in [13]. That is, the points will be represented by elements of $GF(q^2) \cup \{\infty\}$, and the circles by 1-dimensional subspaces over $GF(q)$ of certain

2×2 matrices. More explicitly, a circle will be represented by a matrix of the form $\begin{pmatrix} \alpha & a \\ b & -\alpha^q \end{pmatrix}$, where $\alpha \in GF(q^2)$, $a, b \in GF(q)$, and $\alpha^{q+1} + ab \neq 0$. Two such matrices represent the same circle if and only if one is a $GF(q)$-scalar multiple of the other.

It should be noted that the unique inversion associated with the circle $C = \begin{pmatrix} \alpha & a \\ b & -\alpha^q \end{pmatrix}$ is the semilinear fractional collineation $z \rightarrow \dfrac{\alpha z^q + a}{bz^q - \alpha^q}$ for all $z \in GF(q^2) \cup \{\infty\}$, where the usual conventions concerning ∞ are invoked. By definition this inversion fixes the points of the circle C and moves all other points of $M(q)$ in orbits of size two. Thus the $q + 1$ points of C are simply those points satisfying the equation $z = \dfrac{\alpha z^q + a}{bz^q - \alpha^q}$. We will alternately represent a circle in $M(q)$ by a matrix of the above type or by the set of points incident with it.

One advantage of this model is the ease with which images of circles under various maps can be computed. For instance, let $R = \begin{pmatrix} \alpha & \beta \\ \gamma & \delta \end{pmatrix}$ where $\alpha, \beta, \gamma, \delta \in GF(q^2)$ with $\alpha\delta - \beta\gamma \neq 0$, and let $\phi_R : z \rightarrow \dfrac{\alpha z + \beta}{\gamma z + \delta}$ for all $z \in GF(q^2) \cup \{\infty\}$. It is easy to see that the linear fractional map ϕ_R is a collineation of $M(q)$. Moreover, if C is any matrix representing a circle of $M(q)$ as described above, it is shown in [13] that $\phi_R(C) = RCR'^{*}$, where R' denotes the matrix obtained by raising each entry of R to the qth power and R^* denotes the classical adjoint of R.

A second advantage of this model is the associated collection of computational tools for easily determining the intersection pattern of a given set of circles. For this purpose we restrict to odd q, and let C and D be matrices representing distinct circles of $M(q)$. Define
$h(C,D) = \|C + D\| - \|C\| - \|D\|$, where $\| \ \|$ denotes the determinant function. Then define $C \times D = (\frac{1}{2}h(C,D))^2 - \|C\| \, \|D\|$. The following result is found in [13], and will be stated here without proof.

Lemma (1): Using the above notation, C and D represent circles that are disjoint, tangent, or secant accordingly as $C \times D$ is a nonzero square, zero, or a nonsquare in $GF(q)$. □

We now exploit these computational tools to construct q-nests of circles in $M(q)$ for each odd prime q.

3. THE EXISTENCE QUESTION

For the remainder of this paper we restrict to odd primes (rather than prime powers) q. Using the above-mentioned correspondence between the lines and reguli of a regular spread in $PG(3,q)$ and the points and circles of a miquelian plane $M(q)$, we construct a nest of size q by working in $M(q)$. To accomplish this we distinguish the cases $q \equiv 1 \pmod 4$ and $q \equiv 3 \pmod 4$.

Suppose first that $q \equiv 3 \pmod 4$. Let β denote a primitive element of $GF(q^2)$, and let $\epsilon = \beta^{(q+1)/2}$. Then $\epsilon^{q-1} = -1$ and $w = \epsilon^2 = \beta^{q+1}$ is a primitive element of the subfield $GF(q)$. Let $R = \begin{pmatrix} 1 & \epsilon \\ 0 & 1 \end{pmatrix}$ and let $\theta = \phi_R$ be the linear fractional collineation of $M(q)$ associated with R. Let $D = \begin{pmatrix} \epsilon & w-1 \\ 1 & \epsilon \end{pmatrix}$ represent a circle in $M(q)$. Using the techniques described above, it is easy to compute that

$$\theta^s(D) = \begin{pmatrix} (s+1)\epsilon & (s+1)^2 w - 1 \\ 1 & (s+1)\epsilon \end{pmatrix}$$

for $s = 0, 1, 2, \ldots, q-1$. Clearly the order of the map θ is q, as q is a prime. Letting $D_s = \theta^{s-1}(D)$ for $s = 1, 2, 3, \ldots, q$, we compute that $D_s \times D_t = \frac{1}{4}(s-t)^2 w\,[(s-t)^2 w - 4]$ for $s \neq t$.

Since $(s-t)^2 w - 4$ is a nonzero square for precisely $(q+1)/2$ nonzero choices of $s-t$ (see theorem (67) of [10], for instance), we see that D_s is secant to precisely $(q+1)/2$ D_t's from lemma (1) and the fact that w is a nonsquare in $GF(q)$. But each circle has precisely $q+1$ points. Thus, if we can show that no point lies on 3 D_i's, we will have shown that $\{D = D_1, D_2, \ldots, D_q\}$ is a q-nest of circles.

To that end suppose that $z \in D_s \cap D_t \cap D_u$ for distinct subscripts s, t, u. It is easy to see that $0, \infty \notin D_i$ for $i = 1, 2, \ldots, q_s$ and hence $z \in GF(q)^{\#}$. Using the inversions associated with each circle, we obtain

$$z = \frac{s\epsilon z^q + s^2 w - 1}{z^q + s\epsilon} = \frac{t\epsilon z^q + t^2 w - 1}{z^q + t\epsilon} = \frac{u\epsilon z^q + u^2 w - 1}{z^q + u\epsilon}.$$

Letting $x = z^q$ and solving simultaneously, we obtain $x^2 + (s+t)\epsilon x + stw + 1 = 0$ and $x^2 + (t+u)\epsilon x + tuw + 1 = 0$. This implies that $s = u$, a contradiction. Hence the existence of a q-nest has been established for all primes $q \equiv 3 \pmod 4$.

It should also be noted that θ fixes the point ∞ and leaves invariant each circle of the pencil

$$\mathbf{L}_0 = \left\{\mathbf{C}_0 = \begin{pmatrix} 1 & 0 \\ 0 & -1 \end{pmatrix}\right\} \cup \left\{\mathbf{C}_a = \begin{pmatrix} a & 1 \\ 0 & -a \end{pmatrix} : 0 \neq a \in GF(q)\right\}$$

with carrier ∞. We again easily compute that $\mathbf{C}_0 \times D = 1$ and $\mathbf{C}_a \times D = a^2 + \frac{1}{4}$. Since there are $(q-3)/2$ choices for $0 \neq a \in GF(q)$ such that $a^2 + \frac{1}{4}$ is a nonzero square (see theorem (67) of [10], for instance), the corresponding circles $\mathbf{C}_a$ together with $\mathbf{C}_0$ form a partial pencil of $(q-1)/2$ circles which are disjoint from D by lemma (1). Hence these circles are disjoint from each circle of the q-nest constructed above, since the q-nest is nothing other than the orbit of D under θ. A trivial counting argument now shows that the $q^2 + 1 - q(q+1)/2 = (q^2-q+2)/2$ points of $M(q)$ not incident with the circles in our q-nest are covered by the $(q-1)/2$ circles in the above partial pencil with carrier ∞.

We now turn our attention to the case $q \equiv 1 \pmod 4$. Let $\theta = \phi_R$ be the same linear fractional collineation of $M(q)$ as above, but this time let D be the circle represented by $\begin{pmatrix} \epsilon & 0 \\ 1 & \epsilon \end{pmatrix}$. Then $\mathbf{C}_0 \times D = w$ and $\mathbf{C}_a \times D = a^2 w + \frac{1}{4}$, and lemma (1) implies that D is once again disjoint from $(q-1)/2$ circles in the pencil $\mathbf{L}_0$ (none of which is $\mathbf{C}_0$).

Computing as above we obtain $\theta^s(D) = \begin{pmatrix} (s+1)\epsilon & s(s+2)w \\ 1 & (s+1)\epsilon \end{pmatrix}$ for $s = 0, 1, 2, \ldots, q-1$. Again letting $D_s = \theta^{s-1}(D)$ for $s = 1, 2, 3, \ldots, q$, we easily compute $D_s \times D_t = \frac{1}{4}(s-t)^2 w^2 \, [(s-t)^2 - 4]$ for $s \neq t$. Since theorem (67) of [10] implies that $(s-t)^2 - 4$ is a nonzero square in $GF(q)$ for $(q-5)/2$ nonzero choices of $s - t$ and since -4 is a nonzero square in $GF(q)$, we see that $(s-t)^2 - 4$ is a nonsquare for $(q-1)/2$ (nonzero) choices of $s - t$. Hence by lemma (1) each D_s is disjoint from $(q-5)/4$ D_t's, secant to $(q-1)/2$ D_t's, and tangent to 2 D_t's. Just as in the previous case it can easily be shown that no point is incident with 3 distinct D_t's, and hence $\{D_1, D_2, ..., D_q\}$ is a q-nest of circles by elementary counting. We thus have the following theorem.

Theorem (2): Let q be any odd prime, and let Ω be a regular spread of $PG(3,q)$. Then Ω contains a q-nest of reguli as defined above. Moreover, this q-nest can be chosen so that if U denotes the $q(q+1)/2$ distinct lines contained in the reguli of this nest, then the $(q^2-q+2)/2$ lines of $\Omega - U$ constitute a partial pencil of $(q-1)/2$ reguli.

4. THE REPLACEMENT QUESTION

In this section we show that the partial spreads $U \subseteq \Omega$ of theorem (2) are always replaceable. More explicitly, if $\{R_1, R_2, ..., R_q\}$ is a q-nest of reguli such that $U = \bigcup_{i=1}^{q} R_i$, we will exhibit a partial spread of $\sum = PG(3,q)$ covering the same points as U and consisting of $(q+1)/2$ lines from the opposite regulus R_i^{opp} for each $i = 1, 2, .., q$.

To this end we again use the correspondence established in [3] between the points and circles of $M(q)$ and the lines and reguli of Ω. Letting $\epsilon = \beta^{(q+1)/2}$ as defined in the previous section, every element z of $GF(q^2)$ can be uniquely expressed as $a + b\epsilon$ for some $a, b \epsilon GF(q)$. Representing the points, lines, and planes of $PG(3,q)$ by the 1–, 2–, and 3-dimensional subspaces of a 4-dimensional vector space over $GF(q)$, the point $z = a + b\epsilon$ of $M(q)$ corresponds to the line $l_{(a,b)} = <(a,wb,1,0), (b,a,0,1)>$ and the point ∞ of $M(q)$ corresponds to the line $l_\infty = <(1,0,0,0), (0,1,0,0)>$. As before $w = \epsilon^2 = \beta^{(q+1)}$ is a primitive element of $GF(q)$. Then $\Omega = \{l_\infty\} \cup \{l_{(a,b)} : a, b \epsilon GF(q)\}$ is a regular spread of $PG(3,q)$ (see [3]). Once again we distinguish the cases $q \equiv 1 \pmod 4$ and $q \equiv 3 \pmod 4$.

Suppose first that $q \equiv 3 \pmod 4$. Using the notation established above for $M(q)$, $D_q = \theta^{q-1}(D) = \begin{pmatrix} 0 & -1 \\ 1 & 0 \end{pmatrix}$ and hence consists of the points $z \in GF(q^2)$ satisfying $z^{q+1} = -1$. That is, $D_q = \{a + b\epsilon : b^2 w = a^2 + 1\}$. Pulling back to $PG(3,q)$, we obtain the regulus $R_q = \{l_{(a,b)} : b^2 w = a^2 + 1\}$. To describe the other reguli of our q-nest, recall that $\theta : a + b\epsilon \to a + (b+1)\epsilon$ and $\theta : \infty \to \infty$ as a collineation of $M(q)$. Defining T_1 to be the collineation of $PG(3,q)$ defined homogeneously by

$T_1 : (x_1,x_2,x_3,x_4) \to (x_1+x_4,x_2+wx_3,x_3,x_4)$, we see that

$T_1 : (a,wb,1,0) \to (a,w(b+1),1,0)$ and $T_1 : (b,a,0,1) \to (b+1,a,0,1)$. Thus $T_1 : l_{(a,b)} \to l_{(a,b+1)}$, and similarly $T_1 : l_\infty \to l_\infty$. Hence T_1 is a collineation of the regular spread Ω which is a pre-image of θ under the canonical map π taking the collineation group of Ω onto the collineation group of $M(q)$. It is easy to see that the projective order of T_1 is q and

$T_1^i : (x_1,x_2,x_3,x_4) \to (x_1+ix_4,x_2+iwx_3,x_3,x_4)$ for $i = 1, 2, 3, \ldots, q-1$. Hence, reading the subscripts modulo q, we obtain

$R_i = T_1^i(R_0) = T_1^i(R_q) = \{l_{(a,b)} : w(b-i)^2 = a^2 + 1\}$ for $i = 0, 1, 2, \ldots, q-1$ as the reguli of our q-nest.

If $l_{(a,b)}$ is any line of R_0, we know there is a unique $i \not\equiv 0 \pmod q$ such that $l_{(a,b)} \in R_i$. A simple computation shows that this means $w(b-i)^2 = a^2 + 1 = wb^2$ and hence $i \equiv 2b \pmod q$. It should be noted that $l_{(a,b)} \in R_0$ implies that $b \neq 0$ since $q \equiv 3 \pmod 4$ and $a \neq 0$ since w is a nonsquare of $GF(q)$. One similarly shows that $R_0 \cap R_{2b} = \{l_{(a,b)}, l_{(-a,b)}\}$.

Temporarily fix $a, b \in GF(q)$ with $wb^2 = a^2 + 1$. Thus $l_{(a,b)}$ is a line of R_0 and $P = \langle(a,wb,1,0)\rangle$ is some point on this line. If L denotes the unique line of R_0^{opp} containing P, an easy computation shows that $L = \langle(a,wb,1,0), (0,1,b,a)\rangle$. In fact, $Q = \langle(0,1,b,a)\rangle = L \cap l_{(-a,b)}$. A simple coordinate argument now shows that $L \cap T_1^i(L) = \phi$ for $i = 1, 2, \ldots, q-1$. Thus, given a line of R_0^{opp} and letting T_1 act on that line, we obtain a set of q mutually skew lines of Σ, one from each R_i^{opp} for $i = 0, 1, 2, \ldots, q-1$.

We now more carefully look at $\ker(\pi)$, where π is the canonical map taking $\mathrm{Aut}(\Omega)$ onto $\mathrm{Aut}(M(q))$. Thus $\ker(\pi)$ consists of the collineations of Σ leaving invariant every line of Ω. As before let β be a primitive element of $GF(q^2)$ and let $w = \beta^{q+1}$. Choose $s, t \in GF(q)$ such that the minimal polynomial of β over $GF(q)$ looks like $f(x) = (x-s)^2 - w^{-1}t^2$. The irreducibility of $f(x)$ allows for such a choice of s and t, and also forces that $t \neq 0$. Treating $GF(q^2)$ as a 2-dimensional vector space over $GF(q)$, consider the linear operator on $GF(q^2)$ defined by $z \to \beta z$. Relative to the basis $\{1, \frac{1}{t}(\beta-s)\}$, the matrix representation for this operator is $A = \begin{pmatrix} s & w^{-1}t \\ t & s \end{pmatrix}$. The projective order of this operator, treated as a collineation of $PG(1,q)$, is $q+1$.

Next let $\overline{T}_2$ denote the collineation of $PG(3,q)$ induced by the matrix $\left(\begin{array}{c|c} A & 0 \\ \hline 0 & A \end{array}\right)$ relative to our original basis. Since

$$\left(\begin{array}{c|c} A & 0 \\ \hline 0 & A \end{array}\right)\begin{pmatrix} a \\ wb \\ 1 \\ 0 \end{pmatrix} = s\begin{pmatrix} a \\ wb \\ 1 \\ 0 \end{pmatrix} + t\begin{pmatrix} b \\ a \\ 0 \\ 1 \end{pmatrix}$$

and

$$\left(\begin{array}{c|c} A & 0 \\ \hline 0 & A \end{array}\right)\begin{pmatrix} b \\ a \\ 0 \\ 1 \end{pmatrix} = w^{-1}t\begin{pmatrix} a \\ wb \\ 1 \\ 0 \end{pmatrix} + s\begin{pmatrix} b \\ a \\ 0 \\ 1 \end{pmatrix},$$

$\overline{T}_2$ is a collineation of order $q+1$ leaving invariant each line of the regular spread Ω. Thus $T_2 = (\overline{T}_2)^2$ is a collineation of order $(q+1/2$ in $\ker(\pi)$. Since T_2 commutes with T_1 and $\gcd(q,(q+1)/2) = 1$, we obtain a collineation $T = T_1 T_2$ of Ω whose order is $q(q+1)/2$. We now claim that the orbit under T of any line if R_0^{opp} is a replacement set for U. Since the lines of such an orbit consist of $(q+1)/2$ lines from R_i^{opp} for $i = 0, 1, 2, \ldots, q-1$, it suffices to show that these lines are pairwise skew.

To that end let R_0 and R_i be any two distinct reguli of our q-nest that share a common line. Previous work shows that R_0 and R_i are secant with $R_0 \cap R_i = \{l_{(a,b)}, l_{(-a,b)}\}$, where $i \equiv 2b \pmod q$. As before $a \neq 0$, $b \neq 0$, and $wb^2 = a^2 + 1$. Let $L = <(a,wb,1,0), (0,1,b,a)>$. As indicated above, L is a line of

R_0^{opp}, $P = L \cap l_{(a,b)} = \langle(a,wb,1,0)\rangle$,
$Q = L \cap L_{(-a,b)} = \langle(0,1,b,a)\rangle$, and $L \cap T_1^i(L) = L \cap T^{2b}(L) = \phi$. In fact,
$P' = T_1^{2b}(L) \cap l_{(a,b)} = \langle(2a^2+1, 2abw, a, bw)\rangle$ and
$Q' = T_1^{2b}(L) \cap l_{(-a,b)} = \langle(b,-a,0,1)\rangle$.

Suppose by way of contradiction that some nonzero power of T_2, say T_2^k, mapped P' onto P. As the matrix representation for T_2^k looks like $\left(\begin{array}{c|c} A^{2k} & 0 \\ \hline 0 & A^{2k} \end{array}\right)$, this would force $A^{2k}\begin{pmatrix} a \\ bw \end{pmatrix} \sim \begin{pmatrix} 1 \\ 0 \end{pmatrix}$, where $\sim$ denotes that the two vectors are $GF(q)$-scalar multiples of one another. A simple induction argument shows that A^{2k} has the form $\begin{bmatrix} y_1 & w^{-1}y_2 \\ y_2 & y_1 \end{bmatrix}$, where $y_2 \neq 0$ as $k \neq 0$. Hence we must have $y_2 a + y_1 bw = 0$, and therefore $wb^2 = a^2 + 1 = (y_1^2 w^2 b^2 + y_2^2) / y_2^2$. This implies that
$w^2(y_1^2 - w^{-1}y_2^2)\, b^2 = -\, y_2^2$ is a nonsquare in $GF(q)$, since $q \equiv 3 \pmod 4$. But $y_1^2 - w^{-1} y_2^2 = \det(A^{2k}) = (\det A)^{2k}$ is necessarily a square in $GF(q)$, yielding a contradiction. One similarly shows that no power of T_2 maps Q' onto Q, and hence no line in the orbit of $T_1^{2b}(L)$ under T_2 meets L. The transitivity inherent in the problem and the commutativity of T_1 with T_2 now implies that the orbit of L under $T = T_1 T_2$ is a replacement set for U.

Finally, when $q \equiv 1 \pmod 4$, a completely analogous situation occurs. This time $D_q = D_0 = \begin{pmatrix} 0 & -w \\ 1 & 0 \end{pmatrix} = \{a + b\epsilon : a^2 + w = wb^2\}$, and hence $R_i = \{l_{(a,b)} : a^2 + w = w(b-i)^2\}$ for $i = 0, 1, 2, \ldots, q-1$. If $l_{(a,b)}$ denotes any line of R_0, then necessarily $b \neq 0$ and R_{2b} is the other regulus of the q-nest containing $l_{(a,b)}$. Moreover,

$$R_0 \cap R_{2b} = \begin{Bmatrix} l_{(a,b)} & \text{if } a = 0 \\ l_{(a,b)},\ l_{(-a,b)} & \text{if } a \neq 0 \end{Bmatrix}.$$

If $l_{(a,b)}$ is a line of R_0 chosen so that $a \neq 0$ and L is the unique line of R_0^{opp} passing through the point $\langle(a,wb,1,0)\rangle$ of $l_{(a,b)}$, then $L = \langle(a,wb,1,0), (0,w,b,a)\rangle$ where $\langle(0,w,b,a)\rangle = L \cap l_{(-a,b)}$. Similar arguments to those given above show that the orbit of L under $T = T_1 T_2$ is a replacement partial spread for U in this case as well. Hence we have the following theorem.

Theorem (3). Let q be any odd prime, and let U denote the $q(q+1)/2$ distinct lines of a regular spread Ω that are contained in a q-nest $\{R_0, R_1, \ldots, R_{q-1}\}$ constructed as in theorem (2). Then U is replaceable. Moreover, a replacement partial spread V can be obtained as the orbit of any line in R_0^{opp} under a collineation of Ω whose order is $q(q+1)/2$. The set V consists of $(q+1)/2$ lines from R_i^{opp} for $i = 0, 1, 2, \ldots, q-1$. □

Corollary. For any odd prime q a nondesarguesian translation plane of order q^2 can be constructed from the spread $(\Omega - U) \cup V$ of theorem (3). □

For $q = 5$ the spread obtained via theorem (3) is equivalent to the spread $\mathbf{S}_4$ found in a computer search by Oakden [12]. It was a new plane at the time it was found. For $q = 7$ the spread obtained is most likely equivalent to the one recently found by Cohen and Ganley [8] using Chebyshev polynomials. However, for $q \geq 11$ the translation planes obtained here appear to be new.

5. DERIVATION AND COLLINEATION GROUPS

In this section we discuss the reguli contained in the spreads constructed above, some derivations of these spreads, and the orbit structure of the spreads under their collineation groups.

Theorem (4). If $S = (\Omega - U) \cup V$ is a spread obtained via theorem (3), then S contains at least q reguli.

Proof. From theorem (2) we know that the lines of $\Omega - U$ comprise a partial pencil of $(q-1)/2$ reguli with carrier l_∞. Our work above shows that V may be taken to be the orbit under T of any line in R_0^{opp}. An easy computation shows that $L = \langle (a,wb,1,0), (-b,-a,0,1) \rangle$ is a line of R_0^{opp}, where $wb^2 = a^2 + 1$ if $q \equiv 3$ (mod 4) and $wb^2 = a^2 + w$ $(a \neq 0)$ if $q \equiv 1$ (mod 4). In fact, L contains the points $\langle (1,0,-a,-wb) \rangle$ and $\langle (0,1,b,a) \rangle$ if $q \equiv 3$ (mod 4), while L contains the points $\langle (1,0,-aw^{-1},-b) \rangle$ and $\langle (0,w,b,a) \rangle$ if $q \equiv 1$ (mod 4).

As shown above L, $T_1(L)$, $T_1^2(L)$, ..., $T_1^{q-1}(L)$ is a set of q skew lines in V. We claim that this set, together with l_∞, forms a regulus. To see this simply let
$M = <(a,wb,1,0), (a,w(b+1),1,0)>$,
$N = <(-b,-a,0,1), (1-b,-a,0,1)>$, and

$$J = \begin{cases} <(0,1,b,a)< (a,wb+1,b,a)> & \text{if } q \equiv 3 \pmod 4 \\ <(0,w,b,a), (a,w(b+1),b,a)> & \text{if } q \equiv 1 \pmod 4 \end{cases}.$$

More easy calculations show that M, N, and J are three skew lines that meet each of l_∞, L, $T_1(L)$, ..., $T_1^{q-1}(L)$, and therefore the latter set of $q+1$ skew lines forms a regulus R of S. Since $T_2(l_\infty) = l_\infty$ and $T_2(V) = V$, we obtain that $T_2^i(R)$ is a regulus of S for $i = 0, 1, 2, ..., (q-1)/2$. Hence S contains at least $(q-1)/2 + (q+1)/2 = q$ reguli. It should be remarked that all of these reguli contain the line l_∞, and any two of them share only that line. □

When $q = 5$ or $q = 7$, it has been verified that the above reguli are the only ones contained in S, and it is conjectured that this is always the case. Reversing any one of these reguli in S when $q = 5$ yields a spread S' that contains exactly one regulus. The spread S' is not subregular and, in fact, is isomorphic to Oakden's spread S_9 (see [12]). When $q = 7$ one of the three reguli contained in $\Omega - U$ is special. It corresponds to the circle C_0 as defined previously. Whenever $q \equiv 3 \pmod 4$, S will always contain such a "special" regulus. Reversing this regulus when $q = 7$ again yields a non-subregular spread $\bar{S}$ containing exactly one regulus. Reversing any one of the other six reguli of S yields a spread inequivalent to $\bar{S}$, although still non-subregular and containing exactly one regulus. It appears as if the six spreads so obtained are all equivalent, but this has not been completely verified.

Finally we discuss the orbit structure of S under its collineation group. This is equivalent to looking at the orbit structure of the line at infinity under the translation complement of the corresponding translation plane. As shown before, the collineation T_1 maps each of the $(q-1)/2$ reguli of $\Omega - U$ upon itself by fixing the line l_∞ and moving the remaining q lines in a single orbit. We also know that V itself is a single orbit under $T = T_1 T_2$. Thus, using only the collineations T_1 and T_2 of S, we obtain one orbit of size 1, $(q-1)/2$ orbits of size q, and one orbit of size $q(q+1)/2$.

When $q = 7$ these orbits can be further combined by using two other collineations of S, one being an involution induced by the matrix $\left(\begin{array}{c|c} I & 0 \\ \hline 0 & -I \end{array}\right)$ and the other being an element of order 3 found by ad hoc methods. With the addition of these collineations S is reduced to one orbit of size 1, one orbit of size 7, and one orbit of size 42. It is because of this orbit structure that we feel this particular spread is probably equivalent to the one found by Cohen and Ganley [8]. The one orbit of size 7 corresponds to the "special" regulus of $\Omega - U$ mentioned above. The other two orbits of size 7 were combined by the mapping $\left(\begin{array}{c|c} I & 0 \\ \hline 0 & -I \end{array}\right)$, and then this orbit was combined with the orbit of size 28 by the ad hoc collineation of order 3.

It should be mentioned that this collineation of order 3 for $q = 7$ is not inherited from the collineation group of the regular spread Ω. We believe that this sort of behavior cannot happen for large q. The mapping corresponding to $\left(\begin{array}{c|c} I & 0 \\ \hline 0 & -I \end{array}\right)$, however, is a collineation of S whenever $q \equiv 3 \pmod 4$.

<u>Proposition</u> <u>(5).</u> Let $q \equiv 3 \pmod 4$ be a prime, and let S be a spread of $\sum = PG(3,q)$ obtained via theorem (3). Then the collineation T_3 of $\sum$ induced by the matrix $\left(\begin{array}{c|c} I & 0 \\ \hline 0 & -I \end{array}\right)$ is a collineation of S. Moreover, T_3 leaves invariant one regulus of $\Omega - U$ and pairs off the remaining $(q-3)/2$ reguli of $\Omega - U$.

<u>Proof.</u> Straightforward computations show that $T_3 : l_\infty \to l_\infty$ and $T_3 : l_{(a,b)} \to l_{(-a,-b)}$. Thus T_3 is an involution in Aut (Ω). Since $\Omega - U = \{l_{(a,b)} : a^2 + 1 \text{ is a square in } GF(q)\}$, $T_3 : \Omega - U \to \Omega - U$. Let $L = \langle (a,wb,1,0), (-b,-a,0,1) \rangle$ be a line of R_0^{opp}, where $wb^2 = a^2 + 1$. Then V may be taken to be the orbit of L under $T = T_1 T_2$. A simple computation shows that $T_3(L) = \langle (-a,-wb,1,0), (b,a,0,1) \rangle$. Since T_2^k has the form $\left(\begin{array}{c|c} \begin{matrix} x & w^{-1} \\ 1 & x \end{matrix} & 0 \\ \hline 0 & \begin{matrix} x & w^{-1} \\ 1 & x \end{matrix} \end{array}\right)$ for $k = 1, 2, \ldots, (q-1)/2$ and since $T_2^{(q+1)/4}$ is an involution, it

must be the case that $T_2^{(q+1)/4}$ has the form $\left(\begin{array}{cc|cc} 0 & w^{-1} & 0 & \\ 1 & 0 & & \\ \hline 0 & & 0 & w^{-1} \\ & & 1 & 0 \end{array}\right)$. Thus $T_2^{(q+1)/4} : L \to T_3(L)$. Since $T_3\, T_2 = T_2\, T_3$ and $T_3\, T_1\, T_3 = T_1^{-1}$, this implies that $T_3 : V \to V$ and therefore T_3 is an involution in Aut(S).

From our work in section 3 we know that the $(q-1)/2$ reguli of S contained in $\Omega - U$ are $\overline{R}_0 = \{l_\infty\} \cup \{l_{(a,b)} : a = 0\}$ and $\overline{R}_d = \{l_\infty\} \cup \{l_{(a,b)} : a = -1/2d\}$, where $d \in GF(q)^{\#}$ such that $d^2 + \frac{1}{4}$ is a nonzero square in $GF(q)$. As $T_3 : \overline{R}_0 \to \overline{R}_0$ and $T_3 : \overline{R}_d \to \overline{R}_{-d}$, the result now follows immediately. $\square$

We therefore see that when $q \equiv 3 \pmod 4$, the orbit structure of S under the collineation group of S generated by T_1, T_2, and T_3 consists of one orbit of size 1, one orbit of size q, $(q-3)/4$ orbits of size $2q$, and one orbit of size $q(q+1)/2$. It is not currently known if Aut $(S) = \langle T_1, T_2, T_3 \rangle$ for $q \equiv 3 \pmod 4$ and $q > 7$, or if any further contraction of orbits is possible in this case. For $q \equiv 1 \pmod 4$ it is easy to see that $T_3 \notin$ Aut (S) since $T_3(V) \neq V$. However this problem can be easily overcome.

Proposition (6). Let $q \equiv 1 \pmod 4$ be a prime. Using the notation established above, $\overline{T}_3 = T_3\, T_2$ is a collineation of $S = (\Omega - U) \cup V$ that pairs off the $(q-1)/2$ reguli contained in $\Omega - U$.

Proof. Recall that $(\overline{T}_2)^2 = T_2$. Just as in the proof of proposition (5), we see that $\overline{T}_3 : \Omega - U \to \Omega - U$, $\overline{T}_3\, T_2 = T_2\, \overline{T}_3$, and $\overline{T}_3\, T_1 = T_1^{-1}\, \overline{T}_3$. As shown above $\overline{T}_2^{(q+1)/2}$ has the form $\left(\begin{array}{cc|cc} 0 & w^{-1} & 0 & \\ 1 & 0 & & \\ \hline 0 & & 0 & w^{-1} \\ & & 1 & 0 \end{array}\right)$, and therefore $T_2^{(q+3)/4} = \overline{T}_2^{(q+3)/2} : L \to \overline{T}_2(T_3 L) = \overline{T}_3(L)$. Thus $\overline{T}_3$ is a collineation of S since $\overline{T}_3 : V \to V$.

As discussed in section 3, the $(q-1)/2$ reguli contained in $\Omega - U$ are $\overline{R}_d = \{l_\infty\} = \{l_{(a,b)} : a = -1/2d\}$, where $d \in GF(q)^\#$ such that $d^2w + \frac{1}{4}$ is a nonzero square in $GF(q)$. Since $\overline{T}_3 : \overline{R}_d \to \overline{R}_{-d}$, the result follows. □

Hence, when $q \equiv 1 \pmod 4$, the collineation group $<T_1, T_2, \overline{T}_3>$ of S partitions S into one orbit of size 1, $(q-1)/4$ orbits of size $2q$, and one orbit of size $q(q+1)/2$.

R. D. Baker
G. L. Ebert
University of Delaware
Newark, DE 19716

REFERENCES

[1] L. M. Abatangelo, A translation plane of order 81 and its full collineation group, *Bull. Austral. Math. Soc.* 29 (1984), 19-34.
[2] J. André, Uber nicht-Desarguessche Ebenen mit transitiver Translationgruppe, *Math. Z.* 60 (1954), 156-186.
[3] R. H. Bruck, Construction problems of finite projective planes, Conference on Combinatorics, Univ. North Carolina, 1967.
[4] R. H. Bruck and R. C. Bose, The construction of translation planes from projective spaces, *J. Alg.* 1 (1964), 85-102.
[5] A. A. Bruen, Inversive geometry and some translation planes I, *Geom. Ded.* 7 (1977), 81-98.
[6] A. A. Bruen and J. A. Thas, Flocks, chains, and configurations in finite geometries, *Accad. Naz. Lincei* (1975), 744-748.
[7] M. Capursi, A translation plane of order 11^2, *JCT A* 35 (1983), 289-300.
[8] S. D. Cohen and M. J. Ganley, Some classes of translation planes, *Quart. J. Math. Oxford* (2) 35 (1984), 101-113.
[9] P. Dembowski, *Finite Geometries,* Springer-Verlag, New York, 1968.
[10] L. E. Dickson, *Linear Groups,* Dover Publications (reprint), New York, 1958.
[11] G. Korchmaros, A translation plane of order 49 with non-solvable collineation group, *J. Geom.* 24 (1985), 18-30.
[12] D. J. Oakden, The regulus-containing spreads of $PG(3,5)$, part of Ph.D. Thesis, Univ. of Toronto, 1974.
[13] W. F. Orr, The miquelian inversive plane $IP(q)$ and the associated projective planes, Ph.D. Thesis, Univ. of Wisconsin, 1973.
[14] G. Raguso and B Larato, Piani di traslazione di ordine $l3^2$, *Riv. Mat. Univ. Parma* (4) 10 (1984), 223-233.

Annals of Discrete Mathematics 37 (1988) 21–30

QUASIGROUPS AND GROUPS ARISING FROM CUBIC SURFACES

by Professor Lucien BENETEAU

Université Paul Sabatier, U.E.R. M.I.G.
118 Route de Narbonne F-31062 TOULOUSE Cedex (France)

We shall discuss Manin's structure theorems about cubic hypersurfaces. Under some conditions, a suitable factor of the set of non-singular points of a projective cubic hypersurface may be endowed with a structure of symmetric terentropic quasigroup in which any square is idempotent ; it is therefore isotopic to an exponent 6 Commutative Moufang Loop. This generalizes the classical construction of an abelian group from a plane curve. But in the case of a surface there are still several open questions. Some properties of the terentropic quasigroups and of the cubic quasigroups are stated. Several concrete examples are presented. The multiplication groups of the cubic quasi-groups are described.

1. INTRODUCTION

1.1. Let E be the set of non-singular points of an absolutely irredictible cubic hypersurface V of $\mathbb{P}_n(k)$. In case V is a curve, one defines classifically a symmetric quasigroup (E,o) where xoy is, so to speak, the third point of intersection of the line through x and y (in case $x=y$, take the tangent line). By setting $x+y=uo\,(x\,o\,y)$ one obtains an abelian group. This report is designed to give an elementary approach to Manin's generalization to the case dim $V \geq 2$: in a suitable factor-set $\overline{E} = E/R$ one obtains then a symmetric quasigroup satisfying $(a\,o\,x)\,o\,(a\,o\,y) = (a\,o\,a)\,o\,(x\,o\,y)$ and $(x\,o\,x)\,o\,(x\,o\,x) = (x\,o\,x)$. It follows that all the fibres $A = p^{-1}(e)$ of the projection $x \to p(x) = x\,o\,x$ are isomorphic elementary abelian 2-groups, that $\overline{E}^2 = \mathrm{Im}(p)$ is a distributive Steiner quasigroup and that $(\overline{E},o)$ splits as a direct product of A by $(\overline{E}^2, o)$. Here the previously defined addition makes $\overline{E}$ into an exponent 6 commutative Moufang Loop, direct product of A by the exponent 3 loop $(\overline{E}^2, +)$. But are there examples in which $(\overline{E}, +)$ is not an abelian group ?... This is still an open problem.

1.2. Section 2 yields a simple algebraic presentation of the cubic quasigroups : they are defined as the totally symmetric quasigroups satisfying $ax.ay = a^2.xy$. As a consequence of the Moufang theorem, these quasigroups can also be defined as the symmetric terentropic quasigroups. Now the terentropic quasigroups can be shown to be locally nilpotent with respect to their lacking property, the entropic law. We just give the principle of the proof. The multiplication groups of the various cubic quasigroups isotopic to a given Commutative Moufang Loop are described ; they are all isomorphic. Section 3 is devoted to those cubic quasigroups that arise from cubic hypersurfaces. We recall the framework of Manin's construction and then turn to the description of $(\bar{E},o)$ we sketched in 1.1.

2. DEFINITIONS AND STRUCTURE THEOREMS

2.1. Let $(Q,\cdot)$ be a set Q provided with a binary operation $x,y \to x\cdot y$. One says that $(Q,\ )$ is a *symmetric quasigroup* if any equality of the form $x\cdot y = z$ remains true under all the permutations of x,y,z (equivalently $xy = yx$ and $x\cdot xy = y$). Let us set $x^2 = x\cdot x$.

Proposition 2.2 : Any symmetric quasigroup satisfies all or none of the following identities : (i) $ax\cdot ay = a^2\cdot xy$; (ii) $(ax\cdot y)\,a = a^2y\cdot x$ (iii) $a\cdot xy = ax\cdot a^2y$.

Proof : From (i) it follows that $y^2\cdot(ax\cdot ay) = y^2\cdot(a^2\cdot xy)$ and since $y\cdot ay = a$ and $y\cdot xy = x$ one gets $(y\cdot ax)a = ya^2\cdot x$, equivalent to (ii). Further in (ii) $(ax\cdot y)a = a^2y\cdot x$, if one sets $ax = X$ one obtains $Xy\cdot a = a^2y\cdot aX$, equivalent to (iii). By multiplying both members of (iii) by ax one gets $ax\cdot(a\cdot xy) = a^2y$, equivalent to (i).

In what follows $(Q,\cdot)$ is a set endowed with a multiplication.

Definitions 2.3. : (i) $(Q,\cdot)$ is a *cubic quasigroup* iff $(Q,\cdot)$ is a symmetric quasigroup such that $ax\cdot ay = a^2\cdot xy$ identifically. (ii) The multiplication is said to be *entropic* (or Abelian, metabelian, medial) if $ab\cdot xy = ax\cdot by$ identically. (iii) The multiplication is *distributive* (resp. *idempotent*) iff $a\cdot xy = ax\cdot ay$ (resp. $a^2 = a$) identically.

It is clear that entropic symmetric quasigroups are a special case of cubic quasigroups. Further a cubic quasigroup is distributive iff it is idempotent.

2.4. First examples : (1) Elementary abelian 2-groups are the only groups with symmetric multiplication. They are entropic cubic quasigroups. (2) Let $(A,+)$ be an elementary abelian 3-group. The mid-point law $x,y \to -x-y$ makes A into a distributive entropic cubic quasigroup. Any distributive entropic cubic quasigroup arises in this manner. (3) Let G be an exponent 3 group. The underlying set of G may be provided with a distributive cubic quasigroup structure by the new binary operation $x,y \to x \circ y = xy^{-1}x$.

2.5. There is a canonical kinship between entropic cubic quasigroups and abelian groups. In what follows, *to any element u of a cubic quasigroup* $(Q,\cdot)$ *we attach the operation* : $x,y \to x \underset{u}{*} y = u\cdot xy$. One checks that if $(Q,\cdot)$ is entropic, $(Q,\underset{u}{*})$ is an abelian group which, up to isomorphism, does not depend on the choice of u. Conversely if $(A,+)$ is an abelian group, for any c in A the operation $x \circ y = c - x - y$ makes A into an entropic cubic quasigroup. This correspondence can be extended to a canonical correspondence between cubic quasigroups and Commutative Moufang Loops (CMLs). Recall that a CML is a loop $(Q,*)$ whose multiplication obeys the identity : $(a*a)*(x*y) = (a*y)*(a*y)$. In such a loop the *center* $Z(Q,*)$ is the set of elements c satisfying $(x*y)*c = x*(y*c)$ for any x and y in Q.

Theorem 2.6 : Let $(Q,\cdot)$ be a cubic quasigroup. For any u in Q, $(Q,\underset{u}{*})$ is a CML whose identity element is u. Its center contains u^2. The inverse of any element x is : $\underset{u}{-}x = u^2x$ and $x\cdot y = u^2\underset{u}{-}(x\underset{u}{*}y)$. Besides for any v in Q, $(Q,\underset{u}{*}) \simeq (Q,\underset{v}{*})$. □

Remark : Manin stated similar facts, but his definition of "cubic hypersurface quasigroup" is formally different, so an explicit proof is needed here. The equivalence between the two definitions appears further on.

Proof : In the loop $(Q,\underset{u}{*})$, for any x, $x\underset{u}{*}u = u.xu = x$ and $x\underset{u}{*}u^2x = u(x\cdot u^2x) = u\cdot u^2 = u$. The identity $(a*x)*(a*y) = (a*a)*(x*y)$ follows from $ax\cdot ay = a^2\cdot xy$. Further $u^2\underset{u}{-}(x\underset{u}{*}y) = u^2\underset{u}{*}(u^2\cdot(u\cdot xy)) = u\cdot(u^2(u^2(u\cdot xy)) = xy$. The center $Z(Q,\underset{u}{*})$ is the set of elements c satisfying $(u\cdot xy)\cdot c = (u\cdot cy)\cdot x$ for any x and y in Q ; it contains u^2 since $u\cdot xy = u^2y\cdot ux$. Lastly the mapping $\tau : x \to x\underset{u}{*}v = u\cdot vx$ is an isomorphism from $(Q,\underset{u}{*})$ onto $(Q,\underset{v}{*})$ since $\tau(x)\underset{v}{*}\tau(y) = v\cdot((u\cdot vx)(u\cdot vy)) = v\, u^2\cdot(v^2\cdot xy) = vu^2\cdot xy = u(v\cdot(u\cdot xy)) = \tau(x\underset{u}{*}y)$ (use identities (i) and (iii) of 2.2).

Definition : The loop $(Q,\underset{u}{*})$ is to be referred to as *the CML related to* $(Q,\cdot)$ *with origin* u. Up to isomorphism it does not depend on the choice of u, and one may speak of the CML related to $(Q,\cdot)$; it is clearly the unique CML isotropic to $(Q,\cdot)$. It is an abelian group iff (Q,) is entropic.

There is a converse to theorem 2.6. Its verification is straightforward :

Proposition 2.7 : If $(Q,+)$ is a CML with identity element e, for any central element c of $(Q,+)$ the binary operation $x,y \to x \circ_c y = c-x-y$ turns Q into a cubic quasigroup. The CML with origin e related to $(Q, \circ_c)$ coincides with $(Q,+)$ since $e \circ_c e = c$ and $x + y = e \circ_c (x \circ_x y)$.

The link between cubic quasigroups and CMLs recalls the link between affine spaces and vector spaces, with a remarkable difference : non-isomorphic cubic quasigroups can have the same related CML $(Q,+)$. (see the next examples).

Lemma 2.8 : Let u be an arbitrary element of a cubic quasigroup $(Q,\cdot)$ and S a subset of Q. The subquasigroup of $(Q,\cdot)$ generated by $\{u\} \cup S$ coincides as a set with the subloop of $(Q,_u)$ generated by $\{u^2\} \cup S$.

Proof : A non-empty subset H of Q either satisfies both the two following properties, or neither : (i) H contains u and is closed under $x\cdot y$; (ii) H contains u^2 and is closed under $x\underset{u}{*}y = u(x.u^2y)$. □

Corollary 2.9 : A symmetric quasigroup is cubic iff any three elements generate an entropic subquasigroup.

Proof : If $(Q,\cdot)$ is cubic then the subquasigroup H generated by any three elements u, v, w coincides with the subloop of $(Q,\underset{u}{*})$ generated by u^2, v, w. Since u^2 is central, $(H,\underset{u}{*})$ is an abelian group by the Moufang theorem [Bruck 51b, 58], hence $(H,\cdot)$ is entropic. The converse is trivial. □

Remark : This shows that the cubic quasigroups are exactly the symmetric terentropic quasigroups ; they are the CH-quasigroups in Manin's terminology.

2.10. Further examples :

(1) Consider the rank (n+1) torsion free abelian group $\mathbb{Z}^{n+1}$. Let $B = \{e_i | i = 1,2,\ldots n+1\}$ be its canonical basis ; $x,y \to x \circ y = e_1 - x - y$ turns the underlying set of $\mathbb{Z}^{n+1}$ into an entropic cubic quasigroup. $(\mathbb{Z}^{n+1},\circ)$ is the free entropic cubic quasigroup on n+1 generators (it is freely generated by $\{0\} \cup (B \setminus \{e_1\})$). By replacing $\mathbb{Z}^{n+1}$ by the direct product $L_n \times \mathbb{Z}$ where L_n is the free CML on n generators, one obtains the free cubic quasigroup on n+1 generators.

(2) Let $(L_3,+)$ be the free exponent 3 CML on 3 generators (see chapters I and III), and c one of the two non-trivial central elements of L_3. Both the binary operations $x \cdot y = -x - y$ and $x \circ y = c - x - y$ provide L_3 with cubic quasigroup structure. But $(L_3,\cdot)$ is distributive, while $(L_3,\circ)$ has no idempotent elements.

(3) Let N_3 be the set-product $\mathbb{Z}_3 \times \mathbb{Z}_3 \times \mathbb{Z}_9$, where $\mathbb{Z}_n = \mathbb{Z} / n\mathbb{Z}$. A typical element of N_3 is $X = (x_1, x_2, x_3)$ with $x_1, x_2 \in \mathbb{Z}_3$ and $x_3 \in \mathbb{Z}_9$. One may define the following two binary operations on N_3 :

$$X,Y \to X * Y = (x_1 + y_1,\ x_2 + y_2,\ x_3 + y_3 + 3(x_1 - y_1)(x_2 y_3 - x_3 y_2))$$

$$X,Y \to X \cdot Y = -(X * Y) = (-x_1 - y_1,\ -x_2 - y_2,\ -x_3 - y_3 + 3(y_1 - x_1)(x_2 y_3 - x_3 y_2))$$

It turns out that $(N_3,*)$ is a CML whose center Z consists of the three elements of the form : $(0,0,3\varepsilon)$ where $\varepsilon \in \{0, \pm 1\}$. Any element of Z being a 3-power, there is just one cubic quasigroup isotropic to $(N_3,*)$, namely $(N_3,\cdot)$.

(4) Let G be a group whose set of involutions R generates G. Assume that $(xy)^3 = 1$ holds for any two elements of R (one says then that G is a *Fischer group* in the restricted sense). For any fixed element in R, the set aR may be provided with an exponent 3 CML structure by the operation : $x,y \to x^2yx^2$. Any exponent 3 CML arises in this way. The same set may be endowed with a distributive cubic quasigroup structure by the binary operation : $x,y \to xy^2x$.

2.11 By definition, a *terentropic quasigroup* is a quasigroup in which any three elements generate an entropic subquasigroup. We have shown that the cubic quasigroups are exactly the symmetric terentropic quasigroups. Let us discuss briefly the algebraic properties of CMLs and terentropic quasigroups.

In a CML any two elements generate an abelian group but in general the associativity is not satisfied. To a certain extent, CML theory is like group theory with the rôle of commutativity and associativity exchanged. Just as the deviation of the commutativity of a group is measured by the commutator, the deviation of a CML from associativity is measured by the *associator*

$$(x, y, z) = ((x+y)+z) - ((x+(y+z)).$$

The derived subloop D is the subloop generated by all the associators. The centre Z, as we have seen, is the set of the elements z that behave "associatively" with respect to any pair x, y of elements of Q : $(x,y,z) = 0$ (identity element). Both D and Z are normal subloops. Of course D is the smallest normal subloop by which the quotient Q/D is an abelian group. If Q is associative, one says that Q is centrally nilpotent of class 1. If Q is not associative, but if its central quotient Q/Z is an abelian group, one says that Q is centrally nilpotent of class 2. More generally one may define recursively the fact that "Q is centrally nilpotent of class $k > 2$", by deciding that it is true iff Q/Z is centrally nilpotent of class $(k-1)$.

Despise several analogies, the theory of CMLs display far more features than the theory of groups.

1 - The first sign of this is the identity :

$$n(x,y,z) = (n+3m)(x,y,z)$$

which holds for any two integers n and m. As a consequence, both the central factor and the derived subloop have exponent 3, since $(x,y,3z) = 0 = 3(x,y,z)$.

2 - But above all there is the deep Bruck-Slaby Theorem : any CML on n generators, $n > 2$ is centrally nilpotent of class less than n.

As a consequence, the CMLs enjoy very strong properties which are to be compared to those of locally nilpotent groups having a locally finite derived subgroup.

2.12. Now each terentropic quasigroup $(Q,\cdot)$ is isotropic to an essentially unique CML, that may be constructed by taking an arbitrary element e from Q and by endowing the set Q with the binary multiplication :

$$x,y \longrightarrow x+y = (x/e).((e/e)\backslash y)$$

where x/u and $u \backslash x$ are defined as usual by :

$$(x/u)\cdot u = x = u\cdot(u\backslash x).$$

Conversely if $(Q,+)$ is a CML, one may recover all the terentropic quasigroups isotopic to $(Q,+)$ by considering binary operations of the form :

$$x\cdot y = c + f(x) + g(y)$$

where c is a central element of $(0,+)$ and f,g are commuting automorphisms of

$(Q,+)$ such that, for any x in Q, $f(x)+x$ and $g(x)+x$ are central elements. When $(Q,+)$ is an abelian group, then $(Q,\cdot)$ is entropic, and conversely. Further if $(Q,\cdot)$ is generated by n elements with $n > 1$ then the isotropic CML can be generated by $(n-1)$ elements "up to central elements". As an important consequence one may generalize Soublin's result about the nilpotence of distributive quasigroups :

Theorem 2.12 : Any n-generated terentropic quasigroup with $n > 2$ is centrally nipotent of class less than $(n-1)$ with respect to the entropic law.

Therefore several properties of the CMLs carry over to the terentropic quasigroups. As an intermediary result in the proof, one obtains the following by-product :

Proposition 2.13 : In any CML $(G,+)$ the minimal subsets that generate G up to central elements form the family of the bases of a matroid in the underlying set G.

2.14. Let us turn now to the study of the multiplication group $M = \mathrm{Mlt}(G,\cdot)$ of any cubic quasigroup $(G, \ )$. Though several pairwise non-isomorphic cubic quasigroups can be isotopic to the same CML, they all have the same multiplication group. Let us say that a "reflexion" of a $\mathrm{CML}(G,+)$ is a permutation σ of G of the form : $x \to d-x = x'$, where d is any fixed element of G ; d is then unique and $x' = \sigma(x)$ is characterized by $x+x' = d$; notation : $\sigma = \sigma_d$. Designate by $\mathcal{R}$ the set of all the reflexions and set $\widetilde{M} = \langle \mathcal{R} \rangle$. The subgroup generated by the squares of elements of $\widetilde{M}$ is denoted by $\widetilde{M}^{(2)}$.

Proposition 2.15 : $\widetilde{M}$ is the multiplication group of every cubic quasigroup isotopic to $(G,+)$. $\mathcal{R}$ is a union of complete conjugacy classes. Any element of $\widetilde{M}$ is a product of pairwise distinct elements from $\mathcal{R}$. The derived subgroup $\widetilde{M}'$ equals $\widetilde{M}^{(2)}$ and is contained in $M = \mathrm{Mlt}(G,+)$. Whenever $(G,+)$ is not an elementary abelian 2-group, $[\widetilde{M} : M] = 2$. □

Proof : A cubic quasigroup isotopic to $(G,+)$ has a multiplication of the form $x \cdot y = c + f(x) + g(x)$ where $f = -\mathrm{Id} = g$, thus $\mathrm{Mlt}(G,\cdot) = \langle -\mathrm{Id} \rangle M$. If $x' = \sigma_d(x)$, $x + x' = d$ and $L_a(x) + L_a(x') = (a+x) + (a+x') = 2a + d = \delta$, hence $L_a \sigma_d L_a^{-1} = \sigma_\delta$. This shows that $\mathcal{R}$ is stable under conjugation by any element   . The remainder of the statement is a standard consequence of the fact that $\widetilde{M}$ is generated by a normal subset of involutions. In case $(G,+)$ is not a elementary abelian 2-group, $-\mathrm{Id}$ is an order 2 element, and thus it cannot belong to $\mathrm{Inn}(G,+)$. □

3. GEOMETRICAL MOTIVATIONS

3.1. In what follows k is a field whose algebraic closure is $\bar{k}$ and $\mathbb{P}_n(k)$ is the n-dimensional projective space over k. Let V be an absolutely irreductible cubic hypersurface defined over k : the points of V are those whose homogeneous coordinates satisfy an equation of the type $F(X_0, X_1, \ldots X_n) = 0$ where F is a homogeneous form of the third degree over k which is assumed to be irreductible in $\bar{k}[X_0, X_1, \ldots X_n]$. Denote by $E = V_r(k)$ the set of non-singular k-points of V. By definition three points x, y, z of E are said to be *collinear* iff there exists a line such that :

- either $x, y, z \in \ell \subset V$
- or $\ell \cdot V = x + y + z$ (intersection cycle ; each point turning up equally often as its intersection multiplicity).

The so-defined three-place relation of collinearity L(x, y, z) is clearly invariant under any permutation of x, y, z, and for any x, y in E there exists at least one point z in E such that L(x, y, z). We must now distinguish between the case of a curve and the case dim V ≥ 2.

3.2. The case n = 2 (i.e. dim V = 1)

Proposition : If V is absolutely irreductible cubic plane curve of $\mathbb{P}_2(k)$ then : (i) V admits at most one singular point (either E = V or $E = V \setminus \{s\}$ where s is the unique singular point) ; (ii) no line is contained in V ; (iii) for any two (not necessarily distinct) points x and y from E there is only one point z in E such that L(x,y,z), and only one line satisfying $\ell \cdot V = x + y + z$; (iv) the operation $x, y \to x \circ y = z$ turns E into an entropic cubic quasigroup : (v) each operation of the form $x, y \to x \underset{u}{*} y = u \circ (x \circ y)$, where u is an arbitrary fixed element of E, turns E into an abelian group (Lamé's theorem). □

The foregoing facts are mostly classical, though part (iv) is very seldom explicity stated in the literature.

3.3. From now on *assume* $n > 2$, so that dim $V = n - 1 \geq 2$ (the conclusions are not valid for a curve). We also assume that k is infinite (this ensures the existence of some points). One observes at once that parts (i) (ii) and (iii) of proposition 3.2 fail to hold here. In order to recover a uniquely defined binary operation one must replace $E = V_r(k)$ by a suitable factor set $\bar{E} = E/R$. In the sequel R designates a binary relation on E, say $R \subset E \times E$; we write xRy for $(x,y) \in R$.

Definition : R is said to be an *admissible relation* if R is an equivalence relation on E such that for any five points x, y, z, y', z' from E such that L(x, y, z), L(x, y', z') and yRy' one has zRz'. In the next statements R is any given admissible relation on E, and a typical element of the factor set $\bar{E} = E/R$ is $X = \bar{x}$ where $x \in E$.

Lemma 3.4 : Given any two elements X and Y from $\bar{E}$ the set :

$$Z = \{z \in E \mid \exists\, x \in X, \exists\, y \in Y, L(x, y, z) \text{ holds}\}$$

is an equivalence class modulo R. The operation $X, Y \to Z = X \circ Y$ makes $\bar{E}$ into a symmetric quasigroup. □

Proof : Let x_0, y_0 be representatives of X and Y respectively. There exists z_0 in E such that $L(x_0, y_0, z_0)$ holds. Consider any triple of collinear points x, y, z with $x \in X$ and $y \in Y$. Let z' be a point from E such that $L(x_0, y, z')$ holds. Since $y_0 R y$ the admissibility implies that $z_0 R z'$. Likewise $x_0 R x$ implies $z' R z$, hence $z' \in \bar{z}_0$, namely $Z \subset \bar{z}_0$. Conversely if $t \in \bar{z}_0$, then any point y' such that $L(x, y', t)$ holds belongs to Y, so that $Z = \bar{z}_0$. The initial inclusion was sufficient to ensure the existence of an operation o on E such that $L(x, y, z) \Longrightarrow \bar{x} \circ \bar{y} = \bar{z}$. This operation is of course symmetric.

3.5. Let us come back to the hypothesis of 3.1 and 3.3. For any $x \in E$ denote by T_x the tangent hyperplane at x with V. Let $C_x = T_x \cap V$. Of course C_x is a (possibly degenerate) cubic hypersurface in T_x. Following Manin we shall say that a *point* x from E is *of general type* if the following two conditions are satisfied :

(1) C_x is geometrically irreductible and reduced ;

(2) there exists a line ℓ containing x and another point y from $C_x \setminus \{x\}$ which is not completely contained in C_x (this later condition means that x is not "conical" in C_x). Denote by $\mathcal{G}(V)$ the (possibly empty) open set of all the points of general type of E. For any x in E there exists a unique birational map $t_x : V \to V$ such that $t_x(y)$ is defined for every $y \in V \setminus C_x$ and such that $L(x, y, t_x(y))$ holds (Manin, [5], II, 12-13).

Theorem 3.6. (*Manin's theorem*) : Let V be an absolutely irreductible cubic hypersurface of dimension ≥ 2 over an infinite field k. Let $E = V_r(k)$ be the set of its non-singulat k-points. Assume that $\mathcal{G}(V) \neq \phi$. Consider some admissible relation R on E and let $(\bar{E}, o)$ be the corresponding symmetric quasigroup. Then every class modulo R is dense in the Zariski topology and admits a representative in $\mathcal{G}(V)$. Further $(\bar{E}, o)$ is a cubic quasigroup in which $(x^2)^2 = x^2$. □

Remark 3.7. $\bar{E}$ *can be non-trivial only* if $k \neq \bar{k}$ (if k is algebraically closed there is just one class [Manin 74, II.1.3.1. (i)]). But *it is an open question whether* $(\bar{E}, o)$ *can be non-entropic in the general case* (see [Manin, [5], II, Problem 11.11]).

3.8. Let (Q,) be a cubic quasigroup. Since $x^2 \cdot y^2 = (xy)^2$ *the square mapping* $x \to p(x) = x^2$ *is an endomorphism of* $(Q, \cdot)$. Hence its image $Q^2 = \{x^2 \mid x \in Q\}$ is a subquasigroup and the set Idemp$(Q, \cdot)$ of idempotent elements of $(Q, \cdot)$ is either empty or a subquasigroup.

Proposition 3.9. The three following conditions are equivalent : (i) all the squares are idempotent ($(x^2)^2 = x^2$) ; (ii) $(Q, \cdot)$ admits an idempotent and its related CML has exponent 6. (iii) $(Q, \cdot)$ is a direct product $A \quad D$ of an elementary abelian 2-group A by a distributive cubic quasigroup D. When these conditions are satisfied the factors of the decomposition in (iii) are essentially unique, and more precisely :

- $D \simeq Q^2 = \mathrm{Idemp}(Q, \cdot)$
- all the fibres of p are groups isomorphic to A. □

Proof : Each of the conditions (i) (ii) (iii) ensures the existence of at least one idempotent e. Let us fix e for all the proof and set $x + y = e \cdot xy$. Since $e^2 = e$, $xy = -x - y$ and $x^2 = -2x$.

Starting from (i) $x^2 \cdot x^2 = x^2$, one deduces : $2x^2 = x^2 + x^2 = e \cdot x^2 x^2 = e \cdot xx = 2x$, hence $-4x = 2x$ and $6x = e$. If one assumes that $(Q, \cdot)$ satisfies (ii), then $p(x) = x^2 = -2x$ is an idempotent endomorphism of $(Q,+)$ and any x can be written uniquely as a sum of the form $x = a_x + d$, where $d \in \mathrm{Im}(p)$ and $a_x \in \mathrm{Ker}(p)$ (in fact : $d = x^2$ and $a_x = x - x^2$). Since a CML without order 3 elements must be an abelian group, $(A,+)$ is an elementary abelian 2-group and $\forall\, a, b \in A$, $a + b = -(a+b) = a \; b$ (the operations $(+)$ and $(\cdot)$ coincide in A). Since $y^2 = y$ for any y in Q^2, $(Q^2, \cdot)$ is a distributive cubic quasigroup. Further for $x = a_x + x^2$ and $y = b_y + y^2$ we have $xy = -x - y = a_x - x^2 - y^2 = a_x b_x + x^2 y^2$, whence $x \to (a_x, x^2)$ is an isomorphism from $(Q, \cdot)$ onto $(A, \cdot) \times (Q^2, \cdot)$. Lastly if one assumes (iii), if $x = (a,d)$, $a \in A$, $d \in D$, then $x^2 = (0_A, d) = (x^2)^2$, so that $x^2 = x$ iff $a = 0_A$. □

The preceding proposition has two immediate consequences, both of geometrical interest.

Corollary 3.10 : If $(Q, \cdot)$ is a cubic quasigroup in which $(x^2)^2 = x^2$, then using the notations of XI.3.9 its multiplication group is isomorphic to the direct product of A by the centerless Fischer group $\mathrm{Mlt}(Q^2, \cdot)$. □

Corollary 3.11 : If $(Q, \cdot)$ is a cubic quasigroup in which $(x^2)^2 = x^2$ identically, then $(Q, \cdot)$ is non-entropic iff Q^2 is non-entropic, or equivalently iff Q^2 contains an 81-order subquasigroup isomorphic to $(L_3, \cdot)$). □

We just mention here that CMLs arose in combinatorial contexts. There is a number of papers devotes to the Hall triple systems ; now it is well-known that these special Steiner triple systems are merely another way to describe the exponent 3 CMLs (or equivalently, the distributive totally symmetric quasigroups).

REFERENCES

[1] Beneteau L., Free Commutative Moufang Loops and Anticommutative Graded Rings, Journ. Alg. 67 (1980) pp. 1-35 ; MR 82c 20 118.

[2] Beneteau L., Contribution à l'étude des Boucles de Moufang commutatives et des espaces apparentés (Algèbre, Combinatoire, Géométrie), Thèse d'Etat, Univ. de Provence, Marseille (1981).

[3] Bruck R.H., A survey of binary systems, Springer, Berlin, New-York, (1958) ; M.R. 20 # 76.
[4] Kepta T. Structure of triabelian quasigroups, Comment. Math. Univ. Carolinae 17 (1976).
[5] Manin Yu.I., Cubic forms ; algebra, geometry, arithmetics, North-Holland Publishing Company, Amsterdam, London, New York, 1974.
[6] Soublin J.P., Etude algébrique de la notion de moyenne, Journ. Math. pures et appl., Sér. 9, 50, (1971) pp. 53-264 ; M.R. 45 # 436, # 437 et # 438.

Annals of Discrete Mathematics 37 (1988) 31–42

BLOCKING SETS IN THE LARGE MATHIEU DESIGNS, I : THE CASE S(3,6,22)

Luigia BERARDI

Dipartimento di Ingegneria Elettrica. Universita' de L'Aquila, Italia

A classification of the blocking sets in S(3,6,22) is given. Each of them is characterized by incidence properties.

1. INTRODUCTION

Denote by $(S,\mathcal{B})$ a pair where S is a v-set of elements called points and $\mathcal{B}$ is a family of subsets of S called blocks. The pair $(S,\mathcal{B})$ is called a Steiner system S(t,k,v), where t,k,v are integers with $2\leq t<k<v$, if:

i) every block contains k points,

ii) every t-subset of S is contained in exactly one block of $\mathcal{B}$.

Denote by r_s (s=0,1,...,t) the constant number of blocks that contain a fixed s-subset of S. Then:

$$r_s = \binom{v-s}{t-s}\Big/\binom{k-s}{t-s} \qquad s=0,1,\dots,t. \tag{1.1}$$

Consider a fixed $x\in S$ and the family $\mathcal{B}_x = \{B-\{x\},\ x\in B\in\mathcal{B}\}$; the pair $(S-\{x\}, \mathcal{B}_x)$ is a Steiner system S(t-1,k-1,v-1) called the contraction of $(S,\mathcal{B})$ at point x.

We don't know examples of Steiner systems with $t>5$. There are two very special types of Steiner systems with t=5, namely S(5,6,12) and S(5,8,24) which toghether with their contractions are respectively called the little and the large Mathieu designs. These systems are uniquely determined by their parameters. They have been the subject of several studies for the following reasons: the authomorphism groups of S(5,6,12), S(4,5,11), S(5,8,24), S(4,7,23) S(3,6,22) are respectively Mathieu's five classical groups M_{12}, M_{11}, M_{24}, M_{23}, M_{22} i.e. the first five sporadic groups discovered in 1860, cf. [6], [10].

My purpose is to contribute to the study of these designs with regard to their blocking sets.

A blocking set in a Steiner system is a set C such that every block intersects C but no block is contained in C. Note that the complement of a blocking set is a blocking set too. A blocking set C is said to be reducible if there exists a point $x\in C$ such that $C-\{x\}$ is a blocking set. Otherwise, C is said to be irreducible.

In [1] we obtained a very simple characterization for blocking sets in S(5,6,12) and their contractions. The blocking sets in S(5,6,12) are the 6-sets which are not blocks. The system S(4,5,11) does not contain blocking sets, while in the inversive plane S(3,4,10), of order 3, the blocking sets are the union of two 2-secant blocks in which we delete a common point, and with symmetric difference distinct from a block (cf. [5]). Finally, as proved in [8], the system S(2,4,9), that is AG(2,3), does not contain blocking sets.

In [2] we obtained the characterization of blocking sets in S(2,5,21) $\simeq$ PG(2,4), that is the contraction of S(3,6,22).

The aim of this paper is to give a characterization of blocking sets in S(3,6,22) that is the second contraction of the big system S(5,8,24). In S(3,6,22) we have the following examples of blocking sets:

EXAMPLE (a). (F-Fano sets). In S(3,6,22) there exists a 7-set of type (1,3), i.e. every block is 1-secant or 3-secant. The set $\{1,2,3,8,15,16,19\}$, with regard to the pattern of S(3,6,22) given in Section 2 (Constr. I) is one of them. In Section 3 we shall characterize these sets that we call Fano sets.

EXAMPLE (b). (D-sets). Suppose B, B' are two blocks in $S(3,6,22)$ with $B \cap B' = \emptyset$. Fix $u \in B$, $v \in B'$. The set

$$D := (B - \{u\}) \bigcup (B' - \{v\})$$

is an irreducible blocking set. (See next 4.1).

Example (c). (E-sets). In $S(3,6,22)$ we consider the 11-sets

$$E := F \cup B - \{x\}$$

where B is a 1-secant block of the Fano set F at the point x. In section 7 we shall prove that such sets are sets of type $(1,3,5)$. They are irreducible blocking sets.

In this paper we prove that the above examples are the irreducible blocking sets of $S(3,6,22)$. Namely I am going to prove.

1.1 THEOREM. Denote by C a blocking set in $S(3,6,22)$. Then $|C| \geq 7$, and

(a) $|C| = 7$ implies that C is a Fano set, (section 3).

(b) $|C| = 8$ or 9 implies $C = F \cup X$ with $X \cap F = \emptyset$ and $|X| = 1$ or $|X| = 2$, respectively (section 4 and 5).

(c) $|C| = 10$. Then either $C = D$ (irreducible case) or $C = F \cup X$, where X is a 3-set with $F \cap X = \emptyset$, such that the block containing X is 1-secant to F. (Section 6)

(d) $|C| = 11$. Then C is one of the following sets
   (i) C is an E-set i.e. a set of type $(1,3,5)$, (irreducible case).
   (ii) $C = D \cup \{w\}$, $w \notin D$.
   (iii) $C = F \cup X$ where X is a 4-set with $F \cap X = \emptyset$ such that every block containing a 3-set of X is 1-secant to F. (Section 7).

(e) $|C| \geq 12$ implies that C is reducible and C is the complement of a blocking set considered above. (Section 8).

The classification of blocking sets in $S(4,7,23)$ and in $S(5,8,24)$ will be appear in two next papers.

2. RESULTS AND LEMMAS

Denote by C a c-set in $S(t,k,v)$. Following G.TALLINI (cf. [8]), we denote by t_i $(i=0,1,\ldots,k)$ the number of blocks that are i-secant C. We have:

$$\sum_{i=0}^{k} \binom{i}{s} t_i = \binom{c}{s} r_s \qquad (s=0,1,\ldots,t). \tag{2.1}$$

In the case of $S(3,6,22)$ we have $r_3 = 1$, $r_2 = 5$, $r_1 = 21$, $r_0 = 77$ and (2.1) becomes:

$$\begin{cases} t_0 + t_1 + t_2 + t_3 + t_4 + t_5 + t_6 = 77 \\ t_1 + 2t_2 + 3t_3 + 4t_4 + 5t_5 + 6t_6 = 21c \\ 2t_2 + 6t_3 + 12t_4 + 20t_5 + 30t_6 = 5c(c-1) \\ 6t_3 + 24t_4 + 60t_5 + 120t_6 = c(c-1)(c-2) \end{cases} \tag{2.2}$$

There are two different patterns of $S(3,6,22)$ to which we shall refer to later on.

CONSTRUCTION I. Let us consider TODD's table of S(5,8,24), given in [10] . If we contract at points $0,\infty$ we obtain the following list of blocks of S(3,6,22) for the set $\{1,2,\dots,22\}$.

5 6 7 13 16 17	2 3 4 8 9 21	1 2 3 5 14 17	3 4 5 12 13 18
5 6 8 12 14 21	2 3 6 12 16 26	1 2 4 13 16 22	3 4 6 7 14 22
5 7 11 14 18 19	2 3 7 11 13 15	1 2 6 7 19 21	3 4 10 15 16 17
5 8 9 10 17 18	2 3 10 18 19 22	1 2 8 11 12 18	3 5 6 9 15 19
5 10 13 14 15 22	2 4 5 6 10 11	1 2 9 10 15 20	3 5 7 10 20 21
5 11 12 15 17 20	2 4 7 17 18 20	1 3 4 11 19 20	3 5 8 11 16 22
6 7 8 9 11 20	2 4 12 14 15 19	1 3 6 8 10 13	3 6 11 17 18 21
6 7 10 12 15 18	2 5 7 9 12 22	1 3 7 9 16 18	3 7 8 12 17 19
6 9 10 16 21 22	2 5 8 13 19 20	1 3 12 15 21 22	3 8 14 15 18 20
6 10 14 17 19 20	2 5 15 16 18 21	1 4 5 7 8 15	3 9 10 11 12 14
6 11 12 13 19 22	2 6 8 15 17 22	1 4 6 9 12 17	3 9 13 17 20 22
7 8 13 18 21 22	2 6 9 13 14 18	1 4 10 14 18 21	3 13 14 16 19 21
7 9 14 15 17 21	2 7 8 10 14 16	1 5 6 18 20 22	4 5 9 14 16 20
7 15 10 19 20 22	2 9 11 16 17 19	1 5 9 11 13 21	4 5 17 19 21 22
8 9 12 13 15 16	2 11 14 20 21 22	1 5 10 12 16 19	4 6 8 16 18 19
8 10 11 15 19 21	2 10 12 13 17 21	1 7 10 11 17 22	4 6 13 15 20 21
9 12 18 19 20 21	1 6 11 14 15 16	1 8 9 14 19 22	4 7 9 10 13 19
10 11 13 16 18 20	1 7 12 13 14 20	1 13 15 17 18 19	4 7 11 12 16 21
12 14 16 17 18 22	1 8 16 17 20 21	4 9 11 15 18 22	
4 8 10 12 20 22	4 8 11 13 14 17		

CONSTRUCTION II (Lüneburg [7]). Let us consider the point-set of PG(2,4) and a new point ∞. In S, with $|S|=22$, we give the following families of blocks:
a) Set $\{L\cup\{\infty\}\}$, where L is a line of PG(2,4).
b) A class of 56 hyperovals constructed in the following way. Let $\mathcal{H}$ be the set of the 168 hyperovals of PG(2,4). If $H_1,H_2\in\mathcal{H}$, we say that

$$H_1 \sim H_2 \qquad \text{iff} \qquad |H_1\cap H_2| = 0,2 \text{ or } 6 \ .$$

The relation $\sim$ is an equivalence relation. We have exactly 3 equivalence classes, each of which contains 56 hyperovals. We can assume each of these 3 classes as the class of 56 blocks.

We recall the following

2.1 RESULT (W.Jónsson [6], 3.2, 3.4, 3.5).Let B,B' be two blocks of S(3,6,22). Then either $|B\cap B'|=0$ or $|B\cap B'|=2$.For any fixed B there are 16, or 60 blocks B' such that $B\cap B'=\emptyset$ or $|B\cap B'|=2$, respectively . Moreover, if x is a fixed point, $x\notin B$, there are 6 blocks B' with $x\in B'$ and $B\cap B'=\emptyset$ and 10 blocks with $x\notin B'$ and $B\cap B'=\emptyset$.

The following lemmas will be useful throughout the paper.

2.2 LEMMA. Let B,B' be two blocks of S(3,6,22) with $|B\cap B'|=2$. Let x be a point with $x\notin B\cup B'$. The set $B\cup B'\cup\{x\}$ has two external blocks.

PROOF. Step 1. First, we prove that $B\cup B'$ has exactly 4 external blocks. Denote by t_i the characters of $B\cup B'$. We have $t_6=2$. Moreover, $t_5=t_1=0$, since each block intersecting $B\cup B'$ is at least 2-secant and at most 4-secant $B\cup B'$. The system of characters has the solution: $t_6=2$, $t_5=0$, $t_4=12$, $t_3=32$, $t_2=27$, $t_1=0$, $t_0=4$.

Step 2. Denote by E_1,E_2, E_3, E_4 the four external blocks to $B\cup B'$. These blocks are 2-secant two by two. In fact, if $E_i\cap E_j=\emptyset$, every other block which is different from them, should intersect at least B or B', a contradiction. Then $E_i \cap E_j \neq \emptyset$, that is $|E_i \cap E_j|=2$.

Step 3. Blocks E_h and E_k are 4-secant $E_i \cup E_j$. In fact, in S(3,6,22) there are only two points x,y outside to $B \cup B' \cup E_i \cup E_j$. Moreover E_h and E_k are external to $B \cup B'$. Consequently, they must have 4 points in $E_i \cup E_j$ and $E_h \cap E_k = \{x,y\}$.

Step 4. Finally, we prove that set $B \cup B' \cup \{x\}$ has two external blocks. Step 3 implies $(E_i \cup E_j) \cap E_h \cap E_k = \emptyset$. So, two of four external blocks to $B \cup B'$ pass through each point x outside $B \cup B'$.

Then the assertion is proved.

2.3 LEMMA. Let B,B' be two blocks with $|B \cap B'|=2$. Denote by W the complement point-set of $B \cup B'$, and by $\mathscr{E}=\{E_1, E_2, E_3, E_4\}$ the set of external blocks to $B \cup B'$. Then the pair $(W, \mathscr{E})$ is an 1-(12,6,2) design.

PROOF. Let E_1, E_2 be two external blocks to $B \cup B'$. Then $|E_1 \cap E_2|=2$,necessarily. Put $\{y,z\} := W - E_1 \cup E_2$. The block E_3 (and then E_4) must have 2 points of W outside $E_1 \cup E_2$, and so it contains x and y. Thus, each point of W is incident to two blocks of $\mathscr{E}$, and $(W, \mathscr{E})$ is a 1-(12,6,2) design.

2.4 LEMMA. Let x,y be two points of S(3,6,22). Denote by B_1, B_2, B_3 three blocks through x,y. Then the set $C=B_1 \cup B_2 \cup B_3$ has these characters: $t_6=7$, $t_4=56$, $t_2=14$, $t_0=0$, $t_1=t_3=t_5=0$, in other words C is of type (2,4,6).

PROOF. It is clear that $t_1=t_3=t_5=0$. Then (2.2) implies the assertion.

2.5 REMARK. The complement of the set C defined in 2.4 is a set of type (0,2,4).

2.6. LEMMA. The set of four blocks which pass through two fixed points has these characters: $t_6=20$, $t_4=24$, $t_2=1$, $t_0=t_1=t_3=0$, $t_5=32$.

PROOF. Similar to 2.4.

2.7 LEMMA. The 12-set of two disjoint blocks B,B' of S(3,6,22) is of type (2,4,6).

PROOF. Set $B \cup B'$ has $t_1=t_3=0$, $t_6=2$,since each other block intersects B (or B') in 0 or 2 points. From (2.2) we obtain $t_0= t_5=0$ and $t_2=30$, $t_4=45$.

2.8 LEMMA. Let B,B' be two blocks of S(3,6,22) with $B \cap B'=\{p,q\}$. Fix $u \in B \setminus B'$ and $a \in B' \setminus B$. Denote by B'', B''' the blocks through p,u,a and q,u,a respectively. The set $\bar{H} := B \cup B' \cup B'' \cup B'''$ is the complement of a block.

PROOF. Put $I= \{p,q,u,a\}$ and $\mathscr{H}=\{B,B',B'',B'''\}$ .We investigate the characters of $\bar{H}$. Obviously $t_1=0$. We prove that $t_3=0$. If there exists a block B_0 3-secant to $\bar{H}$, then B_0 contains at least one element of I, otherwise B_0 should be 1-secant to a block of $\mathscr{H}$. If B_0 contains only one element of I,B_0 is 1-secant to one block of $\mathscr{H}$, since 3 blocks pass through each point of I and B_0 should contain two other points on two of these 3 blocks. So B_0 contains at least 2 points of I. If B_0 contains 2 points of I and one outside I, B_0 is 2-secant to three blocks of $\mathscr{H}$ (those containing the two elements of I and the one which the third point is on) and 1-secant to one block of $\mathscr{H}$(that contains only one of the 2 points of I said above). So B_0 should have 3 points of I. But it is impossible, since through any 3 elements of I there is one block of $\mathscr{H}$, which is 6-secant to $\bar{H}$. Consequently $t_3=0$.

We prove that $t_5=0$. If B_0 is 5-secant to the set $\bar{H}$, then B_0 contains at least a point of I, otherwise B_0 should be 1-secant to one block of $\mathscr{H}$. B_0 cannot contain only one point of I, since B_0 should contain another point on each of the 3 blocks through this point and only one point on the remaining block of $\mathcal{H}$. The block B_0 cannot contain two points of I, since, in this case, B_0 could be 4-secant to $\bar{H}$. B_0 cannot contain 3 points of I, since B_0 should be a block of $\mathcal{H}$, which is 6-secant to $\bar{H}$. Hence $t_5=0$. Then $\bar{H}$ is a 16-set with $t_1=t_3=t_5=0$. By (2.2) we obtain $t_0=1$, $t_2=2$, $t_4=60$, $t_6=16$. So $\bar{H}$ has one external block, exactly.

3. BLOCKING SETS OF MINIMAL CARDINALITY

In [1] we proved that if C is a blocking set in S(3,6,22), then

$$7 \leq |C| \leq 15. \tag{3.1}$$

In this section we deal with blocking sets of seven points. We shall prove Theorem 1.1 (a). In view of example (a), there exists a blocking set of minimal cardinality 7. By example (a) the following is a non-empty definition:

3.1 DEFINITION. A set F of points in S(3,6,22) of type (1,3) is called a Fano set. We prove that:

3.2 THEOREM. The blocking sets in S(3,6,22) of minimal size are exactly the Fano sets.

PROOF. Suppose F is a Fano set, i.e. a set of type (1,3) in S(3,6,22). By (2.2) we obtain $t_1=42$, $t_3=35$, $c=7$. So a Fano set is a blocking set of minimal size in view of (3.1).

Suppose now that C is a blocking set with $|C|=7$. By (2.2) with $t_0=t_6=0$, $c=7$ we obtain:

$$t_4=t_5=t_2=0, \quad t_3=35, \quad t_1=42.$$

So, the assertion follows.

We recall that in PG(2,4) there are two classes of sets of type (1,3): the Fano (or Baer) subplanes and the Hermitian arcs. In S(3,6,22) only the class of Fano sets is.

Next, let us look at the connection between a Fano subplane in PG(2,4) and a Fano set of S(3,6,22).

In PG(2,4) we consider a conic $\mathcal{C}$, a tangent T at a point t of $\mathcal{C}$ and the knot k of $\mathcal{C}$ which also lies on T. The set $\mathcal{C} \cup T-\{t\}$ is a blocking set of PG(2,4) which is reducible since the set $\mathcal{C} \cup T-\{t,k\}$ is a blocking set (of 7 points) too. Obviously, each Fano subplane can be constructed in this way. We prove:

3.3 THEOREM. Let F(2) be a Fano subplane of PG(2,4). Denote by T a 3-secant line of F(2). The symmetric difference set $\mathcal{K}:=T \,\triangle\, F(2)$ is a hyperoval of PG(2,4). Vice versa, if $\mathcal{K}$ is a 6-arc of PG(2,4) and L is a 2-secant line of $\mathcal{K}$, then set $\mathcal{K} \,\triangle\, L$ is a Fano subplane.

PROOF. We prove that any line M has either zero or two points in $\mathcal{K}$. Suppose that M is 1-secant to F(2). If M intersects F(2) at a point of $F(2) \cap T$, then M is 0-secant to $\mathcal{K}$. If M intersects F(2) at a point outside T, then M is 2-secant to $\mathcal{K}$. If M=T, then M is 2-secant to $\mathcal{K}$ by construction. If $M \neq T$ is 3-secant to F(2), then M is 1-secant to T at a point of $T \setminus F(2)$, since two 3-secants of F(2) intersect at a point of F(2). Consequently M is 2-secant to $\mathcal{K}$.

Conversely, let $\mathcal{K}$ be a 6-arc of PG(2,4) and L a 2-secant of $\mathcal{K}$. Put $H := \mathcal{K} \,\triangle\, L$. Line L is 3-secant to H. If M is a 0-secant line of $\mathcal{K}$, then M is 1-secant to L at a point of $L \setminus \mathcal{K}$. Suppose that M is a 2-secant line of $\mathcal{K}$. If M intersects $\mathcal{K}$ at a point of $L \cap \mathcal{K}$, then M is 1-secant of H; if M intersects $\mathcal{K}$ at two points of $\mathcal{K} \setminus L$, then M is 3-secant to H. So $\mathcal{K} \triangle L$ is a Fano subplane.

Let F(2) be a Fano subplane and $\mathcal{A}$ the family of hyperovals of PG(2,4) constructed by starting from any 3-secant line to F(2). We call $\mathcal{A}$ the family of the 6-arcs associated with F(2). Clearly, each 6-arc determines one subplane F(2), but one subplane F(2) fixes a family $\mathcal{A}$ of 6-arcs. We prove.

3.4 THEOREM. Let F(2) be a Fano subplane of PG(2,4) and $\mathcal{A}$ the family of the 6-arcs associated with F(2). Any two 6-arcs of $\mathcal{A}$ have two common points.

PROOF. Let T_1 and T_2 be two lines of PG(2,4) which are 3-secant to F(2). Denote by $\mathcal{K}_1 = T_1 \triangle F(2)$ and $\mathcal{K}_2 = T_2 \triangle F(2)$ the two arcs of $\mathcal{A}$ constructed by starting from T_1 or T_2, respectively. It is very easy to prove that $\mathcal{K}_1 \cap \mathcal{K}_2 = F(2) - (T_1 \cup T_2)$. So $|\mathcal{K}_1 \cap \mathcal{K}_2| = 2$.

We note that each Fano set of $\underline{S}$=(3,6,22) is a Fano subplane in the contraction at a point $x \notin F$. We ask about the vice versa.

For the vice versa we work in PG(2,4). We consider a Fano subplane F(2) and the family $\mathcal{A}$ of the 6-arcs associated with F(2). Note that one of these 6-arcs is a block if and only if each 6-arc of $\mathcal{A}$ is a block, since by 3.4 two 6-arcs of $\mathcal{A}$ are in the same Lüneburg class.

Now we are ready to characterize the Fano subplane which are Fano sets too. We prove:

3.5 THEOREM. A Fano subplane F(2) in the contraction of $\underline{S}$=S(3,6,22) at a point $x \notin F(2)$ is a Fano set of $\underline{S}$ if and only if the family $\mathcal{A}$ of 6-arcs associated with F(2) does not contain blocks of $\underline{S}$.

PROOF. Let F a Fano set. We consider the contraction of S at $x \notin F$. Since F is of type (1,3), it follows that F is a Fano subplane in PG(2,4).

Denote by F(2) a Fano subplane of PG(2,4). So there is a conic $\mathcal{C}$, a point $t \in \mathcal{C}$, a 1-secant line T to $\mathcal{C}$ at point t, such that $F(2) = \mathcal{C} \cup T - \{t,k\}$, where k is the knot of $\mathcal{C}$. Denote by $\mathcal{K} := \mathcal{C} \cup \{k\}$ one of the 6-arcs associated with F(2). Note that if $\mathcal{K}$ is a block of S(3,6,22), set F(2) cannot be a Fano set, since $\mathcal{K}$ should be a 4-secant block. Assume that $\mathcal{K}$ is not a block of S(3,6,22). So, by 3.4, no 6-arc of $\mathcal{A}$ can be a block. Each block B of S(3,6,22) which does not pass through pole ∞, has 1 or 3 common points with $\mathcal{K}$, since $\mathcal{K}$ and B are two 6-arcs which are in two different Lüneburg's classes.

a) Suppose that B has 1 common point with $\mathcal{K} - \{t,k\}$. Block B intersects block $T \cup \{\infty\}$ in 0 or 2 points, since they are blocks of S(3,6,22). So B intersects F(2) in 1 or 3 points.

b) Suppose that $t \in B$ and $k \notin B$. Then B contains another point x of $T - \{t,k\}$ (since $T \cup \{\infty\}$ is a block) and 0 or 2 points of $\mathcal{K} - \{t,k\}$. Then B is 1 or 3-secant to F(2).

c) Now suppose that $k \in B$ and $t \notin B$. Since $\mathcal{C}' = \mathcal{C} - \{t\} \cup \{k\}$ is a conic and T is a tangent line of $\mathcal{C}'$ at k, we can change t with k. So all is as in point b).

d) Suppose that $t,k \in B$. Then B is 0-secant to $T - \{t,k,\infty\}$ and 1-secant $\mathcal{K} - \{t,k\}$.

e) Finally, suppose that B has 3 common points with $\mathcal{K} - \{t,k\}$. We shall prove that $B \cap (T \cup \{\infty\}) = \emptyset$. On the contrary, suppose that $B \cap (T - \{t,k,\infty\}) = \{x,y\}$. Consider the lines xz and yz, where z is a point of $\mathcal{C} \cap B$. These two lines do not contain k, so they are secant to $\mathcal{C}$. Then one of them, say xz, contains two points of $B \cap (\mathcal{C} - \{t,k\})$. So the blocks B and $xz \cup \{\infty\}$ should have 3 common points, a contradiction.

4. BLOCKING SETS WITH EIGHT POINTS

In this section we shall prove point (b) of 1.1 for the blocking sets C with $|C|=8$. Moreover, we shall prove that the 10-set D defined in Example (b) of Introduction is an irreducible blocking set.

4.1 LEMMA. Let B,B' be two blocks of S(3,6,22) with $B \cap B' = \emptyset$. Let u,v be two points with $u \in B$, $v \in B'$. Then the 10-set

$$D := (B - \{u\}) \cup (B' - \{v\}).$$

is an irreducible blocking set.

PROOF. Set D is a blocking set by 2.7. Now we prove that D is irreducible. Fix a point $x \in B$. Four blocks different from B pass through u,x. Only 3 of these 4 blocks intersect B'. So, each point x is incident to a block which is 1-secant to D. If $x \in B'$, the reasoning is the same.

4.2 THEOREM. In S(3,6,22) the blocking sets with 8 points are all reducible. They are exactly the sets $F \cup \{x\}$, being F a Fano set and x a point with $x \notin F$.

PROOF. Let C be a blocking set of S(3,6,22) with $|C|=8$. Since C is a blocking set, we have $t_0=t_6=0$. Then we obtain by (2.2):

$$t_4=7-4t_5 \quad , \quad t_3=28+6t_5 \quad , \quad t_2=14-4t_5 \quad , \quad t_1=28+t_5 \quad .$$

From $t_4 \geq 0$, it follows $0 \leq t_5 \leq 1$. We prove that $t_5=0$.

Suppose $t_5=1$. Let B be the 5-secant block to C. Denote by B' the block through the point of $B \setminus C$ and two points of C-B. C is contained in the union of $B \cup B'$ with a point. So C is not a blocking set by 2.2. Then $t_5 \neq 1$, i.e. $t_5=0$.

Since $t_5=0$, set C is also a blocking set in the contraction of S(3,6,22) with respect to a point b outside C. Then C is either the set $F \cup \{x\}$, where F is a Fano subplane and $x \notin F$, or the triangle Δ, [2]. We recall that in PG(2,4), contraction of S(3,6,22), the triangle Δ is defined as follows. Let L,M be two lines with $L \cap M = \{p\}$. Fix $u \in L$, $v \in M$. Fix a point $w \in uv$ with $w \neq u,v$. Then $\Delta := (L-\{u\}) \cup (M-\{v\}) \cup \{w\}$. Suppose $C = \Delta$ in PG(2,4). It follows that $B_1 := L \cup \{b\}$ and $B_2 := M \cup \{b\}$ are two blocks of S(3,6,22) with $B_1 \cap B_2 = \{b,p\}$. Since C is contained in $B_1 \cup B_2 \cup \{w\}$, C has external blocks in view of 2.2. So, since C is reducible, there exists a point x such that $C-\{x\}$ is a Fano subplane of PG(2,4). Finally, we prove that such Fano subplane is a Fano set of S(3,6,22), necessarily. In PG(2,4) we have $F = \mathscr{C} \cup T - \{t,k\}$, where $\mathscr{C}$ is a conic with knot k and T is a 1-secant line to $\mathscr{C}$ at $T \in \mathscr{C}$. If F is not a Fano set of S(3,6,22), then $\mathscr{C} \cup \{k\}$, by 3.3, is a block of S(3,6,22). In this case the blocking set C is contained in the union of two blocks $\mathscr{C} \cup \{k\}$ and $T \cup \{b\}$, a contradiction by 2.2.

5. BLOCKING SETS WITH NINE POINTS

In this section we shall prove point (b) of 1.1 for the blocking sets C with $|C|=9$. This is a consequence of some lemmas. We also explain the blocking set T defined in 5.3 which will appear in the next 6.2. The blocking sets N_0, N_1 defined in this Section (cf. 5.2, 5.4) are only ausiliary sets and they appear neither in 1.1 nor in other Sections.

5.1 LEMMA. Let B,B' be two blocks of S(3,6,22) with $|B \cap B'| = 2$. Denote by a,b two points of $B' \setminus B$. Let $\mathscr{E} = \{E_1, E_2, E_3, E_4\}$ be the set of the four blocks external to $B \cup B'$.

(a) There exist two blocks S_1, S_2 which are 2-secant to $B \cup B'$ at a and b.

(b) For every $x \in S_1$ (or S_2), with $x \neq a,b$ the two blocks E_i, E_j of $\mathscr{E}$ through x have the other common point y on S_1 (or S_2).

(c) One of the two points outside $B \cup B' \cup E_1 \cup E_2$ is in S_1 and the other in S_2.

PROOF. (a) Four blocks different from B' pass through a and b. Exactly two of these are 2-secant to $B \setminus B'$, so the other two intersect $B \cup B'$ only at a,b.

(b) Suppose $x \in S_1 - \{a,b\}$. Let E_1, E_2 be the two blocks of $\mathscr{E}$ through x (cf. 2.3) and denote by y the other point of $E_1 \cap E_2$.

Step 1. First we prove that $y \notin S_1 - \{a,b\}$. Assume on the contrary that $x,y \in S_1 - \{a,b\}$. Then the three blocks E_1, E_2, S_1 pass through x,y. Each of these blocks has no common point with B. This is a contradiction, since in the contraction with respect to x (or to y) there would be three lines of PG(2,4) through y (or x) external to 6-arc B.

Step 2. Now we prove that $y \in S_2 - \{a,b\}$. Assume the contrary. In the contraction with respect to x, the sets $E_1 - \{x\}$ and $E_2 - \{x\}$ are two lines through y, while $S_1 - \{x\}$ is a line which intersects 6-arc S_2 at a and b. The lines $E_1 - \{x\}$ and $E_2 - \{x\}$ intersect S_2 at points different from a and b. Since there cannot be 4 lines containing y and secant the 6-arc S_2, necessarily $y \in S_1 - \{x,a,b\}$. This is a contradiction in view of Step 1.

(c) The proof follows by counting arguments.

5.2 LEMMA. Use the same notations of 5.1. Fix a point $u \in B \setminus B'$ and denote by z the only point of S_2 outside $B \cup B' \cup E_1 \cup E_2$. Then

$$N_1 := B \cup B' - \{a,b,u\} \cup \{x,z\} \tag{5.1}$$

is a blocking set with nine points.

PROOF. The blocks that are not elements of $\mathscr{E} \cup \{S_1, S_2\}$ intersect N_1, since they intersect $B \cup B'$. Moreover, E_1, E_2, S_1 contain x, and E_3, E_4, S_2 contain z. So each block intersects N_1.

Suppose that there exists a block 6-secant to N_1. This block necessarily contains x and z; moreover, it is 2-secant to $B \setminus B'$ and $B' \setminus B$ at points of N_1. Consequently, it should be 1-secant to S_1 and S_2, a contradiction.

We shall use the following Lemma in the sequel.

5.3 LEMMA. Let B,B' be two blocks of S(3,6,22) with $|B \cap B'| = 2$. Fix $u \in B \setminus B'$ and $a \in B' \setminus B$. There are two points x,y outside $B \cup B'$ such that:

$$T := [B \cup B' - \{u,a\}] \cup \{x,y\} \tag{5.2}$$

is a reducible blocking set with ten points.

PROOF. Each block which intersects $B \cup B'$, intersects T. There are 4 blocks E_1, E_2, E_3, E_4 external to $B \cup B'$. In view of 2.2 (cf. Step 2), if $x \in E_1 \cap E_2$, $y \in E_3 \cap E_4$, then T has no external block.

Suppose that B_0 is a 6-secant block to T, necessarily unique; B_0 intersects $B' \setminus B$ at two points, $B \setminus B'$ at two points and must contain x and y. So, $|E_1 \cap B_0| = 1$, a contradiction.

Finally, we prove that T is reducible. Denote by p and q the points of $B \cap B'$. Let B'',B''' be the blocks through p,u,a and q,u, a respectively. By Lemma 2.8 the complement of set $B \cup B' \cup B'' \cup B'''$ is a block H. Since both x,y cannot be in H (otherwise there exists an external block to T by (2.2), then one of them lies in B'' or in B'''. If, for example, $x \in B''$, then $T - \{p\}$ is a blocking set. So, T is reducible.

5.5 LEMMA. Let B,B',B'' be three blocks with $|B \cap B' \cap B''| = 1$. Fix a point z of $B'' \setminus (B \cup B')$. Denote by E_1, E_2 the two external blocks to $B \cup B' \cup \{z\}$. Let y be a point of $E_1 \cap E_2$. Then

$$N_0 := \left[(B \cup B') \setminus B''\right] \cup \{y,z\} \tag{5.3}$$

is a blocking set with nine points.

PROOF. Put $B \cap B' \cap B'' = \{u\}$ and $(B \cup B') \cap B'' = \{u,v,w\}$. The blocks intersecting $B \cup B' \setminus B''$, intersect N_0. The block through u,v,w, i.e. B'', contains z. Each block through u and one of points v,w intersects N_0. Finally, blocks E_1 and E_2 contain y.

Suppose that B_0 is a 6-secant block to N_0. Then B_0 contains y,z and is 4-secant to $(B\setminus B')\cup(B'\setminus B)$ at points of N_0. So $|B\cap B''| = 1$, a contradiction.

Now we give a characterization of blocking sets with 9 points, in terms of N_0 and N_1.

5.6 LEMMA. Let C be a blocking set of S(3,6,22) with $|C| = 9$. Then $C=N_0$ or $C=N_1$, where N_0 and N_1 are defined in (5.3) and (5.1).

PROOF. First, we prove that there is at most one 5-secant block to C. On the other hand, suppose that B_1, B_2 are two of them. Obviously $B_1\cap B_2\neq\emptyset$. It follows that C is contained either in $B_1\cup B_2$ or in the union of B_1, B_2 and one point. This is a contradiction by 2.2.

Suppose that B is a 5-secant block to C. If we put $t_0=t_6=0$, $t_5=1$, $|C|=9$ in (2.1), we obtain $t_4=12$. Let B' be a 4-secant block to C. We prove that $B\cap B'$ contains two points of C. Assume the contrary. Then either $B\cap B'=\emptyset$ or $B\cap B'$ contains one point of C and one point outside C. In the first case C is strictly contained in the blocking set D (cf. 4.1), which is irreducible, a contradiction. The other case is impossible in view of 2.2.

Put $\{u\} = B\setminus C$, $\{a,b\}= B'\setminus C$ and $\{x,z\} = C\setminus(B\cup B')$. Denote by S_1, and S_2 the two blocks through a and b, which are 2-secant to $B\cup B'$. So x,z are necessarily as in (5.1). Consequently, $C=N_1$.

Suppose now that there is no 5-secant block to C. Let B be a 4-secant block to C. In the contraction PG(2,4) of S(3,6,22) at a point u of $B\setminus C$, set C is also a blocking set. The line $L=B-\{u\}$ is a 4-secant of C in PG(2,4). So in PG(2,4) set C cannot be a hermitian arc and it is necessarily formed by $(L-\{v\})\cup(M-\{w\})\cup\{z,y\}$ where L,M are lines, $v\in L$, $w\in M$, z,v,w are collinear, $y\notin L\cup M$ (cf. [2], Theor. 2.3). In S(3,6,22) we consider the blocks $B=L\cup\{u\}$, $B'=M\cup\{u\}$ and point z lying on block B" through u,v,w. So C verifies the hypothesis of 5.5 and point y must be fixed as in (5.3). Then $C=N_0$ and our lemma is completely proved.

5.7 THEOREM. Denote by F a Fano set in S(3,6,22). Suppose that $x,y\notin F$. Then the set $N :=F\cup\{x,y\}$ is a reducible blocking set with 9 points. Moreover, if B is a block such that $|B\cap F| =3$ and $x,y\in B$, then $N=N_1$, otherwise $N=N_0$.

PROOF. Set F is of type (1,3), consequently N is a reducible blocking set. Suppose that x,y are on a 3-secant block B of F. Then B is 5-secant to N, and in view of 5.6 $N=N_1$. If x,y are not in the same 3-secant block of F, then there is no 5-secant block to N. Then $N=N_0$ by 5.6.

So, the case 1.1(b), $|C| = 9$ is a consequence of the above theorems.

6. BLOCKING SETS WITH TEN POINTS

In this section we shall prove point (c) of 1.1. If C is a blocking set with $|C| = 10$, we have $t_0=t_6=0$ and we obtain by (2.2):

$$t_1=8+t_5 , \quad t_2=8+6t_5 , \quad t_3=33-4t_5 , \quad t_4=28-4t_5 .$$

These imply $t_5\leq 7$. So we have many distinct possibilities for t_5.

We begin with the following

6.1 THEOREM. Let C be a blocking set in S(3,6,22) with $|C| =10$. Then $t_5=2$ if and only if C is the irreducible blocking set D defined in 4.1.

PROOF. If $t_5=2$, we have $t_1=10$, $t_2=20$, $t_3=25$, $t_4=20$. If the two blocks B,B' that are 5-secant to C have no common point, then C=D (defined by 4.1). On the other hand, if $|B\cap B'|=2$, we have a contradiction. In fact both 2 common points must be in C, otherwise C would be contained in the union of two blocks and a point, a contradiction by 2.2. Then C is contained in $B\cup B'\cup\{x,y\}$, where $x,y\notin B\cup B'$. There are at most 4 distinct blocks through x,y that are 4-secant to C.(Denote by u_i the characters of blocks through x and y. We have $u_2+u_3+u_4=5$, $u_3+2u_4=8$, so $u_4\leq 4$).The other 4-secant blocks to C are either those 4-secant to C at points of $B\cup B'$ or those 3-secant to C at points of $B\cup B'$ and containing x or y.Count the 4-secant blocks to C at points of $B\cup B'$ Fix two points p,q of C in $B'\setminus B$ in $\binom{3}{2}$ ways. Denote by u_i the characters of blocks through p and q. We have: $u_2+u_3+u_4+u_5=5$, $u_3+2u_4+3u_5=6$, $u_5=1$. So, $u_4\leq 1$. Then the required blocks are at most $\binom{3}{2}$.
Count the 3-secant blocks to C at points of $B\cup B'$ and containing one point of $\{x,y\}$.
These blocks can be either those containing one point of $B\cap B'$, which are at most 6, or those containing the point v of $B\setminus C$ (or u of $B'\setminus C$) one point z of $B\cap C$ (or $B'\cap C$) and two points of $B'\cap C$ (or $B\cap C$). Denote by u_i the characters of blocks through v,z with respect to set $(B\cup B')\cap C$. We have $u_1+u_2+u_3+u_5=5$, $u_2+2u_3+4u_5=7$, $u_5=1$. It follows $u_3\leq 1$. Consequently these last 4-secant blocks to C are at most 6. Then $t_4\leq 4+3+6+6=19$, a contradiction.

Vice versa is trivial.

6.2 THEOREM. Let C be a blocking set with $|C|=10$ in S(3,6,22). If $t_5\neq 2$, then $C=F\cup X$, where $|X|=3$, $X\cap F=\emptyset$ and the block through X is 1-secant F.

PROOF. If $t_5\geq 3$, let B,B' be two 5-secant blocks to C. It is $|B\cap B'|=2$, since if $B\cap B'=\emptyset$, then it would be $t_5=2$. Moreover $B\cap B'\subset C$, otherwise C would have external blocks. So, C is contained in $B\cup B'\cup\{x,y\}$ with $x,y\notin B\cup B'$. The points x,y are necessarily as in (5.2) and C=T. Since T is reducible the assertion is proved. If $t_5\leq 1$, then $t_1\leq 9$, so C is reducible,and the assertion is also true in this case.

7. BLOCKING SETS WITH ELEVEN POINTS

In this section we shall prove point (d) of 1.1. We know that in S(3,6,22) there are 11-sets of type (1,3,5) (cf. Example (c), Introduction). If $|C|=11$ and $t_0=t_6=0$, we have by (2,2):

$$t_4=t_2=4(11-t_5)\ ,\quad t_3=6t_5-11,\quad t_1=t_5\ ,\tag{7.1}$$

necessarily with $2\leq t_5\leq 11$. We begin by proving:

7.1 LEMMA. A blocking set C in S(3,6,22) with $|C|=11$ is irreducible if and only if C is an 11-set of type (1,3,5).

PROOF. Suppose C is irreducible. Then $t_5=t_1\geq|C|=11$, and so $t_5=11$ and $t_4=t_2=0$. This means C is of type (1,3,5).

Now we prove that an 11-set C of type (1,3,5) is an irreducible blocking set. Assume the contrary and suppose that no tangent passes through a point $x\in C$. Then the blocks through x are either 3-secant or 5-secant C. In the contraction with respect to x the set $C-\{x\}$ is of type (2,4). Denote the characters of the 10-set $C-\{x\}$ with respect to lines of PG(2,4) by y_i. We should have: $y_2+y_4=21$, $2y_2+4y_4=50$, $2y_2+12y_4=90$, a contradiction.

7.2 REMARK. In S(3,6,22) the complement of an 11-set of type (1,3,5) is also an 11-set of type (1,3,5).

7.3 REMARK. Each 11-set of type (1,3,5) is an example of blocking set which is irreducible with its complement.

7.4 THEOREM. Let F be a Fano set of S(3,6,22). Denote by B a block such that $F \cap B = \{x\}$. In S(3,6,22) the sets of type (1,3,5) are exactly the sets E defined by $E := F \cup B - \{x\}$.

PROOF. First, we prove that set E is of type (1,3,5). The block B is 5-secant to E. Each block B' with $x \notin B'$ is m-secant F and n-secant B with m=1,3 and n=0,2. Hence B' is (m+n)-secant to E with m+n=1,3,5. Each block $B' \neq B$ with $x \in B'$ is 1-secant to $B \cap E$ and m-secant to $F \cap E$ with m=0,2. So B' is (m+n)-secant to E with m+n= =1,3.

Let C be a set of type (1,3,5). We have $t_5 = 11$. Denote by B a 5-secant block to C and by x the point of $B \setminus C$. We prove that the set $H := (C \setminus B) \cup \{x\}$ is a Fano set. Let B' be a block with $x \notin B'$. If B' is 5-secant to C, then $|B \cap B'| = 2$, necessarily. Assume the contrary, i.e. $B \cap B' = \emptyset$. Then C should be the union of two disjoint blocks B,B', each without one point, with a point w outside to $B \cup B'$. Then C should be a reducible blocking set, since $C - \{w\}$ should be blocking set D defined in 4.1. Hence $|B \cap B'| = 2$. Consequently B' is 3-secant to $C \setminus B$. If B' is 3-secant to C, then B' is either 1-secant or 3-secant to $C \setminus B$. If B' is 1-secant to C, then B' is 1-secant to $C \setminus B$ since B' cannot be 1-secant to B.

Let B' be a block different from B with $x \in B'$. Then B' intersects B at another point w. We claim that B' cannot be 5-secant to C. Assume the contrary. Then C should be the union of $B \cup B' - \{x\}$ with 2 points $a, b \notin B \cup B'$. The block through a,b,x necessarily contains w, otherwise it should be 4-secant to C. Then, there is at least a block containing a,b, which has an even number of points in C, a contradiction. Hence $B' (\neq B)$ is m-secant C with m=1,3. Consequently B' is n-secant to $C \setminus B$ with n=0,2, and then s-secant to $(C \setminus B) \cup \{x\}$ with s=1,3.

Finally we prove

7.5 THEOREM. Suppose C is reducible with $|C| = 11$. Then 1.1 (d) holds.

PROOF. If C is reducible, then there exists a point $w \in C$ such that $C - \{w\}$ is a blocking set of ten points. Then we necessarily have the following two cases.

(i) $C - \{w\} = D$, $w \notin D$.

(ii) $C - \{w\} = F \cup \{x,y,z\}$, where the block through x,y,z is 1-secant to F.

In this case we require that whenever we take 3 points in $\{x,y,z,w\}$, a block through them is 1-secant to F.

8. BLOCKING SETS WITH MORE THAN ELEVEN POINTS

All the blocking sets with 7,8,9,10,11 points have been characterized in S(3,6,22). Now, we prove the following

8.1 THEOREM. Denote by C an irreducible blocking set in S(3,6,22). Then $|C| \leq 11$. The equality holds if and only if C is an 11-set of type (1,3,5).

PROOF. By 7.1 it is sufficient to prove that if $|C| \geq 12$, then C is reducible. This is trivial if $|C| = 15$, since C is a set of type (3,5). In the other cases: $|C| = 12$ implies $t_2 = 60 - 4t_5$, $t_1 = t_5 - 8$, then $t_5 \leq 15$ and $t_1 \leq 7 < |C|$. $|C| = 13$ implies $t_2 = 84 - 4t_5$, $t_1 = t_5 - 17$, then $t_5 \leq 21$ and $t_1 \leq 4 < |C|$. Finally, $|C| = 14$ implies $t_2 = 119 - 4t_5$, $t_1 = t_5 - 28$. So, $t_5 \leq 29$ and then $t_1 \leq 1 < |C|$.

In conclusion, we note that all blocking sets in this section are characterized by their complement.

REFERENCES

[1] L.Berardi-F.Eugeni-O.Ferri, Sui blocking sets nei sistemi di Steiner. Boll. Un. Mat. Ital. Sez. D, 1 (1984), 141-164.

[2] L.Berardi-F.Eugeni, Blocking sets in the projective plane of order four. In these Proceedings.

[3] T.Beth, D.Jungnickel, H.Lenz, Design theory. Wissenschaftsverlag, Mannheim-Wien-Zurich, 1985.

[4] P.J.Cameron, J.H. Van Lint, Graphs, codes and Designs. London Math. Soc. 43 (1980), Cambridge University Press.

[5] F.Eugeni,E.Meyer, Blocking sets of index two. In these Proceedings.

[6] W.Jónsson, On the Mathieu groups M_{22} , M_{23} , M_{24} and the uniqueness of associated Steiner systems. Math. Z. 125 (1972), 193-214.

[7] H.Lüneburg, Transitive Erweiterungen endlicher Permutations-gruppen, Lectures Notes Math. 84, Springer-Verlag, Berlin (1969).

[8] G.Tallini, Blocking sets nei sistemi di Steiner e d-blocking sets in PG(r,q) ed AG(r,q)., Quad. Sem. Geom. Comb. Fac. Ing. L'Aquila, 3 (1983).

[9] G.Tallini, On blocking sets in finite projective and affine spaces. In these Proceedings.

[10] J.A.Todd, A representation of the Mathieu group M_{24} as a collineation group. Ann. Mat. Pura ed Appl., 71 (1966), 199-238.

Luigia BERARDI
Dipartimento di Ingegneria Elettrica
Facolta` di Ingegneria - L'Aquila
Italia.

Annals of Discrete Mathematics 37 (1988) 43–50

BLOCKING SETS IN THE PROJECTIVE PLANE OF ORDER FOUR

Luigia BERARDI and Franco EUGENI

Dipartimento di Ingegneria Elettrica.Universita' de L'Aquila (Italia)
Istituto di Scienze. Universita' di Chieti (Italia)

We give a complete characterization of blocking sets in PG(2,4).

1. INTRODUCTION

A blocking set in a finite incidence structure $(P, \mathscr{B}, I)$ is a subset S of P such that every block of $\mathscr{B}$ is incident to S and to P-S at least in one point.

The definition implies that if S is a blocking set, then P-S is a blocking set too. A blocking set is said to be irreducible if $\forall x \in S$ the set $S-\{x\}$ is not a blocking set.

The concept of blocking set was introduced, following an idea of Von Newmann and Morgenstern ([20], footonote 3, p. 469), with the name blocking coalitions by M.Richardson in 1956 [16], in relation to the so-called finite projective games (see also [12]). In 1966 J.Di Paola wrote two papers on these questions with the first results in projective planes of small order and in the dual of a Steiner system, [8],[9]. In 1970 A.Bruen introduced the word "blocking set" and began a sistematic study. Since then several authors have contributed to this study. Recently it has been proved that the existence of blocking sets in projective or affine spaces of high dimension whose order is large enough (cf. [3], [7], [14]).A useful list can be found in [19].We recall:

1.1 RESULT (Bruen [4],[5]). Let S be a blocking set in a projective plane of order q.Then: $q+\sqrt{q}+1 \leq S \leq q^2-\sqrt{q}$. Moreover, equality on the left hand-side holds if and only if S is a Baer subplane. (On the right hand-side iff S is the complement of a Baer subplane).

1.2 RESULT (Bruen-Thas [6],Tallini [17], [19]). Let S an irreducible blocking set in a projective plane of order q. Then $S \leq q\sqrt{q}+1$, moreover the equality holds if and only if S is a hermitian arc.

Suppose $q=p^h$, p a prime, $q>2$. We define the function m(q) as follows: $m(q)=\sqrt{q}$, if q is a square; $m(q)=(q+1)/2$, if q is a prime and $m(q)=p^{h-d}$, if $h>1$ is odd and where d denotes the greatest divisor of h, different from h. Then

1.3 RESULT (Berardi-Eugeni [1]). For any integer k with $q+m(q)+1 \leq k \leq q^2-m(q)$, there exists a blocking set in PG(2,q) having exactly k elements.

Many informations about sets of PG(2,q), q even, intersected by a line in 0, q/2 - 1, q/2, q/2+1 or q points, can be found in [10].

The aim of this paper is to study the blocking sets in PG(2,4). Why just this case? We have several reasons.

a) If q=2, there is no blocking set in PG(2,2); if q=3 we have in PG(2,3) only two blocking sets up to isomorphism (see [17]). So q=4 is really the first case; moreover for any k with $7 \leq k \leq 14$ we have at least one blocking set with k points, by 1.3.

b) The space PG(3,4) is the only 3-dimensional case in which the existence of blocking sets is an open problem. We hope that the study of blocking sets of PG(3,4) can depend on their sections with a plane that are the blocking sets of PG(2,4).

c) In [2] and [24] all blocking sets in the little Mathieu systems

S(5,6,12) and its contractions have been studied. We intend to classify all blocking sets in the large Mathieu systems S(5,8,24), S(4,7,23), S(3,6,22) and S(2,5,21) i.e. PG(2,4). Each of these case is a special case and PG(2,4) is only the first. The study of blocking sets in other big Mathieu systems will be the aim of our next papers.

In what follows we use the following notations.Denote by F the Fano (Baer) subplane of PG(2,4). Fix three non-concurrent lines L,M,N.Put $p=L\cap M$, $u\in M/L$, $v\in M/L$, and suppose $u,v,w\in N$. Put

$$\Delta :=(L-\{u\})\cup(M-\{v\})\cup\{w\} .$$

The line N is called the short side of Δ. Denote by H a triangle in PG(2,4) without vertices; all hermitian arcs are isomorphic to H.

In this paper we prove the following

1.4 THEOREM. In PG(2,4) the blocking sets S, $|S|=k$, with necessarily $7\leq k\leq 14$, are exactly the following:
(a) if k=7, then S=F;
(b) if k=8, then either $S=F\cup\{x\}$ with $x\notin S$, or $S=\Delta$ which is irreducible;
(c) if k=9, then either S=H (irreducible), or $S=\Delta\cup\{x\}$ with $x\notin\Delta$;
(d) if k=10, then $S=\Delta\cup\{x,y\}$, where x,y are not vertices and the following condition holds:if x,y,w are collinear then either x,y,w,p or x,y,w,u,v are collinear.
(e) Moreover, the blocking sets with $|S|\geq 11$ are exactly the complements of those defined above.

2. PROOF OF THE THEOREM

By Results 1.1, 1.2, 1.3 it follows that if S denotes a blocking set in PG(2,4), then for any k with $7\leq k\leq 14$ there is at least one blocking set. We note that it is enough to classify the blocking sets S with $7\leq|S|\leq 10$, since the others are the complementary point-sets. Moreover, we note that by 1.1 it follows 1.4 (a) and by 1.2 the first part of 1.4 (c) has been proved.

In the following Remark, a particular construction of a Fano subplane F of PG(2,4) is given.

2.1 REMARK. Let C be a (q+1)-arc in PG(2,q), $t\in C$ a point and T a line 1-secant of C at T. The set $B:=C\cup T-\{t\}$ is an irreducible blocking set if q is odd, cf.[17]. If q is even and the knot of C is denoted by k, the set $B-\{k\}$ is a blocking set. In the case of PG(2,4) such a blocking set is the Fano plane since it contains 7 points. Moreover, B and $B\cup T-\{k\}$ are clearly isomorphic, since $C\cup\{k\}-\{t\}$ is a conic (cf.[11], 8.4.1).

In the following we use the notion of characters of a k-set K. We denote by x_i the number of lines of PG(2,4) which are i-secant of set K. It is well known [17] that the following relations hold

$$(2.1)\quad\begin{cases} x_0+x_1+x_2+x_3+x_4+x_5=21\\ x_1+2x_2+3x_3+4x_4+5x_5=5k\\ 2x_2+6x_3+12x_4+20x_5=k(k-1).\end{cases}$$

Now we prove 1.4 (b), namely:

2.2 THEOREM. In π=PG(2,4) there are only two types (under up isomorphism) of blocking sets with 8 points:
(a) the set Δ, that is irreducible,
(b) the set $F\cup\{x\}$, where F is a Fano subplane and $x\notin F$.

PROOF - Let C be a blocking set in with $|C|=8$. By (2.1) we have:

$$x_1+x_2+x_3+x_4 = 21,\quad x_2+2x_3+3x_4 = 19,\quad x_3+3x_4 = 9\ ,\ \text{so}\ x_4\leq 3\ .$$

Step 1. We prove that $x_4=2$ or $x_4=1$.

Suppose $x_4=0$. It follows that $x_3=9$, $x_2=1$, $x_1=11$. Denote by u_i the number of lines through a fixed point of C that are i-secant of C. We have: $u_1+u_2+u_3=5$, $u_2+2u_3=7$. So, $u_2\neq 0$ for any point of C, then $x_2>1$, a contradiction.

Suppose $x_4=3$. Two 4-secant lines L,M are necessarily incident at a point of C, otherwise C is not a blocking set. Moreover, the third 4-secant intersects L and M, and so $|C|\geq 9$, a contradiction.

Step 2. We prove that $x_4=1$ implies $C=F\cup\{x\}$.

If $x_4=1$, then $x_3=6$, $x_2=4$, $x_1=10$. Denote by T the 4-secant of C and by t the point of T outside C. Put $C_1:=C/T$.

We note that $C' := C_1\cup\{t\}$ is a 5-arc of π.

(Each line $\neq T$ through t intersects C, and then C_1, at exactly one point, and so such lines are 2-secant of C').

Fix $p\in T$ with $p\neq t$. Denote by u_i the number of lines through p and i-secant of C. We have $u_1+u_2+u_3+u_4=5$, $u_2+2u_3+3u_4=7$. We have $u_4=1$ and $u_3=2$, (in fact $x_4=1$ implies $u_4=1$. If $u_3=0$ or 1 then $\forall p\in T-\{t\}$ we have at most 4 lines that are 3-secant of C, since each 3-secant intersects $T-\{t\}$, while $x_4=1$ implies $x_3=6$). Then $u_2=0$ and $u_1=2$.

Consequently, any line $\neq$ T through a point p intersects C at 1 or 3 points and C_1 at 0 or 2 points. The set $C' = C_1\cup\{t\}$ is of type (0,1,2) i.e. C' is a 5-arc of π. By [11] (cf.8.4.1) C' is a conic and $C=F\cup\{x\}$.

Step 3. Finally, we prove that $x_4=2$ implies $C=\Delta$.

The two 4-secants L,M are necessarily incident at a point of C. Put $u=L\setminus C$ and $v\in M\setminus C$. Since the line containing u and v must contain a point $w\in C$, it follows $C=\Delta$. □

We deal now with the blocking sets of cardinality 9 in π. We use the notations of 1.4 and 2.1. In π a hermitian arc H is an irreducible blocking 9-set and the 9-set $\Delta\cup\{x\}$, where x is not a vertex of Δ, is clearly a reducible blocking set. Note that in π the set $(F\cup\{y\})\cup\{x\}$ is a special case of the 9-set $\Delta\cup\{x\}$, obtained if x is collinear with p and w.

We prove that

2.3 THEOREM. In $\pi=PG(2,4)$ there are (under up isomorphism) only two blocking sets of 9 points: the hermitian arc and $\Delta\cup\{x\}$ with $x\notin\Delta$.

PROOF. Let C be a blocking set in π with $|C|=9$. Denote by x_i the characters of C. By (2.1), we have:

$$x_1+x_2+x_3+x_4=21\ ,\quad x_2+2x_3+3x_4 = 24\ ,\quad x_3+3x_4 = 12\ .\ \text{So},\ x\leq 4.$$

Step 1. Suppose $x_4=0$, then C is a set of type (1,3) and by [11], cf. 13.4.6, set C is a hermitian arc, since $|C|=9$.

Step 2. Suppose $x_4=1$, then $x_3=9$, $x_2=3$, $x_1=8$. Denote by $w\in C$ a point of the line L which is 4-secant of C. We have:

$$u_1+u_2+u_3+1 = 5\qquad ,\qquad u_2+2u_3+3 = 8,$$

where u_i is the number of i-secant lines through $w\in L$. These imply

$$u_1= u_3-1\qquad ,\qquad u_2= 5 - 2u_3\ .$$

It follows that $u_3=1$ or 2. In both these cases $u_2\geq 1$, and so there is at least one 2-secant through each of the four points $w \in C \cap L$. So $x_2 \geq 4$, a contradiction.

Step 3. Now we prove that $x_4=2$ implies $C=\Delta \cup \{x\}$, where either x is a point on the short side of Δ or x is collinear with p and w.

Suppose $x_4=2$, then $x_3=6=x_2$, $x_1=7$. Since $x_1=7<|C|$, there is a point x which does not lie on a tangent line. So $C-\{x\}$ is a blocking set of 8 points, then $C-\{x\}=\Delta$ or $F \cup \{y\}$, $y \notin F$.

Suppose $C-\{x\}=\Delta$. Then x lies on the short side of Δ with $x \notin \{u,v,w\}$ (otherwise the line xw is a third 4-secant). If we complete Δ with x in such a way, we have a blocking set with $x_4=2$.

Suppose $C-\{x\}=F \cup \{y\}$. Then $x \notin F \cup \{y\}$, since C contains no line. If we complete $F \cup \{y\}$ with the fixed point x, we have a blocking set with $x_4=2$. Since $F \cup \{x,y\}$ is the same as $\Delta \cup \{x\}$ when x,w and p are collinear, the assertion is proved.

Step 4. Now we prove that $x_4=3$ implies that $C=\Delta \cup \{x\}$, where $x \notin uv$ and x,p,w are not collinear.

Suppose $x_4=3$. Then $x_3=3$, $x_2=9$, $x_1=6$. Since $x_1=6<|C|$, there is a point x which does not lie on a tangent line. So, $C-\{x\}$ is a blocking set of 8 points.

Suppose $C-\{x\}=F \cup \{y\}$. Then (cf. Step 3 above) set C has $x_4=2$ necessarily, a contradiction.

Suppose $C-\{x\}=\Delta$. Then $x \notin uv$ and x,p,w are not collinear (otherwise, by above Step 3, $x_4=2$, a contradiction). It follows that the line xw intersects $C=\Delta \cup \{x\}$ at two points of C, since it does not contain p, then xw is the third 4-secant.

Step 5. Finally, we prove that $x_4=4$ is impossible.

Two 4-secant lines L,M have necessarily one common point in C. So, a third 4-secant intersects L and M at two points of C.

If three 4-secants are concurrent, then $|C| \geq 10$, a contradiction.

Suppose that the third 4-secant N of C intersects L and M at two distinct points of C. The fourth 4-secant has at most three points of C in common with L,M,N. Consequently $|C| \geq 10$, a contradiction.

So, 2.3 is completely proved.

Now we deal with blocking sets in π with 10 points.

If x,y are points of π (with the exception of just a few positions for x and y) the set $\Delta \cup \{x,y\}$ is a blocking set of 10 points. For example, if z is a vertex of a hermitian arc H, then the blocking 10-set $H \cup \{z\}$ is a special case of set $\Delta \cup \{x,y\}$, obtained by taking x,y on the short side of Δ. Moreover, if z does not belong to a side of H, set $H \cup \{z\}$ is isomorphic to an appropriate $\Delta \cup \{x,y\}$ with x,y on the short side. This is easy to prove.

We conclude this section with the following

2.4 THEOREM. In $\pi=PG(2,4)$ the blocking sets of 10 points (necessarily reducible) are exactly the sets $\Delta \cup \{x,y\}$, where x,y are not vertices of Δ and the following condition holds: if x,y,w are collinear, then the line xy either coincides with the short side of Δ or it contains p.

PROOF. Let C be a blocking set in π with $|C|=10$. Denote by x_i the characters of C. By (2.1) we obtain $x_1=8-x_4$, $x_2=3(x_4-1)$, $x_3=16-3x_4$. So, $1 \leq x_4 \leq 5$. It is obvious that if two 4-secant L,M exist, they intersect at a point of C (otherwise $|C| \geq 11$). Put $p:=L \cap M$, $u:=L \setminus C$, $v:=M \setminus C$; denote by w a point of C on the line uv. Then the triangle $\Delta=L \cup M \cup \{w\}-\{u,v\}$ is contained in C. Put $\{x,y\}:=C \setminus \Delta$.

Step 1. The case $x_4=1$ is not possible.

If $x_4=1$, then $x_3=13$, $x_2=0$, $x_1=7$. So, there exists a point $x \in C$ through which no tangent passes. Consequently, if u_i is the number of i-secant of C through x, we have $u_1=u_2=0$, and so $u_3+u_4=5$ and $2u_3+3u_4=9$, which imply $u_4=-1$, a contradiction.

Step 2. The case $x_4=2$ implies that $C=\Delta\cup\{x,y\}$, where x,y lie on the short side uv of Δ.

Suppose $x_4=2$. Denote by u_i the number of lines through $p=L\cap M$ and i-secant of C. We have $u_1+u_2+u_3+2 = 5$ and $u_2+2u_3+6 = 9$, from which $u_1=u_3$ and $u_2=3-2u_3$, consequently $u_3=1$ or 0.

Suppose $u_3=1$, then $u_1=u_2=1$. If x is a point on pw, the line py is the 2-secant. In this case y is on the short side; otherwise yw is a third 4-secant. But in this case xy is also a 4-secant, and this is a contradiction.

This leaves $u_3=0$, which implies $u_1=0$, $u_2=3$. In this case pw, px, py are three distinct lines. Moreover, x and y necessarily lie on the short side, otherwise we have $x_4>2$. So, the assertion of Step 2 is proved.

Step 3. The case $x_4=3$ implies that $C=\Delta\cup\{x,y\}$, where x and y are points that satisfy one of the following conditions:

(i) $x,y\in pv$

(ii) $x\in uv$, $y\in px$

(iii) $x\in wp$, $y\in xv$ (or xu).

If $x_4=3$, then $x_3=7$, $x_2=6$, $x_1=5$. Denote by N the third secant. It results $v = N\cap uv$. We have two possibilities: L,M and N are concurrent in p or not.

First, suppose $p = L\cap M\cap N$. The points of C are all on $L\cup M\cup N$, so $w,x,y\in N$ since they are not in $L\cup M$. So $C=\Delta\cup\{x,y\}$ with condition (i).

Next, suppose $p=L\cap M\notin N$. The points of C are in $L\cup M\cup N$.
Put $y := C\cap(N\setminus(L\cup M)) - \{w\}$ and $x := C - (\Delta\cup\{y\})$. Denote by u_i the number of lines which are i-secant of C and contain p. As in the case of Step 2, we have:

$$\text{(I)}\quad u_3=u_1=0\ ,\ u_2=3 \quad\text{or}\quad \text{(II)}\quad u_3=u_2=u_1=1.$$

The case (I) implies a contradiction. (Since in $\Delta\cup\{y\}$ the lines pw and py are 2-secant of C, the point x lies on the fifth line through p. In any case xw or xy is an ulterior 4-secant).

Finally, suppose that (II) holds. Since in $\Delta\cup\{y\}$, the lines pw and py are 2-secant of C, the point x lies on one of them. If $x\in py$, necessarily $x\in uv$; otherwise xw is a 4-secant. This is the case (ii). If $x\in pw$, the line xy necessarily contains either u or v, otherwise xy should be a 4-secant. This is the case (iii). So, the proof of Step 3 is finished.

Step 4. The case $x_4=4$ implies that $C=\Delta\cup\{x,y\}$, where x,y are points which satisfy one of the following conditions:

(1) $p\in xy$, $w\notin xy$, $x,y\notin uv$;

(2) $u\in xu$ (or $v\in xy$) , $x,y\notin\Delta$;

(3) $x\in uv$, $w,p\notin xy$;

(4) $x\in wp$, $u,v\notin xy\neq pw$, $y\notin uv$.

Suppose $x_4=0$ and denote by u_i the number of i-secant containing p. As in Steps 2 and 3, we have either (a) $u_1=u_3=0=u_2$ or (b) $u_1=u_2=u_3=1$.

In the case (a) the lines pw, px, py,L,M are all distinct. It is trivial that either xy contains u or $x\in uv$; otherwise $x_4>4$. So, we are in situations (2) and (3) for x and y.

In the case (b) we necessarily have either $p\in xy\neq pw$ or $x\in pw$ with y on one of two remaining lines through p. Moreover, yx contains neither u nor v and $y\notin uv$. This corresponds to situations (1) and (4) for x and y. So, Step 4 is proved.

Step 5. The case $x_4=5$ is impossible.

If $x_4=5$, we have $x_3=1$, $x_2=12$, $x_1=3$. Let R be the only 3-secant, take $h \in R \cap C$. If u_i are the number of i-secant containing h, we have: $u_1+u_2+1+u_4=5$, $u_2+2+3u_4=9$ and then $u_1=2u_4-3$, $u_2=7-3u_4$, from which $u_4=2$, $u_2=u_1=1$.
Consequently, each of three points on $C \cap R$ is incident with exactly two 4-secant. Then $x_4 \geq 6$, a contradiction.

Step 6. As a last remark, we note that for $\Delta \cup \{x,y\}$ every case is presented in the above steps, except the following: $xy \neq uv$, $w \in xy$, $p \notin xy$. Note that in this exceptions xy is entirely contained in $\Delta \cup \{x,y\}$ which, in this case, is not a blocking set. Since this case is excluded by the hypothesis of the theorem, the assertion 2.4 is completely proved.

3. FINAL REMARKS

In this Section we deal about other questions connected with blocking sets. We recall that an n-fold blocking set in an incidence structure is the union of n disjoint blocking sets. It is well known (cf. [11]) that in PG(2,q), q a square, there exists a Baer partition in $q-\sqrt{q}+1$ Baer subplanes. Such a partition, when q=4, is a 3-fold blocking set that we call a Fano partition.

3.1 REMARK. Denote by $\mathcal{B}$ an n-fold blocking set in PG(2,4), then $n \leq 3$. Namely:
(a) n=2, and $\mathcal{B}$ is the union either of one blocking set and its complement or of two blocking sets contained in each of them.
(b) n=3, and $\mathcal{B}$ is a Fano partition.

Another matter is the open problem of the existence of blocking sets in projective or affine spaces of small order. Recently Rajola [25] has given an example of blocking set in PG(3,5). So, when $q \leq 5$, the problem is open in PG(r,5) $(r \geq 4)$, AG(r,5) $(r \geq 3$, cf. [26]) and in PG(r,4) $(r \geq 3)$. In AG(r,4) $(r \geq 3)$ no blocking set there exists. We hope that the classification of blocking sets in PG(2,4) could be useful for this problem. Moreover we aim to give a complete classification of blocking sets in projective or affine spaces of order $q \leq 9$. See the last part of [19] in these Proceedings.

The last problem that we want to remark in this Section is the problem of the index (cf. [22], [24]). Denote by B a blocking set in an incidence structure. The index i=i(B) of B is the smallest number of blocks such that the union of them contains B. In [22], it has been proved that in a projective of order q plane we have $3 \leq i(B) \leq q+1$ for any blocking set. In PG(2,4), with regard to 1.4 Theorem of Section 1, we have the following situation.

(a) A Fano subplane F, set $F \cup \{x\}$ with x on a 3-secant, a triangle Δ, a hermitian arc H and the blocking sets of type either $\Delta \cup \{x\}$ or $\Delta \cup \{x,y\}$ with x,y w collinear, have all index three.

(b) Either $F \cup \{x\}$ with x outside one 3-secant or $\Delta \cup \{x,y\}$ with x,y,w non--collinear have index four.

It remains to classify the index of the complements of above considered blocking sets which are reducible. This is a trivial exercise, we remark only that one of them, namely the complement of the Fano subplane has index 5 necessarily, so each index is possible.

REFERENCES

[1] L.Berardi-F.Eugeni, On the cardinality of blocking sets in PG(2,q). J. of Geometry 22 (1984), 5-14.

[2] L.Berardi-F.Eugeni-O.Ferri, Sui blocking sets nei sistemi di Steiner. Boll.Un.Mat.Ital., sez. D, 1 (1984), 141-164.

[3] A.Beutelspacher-F.Eugeni, On blocking sets in projective and affine spaces of large order. Communicated to "Grundlagen der Geometrie" Oberwolfach, 27-10/2-11-1985. Rend.di Mat. Roma, to appear.

[4] A.Bruen, Blocking sets in finite projective planes. SIAM J.Appl. Math., 21 (1971), 380-392.

[5] A.Bruen, Baer subplanes and blocking sets. Bull.Amer.Math.Soc.,76 (1970), 342-344.

[6] A.Bruen-A.Thas, Blocking sets. Geom. Dedicata, 6 (1977), 193-203.

[7] P.Cameron-F.Mazzocca, Bijections which preserve blocking sets. Geom. Dedicata, to appear.

[8] J.Di Paola, On a restricted class of block design games., Canad. J.Math., 18 (1966), 225-236.

[9] J.Di Paola, On minimum blocking coalitions in small projective plane games. SIAM J. Appl. Math., 17 (1969), 378-392.

[10] M. de Resmini, On k-sets of class [0,q/2-1, q/2 q/2+1,q] in a plane of even order q. Europ.J.Combin., 6 (1985), 303-315.

[11] J.W.P.Hirschfeld, Projective geometries over finite fields. Clarendon Press, Oxford, 1979.

[12] A.J.Hoffman-M.Richardson, Block design games. Canad. J. Math., 13 (1961), 110-128.

[13] W.Jónsson, On the Mathieu groups M_{22}, M_{23} , M_{24} and the uniqueness of associated Steiner systems. Math. Z. 125 (1972), 193-214.

[14] F.Mazzocca-G.Tallini, On the non existence of blocking-sets in PG(n,q) and AG(n,q), for all large enough n. Simon Stevin, 1 (1985), 43-50.

[15] M.Richardson, On finite projective games. Proc. Amer. Math. Soc., 7, (1956), 458-465.

[16] G.Tallini, Problemi e risultati nelle geometrie di Galois. Relaz. n.30, (1973) Ist. Mat. Univ. Napoli.

[17] G.Tallini, k-insiemi e blocking sets in PG(r,q) e in AG(r,q). Sem. Geom. Comb. Fac. Ing. Univ. L'Aquila, 1 (1982).

[18] G.Tallini, Blocking sets nei sistemi di Steiner e d-blocking sets in PG(r,q) ed AG(r,q). Sem.Geom.Comb.Fac. Ing. Univ.L'Aquila, 3 (1983).

[19] G.Tallini, On blocking sets in finite projective and affine spaces. In these Proceedings.

[20] J.A.Tood, A representation of the Mathieu group M_{24} as a collineation group. Ann. Mat.Pura e Appl., 71 (1966), 199-238.

[21] J.Von Neumann-O.Morgenstern, Theory of Games and Economic Behavior. 3d. ed., Princeton University Press, 1953.

Supplementary references on Section 3.

[22] A.Beutelspacher-F.Eugeni, Sui blocking sets di dato indice con particola-

re riguardo all'indice tre. Boll.Un.Mat.Ital., 4-A(1985), 441-450.

[23] A.Beutelspacher-F.Eugeni, On n-fold blocking sets. Annals of Discrete Mathematics 30 (1986), 31-38. Proceedings of International Conference "Combinatorics '84"-Bari.

[24] F.Eugeni, E. Mayer, On blocking sets of index two, In these Proceedings.

[25] A.Rajola, Un esempio di blocking set in PG(3,q) per ogni $q \geq 5$. Quad.n.57 (1985), Sem. Geom. Comb. Univ. Roma "La Sapienza". In these Proceedings.

[26] C.O'Keefe-A.Venezia, Blocking sets in AG(r,5). Quad. n. 56, Sem. Geom. Comb. Univ. Roma "La Sapienza".

L.Berardi
Dipartimento di Ingegneria Elettrica
Seminario di Geometria Combinatoria
L'Aquila (Italy)

F.Eugeni
Istituto di Scienze
Facolta' di Architettura
Chieti (Italy)

Annals of Discrete Mathematics 37 (1988) 51–58

KALAHARI AND THE SEQUENCE "SLOANE NO. 377"

DIETER BETTEN

When playing Kalahari (Kalaha, Owari, Bohnenspiel) a move consists in taking all pebbles from a playing mould and distributing them one after the other to the following moulds. If the last pebble falls into one's own counting mould, then another move is allowed. We observe that for any number of pebbles there is exactly one position where all pebbles can be removed by iteration. So we get for every natural number a unique presentation as sum of natural numbers. The sequence of those numbers for which the length of the corresponding sum representation is increasing is the sequence No. 377 in Sloane's book. This sequence was constructed by Erdös and Jabotinsky by some sieving process. Erdös, Jabotinsky and David described the growth of this sequence. We consider the rectangular scheme of all summands, having in the n'th row the sum representation for the number n. David found arithmetic progressions in the rows and described them. In the following note we study the columns of the scheme and show: the m'th column is periodic of length l.c.m. $\{2,3,\ldots,m\}$.

We go over from the original Kalahari playing board

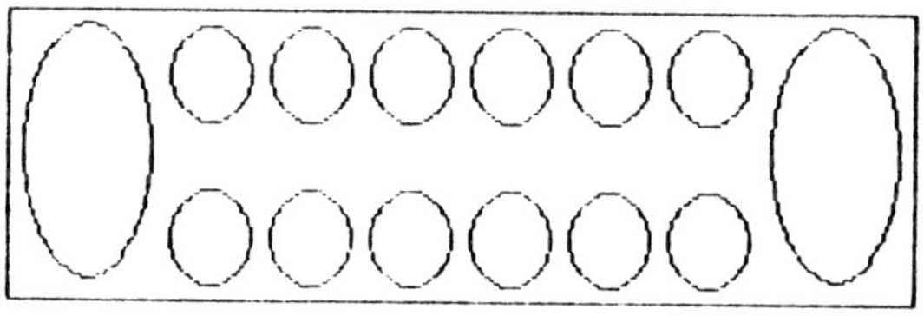

to a simplified version for one person

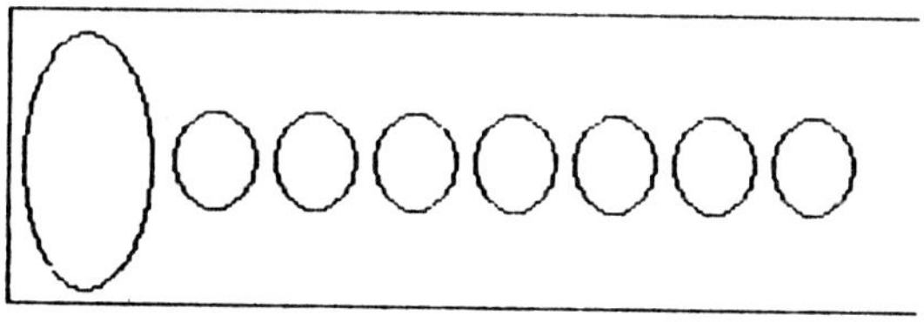

having one counting mould and countably many playing moulds, numbered by 2,3,4,... . We play from right to left, now. Only such moves are allowed for which the last pebble falls into the counting mould.

Observation 1: For every n there exists exactly one position with n pebbles, where all pebbles can be removed by iteration.
Proof: Begin at the end and play backwards.

Definition 1: Let $a_{n,m}$, $n = 0,1,2,\dots$, $m = 2,3,4,\dots$, be the number of pebbles lying in the m-th mould for the n-position.

In figure 1 we give the scheme of the $a_{n,m}$. The bracket in the n'th row indicates the mould which must be taken when carrying out the move.

Note: There are positions where pebbles could be moved from more than one mould for the last pebble to fall into the counting mould (for instance $n = 8$, $m = 3$ and $m = 5$); however, the move has to be started from the mould with minimal m in order to lead to the end.

Proposition 1: The rectangular scheme $\{a_{n,m}\}$ can be defined by induction on n as follows: $a_{0,m} = 0$ for all $m \geq 2$. Given $a_{n,m}$, $m \geq 2$, for fixed $n \geq 0$ then

$$a_{n+1,m} = \begin{cases} a_{n,m} - 1 & \text{if} \quad m < \mu \\ m - 1 & \text{if} \quad m = \mu \\ a_{n,m} & \text{if} \quad m > \mu \end{cases}$$

where $\mu(n)$ = minimal m with $a_{n,m} = 0$.

Proof: This is exactly what is being done when playing backwards.

Proposition 2: A direct description of the n-th row can be given by induction on m : $a_{n,1} = 0$ and $a_{n,m}$ = remainder when dividing $n - \sum_{\mu=1}^{m-1} a_{n,\mu}$ by $m (m \geq 2)$.

Proof: By induction on n using proposition 1.

Definition 2: Let $\{a_k\}$, $k = 1,2,3,\dots$, be the sequence of those n for which the length of the row is increasing. By figure 1 the first members of this sequence are 1,2,4,6,10,12,18,22,30,34,... .

Looking in Sloane's book [3], one finds that this sequence might be the sequence No. 377 given there. This sequence is constructed in [1,2] by a sieving process in the following way: Write in the first column all natural numbers. To get the second column cross out 1 and every second number. From the remaining numbers cross out the first and every third to get the third column. Then cross out the first and every fourth number and so on. The resulting sequence is the sequence No. 377 (figure 2). So in figure 2 a bracket around a number means that this number has to be deleted in the following column.

<u>Observation 2:</u> The sequence of definition 2 coincides with the sequence No. 377. The places of brackets in figure 2 coincide with the places of brackets in figure 1.

	m = 2	3	4	5	6	7	8	9	10	11	12	13
n= 1	(1)											
2	0	(2)										
3	(1)	2										
4	0	1	(3)									
5	(1)	1	3									
6	0	0	2	(4)								
7	(1)	0	2	4								
8	0	(2)	2	4								
9	(1)	2	2	4								
10	0	1	1	3	(5)							
11	(1)	1	1	3	5							
12	0	0	0	2	4	(6)						
13	(1)	0	0	2	4	6						
14	0	(2)	0	2	4	6						
15	(1)	2	0	2	4	6						
16	0	1	(3)	2	4	6						
17	(1)	1	3	2	4	6						
18	0	0	2	1	3	5	(7)					
19	(1)	0	2	1	3	5	7					
20	0	(2)	2	1	3	5	7					
21	(1)	2	2	1	3	5	7					
22	0	1	1	0	2	4	6	(8)				
23	(1)	1	1	0	2	4	6	8				
24	0	0	0	(4)	2	4	6	8				
25	(1)	0	0	4	2	4	6	8				
26	0	(2)	0	4	2	4	6	8				
27	(1)	2	0	4	2	4	6	8				
28	0	1	(3)	4	2	4	6	8				
29	(1)	1	3	4	2	4	6	8				
30	0	0	2	3	1	3	5	7	(9)			
31	(1)	0	2	3	1	3	5	7	9			
32	0	(2)	2	3	1	3	5	7	9			
33	(1)	2	2	3	1	3	5	7	9			
34	0	1	1	2	0	2	4	6	8	(10)		
35	(1)	1	1	2	0	2	4	6	8	10		
36	0	0	0	1	(5)	2	4	6	8	10		
37	(1)	0	0	1	5	2	4	6	8	10		
38	0	(2)	0	1	5	2	4	6	8	10		
39	(1)	2	0	1	5	2	4	6	8	10		
40	0	1	(3)	1	5	2	4	6	8	10		
41	(1)	1	3	1	5	2	4	6	8	10		
42	0	0	2	0	4	1	3	5	7	9	(11)	
43	(1)	0	2	0	4	1	3	5	7	9	11	
44	0	(2)	2	0	4	1	3	5	7	9	11	
45	(1)	2	2	0	4	1	3	5	7	9	11	
46	0	1	1	(4)	4	1	3	5	7	9	11	
47	(1)	1	1	4	4	1	3	5	7	9	11	
48	0	0	0	3	3	0	2	4	6	8	10	(12)

Figure 1

	m = 2	3	4	5	6	7	8	9	10	11	12	13
n= 1	(1)											
2	2	(2)										
3	(3)											
4	4	4	(4)									
5	(5)											
6	6	6	6	(6)								
7	(7)											
8	8	(8)										
9	(9)											
10	10	10	10	10	(10)							
11	(11)											
12	12	12	12	12	12	(12)						
13	(13)											
14	14	(14)										
15	(15)											
16	16	16	(16)									
17	(17)											
18	18	18	18	18	18	18	(18)					
19	(19)											
20	20	(20)										
21	(21)											
22	22	22	22	22	22	22	22	(22)				
23	(23)											
24	24	24	24	(24)								
25	(25)											
26	26	(26)										
27	(27)											
28	28	28	(28)									
29	(29)											
30	30	30	30	30	30	30	30	30	(30)			
31	(31)											
32	32	(32)										
33	(33)											
34	34	34	34	34	34	34	34	34	34	(34)		
35	(35)											
36	36	36	36	36	(36)							
37	(37)											
38	38	(38)										
39	(39)											
40	40	40	(40)									
41	(41)											
42	42	42	42	42	42	42	42	42	42	42	(42)	
43	(43)											
44	44	(44)										
45	(45)											
46	46	46	46	(46)								
47	(47)											
48	48	48	48	48	48	48	48	48	48	48	48	(48)

Figure 2

The numbers $a_{n,m}$ occur also in [1], so the two procedures (Kalahari and Sieving) are in fact related to each other. Erdös and Jabotinski [2] study the growth of the sequence a_k and prove

$a_k = \frac{k^2}{\pi} + O(k^{4/3})$. Another proof for this result is given by David [1]. Looking at the rows of the scheme 1 for great n it can be seen that the row ends with an arithmetic progression of difference 2; this progression is preceeded by an arithmetic progression of difference 4, which in turn is preceeded by an arithmetic progression of difference 6 and so on. Those progressions are studied by David [1]. As an illustration we give here the sum representation for n = 100000 (figure 3). The numbers in brackets give the differences in the arithmetic progressions.

```
100000  =

    0 +  1 +  3 +  1 +  5 +  2 +  4 +  3 +  1 +  1 +  7 +  2 + 10 +  0 +  8 +
 +  9 +  7 + 15 +  1 +  2 + 16 + 13 +  1 + 13 +  9 + 20 + 26 +  2 +  8 + 21 +
 + 13 + 17 +  3 +  6 + 30 +  5 +  3 + 28 +  4 +  9 +  5 + 35 + 15 + 31 + 41 + 45 +
 + 43 + 35 + 21 +  1 + 27 + 50 + 16 + 31 + 43 + 52 +  0 +  1 + 59 + 56 + 50 +
                                                         (58)
 + 41 + 29 + 14 + 62 + 43 + 21 + 65 + 39 + 10 + 50 + 17 + 55 +
           (48)           (44)           (40)      (38)
 + 18 + 54 + 13 + 47 +  2 + .. + 66 + 16 + .. + 46 +
    (36)      (34)       (32)          (30)
 + 21 + .. + 77 + 17 + .. + 69 +  4 + .. + 76 +  5 + .. + 93 +
      (28)          (26)          (24)          (22)
 + 15 + .. + 95 + 10 + .. +100+  7 + .. +103+  1 + .. +113
      (20)         (18)          (16)          (14)
 +  0 + .. +132+  5 + .. +145+  1 + .. +169+  1 + .. +205 +
      (12)          (10)          (8)           (6)
 +  0 + .. +280+  2 + .. +560
      (4)           (2)
```

Figure 3

Wheras David has studied the rows of scheme 1, we now want to look at the columns and show

<u>Theorem:</u> The m'th column of the scheme $\{a_{n,m}\}$ is periodic of length l.c.m. $\{2,3,4,\dots,m\}$.

In order to prove this theorem we define another scheme $\{A_{N,m}\}$, $N \geq 0$, $m \geq 2$, which gives the numbers of vertical repetitions of the digits $a_{n,m}$. We call the scheme $\{A_{N,m}\}$ the quotient scheme of $\{a_{n,m}\}$, see figure 4.

<u>Proposition 3:</u> The columns of the scheme $\{A_{N,m}\}$ are given by induction on m:

$$A_{k(m-1),m+1} = A_{km,m} + A_{km+1,m}$$

$$A_{k(m-1)+i,m+1} = A_{km+1+i,m}$$

$$(k = 0,1,2,\dots, \quad i = 1,2,\dots, m-2) \quad .$$

Proof: We look at the original scheme $\{a_{n,m}\}$ and compare two numbers $\begin{matrix} a_{n,m} \\ a_{n+1,m} \end{matrix}$ with the corresponding numbers $\begin{matrix} a_{n,m+1} \\ a_{n+1,m+1} \end{matrix}$ of the next column. Using proposition 1 we note: $a_{n+1,m} = a_{n,m}$ then also $a_{n+1,m+1} = a_{n,m+1}$; if $a_{n+1,m} = a_{n,m}-1$ then also $a_{n+1,m+1} = a_{n,m+1} - 1$; but if $a_{n,m} = 0$ and $a_{n+1,m} = m-1$ then

$a_{n+1,m+1} = a_{n,m+1}$. So, to get column m+1 from column m in the scheme $\{A_{N,m}\}$ the number of 0's and the number of the following (m-1)'s must be added, all other lengths of repetition are carried over without change.

N	m = 2		3		4		5		6		7	
0	0	1	0	2	0	4	0	6	0	10	0	12
1	1	1	2	2	3	2	4	4	5	2	6	6
2	0	1	1	2	2	4	3	2	4	6	5	4
3	1	1	0	2	1	2	2	6	3	4	4	8
4	0	1	2	2	0	4	1	4	2	8	3	4
5	1	1	1	2	3	2	0	2	1	4	2	8
6	0	1	0	2	2	4	4	6	0	2	1	6
7	1	1	2	2	1	2	3	4	5	6	0	6
8	0	1	1	2	0	4	2	2	4	6	6	4
9	1	1	0	2	3	2	1	6	3	6	5	2
10	0	1	2	2	2	4	0	4	2	4	4	12
11	1	1	1	2	1	2	4	2	1	2	3	6
12	0	1	0	2	0	4	3	6			2	4
13	1	1	2	2	3	2	2	4			1	8
14	0	1	1	2	2	4	1	2				
15	1	1			1	2						
L =	1		1		2		3		12		10	
l =	2		6		12		60		60		420	

Figure 4

Note: In the first row (N = 0) we get the sequence $\{a_k\}$, so Proposition 3 gives another method to construct this sequence.

Definition 3: We define $l(m)$ = length of period of column m of the scheme $\{a_{n,m}\}$, $L(m)$ = length of period of column m of the quotient scheme $\{A_{N,m}\}$.

Proposition 4: Induction rules for the numbers L and l :

(a) $L(m+1) = [m,L(m)]\,\dfrac{m-1}{m}$

(b) $l(m) = [m,L(m)]\,\dfrac{l(m-1)}{L(m)}$

(We abbreviate l.c.m. $\{m,L(m)\} = [m,L(m)]$).

Proof: (a) The procedure of combining $A_{km,m}$ and $A_{km+1,m}$ in column m is periodic of length $[m,L(m)]$. The period $L(m+1)$ is somewhat smaller, because the two numbers $A_{km,m}$ and $A_{km+1,m}$ in column m give only one number in column m+1.

Therefore we must subtract $\frac{[m,L(m)]}{m}$ and get

$$L(m+1) = [m,L(m)] - \frac{[m,L(m)]}{m} = [m,L(m)] \cdot \frac{m-1}{m} .$$

(b) The number $l(m)$ is the sum of the A's in column m from $N = 0$ to $N = [m,L(m)] - 1$. This period consists of $\frac{[m,L(m)]}{L(m)}$ sequences of length $L(m)$ and the sum in each of those sequences is $l(m-1)$. Therefore we get $l(m) = \frac{[m,L(m)]}{L(m)} l(m-1)$.

Lemma: $[m+1, \frac{[2,\ldots,m]}{m}] = \frac{[2,\ldots,m+1]}{m}$ for every $m \geq 2$.

Proof: Compare factors on both sides.

Proof of the theorem:

(1) $L(m) = \frac{[2,\ldots,m-1]}{m-1}$ for all $m \geq 3$.

Proof: This is true for $m = 3$. Suppose that (1) is true for some $m \geq 3$, then by proposition 4(a) and Lemma it follows

$$L(m+1) = [m,L(m)] \frac{m-1}{m} = [m, \frac{[2,\ldots,m-1]}{m-1}] \frac{m-1}{m} =$$
$$= \frac{[2,\ldots,m]}{m-1} \frac{m-1}{m} = \frac{[2,\ldots,m]}{m} .$$

(2) $l(m) = L(m+1)\, m$ for every $m \geq 2$.

Proof: This is true for $m = 2$. If (2) has already been proved for some $m \geq 2$, then using proposition 4(b), the lemma and (1) it follows

$$l(m+1) = [m+1,L(m+1)] \frac{l(m)}{L(m+1)} = [m+1,L(m+1)] \cdot m =$$
$$= [m+1, \frac{[2,\ldots,m]}{m}] \cdot m = \frac{[2,\ldots,m+1]}{m} \cdot m =$$
$$= \frac{[2,\ldots,m+1]}{m+1} (m+1) = L(m+2) \cdot (m+1) ,$$

so, by induction, (2) is proved.

Using (2) and (1) the theorem follows:

$$l(m) = L(m+1)\, m = \frac{[2,3,\ldots,m]}{m} \cdot m = [2,3,\ldots,m] .$$

References:

[1] Y. David: On a sequence generated by a sieving process, Riveon Lematematika 11, 26-31 (1957).

[2] P. Erdös and E. Jabotinsky: On a sequence of integers generated by a sieving process, Indagationes Math. 20, 115-128 (1958).

[3] N.J.A. Sloane: A Handbook of integer sequences, Academic Press, New York 1973.

[4] C.D. Grupp: Brettspiele-Denkspiele, Humboldt Taschenbuchverlag, München 1976.

Mathematisches Seminar der Universität
Olshausenstraße 40, 2300 Kiel 1
Bundesrepublik Deutschland

Annals of Discrete Mathematics 37 (1988) 59–68
© Elsevier Science Publishers B.V. (North-Holland)

ENCIPHERED GEOMETRY.
SOME APPLICATIONS OF GEOMETRY TO CRYPTOGRAPHY

Albrecht BEUTELSPACHER

Siemens AG, ZT ZTI SYS 42, Otto-Hahn-Ring 6,
D-8000 München, West Germany

We present three applications of finite geometry to cryptography. These applications comprise threshold schemes, binary sequences and authentication systems. The main geometric objects are classical projective spaces.

1. MUCH ADO ABOUT NOTHING: INTRODUCTION

There are two major branches in todays cryptography. On the one hand one wants to conceal the content of a message. An algorithm which achieves this aim is called an *encryption algorithm*. On the other hand, encryption does not guarantee the integrity of the message; a bad guy can easily change the message without being able to understand what the message says. This is a serious problem in many applications. (Think for instance of changing the address (i.e. the account number) of an electronic cheque.) The part of cryptography in which one studies techniques in order to controll the integrity of a message is called *authentication*.

For both, encryption and authentication, one needs keys, usually secret keys. These keys have to be generated, administrated, changed, etc. All these procedures are controlled by a "master key". It is clear that the access to this master key is a critical point in the system. In other words, before installing a crypto system, one has to have an efficient access control system.

Consequently, we shall deal with three topics: Access control, encryption and authentication. Our aim is neither to search for a complete survey, nor to present the most frequently used algorithms; we focus on applications of finite geometry!

One could think that new types of problems need new methods for solutions. This is true for quite a few parts of cryptography (such as the idea of public keys). On the other hand, surprisingly enough, many classical mathematical structures and results can be applied in cryptography. In particular, number theory was injected with new life by the applications in the RSA-algorithm and signature schemes.

Here, I would like to show that also classical projective geometry has very clever applications in crypto systems. Clearly, geometry is not the center of applied mathematics; but I want to confirm that some parts of geometry are very close to the most modern and exciting parts of computer science.

2. TWO HEADS ARE BETTER THAN ONE: AN ACCESS CONTROL SYSTEM

Threshold schemes are very clever access control systems which were invented to control access to top secret operations such as changing the master key of a computer security system. A widely used method to solve this problem is the following: The critical operation should only be performed if t or more authorized users agree on it.

In order to handle such a situation, threshold schemes have been introduced in the literature (see for instance [1], [5], [6], [7],[10], [13]).

A t-*threshold scheme* consists of a certain number of pieces of information such that the following properties are fulfilled:

(i) a secret datum x can be retrieved from any t shadows,

(ii) determining x with knowledge of only t-1 of the shadows is impossible.

In *geometric language*, a t-threshold scheme can be described as follows. In a *geometry* (consisting of *points* and *blocks*) one choses a block B and n points $P_1,\ldots, P_n$ on B in such a way that

(i) any t of the n points determine B uniquely,

(ii) through any t-1 or fewer of the n points there are 'many' blocks.

One choses also a set S which intersects B in one point X (or a non-empty set X of points).

The system has to know only the geometry (i.e. the points and blocks) and the point X. If a certain number of shadows (i.e. points) enter the system, it tries to determine the block through these points. If there is no unique block through the given shadows, the process stops. If, on the other hand, there is a unique such block C, the systems computes $P_C := C \cap S$. Only if $P_C = X$, the users get access to the system and may perform the operation in question.

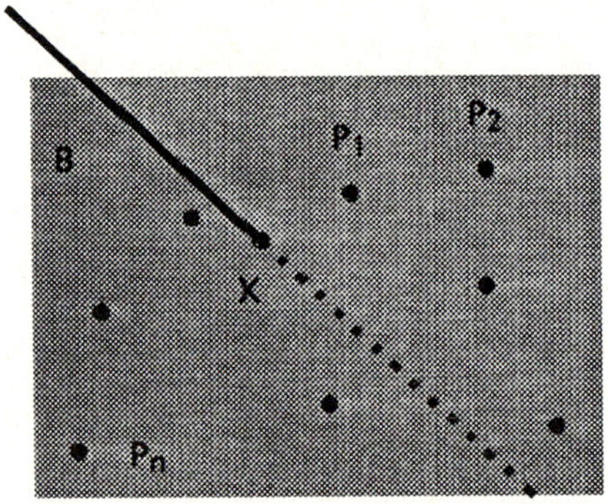

Clearly, such a scheme works equally well, no matter where or who the users are. Also, the system has a certain guarantee that the users are authorized, since the datum X has to be computed. (If an unauthorized user takes part in the process, the system is virtually certain to determine a block C which is different from B; then the system will compute a point P_C different from X.) As a consequence we have that the existence of only one unauthorized user U guarantees that no group of users which U belongs to can perform the critical operation.

Example 1. Curves

One may take as B a randomly chosen *conic* in a Desarguesian projective plane **P** = PG(2,K) over the field K. As S we take a (randomly chosen) tangent to B at a point X. Since the number of conics through four points in general position equals |K| -2 (which can be chosen as big as one wants to), we have a 5-threshold scheme.

If we restrict ourselves to *circles*, we get 3-threshold schemes.

These examples generalize in the following way. A *rational normal curve* in the d-dimensional projective space $\mathbf{P}_s$ has the property that it is determined by any $d+3$ of its points. Moreover, through any $d+3$ points of $\mathbf{P}_s$, no $d+1$ of which are contained in a common hyperplane, there is precisely one rational normal curve (cf. B. Segre [12] and van der Waerden [14]). If $d = 2$, the rational normal curves are precisely the conics.

Example 2. Flats

Consider now an affine or a projective geometry **G** of dimension d.

1. As B we take a (t-1)-dimensional *(flat) subspace*; $P_1,\ldots,P_n$ are n points on B which are in general position and for S we take a (d-t+1)-dimensional subspace which intersects B in just one point X. We get a t-threshold scheme for every positive integer t.

2. Now we consider a variation of the above example. Suppose that we have two types of users which want to get access, say *programs* and *human users*. We would like to design our system in such a way that

(i) Any t of the total number of n users determine the block B uniquely.

(ii) The programs alone, even if their number is bigger than t, do not determine B.

In order to achieve this, we take a (t-1)-dimensional flat B and a (t-2)-dimensional subflat B*. The points which correspond to programs will be chosen as points in general position of B*, whereas the points which correspond to human users are points of B outside B*.

The following picture describes the case $t = 3$.

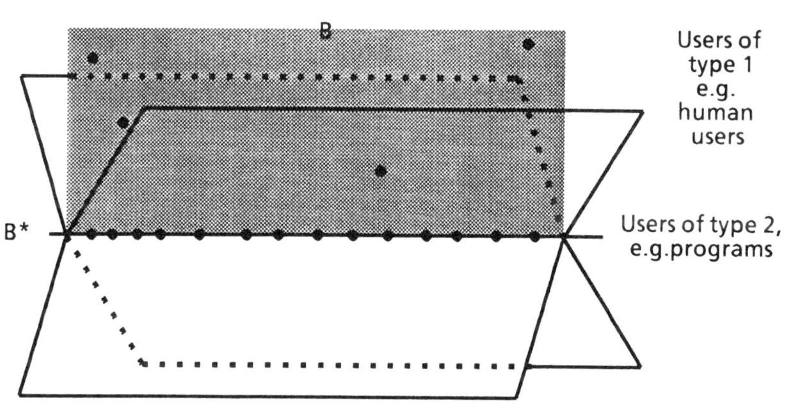

Now we consider the problem where there is a *hierarchy of users* with the property that the levels of the hierarchy correspond to the amount of privileges a certain user has.

A *trivial solution* is to design a threshold scheme for every level seperately and "join" the different systems disjointly. For example, suppose we have users of two types.

The users in group 1 are representated by a 2-threshold scheme (in our example, by a line ℓ), whereas users in group 2 are representated by a 3-threshold scheme (in our situation by a circle C). The line ℓ and the circle C intersect in a point P, which is also the intersectuion of ℓ (or C) by the line S.

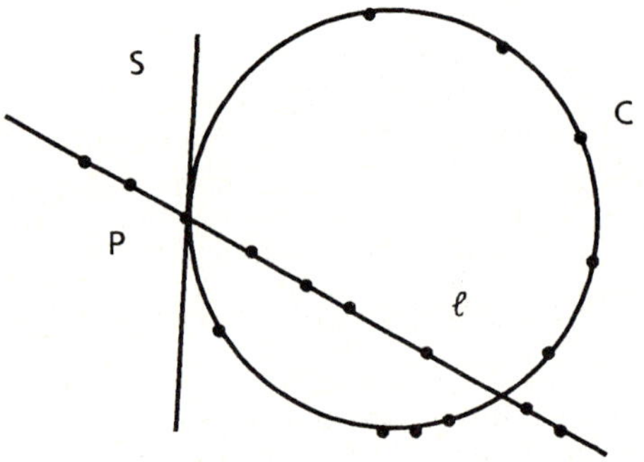

This solution clearly has the disadvantage that no set of users of one group can replace one or more users of another group. This is not the case in the following more *sophisticated* system. In this system, a user i is usually not represented by a point, but by a subspace U_i – in such a way that the dimension of U_i corresponds to the hierarchical level of i. In a military world, we might design our system as follows:

Generals are represented by planes,
colonels by lines,
ordinary soldiers by points,

all within a (say) 5-dimensional space W. The critical operation will be executed if the subspaces which correspond to the users in question span the subspace W. In contrast to the above system, here a (big) number of soldiers might replace a general.

Example 3. Designs

A t-(v,k,λ)-*design* is a geometry **D** consisting of points and blocks (which we always think of sets of points) satisfying the following properties:

(a) D has v points.

(b) Every block of D consists of exactly k points.

(c) through any t points there are exactly λ blocks.

For definitions, background and the theory of block designs see Beth, Jungnickel, Lenz [3], Beutelspacher [4], Hughes and Piper [9].

If $\lambda = 1$, **D** is also called a *Steiner system*. So, by definition, given a block B and n (< k) points on B, the unique block through any t of these points is B. Hence, every Steiner system yields many (usually different) t-threshold schemes.

In a design with $\lambda > 1$, typically, through a certain number s of points there are many blocks. (If $s = t$, this follows from the definition; but in many cases, this is also true for certain values of $s > t$.) But there are certain situations, where for a given number s, any set of s points determine a block uniquely. In other words, any set of s points lies on *at most* one block.

The non-negative integer μ is said to be an *intersection number* of the design **D**, if there are two different blocks of **D** which intersect in precisely μ points. The following statement is obvious:

Denote by **D** *a* t-(v,k,λ)-*design, and fix an arbitrary number* s *wich is bigger than the maximal intersection number of* **D**. *Then through any set of* s *points of* **D** *there is at most one block of* **D**.

As a consequence we have: If s is a number bigger than any intersection number of a given design D such that through some s-1 points there are many blocks, then there exists an s-threshold scheme. This applies in particular to biplanes, i.e. symmetric 2-(v,k,λ) designs.

3. I KNOW SOMETHING YOU DON'T KNOW: ENCRYPTION

Before we can send a message down an "electronic chanel" we have to "encode" it into a binary sequence, that is a sequence of 0's and 1's. The most obvious way (due to Vernam [15]) to encipher such a binary message M consisting of the bits $m_1, m_2, m_3, \ldots$ is adding to it, bit by bit, another binary sequence K (consisting of the bits $k_1, k_2, k_3, \ldots$). The sequence $c_1 = m_1 \oplus k_1$, $c_2 = m_2 \oplus k_2$, $c_3 = m_3 \oplus k_3, \ldots$ (where $\oplus$ denotes binary addition) is called the *ciphertext* and this is the text which is being sent down the chanel.

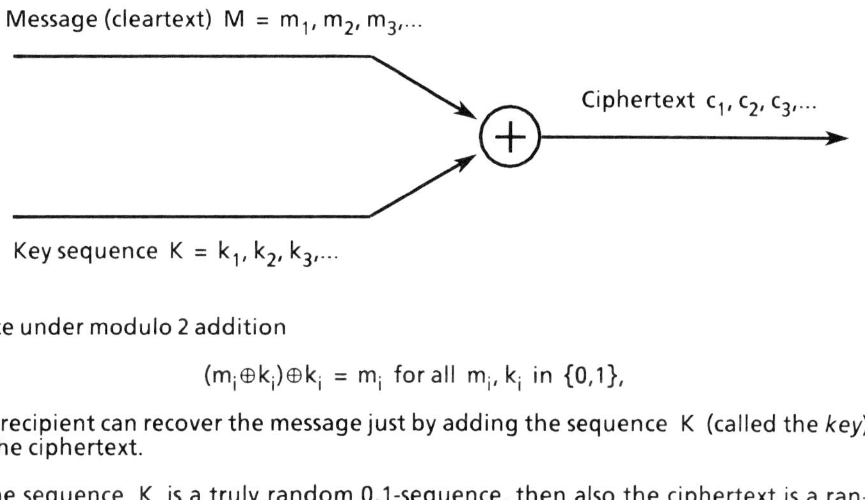

Since under modulo 2 addition

$$(m_i \oplus k_i) \oplus k_i = m_i \text{ for all } m_i, k_i \text{ in } \{0,1\},$$

the recipient can recover the message just by adding the sequence K (called the *key*) to the ciphertext.

If the sequence K is a truly random 0,1-sequence, then also the ciphertext is a random sequence, and so it is "unbreakable"! But the price one has to pay for this super cipher is that the key K is extremely long (its length equals the length of the message) – and, of course, the key has to be forwarded to the receiver before the message is transmitted (otherwise the receiver would be in the same unpleasant situation as an eavesdropper). Moreover, the key has to be transmitted in a secure way (otherwise, an eavesdropper would be in the same pleasant situation as the receiver). Usually, this problem is solved by taking as the key a "pseudo-random" sequence instead of a truly random sequence. A sequence is pseudo-random if it behaves at a first glance as a random sequence, but is determined by very few data. What do we mean by "behaves at a first glance as a random sequence"? The three famous postulates of Golomb (cf. for instance [2]) provide one possible answer to this question. First of all, these postulates deal only with periodic sequences. Let p be the length of the period, and denote by C a generating cycle. The first postulate is very easy to state

P1. *The number of zeros in* C *equals the number of ones.*

If p is an odd number, then this postuilate can never be true in a strict sense. In this case "equals" should be interpreted as "differs by at most 1". A similar interpretation should take place in the following postulates.

By a *gap* we mean a maximal sequence of 0's, whereas a *string* is a maximal sequence of 1's of C. So, the following sequence has two gaps of length 2 and one string of length 1:

Example:

$$C = 011101100101000$$

Now, we can formulate the second postulate:

P2. *For any* i, *the number of gaps of length* i *equals (that is, differ by at most one) the number of strings of length* i.

Clearly, if C is a generating cycle of a periodic sequence, then also the cycle C(a) which is obtained from C by a cyclic shift of a positions is a generating cycle.

If C denotes the cycle in the above example, then

$$C(2) = 11011001010001.$$

For a fixed $a \neq 0$, we denote by A the number of positions in which C and C(a) agree and by D the number of positions in which C and C(a) disagree. (So $D = p - A$.) The number

$$(A\text{-}D)/p = (2A - p)/p$$

is called the *out-of-phase autocorrelation*. Again considering our above example (with $a = 2$) we get $A = 7, D = 8$ and as the out-of-phase autocorrelation

$$(A\text{-}D)/p = -1/15.$$

The third postulate of Golomb reads as follows:

P3. *The out-of-phase autocorrelation is a constant (for* $a \neq 0$*).*

At this point the natural question arises, whether there are sequences satisfying Golombs postulates. Surprisingly enough, most of the known sequences with the above discussed properties arise from projective spaces!

Consider the projective spaces $\mathbf{P} = PG(d,2)$ of dimension d over the field with 2 elements. It is well known that **P** admits a Singer cycle, that is a cyclic group which acts transitively on the points (as well as on the hyperplanes) of **P**. We label the points of **P** by $1,\ldots,v = 2^{d+1}-1$ in such a way that the map σ which maps i onto $i+1 \pmod v$ is a generating element of the Singer cycle. Then we have the following

Theorem. *Let* $C = (a_1,\ldots,a_v)$ *be the incidence vector of a hyperplane* H *of* **P** *with respect to the above defined labeling. (That is* $a_i = 1$ *if* H *is incident with the point* i, *and* $a_i = 0$ *otherwise.) Then the cycle* C *satisfies Golomb's postulates.*

Proof. By definition, the number a of 1's in C equals the number of points in the hyperplane H. So,

$$a = 2^d - 1.$$

On the other hand, the number b of 0's in C is nothing else than the number of points outside H, hence

$$b = 2^d.$$

In other words, **P1** is true.

In order to check **P2** we have to take into consideration that our labeling corresponds to the Singer cycle. Let f be an irreducible polynomial of degree $d+1$ over GF(2). Then the set of points of the projective space **P** can be seen as the set of all non-vanishing polynomials of degree $\leqq$ d. Moreover, the generating element σ of the Singer cycle is just multiplication by x (mod f).

We take as our hyperplane the hyperplane H which is spanned by the points 1, x, $x^2, \ldots, x^{d-1}$. In other words, H consists of all polynomials of degree $\leqq$ d-1. We claim that the corresponding incidence vector C *has one string of length* d, *one gap of length* $d+1$, *and* 2^i *strings and* 2^i *gaps of lengths* $d-1-i$ ($i = 0,1,\ldots,d-2$). Clearly, this will show that **P2** is true.

Since $x^d \notin H$ and $1/x = (f-1)/x \notin H$, there is one string of length d (namely $1,x,x^2,\ldots,x^{d-1}$). Now take a polynomial $h = x^{i+1} + a_i x^i + \ldots + a_1 x + 1$ of degree $i+1$ which has a non-vanishing absolute term. Obviously, there are exactly 2^i such polynomials. It follows that

$$h,\ x \cdot h,\ x^2 \cdot h, \ldots\ x^{d-2-i} \cdot h$$

are points of H. Now, $x^{d-1-i} \cdot h$ and h/x are polynomials of degree d (note that $h/x = h(f-1)/x$, which has degree d, since h has 1 as its absolute term). Hence h gives rise to a string of length d-1-i. In such a way we get at least 2^i strings of length d-1-i. Since we have considered every polynomial of degree $\leqq$ d-1 just once, we have covered every point of H excactly once. So, the considered strings are all possible strings. This proves the first part of our assertion.

The number of gaps can be computed in a similar way by observing that $x^n + h \notin H$ if $h = 0$ or $h \in H$.

For **P3** we again make use of the fact that our labeling is induced by a collineation. From this it follows that not only C is the incidence vector of a hyperplane (namely H), but also C(a). More precisely, C(a) is the incidence vector of $H' = \sigma^a(H)$, which is a hyperplane as well, since σ (and σ^a) is a collineation.

The rest is easy: The number A of positions in which C and C(a) aggree is the number of common 1's plus the number of common 0's; so it is the number of points in H and H' plus the number of points off H and off H'. Consequently,

$$A = 2^{d-1}-1 + 2^{d-1} = 2^d-1.$$

From this it follows

$$D = 2^{d+1}-1 - (2^d-1) = 2^d.$$

Therefore, the out-of-phase autocorrelation is a constant.

Remark. There is only one sequence known which satisfies Golomb's postulates, but which cannot be constructed from a projective space in the avove described way. The interested reader is referred to [11], where further details can be found.

4. MAKING LIFE AS DIFFICULT AS POSSIBLE FOR THE BAD GUY: A PERFECT AUTHENTICATION SYSTEM

In this section we study the following question: How can one guarantee the integrity of a message? Or a little bit more modest: Are there systems which guarantee that the receiver notices whether the message has been changed or not? We illustrate the problem by an example. (The example and the theory can be found in [8].)

The local manager (a bad guy) of a certain gambling casino has to transfer the dollars from the slot machines every night to the owner. The bad guy cheats the owner by reporting the daily takings from the slot machines to be less than they actually are and keeping the rest for himself. But the owner is suspicious and therefore he installes at every slot machine a device which takes as inputs the day's takings M and a (secret) key K and has as its output an *authenticator* (or, a *message authentication code*)

$$A = f(M,K).$$

The bad guy's duty is to transfer not only the dollars, but also the authenticator A. The owner takes the transfered number M' of dollars (which might differ from M) and computes, using the secret key K, the authenticator

$$A' = f(M',K).$$

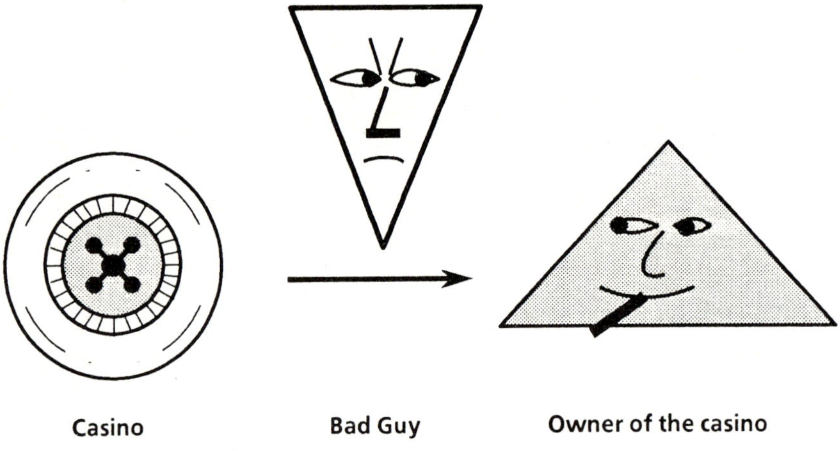

If $M' \neq M$, he is virtually certain that $A' \neq A$. So, if the outcome A' of his computations does not aggree with A, he "knows" that the bad guy is cheating him.

Such a system is called an *authentication system*. Although the bad guy does not know the correct key, it is strongly advisable for the owner that he asks himself the question what chances the bad guy has of guessing the correct authenticator. Of course, the bad guy's chance depends on many things, for instance on the chosen authentication function f. But one principal constraint is the number k of keys. The following theorem shows that there are no authentication systems which are arbitrarily good.

Theorem. (Gilbert, MacWilliams, Sloane [8]) *Let* k *be the total number of keys. Then there is no authentication system in which the bad guy's chance of success is worse than* $1/\sqrt{k}$.

We shall not prove this theorem here. We shall, however, discuss the natural question, whether there are systems which are as ideal as possible, that is systems in which the bad guy is confronted with the unpleasant situation that he has only a chance of $1/\sqrt{k}$ to guess a correct authenticator. This question will be answered in the affirmative by the following

Example. Let **P** be a finite projective plane of order n. Fix one line ℓ_0.

The *messages* are the points on ℓ_0;

the *keys* are all points outside ℓ_0;

the *authenticators* are alle lines different from ℓ_0.

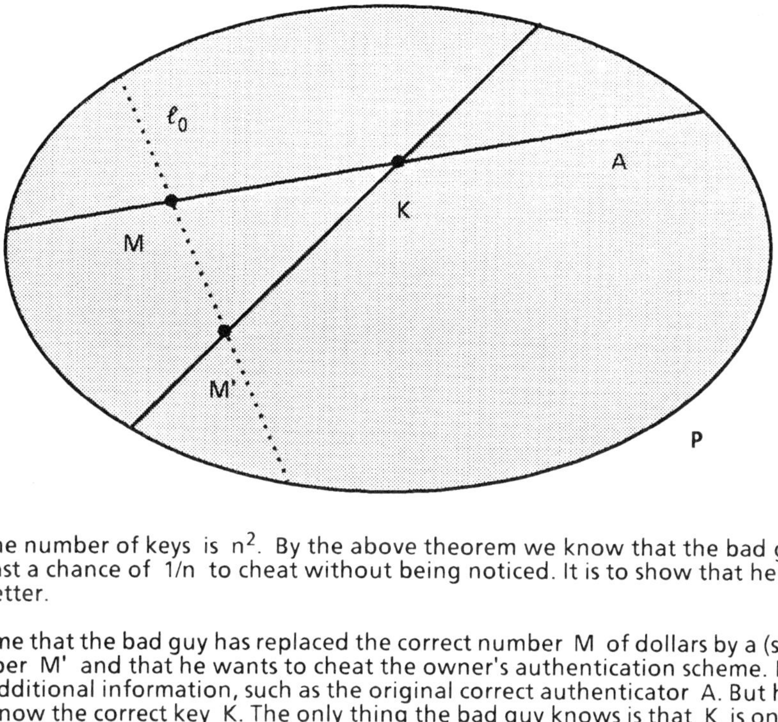

So, the number of keys is n^2. By the above theorem we know that the bad guy has at least a chance of 1/n to cheat without being noticed. It is to show that he can do no better.

Assume that the bad guy has replaced the correct number M of dollars by a (smaller) number M' and that he wants to cheat the owner's authentication scheme. He may use additional information, such as the original correct authenticator A. But he does not know the correct key K. The only thing the bad guy knows is that K is one if the n remaining points on A. On the other hand, any hypothetical authenticator through M' intersects A in a point, which is a hypothetical key. Therefore, the bad guy has only a chance of 1/n to guess the correct key. In other words, the projective-plane-authentication-scheme is as good as one can expect from the theory.

ACKNOWLEDGEMENT

The author would like to thank very much Dr. Jean Georgiades for many helpful discussions and for his strong coffee.

REFERENCES

[1] G.R. Blakley, Safeguarding cryptographic keys. Proc. NCC 48, AFIPS Press, Montvale, N.J., 317-319 (1979).

[2] H. Beker and F.C. Piper, Cipher Systems. Northwood Books, London 1982.

[3] T. Beth, D. Jungnickel and H.Lenz, Design Theory. B.I.-Wissenschaftsverlag, Mannheim - Wien - Zürich 1985.

[4] A. Beutelspacher, Einführung in die endliche Geometrie I. Blockpläne. B.I.-Wissenschaftsverlag, Mannheim - Wien - Zürich 1982.

[5] A. Beutelspacher and K. Vedder, Geometric structures as threshold schemes. To appear.

[6] D. Chaum, Computer systems established, maintained, and trusted by mutually suspicious groups. Memorandum No. UCB/ERL M79/10, University of California, Berkeley, CA, February 22, 1979.

[7] D.E.R. Denning, Cryptography and Data Security. Addison-Wesley 1983.

[8] E.N. Gilbert, F.J. MacWilliams, N.J.A. Sloane, Codes which detect deception. Bell. Syst. Tech. J. **53** (1974), 405-424.

[9] D.R. Hughes and F.C. Piper, Design Theory. Cambridge University Press, 1985.

[10] S.C. Kothari, Generalized Linear Threshold Scheme. Adavances in Cryptology (Proceedings of CRYPTO 84), Lecture Notes in Computer Science **196**, Springer 1985, 231-241.

[11] F.C. Piper and M. Walker, Binary sequences and Hadamard designs.

[12] B. Segre, Lectures on modern geometry. Cremonese, Roma 1961.

[13] A. Shamir, How to share a secret. Comm. ACM Vol. 22(1), 612-613 (1979).

[14] B. van der Waerden, Einführung in die algebraische Geometrie

[15] G.S. Vernam, Cipher printing telegraph systems for secret wire and radio telegraphic communications. J. AIEE **45** (1926), 109-115.

Annals of Discrete Mathematics 37 (1988) 69–74

ON FINITE GRASSMANN SPACES

Paola BIONDI

Dipartimento di Matematica e Applicazioni
Via Mezzocannone 8
80134 Napoli, Italia*

In this paper a characterization of the Grassmann spaces $\Gamma^h(\mathbb{P})$ associated with a finite projective space $\mathbb{P}$ is obtained by means of incidence properties of points and lines only.

1. INTRODUCTION

Let $\mathbb{P}$ be a projective space of dimension at least three. For each integer h, $1 \leq h < \dim \mathbb{P}-1$, $G^h(\mathbb{P})$ and $F^h(\mathbb{P})$ denote the families of h-dimensional subspaces of $\mathbb{P}$ and of pencils of h-dimensional subspaces of $\mathbb{P}$, respectively. Such a pencil consists of all the elements in $G^h(\mathbb{P})$ through a fixed (h-1)-dimensional subspace of $\mathbb{P}$ and which are contained in a given subspace of $\mathbb{P}$ of dimension h+1. The incidence structure $\Gamma^h(\mathbb{P}) = (G^h(\mathbb{P}),F^h(\mathbb{P}))$ is a partial linear space, the h-th Grassmann space associated with $\mathbb{P}$. If $\dim \mathbb{P} < \infty$ and $\mathbb{P}$ is coordinatized by a field, then $\Gamma^h(\mathbb{P})$ is isomorphic to the Grassmann variety representing the h-dimensional subspaces of $\mathbb{P}$.
In relation to special values of h, more characterization of the spaces $\Gamma^h(\mathbb{P})$ are known ([1], [2], [3], [4], [5], [6]).
The spaces $\Gamma^h(\mathbb{P})$, for each h, are characterized in [7], with the help of incidence properties of the maximal subspaces in $\Gamma^h(\mathbb{P})$.
A characterization of the spaces $\Gamma^h(\mathbb{P})$, which only uses incidence properties of points and lines, is obtained, in the special case $\mathbb{P}$ finite and irreducible in [3] as a consequence of a more general theorem.
This paper provides a characterization, in the finite case, of the partial linear spaces (P,L) isomorphic to $\Gamma^h(\mathbb{P})$, by means of incidence properties of points and lines only. This characterization also holds if $\mathbb{P}$ is reducible. Moreover, this result is extendable to the infinite case if the following condition holds : "there exists in (P,L) a maximal subspace of finite rank".

2. INCIDENCE PROPERTIES OF POINTS AND LINES IN $\Gamma^h(\mathbb{P})$

Let (P,L) be a partial linear space ([8]). First of all, we recall some well known basic definitions. Two distinct points $a,b \in P$ are *collinear*, $a \sim b$, if the line (a,b) through them exists. If this is not the case, a and b are *noncollinear*, $a \not\sim b$. Moreover, any point is assumed to be collinear with itself.

*Work supported by National Research Project on "Strutture geometriche, Combinatoria, loro applicazioni" of Italian M.P.I .

A partial linear space is *connected* if for any two distinct points in P there exists a polygonal path in (P,L) joining them. A *subspace* of (P,L) is a subset of P consisting of pairwise collinear points and containing the line through any two of its (distinct) points. A subspace is *maximal* if no subspace exists in (P,L) which properly contains it.
Two points b and c are *of the same type* with respect to a point a, $b \underset{a}{\approx} c$, if both of them are either collinear or noncollinear with a.
Let $a \in P$ and $l \in L$. The point a is *collinear* with l, $a \sim l$, if a is collinear with any point on l, *one-collinear* with l, $a \dashv l$, if it is collinear with exactly one point on l and *noncollinear* with l, $a \nsim l$, if it is collinear with no point on l. If $a \nsim l$, then $\hat{l}_a = \{x \in P : x \sim a, x \sim l\}$.
Let $l, m \in L$. The lines l and m are *collinear*, $l \sim m$, if $x \sim y$ for any $x \in l$ and $y \in m$. If this is not the case, l and m are *noncollinear*, $l \nsim m$.
Finally, let $l, m \in L$. We define a *chain* of length k and with end-lines l and m as an ordered k-tuple $(l_1, \ldots, l_k)$ of elements in L such that : i) $l_1 = l$ and $l_k = m$; ii) $l_i \sim l_{i+1}$, $i=1,\ldots,k-1$, and $l_i \nsim l_{i+2}$, $i=1,\ldots,k-2$. If ii) holds for any i (mod. k) and $l_1 \cap \ldots \cap l_k = \emptyset$, then the chain is called a *cycle*. Denote by $\mathscr{C}$ the set of cycles in (P,L). The chain $(l_1, \ldots, l_k)$ is *minimal* if there exists no chain of length h, h<k, with end-lines l_1 and l_k.
A finite and connected partial linear space will be said to be a *finite Grassmann space* if the following properties hold:

(2.1) $\forall l \in L, \quad \forall a \in P \quad \Rightarrow \quad$ either $a \nsim l$ or $a \dashv l$ or $a \sim l$.

(2.2) $\forall l \in L, \quad \forall a,b,c \in P-l : a \sim l, b \sim l, c \sim l, b \underset{a}{\approx} c \quad \Rightarrow \quad b \sim c.$

(2.3) $\forall l \in L \quad \exists a,b \in P : a \sim l,\ b \sim l$ and $a \nsim b$.

(2.4) $\forall l,m \in L, \quad \forall a \in P : l \sim m, a \nsim l, a \nsim m, \hat{l}_a \neq \emptyset$ and $\hat{m}_a \neq \emptyset \quad \Rightarrow \quad \forall b \in \hat{l}_a \quad \exists c \in \hat{m}_a :$ $b \neq c$ and $b \sim c$.

(2.5) $\forall (l_1, \ldots, l_k) \in \mathscr{C} \quad \Rightarrow \quad k = 2h, \quad h > 2.$

It is easy to verify that, given a projective space $\mathbb{P}$ of dimension at least three, $\Gamma^h(\mathbb{P})$ is connected and satisfies (2.1)-(2.5) ,for each h, $1 \leq h < \dim \mathbb{P} - 1$
From now on, (P,L) denotes a finite Grassmann space.

3. SOME PROPERTIES OF THE SPACE (P,L)

Axioms (2.1) to (2.3) allow to extend prop.s I-VI in [4],sect.2, to (P,L).

PROPOSITION 3.1. If three distinct points a,b,c are pairwise collinear and $c \notin (a,b)$, then there exists a unique maximal subspace containing them.

Proof. By (2.1) and prop.s I and III in [4],sect.2.

PROPOSITION 3.2. Let l be a line and c a point such that $c \nsim l$ and $\hat{l}_c \neq \emptyset$. Then, $\hat{l}_c \in L$ and $\hat{l}_c \cap l = \emptyset$. Moreover, $\hat{l}_c \sim l$ and one of the two maximal subspaces through $\hat{l}_c$ contains c, whereas the other one contains l.

Proof. Denote by V and W the two maximal subspaces through l (see prop.IV in [4],sect.2). By prop.I in [4],sect.2, $\hat{l}_c \subseteq (V \cup W) - l$. Take a point $a \in \hat{l}_c$ and suppose $a \in V$. By (2.4) and prop.V in [4],sect.2, there exists in $V \cap \hat{l}_c$ a point b distinct from a. By (2.1), $(a,b) \subseteq \hat{l}_c$ and, by prop.3.1, a unique maximal subspace through (a,b) and c exists. By prop.V in [4],sect.2, $\hat{l}_c \cap V=(a,b)$. We now prove that $\hat{l}_c \cap (W-l)=\emptyset$. If c were collinear with a point $a' \in W-l$, by the same argument as above there would be in W-l a line (a',b') collinear with c. Denote by V' (W') the maximal subspace through (a,b) and c ((a',b') and c). By prop.III in [4],sect.2, either $V' \cap W' = c$ or $V' \cap W' = l' \in L$, with $c \in l'$. First suppose $V' \cap W' = c$ .Since $a \not\sim (a',b')$ by prop.V in [4],sect.2,and $l \cup \{c\} \subseteq \widehat{(a',b')}_a$, by the same argument as above there exists in W'-(a',b') a line m through c collinear with a . Since $V' \cap W' = c$, the maximal subspace through (a,c) and m is distinct from V'; thus, from prop.V in [4],sect.2, $(a,b) \not\sim m$. Then, the cycle ((a,b),(a,c), m,(a',b'), l) of length 5 exists, a contradiction to (2.5). Now, suppose $V' \cap W' = l'$, with $c \in l'$. In this case, the cycle ((a,b), l',(a',b'), l) of length 4 exists, a contradiction to (2.5). Therefore, $\hat{l}_c = (a,b)$ and the proof is complete.

Proposition 3.2 allows to extend prop.s 3.4 and 3.5 in [2] to (P,L).

PROPOSITION 3.3. Let l and m be two disjoint collinear lines and c a point noncollinear with both of them. If $\hat{l}_c \neq \emptyset$ and $\hat{m}_c \neq \emptyset$, then $\hat{l}_c = \hat{m}_c$.

Proof. Denote by V the maximal subspace through l and m (see prop.s I and III in [4],sect.2) and by L and M the two maximal subspaces through l and m , respectively, different from V (see prop.IV in [4],sect.2).By prop.3.2, $\hat{l}_c$ ($\hat{m}_c$) is a line non-incident with l (m) and belonging to one of the two maximal subspaces through l (m). By (2.4) and prop.3.5 in [2], it cannot occur that both $\hat{l}_c \subset L$ and $\hat{m}_c \subset M$. Furthermore, one can exclude the case where $\hat{l}_c \subset L$ and $\hat{m}_c \subset V$: a contradiction to prop.3.5 in [2] would arise, since $c \notin \hat{l}_c$ and,by prop. 3.2, c belongs to the maximal subspace through $\hat{l}_c$ other than L. Therefore, $\hat{l}_c$ belongs to V. Consequently, $\hat{l}_c \subseteq \hat{m}_c$, so that $\hat{l}_c = \hat{m}_c$.

Propositions 3.2 and 3.3 allow to extend prop.IX in [4],sect.2, and prop.s 3.9 and 3.10 in [2] to (P,L).

PROPOSITION 3.4. For any $l,m \in L$, a chain with end-lines l and m exists.

Proof. Let $l \cap m \neq \emptyset$. If $l \sim m$, then the chain (l,m) of length 2 and with end-lines l and m exists. If $l \not\sim m$, by prop.IX in [4],sect.2, a chain of length 3 and with end-lines l and m exists. Now, suppose $l \cap m = \emptyset$ and consider two points a and b on l and m, respectively. Since (P,L) is connected, there exists an ordered k-tuple $(a_1,\ldots,a_k)$ of points in P such that $a_1=a$, $a_k=b$ and $a_i \sim a_{i+1}$ $i=1,\ldots,k-1$. If k=2, i.e. the line (a,b) exists, by the previous argument a chain with end-lines l and (a,b) and a chain with end-lines (a,b) and m exists. Therefore, we finde an ordered n-tuple of lines, $3 \leq n \leq 5$, with end-lines l and m such that $l_i \sim l_{i+1}$, $i=1,\ldots,n-1$. Obviously, a chain with end-lines l and m can be drawn from this n-tuple. Now, by induction we may assume $k \geq 3$. Then, a chain with end-lines l and (a_{k-1},a_k) exists, either $l \cap (a_{k-1},a_k)=\emptyset$ or $l \cap (a_{k-1}, a_k) \neq \emptyset$. Moreover, a chain with the two incident end-lines (a_{k-1},a_k) and m exists. With the help of this two chains, a chain with end-lines l and m can be easily constructed.

PROPOSITION 3.5. The maximal subspaces in (P,L) are projective spaces of finite dimension.

Proof. Since (P,L) is finite, the statement follows from prop.3.10 in [2].

4. DISTANCE BETWEEN TWO MAXIMAL SUBSPACES IN (P,L)

Let $l,m \in L$. We define the *distance between* l *and* m, $d(l,m)$, as the integer $k-2$, where k is the length of a minimal chain with end-lines l and m (see prop.3.4). Now, let L and M be two maximal subspaces. Set

$$d(L,M) = \{d(l,m)\}_{l\subset L, m\subset M}$$

The integer $d(L,M)$ will be said to be the *distance between* L *and* M.

PROPOSITION 4.1. Let V and W be the two maximal subspaces through a line l. Moreover, let M be a maximal subspace such that $V\cap M = W\cap M = \emptyset$. Then, $d(V,M)= d(W,M)\pm 1$.

Proof. Set $d(V,M)=k$ and $d(W,M)=k'$. Moreover, denote by $(v_1,\ldots,v_{k+2})$ and $(w_1,\ldots,w_{k'+2})$ two chains of length $k+2$ and $k'+2$, respectively, with $v_1\subset V$, $w_1\subset W$ and v_{k+2}, $w_{k'+2} \subset M$. First of all, recall that two lines are collinear iff a maximal subspace through them exists. If $v_1=l$, then $v_2\subset W$. Therefore, $k'\leq k-1$. If $v_1\neq l$, then $v_2 \nsim l$ by prop.V in [4],sect.2. Thus, the chain $(l,v_1,\ldots,v_{k+2})$ of length $k+3$ exists, so that $k'\leq k+1$. In any case, $k'\leq k+1$. By the same argument, $k\leq k'+1$. Therefore, $k-1\leq k'\leq k+1$. In order to prove the statement,it is sufficient to show that $k \neq k'$. By the previous argument, we can assume $v_1,w_1\neq l$. By prop.V in [4],sect.2, $v_1 \nsim w_1$, $v_2 \nsim l$ and $w_2 \nsim l$. If $v_{k+2} \neq w_{k+2}$, then, by prop.V in [4], sect.2, $v_{k+1} \nsim w_{k'+2}$ and $v_{k+2} \nsim w_{k'+1}$. Therefore, the cycle $(v_1,\ldots,v_{k+2},w_{k'+2},\ldots,w_1,l)$ of length $k+k'+5$ exists. By (2.5), $k \neq k'$. Next, assume $v_{k+2} = w_{k'+2}$ and denote by N the maximal subspace through v_{k+2} other than M. Then, v_{k+1} and $w_{k'+1}$ belong to N. Obviously, if $k=0$, then $k'=1$. Now, let $k=1$. If $N\cap V = \emptyset$,then $N\cap W = \emptyset$ by prop.3.5 in [2]. By the previous arguments, $d(N,W)=k'-1$ and $k'-1\neq 0$ $(=d(N,V))$; so $k'\neq 1$. If $N\cap V=a\in P$, with $a\in l$, then $d(W,M)=0$ by prop.IX in [4], sect.2. If $N\cap V=a\in P-l$, then $k'\neq 1$ by prop.3.4 in [2]. Now, by induction we may assume $k\geq 2$. If k' were equal to k, then $N\cap V=N\cap W=\emptyset$ by prop.IX in [4],sect.2 . Moreover, $d(N,V)=d(N,W)=k-1$, which contradicts the hypothesis of induction if $w_{k+1} = v_{k+1}$ and the previous argument if $w_{k+1} \neq v_{k+1}$.

5. CHARACTERIZATION OF THE FINITE GRASSMANN SPACES

Let S_0 be a maximal subspace. Denote by $\mathcal{S}$ the family consisting of S_0, the maximal subspaces meeting S_0 at a unique point, and the maximal subspaces disjoint from S_0 and whose distance from S_0 is even. Furthermore, denote by $\mathcal{T}$ the family consisting of all the other maximal subspaces, i.e. those meeting S_0 at a line and those disjoint from S_0 and whose distance from S_0 is odd.Obviously these two families are non-empty and partition the family of all the maximal subspaces in (P,L).

PROPOSITION 5.1. For any $l\in L$,one of the two maximal subspaces through l belongs to $\mathcal{S}$, whereas the other one belongs to $\mathcal{T}$.

Proof. The statement is obviously true if $l\subset S_0$. Therefore, assume $l\not\subset S_0$. If $l\cap S_0 = a\in P$, the statement follows from prop.IX in [4],sect.2. Now, suppose $l\cap S_0 = \emptyset$ and call V and W the two maximal subspaces through l. If one of them, say V, meets S_0 at a line s, then $W\cap S_0 = \emptyset$ by prop.V in [4],sect.2. Moreover, $d(W,S_0)=0$. Therefore, $V\in \mathcal{T}$ and $W\in \mathcal{S}$. Next, assume that one of V and W,say V, meets S_0 at a unique point a. By prop.V in [4],sect.2, $W\cap S_0 = \emptyset$. Consider a point b on l. By prop.IX in [4],sect.2, the maximal subspace V' trough (a,b) other than V meets S_0 at a line l' and W just at a point b. THen, the chain (l', (a,b),l) exists; thus, $d(W,S_0)\leq 1$. Since $W\cap V'= b\notin l'$, $d(W,S_0)\neq 0$ by prop.3.4 in [2]. Therefore, $V\in\mathscr{S}$ and $W\in\mathscr{T}$. Finally, if V and W are disjoint from S_0, the statement follows from prop.4.1.

From prop.s III and IX in [4],sect.2, and prop.5.1 the next result follows.

PROPOSITION 5.2. For any $S\in\mathscr{S}$ and any $T\in\mathscr{T}$, either $S\cap T = \emptyset$ or $S\cap T\in L$.

PROPOSITION 5.3. The two families $\mathscr{S}$ and $\mathscr{T}$ satisfy the Veblen axiom.

Proof. The proof runs as that one of prop.5.4 in [2].

Propositions 3.1 and 3.5 and the results in this section prove that (P,L) satisfies all the hypotheses of the main theorem in [7]. Therefore, the next theorem holds.

THEOREM I. A finite Grassmann space is isomorphic to $\Gamma^h(\mathbb{P})$, for some finite projective space $\mathbb{P}$ and some integer h.

Remark. Except for prop.3.5, all the propositions in this paper have been proved without taking into account that (P,L) is finite. This hypothesis is necessary only to affirme that one of the maximal subspaces in (P,L), which are projective spaces,has finite dimension.Therefore,if one replace this hypotheses with the axiom A_3 in [7],sect.1, the following result holds.

THEOREM II. A connected partial linear space satisfying (2.1)-(2.5) and A_3 in [7],sect.1, is isomorphic to $\Gamma^h(\mathbb{P})$, for some projective space $\mathbb{P}$.

REFERENCES

[1] Bichara, A. and Tallini, G., On a characterization of the Grassmann manifold representing the planes in a projective space, Ann.Discr.Math. 14 (1982), 129-150.

[2] Biondi, P., A characterization of the Grassmann space representing the 2-dimensional subspaces of a projective space (to appear in Bollettino U.M.I.).

[3] Cohen, A.M., On a theorem of Cooperstein, Europ.J.Combinatorics, 4 (1983), 107-126.

[4] Lo Re, P.M. and Olanda, D., Grassmann spaces, J.Geom.,17 (1981),50-60.

[5] Melone, N. and Olanda, D., A characteristic property of the Grassmann manifold representing the lines of a projective space, Europ.J.Comb., 5 (1984), 323-330.

[6] Tallini, G., ON a characterization of the Grassmann manifold representing the lines in a projective space, Finite geometries and designs (eds.P.J. Cameron et al.), London Math.Soc. Lect.Notes Series n.49, Cambridge University Press (1981), 354-358.

[7] Bichara, A. and Tallini, G., On a characterization of the Grassmann space representing the h-dimensional subspaces in a projective space, Ann. Discr.Math., 18 (1983), 113-132.

[8] Tallini, G., Spazi parziali di rette, spazi polari, geometrie subimmerse, Quaderni Sem.Geom.Combinatorie, Ist.Mat.Univ.Roma n.14 (1979).

Annals of Discrete Mathematics 37 (1988) 75–86

THE REGULAR SUBGROUPS OF THE SHARPLY 3-TRANSITIVE FINITE PERMUTATION GROUPS (*)

To Prof. Guido Zappa on his 70-th birthday

Arrigo BONISOLI

Dipartimento di Matematica
via Campi 213/B
41100 MODENA (Italy)

Every sharply 3-transitive finite permutation group is isomorphic to a certain subgroup G of $P\Gamma L(2,q)$ acting on the projective line $X := GF(q) \cup \{\infty\}$ and containing $PSL(2,q)$ as a subgroup of index at most 2 : using Dickson's classification of the subgroups of $PSL(2,q)$ we determine all the subgroups of G operating sharply 1-transitively on X .

1. The motivations for this paper actually come from geometry; in fact every finite Minkowski plane M determines a sharply 3-transitive set G of permutations on a finite set X (with no loss of generality G is always assumed to contain the identity on X) and conversely every such permutation set determines a finite Minkowski plane . The existence of a *regular* subset R of G (i.e. a subset R of G which is a *sharply 1-transitive* permutation set on X) amounts to the existence of a partition of the set of points of M into pairwise disjoint blocks (in the terminology of block designs we would speak of a *resolution class of blocks*) .

In view of the existence problem for finite Minkowski planes and of possible geometrical characterizations, it seems desirable to have informations about the regular subsets of the known finite sharply 3-transitive permutation sets; so far, every known example of a finite sharply 3-transitive permutation set which contains the identity but is not a group, can be constructed as follows: if p is an odd prime and m is an integer >2 , set $q = p^m$ and let σ be

(*) work done within the activity of G.N.S.A.G.A. and supported by the Italian Ministry of Public Education .

an automorphism of the finite field $GF(q)$ such that $\sigma^2 \neq 1$; define G to be the following subset of $P\Gamma L(2,q)$ in its natural action on the projective line $X := PG(1,q) = GF(q) \cup \{\infty\}$:

$G := \{ x \longmapsto (ax+b)/(cx+d) \ : \ ad-bc$ a square in $GF(q)^* \} \cup$

$\cup \{ x \longmapsto (a\sigma(x)+b)/(c\sigma(x)+d) \ : \ ad-bc$ a non-square in $GF(q)^* \}$.

A family of regular subsets of G, under certain conditions on q, is given in [2] .

The classification of the sharply 3-transitive finite permutation groups goes back to H. Zassenhaus and J. Tits (cfr. [4] or [6, XI 2.6]) and can be stated as follows : if G is a sharply 3-transitive group of permutations on the finite set X, then X can be identified with the projective line $PG(1,q) = GF(q) \cup \{\infty\}$ for some prime power q and either

(1.1) G is $PGL(2,q)$ in its natural action on $PG(1,q)$

or

(1.2) $q = p^{2f}$ for an odd prime p and a positive integer f, and if σ denotes the unique involution in $Aut(GF(q))$ then G is the following subgroup of the group $P\Gamma L(2,q)$ in its natural action on $PG(1,q)$:

$M(p^{2f}) := \{ x \longmapsto (ax+b)/(cx+d) \ : \ ad-bc$ a square in $GF(q)^* \} \cup$

$\cup \{ x \longmapsto (a\sigma(x)+b)/(c\sigma(x)+d) \ : \ ad-bc$ a non-square in $GF(q)^* \}$

(the notation $M(p^{2f})$ is taken from [6]) .

If R is a regular subset of a permutation group G, we may always assume $1 \in R$ (otherwise consider $\rho^{-1}R$ for a fixed element ρ of R) ; it is the aim of this paper to determine all the regular *subgroups* of G when G is one of the groups in (1.1) or (1.2) . Of course the condition that R be a subgroup and not merely a subset is a strong one ; in order to find all the regular *subsets* of G one should investigate all the *transitive* subgroups of G (since a regular subset necessarily generates a transitive subgroup), but this seems to be a harder task .

Our results are summarized in the following

PROPOSITION . The regular subgroups of $PGL(2,p^m)$ can be classified into the following families :

i) a family of $\binom{p^m}{2}$ cyclic subgroups forming a single conjugacy class in $PGL(2,p^m)$;

ii) for p odd, $p^m \neq 3$, a family of $\binom{p^m}{2}$ dihedral subgroups forming a single conjugacy class in $PGL(2,p^m)$; for $p^m = 3$, a unique dihedral normal subgroup of $PGL(2,3)$; if $p^m \equiv -1 \pmod 4$ all such subgroups even lie in $PSL(2,p^m)$ and form there a single conjugacy class ;

iii) for $p = 11$, $m = 1$, a family of 55 subgroups isomorphic to A_4 all of which lie in $PSL(2,11)$ and form there a single conjugacy class ;

iv) for $p = 23$, $m = 1$, a family of 506 subgroups isomorphic to S_4 all of which lie in $PSL(2,23)$ and split there into two conjugacy classes of equal size, while they form a single conjugacy class in $PGL(2,23)$;

v) for $p = 59$, $m = 1$, a family of 3422 subgroups isomorphic to A_5 all of which lie in $PSL(2,59)$ and split there into two conjugacy classes of equal size, while they form a single conjugacy class in $PGL(2,59)$.

No subgroup of $M(p^{2f})$ operates regularly on $PG(1,p^{2f})$.

The basic observation for the proof is that the groups in (1.1) and (1.2) contain $PSL(2,q)$ as a subgroup of index at most 2 , therefore a regular subgroup R is either contained in $PSL(2,q)$ or its intersection with $PSL(2,q)$ has index 2 in R , thus yielding a subgroup of order $(q+1)/2$ of $PSL(2,q)$ operating *semiregularly* on X , i.e. the stabilizer of each point x of X reduces to the identity (cfr. [12, page 8]) ; we use then Dickson's classification of the abstract groups appearing as subgroups of $PSL(2,q)$: we refer to Theorem 4 of [10] or to [5, II 8.27] for such a list and to Theorem 3 of [10] for a list of all the abstract groups which are subgroups of $PGL(2,q)$.

Exhaustive computer search by V. Franceschini showed that every regular subset with identity of $PGL(2,9)$ is a subgroup, while no regular subset exists in $M(9)$, but F. Zironi just announced the existence of regular subsets (no subgroups !) in $M(25)$. The conjecture that every regular subset with identity of $PGL(2,q)$ must be a subgroup has a further piece of evidence in the result of [7] .

We remark that the problem we approached is also motivated by the general search for sharply k-transitive subsets inside n-transitive permutation groups (see for instance [1] and [11]). Of course the main question in this direction is the determination of the finite 2-transitive permutation groups containing a sharply 2-transitive subset : such subsets do not exist in a finite sharply 3-transitive permutation group (even a set) as proved in [9] and also in [11] with a few exceptions. We also point out that the non-existence of regular subsets in M(9) yields an easy proof of the non-existence of sharply 2-transitive subsets in the Mathieu group M_{11} acting sharply 4-transitively on 11 elements : in fact M(9) is the one-point-stabilizer in M_{11}.

We finally metion the article [8] , which contains a method based on character theory for the determination of regular subsets in a transitive permutation group : we have not tried to use this method in our situation.

2. If $PSL(2,p^m)$ contains a regular subgroup R , then this will also be a regular subgroup of $PGL(2,p^m)$, and in case $m = 2f$, p odd, also of $M(p^m)$. We check the list of subgroups of $PSL(2,p^m)$, looking for possible subgroups of order $(p^m + 1)$ and obtain the following possibilities :

(2.1) cyclic subgroups for $p = 2$;

(2.2) dihedral subgroups for p odd ;

(2.3) A_4 for $p = 11$, $m = 1$;

(2.4) S_4 for $p = 23$, $m = 1$;

(2.5) A_5 for $p = 59$, $m = 1$.

Cases (2.1) and (2.2) will be dealt with in the next section. As to case (2.3) , PSL(2,11) does possess subgroups which are isomorphic to A_4 and since the one-point-stabilizers in PSL(2,11) have order $11 \cdot 5$ which is relatively prime with $12 = |A_4|$, we see that each of these subgroups operates regularly on PG(1,11) . Similarly PSL(2,23) resp. PSL(2,59) possesses subgroups which are isomorphic to S_4 resp. A_5 and they all operate regularly on PG(1,23) resp. PG(1,59) . Furthermore, every subgroup of PGL(2,11) resp. PGL(2,23) resp. PGL(2,59) which is isomorphic to A_4 resp. S_4 resp. A_5 is actually contained in PSL(2,11) resp. PSL(2,23) resp. PSL(2,59) ; this is clear for A_4 and A_5 since they have no subgroup of index 2 , while for S_4 it can be seen by the following easy argument . Assume $R \leq PGL(2,23)$, $R \cong S_4$, $R \nleq PSL(2,23)$. Then $R \cap PSL(2,23)$ is isomorphic to A_4 , the unique subgroup of index 2 in S_4 . There is an element ψ of order 4 in R and since PGL(2,23) is 3-transitive we may assume ψ has the form $(\infty,0,1,c) \ldots$, otherwise we replace R by a suitable conjugate R^{γ} in PGL(2,23) ; then $c = 2$ and $\psi : x \longmapsto 2/(2-x)$; the determinant of this mapping is 2 , a square in GF(23) , $2 = 5^2$, and so $\psi \in R \cap PSL(2,23) \cong A_4$, a contradiction since A_4 has no element of order 4 . For the number of given subgroups in cases (2.3) , (2.4) , (2.5) and their partition into conjugacy classes we refer to [3, pages 282-285] .

The other possibility for a regular subgroup R is when p is odd and $R \cap PSL(2,p^m)$ is a semiregular subgroup of order $(p^m+1)/2$ of $PSL(2,p^m)$. Again we look for possible subgroups of order $(p^m+1)/2$ in the list of subgroups of $PSL(2,p^m)$, p odd, and end up with the following possibilities :

(2.6) cyclic subgroups ;

(2.7) dihedral subgroups ;

(2.8) A_4 for $p = 23$, $m = 1$;

(2.9) S_4 for $p = 47$, $m = 1$.

Cases (2.6) and (2.7) will be dealt with in the next section ; as to case (2.7) we only remark here that a dihedral subgroup of order $(p^m+1)/2$ possesses a cyclic subgroup of order $(p^m+1)/4$, yielding $p^m \equiv -1 \pmod 4$ which is never fulfilled in case $m = 2f$. In both cases (2.8) and (2.9) we only have to deal with the linear group $PGL(2,p^m)$ because here $m = 1$. The group PSL(2,23) does possess subgroups which are isomorphic to A_4 and since the one-point-stabilizers in PSL(2,23) have order $23 \cdot 11$ which is relatively prime with $12 = |A_4|$, we see that each such subgroup operates semiregularly on PG(1,23) ; similarly, PSL(2,47) possesses subgroups which are isomorphic to S_4 and they all operate semiregularly on PG(1,47) . Nevertheless, there exists no subgroup of order 24 in PGL(2,23) intersecting PSL(2,23) in A_4 , nor does there exist a subgroup of order 48 in PGL(2,47) intersecting PSL(2,47) in S_4 : we do have direct proofs of these facts, but probably the easiest way to see them is to look at the list of subgroups of $PGL(2,p^m)$, obtaining S_4 as the unique possibility in the first case and no possibilities in the second one ; the first case is then also ruled out once we know that each subgroup R of PGL(2,23) which is isomorphic to S_4 actually lies in PSL(2,23) , as we proved above .

3. By $\Gamma L(2,q)$ we denote as usual the group of semilinear transformations of a 2-dimensional vector space V over GF(q) , $q = p^m$; we may of course assume $V = GF(q^2)$. We denote by $\Phi : \Gamma L(2,q) \longrightarrow Aut(GF(q))$ the group--epimorphism taking every semilinear transformation into its associated field--automorphism : of course $\ker\Phi = GL(2,q)$. If ψ is a semilinear transformation we look at the determinant of its coefficient matrix with respect to a GF(q)-basis of V : if this determinant is a square in GF(q)* then we set $\chi(\psi) = 1$, while we set $\chi(\psi) = -1$ if the determinant is a non-square in GF(q)* ; we thus obtain another group-homomorphism $\chi : \Gamma L(2,q) \longrightarrow \{1,-1\}$. If $\hat{} : \Gamma L(2,q) \longrightarrow P\Gamma L(2,q)$ denotes the canonical epimorphism, then by setting $\Phi(\hat{\psi}) := \Phi(\psi)$ and $\chi(\hat{\psi}) := \chi(\psi)$ we obtain the induced homomorphisms $\Phi : P\Gamma L(2,q) \longrightarrow Aut(GF(q))$ and $\chi : P\Gamma L(2,q) \longrightarrow \{1,-1\}$.

If R is a regular subgroup of PGL(2,q) then every non-identical element of R has no fixed points on the projective line $X := PG(1,q) = GF(q) \cup \{\infty\}$; there is a well-known-partition of the fixed-point-free elements of PGL(2,q) corresponding to the so-called *Singer cyclic subgroups* (see [5, II 8.5]) : we introduce this partition as follows . Let $h(z) = z^2 - ez - b \in GF(q)[z]$ be an irreducible polynomial . To each element $a \in X$ define the mapping $\psi_{a,h(z)}$ (or simply ψ_a when reference to h(z) is clear) by setting $\psi_{a,h(z)}(x) := = (ax+b)/(x+a+e)$ for all $x \in X$. Observe that ψ_∞ is the identity on X and that for $a \in GF(q)$ we have $\det\begin{pmatrix} a & b \\ 1 & a+e \end{pmatrix} = h(-a) \neq 0$ and so in any case the transformation ψ_a belongs to PGL(2,q) . If x , y are in X then it can be easily verified that there exists precisely one element $a \in X$ such that $\psi_a(x) = y$. For $a_1, a_2 \in X$ we have $\psi_{a_1}\psi_{a_2} = \psi_{a_3}$ with $a_3 = (a_1a_2+b)/(a_1+a_2+e)$; furthermore for $a \in X$ we have $\psi_a\psi_{-a-e} = \psi_\infty$. It follows that the subset $C := C(h(z)) := \{ \psi_a : a \in X \}$ is a regular abelian subgroup of PGL(2,q) . We remark that if $k(z) = z^2 - \bar{e}z - \bar{b} \in GF(q)[z]$ is an irreducible polynomial with $k(z) \neq h(z)$ then $C(h(z)) \cap C(k(z)) = \{\psi_\infty\}$. If ψ is a fixed-piont-free element of PGL(2,q) , then in particular $\psi(\infty) \neq \infty$ and thus ψ can be written in the form $\psi(x) = (ax+b)/(x+a+e)$; from $\psi(x) \neq x$ for $x \in GF(q)$ it follows then that the polynomial z^2+ez-b has no roots in $GF(q)$, whence z^2-ez-b is also irreducible over $GF(q)$ and $\psi \in C(z^2-ez-b)$. As a consequence we have that the sets $C(h(z)) \setminus \{\psi_\infty\}$ as $h(z)$ runs over the irreducible quadratic monic polynomials of $GF(q)[z]$ form a partition of the fixed-point-free elements of $PGL(2,q)$. The splitting field of $h(z)$ over $GF(q)$ is $GF(q^2)$, hance $V := GF(q^2) = GF(q)(\zeta)$ where ζ is a root of $h(z)$ and therefore $\{1,\zeta\}$ is a $GF(q)$-basis of V; for any element $c \in GF(q^2)^*$ the transformation $\lambda_c : V \longrightarrow V$, $x \longmapsto cx$ belongs to $GL(2,q)$ and such transformations form a subgroup S of $GL(2,q)$ which is isomorphic to the cyclic multiplicative group $GF(q^2)^*$; if $c = c_1 + c_2\zeta$ then the matrix representing λ_c with respect to the basis $\{1,\zeta\}$ is $\begin{pmatrix} c_1 & c_2 b \\ c_2 & c_1+c_2 e \end{pmatrix}$ and so for $x \in X$ we have $\hat{\lambda}_c(x) = ((c_1/c_2)x+b)/(x+(c_1/c_2)+e)$ showing that $\hat{S} = C(h(z))$; since $\hat{S}$ is the quotient group of S by the subgroup $\{\lambda_b : b \in GF(q)^*\}$, we see that $C(h(z))$ is a cyclic group of order $(q^2-1)/(q-1) = q+1$. [*Remark*: the above arguments hold for any field K and an irreducible quadratic monic polynomial $h(z)$, the group $C(h(z))$ being then isomorphic to $K(\zeta)^*/K^*$, ζ a root of $h(z)$]. Since the number of irreducible quadratic monic polynomials over $GF(q)$ is $\binom{q}{2}$, the above procedure yields $\binom{q}{2}$ regular cyclic subgroups of $PGL(2,q)$ and these account for all regular cyclic subgroups of $PGL(2,q)$, since a generator for such a subgroup has to operate fixed-point-free on X.

We now prove that the normalizer N of S in the group $\Gamma L(2,q) = \Gamma L(V)$ is given by $N = \{x \longmapsto c\,x^{p^i} : c \in GF(q^2)^*,\ 0 \leq i < 2m\}$ (cfr. [5, II 7.3]). A direct check shows that every element of N normalizes S. On the other side assume ψ is a semilinear transformation normalizing S. Then $\psi\,\lambda_c\,\psi^{-1} = \lambda_{\alpha(c)}$ where α is an automorphism of the multiplicative group $GF(q^2)^*$. Since ψ is an automorphism of the additive group $GF(q^2)^+$, we can define $\tau(x) = \psi(x)/\psi(1)$ for $x \in GF(q^2)$ and claim that τ is a field-automorphism of $GF(q^2)$. Clearly τ is an automorphism of the additive group $GF(q^2)^+$; now from $\psi\,\lambda_c\,\psi^{-1}(x) = \alpha(c)\,x$ we get by setting $y := \psi^{-1}(x)$ the equality $\psi(cy) = \alpha(c)\,\psi(y)$, which for $y = 1$ yields $\alpha(c) = \tau(c)$ for all $c \in GF(q^2)^*$ as required. Every field-automorphism of $GF(p^{2m})$ has the form $x \longmapsto x^{p^i}$ for some $i \in \{0,1,\ldots,2m-1\}$; whence $\tau(x) = x^{p^i}$, i.e. $\psi(x) = c\,x^{p^i}$ with $c = \psi(1)$ and so ψ belongs to N.

We have $|N| = (q^2 - 1)2m$ because distinct choices of the coefficient c or of the exponent i yield distinct mappings. If ζ is a primitive element of $GF(q^2)$ we define the mapping $\rho : x \longmapsto \zeta x$; we define further $\varepsilon : x \longmapsto x^q$, $\tau : x \longmapsto x^p$; every element of N admits then a unique representation of the form $\rho^t \varepsilon^j \tau^i$ with $t \in \{0,1,\ldots,q^2-2\}$, $j \in \{0,1\}$, $i \in \{0,1,\ldots,m-1\}$. The group $\hat{N}$ is the normalizer of $\hat{S}$ in $P\Gamma L(2,q)$ and has order $(q+1)2m$; since ζ^{q+1} is a primitive element of $GF(q)$ we see that every element of $\hat{N}$ can be written uniquely in the form $\hat{\rho}^t \hat{\varepsilon}^j \hat{\tau}^i$ with $t \in \{0,1,\ldots,q\}$, $j \in \{0,1\}$, $i \in \{0,1,\ldots,m-1\}$.

The subgroup $M := N \cap GL(2,q)$ is the normalizer of S in $GL(2,q)$ and we have $M = \{x \longmapsto c\,x^{p^i} : c \in GF(q^2)^*,\ i \in \{0,m\}\}$, showing that $|M| \doteq 2(q^2-1)$; the subgroup $\hat{M}$ is the normalizer of $\hat{S}$ in $PGL(2,q)$ and is simply the dihedral subgroup of order $2(q+1)$ generated by $\hat{\rho}$ and $\hat{\varepsilon}$.

The number of distinct conjugates of $\hat{S}$ in $PGL(2,q)$ is $|PGL(2,q) : \hat{M}| = \binom{q}{2}$; this shows that the regular cyclic subgroups of $PGL(2,q)$ form a single conjugacy class in $PGL(2,q)$; from $PGL(2,q) \trianglelefteq P\Gamma L(2,q)$ we see that any conjugate $\hat{T}$ of $\hat{S}$ in $P\Gamma L(2,q)$ lies in $PGL(2,q)$ and since $\hat{T}$ is obviously a regular cyclic subgroup it must coincide with some subgroup $C(h(z))$, thus showing that the subgroups of the form $C(h(z))$ also form a single conjugacy class in $P\Gamma L(2,q)$. If $n > 1$ is a divisor of $q+1$, then each subgroup $C(h(z))$ contains a unique subgroup of order n, which is thus a semiregular cyclic subgroup of order n in $PGL(2,q)$; conversely every semiregular cyclic subgroup of order n in $PGL(2,q)$ is contained in some subgroup $C(h(z))$, and it follows that the semiregular cyclic subgroups of order n of $PGL(2,q)$ form a single conjugacy class both in $PGL(2,q)$ and in $P\Gamma L(2,q)$; consequently the normalizer in $P\Gamma L(2,q)$ resp. in $PGL(2,q)$ of $C(h(z))$ is the same as the normalizer in $P\Gamma L(2,q)$ resp. in $PGL(2,q)$ of any one of its non-trivial subgroups (cfr. [5, II 7.3]).

A regular dihedral subgroup of $PGL(2,q)$ possesses a cyclic subgroup of order $(q+1)/2$, forcing q to be odd, and in such case the cyclic subgroup is a semiregular cyclic subgroup of order $(q+1)/2$ of $PGL(2,q)$, which we may thus assume to be the unique subgroup $\hat{Z} = \langle \hat{\rho}^2 \rangle$ of order $(q+1)/2$ of $\hat{S}$; indeed

$\hat{Z} = \hat{S} \cap PSL(2,q)$, since $\det \rho = \zeta^{q+1}$ (cfr. [5, II 7.3]) and ζ^{q+1} is a primitive element of $GF(q)$, hence a non-square in $GF(q)$. A regular subgroup of $PGL(2,q)$ possessing $\hat{Z}$ as a subgroup of index 2 necessarily normalizes $\hat{Z}$, hence is necessarily contained in the normalizer $\hat{M}$ of $\hat{Z}$ in $PGL(2,q)$. There are exactly three subgroups of $\hat{M}$ containing $\hat{Z}$ as a subgroup of index 2 , namely $\hat{S}$ itself and the two dihedral groups $D^{(1)} := < \hat{Z} , \hat{\rho}\hat{\varepsilon} >$ and $D^{(2)} := = < \hat{Z} , \hat{\varepsilon} >$. As a semiregular group of order $(q+1)/2$ on a set with $q+1$ elements, $\hat{Z}$ yields two orbits of length $(q+1)/2$ on $PG(1,q)$. Let x and y denote the points of $PG(1,q)$ deriving from the elements 1 and ζ^{-1} of $V = = GF(q^2)$ respectively . It is easily seen that x and y lie in distinct orbits of $PG(1,q)$ under the action of $\hat{Z}$. From the transitivity of $\hat{M}$ on $PG(1,q)$ and $|\hat{M}| = 2(q+1)$ it follows the existence of exactly two elements of $\hat{M}$ taking x to y , one of these lies in $\hat{S}$ and the other one lies in exactly one of the subgroups $D^{(1)}$, $D^{(2)}$ and we can conclude that exactly one of the subgroups $D^{(1)}$, $D^{(2)}$ is regular on $PG(1,q)$; from $\rho^{q^2-2}\varepsilon(1) = \zeta^{q^2-2} = \zeta^{-1}$ and $\hat{\rho}^{q^2-2}\hat{\varepsilon} \in D^{(1)}$ we have that $D^{(1)}$ is regular while $D^{(2)}$ is not . We also have that since the semiregular cyclic subgroups of order $(q+1)/2$ of $PGL(2,q)$ form a single conjugacy class in $PGL(2,q)$, so do the regular dihedral subgroups of $PGL(2,q)$ and there are therefore $\binom{q}{2}$ such subgroups for $q > 3$, while we get a unique such subgroup for $q = 3$. If we set $\mu := \rho^{q^2-2}\varepsilon$ we have $D^{(1)} = = < \hat{Z} , \mu >$, $\mu(x) = \zeta^{-1}x^q$ and $\mu^2(x) = \zeta^{-q-1}x$, whence the minimal polynomial of the linear mapping μ over $GF(q)$ is $z^2 - \zeta^{-q-1}$, showing that the determinant of μ over $GF(q)$ is $-\zeta^{-q-1}$; since ζ^{-q-1} is a primitive element of $GF(q)$, hence a non-square in $GF(q)$, we conclude that $\hat{\mu} \in PSL(2,q)$ (i.e. $D^{(1)} \leqq PSL(2,q)$) if and only if -1 is a non-square in $GF(q)$, that is if and only if $q \equiv -1 \pmod 4$. If this is the case then $D^{(1)}$ is the normalizer of $\hat{Z}$ in $PSL(2,q)$; it follows then easily that the regular dihedral subgroups of $PGL(2,q)$ all lie in $PSL(2,q)$ and form there a single conjugacy class .

If $q \equiv -1 \pmod 4$, $q > 3$, then $\hat{Z}$ has a non-trivial subgroup $\hat{W}$ of index 2 ; the normalizer of $\hat{W}$ in $PSL(2,q)$ is precisely the subgroup $D^{(1)}$; there are exactly two dihedral subgroups of order $(q+1)/2$ of $D^{(1)}$ containing $\hat{W}$, namely $E^{(1)} := < \hat{\rho}^4 , \hat{\rho}\hat{\varepsilon} >$ and $E^{(2)} := < \hat{\rho}^4 , \hat{\rho}^3\hat{\varepsilon} >$; both $E^{(1)}$ and $E^{(2)}$ are semiregular on $PG(1,q)$ and these are exactly the subgroups of $PSL(2,q)$ yielding case (2.7) , as $\hat{W}$ runs over the semiregular cyclic subgroups of order $(q+1)/4$

of $PSL(2,q)$. For $\nu = 1,2$ let $F^{(\nu)}$ denote the normalizer of $E^{(\nu)}$ in $PGL(2,q)$; by [3, pages 267-268] we have that, for $q > 7$, $F^{(\nu)}$ is $D^{(1)}$, while, for $q = 7$, $F^{(\nu)}$ is a subgroup of $PSL(2,7)$ which is isomorphic to S_4 ; in either case we can check that every regular subgroup of $F^{(\nu)}$ containing $E^{(\nu)}$ as a subgroup of index 2 is dihedral and falls therefore under those in (2.2) .

4. In this last section we prove that $M(p^{2f})$ has no regular subgroups . Indeed we know from sections 2 and 3 that no regular subgroups of $PSL(2,p^{2f})$ exist when p is odd . Hence if a regular subgroup R of $M(p^{2f})$ exists at all, it must intersect $PSL(2,p^{2f})$ in a semiregular subgroup of order $(p^{2f}+1)/2$ which by section 2 must necessarily be cyclic (case (2.6)) ; with the previous notation let $\hat{Z}$ be such a subgroup and let $\hat{S}$ be the unique regular cyclic subgroup of $PGL(2,p^{2f})$ containing $\hat{Z}$. From $|R:\hat{Z}| = 2$ it follows that R is contained in the normalizer $\hat{U}$ of $\hat{Z}$ in $M(p^{2f})$, for which we have $\hat{U} = \hat{N} \cap M(p^{2f})$ where $\hat{N}$ is the normalizer of $\hat{Z}$ (as well as the normalizer of $\hat{S}$) in $P\Gamma L(2,p^{2f})$. Recalling the definition of the group $M(p^{2f})$ and of the homomorphisms Φ and χ , from $\hat{\varepsilon} = \hat{\tau}^{2f}$ we obtain $\chi(\hat{\varepsilon}) = = \chi(\hat{\tau}^f)^2 = 1$ and have therefore the following possibilities for $\hat{U}$ (with $q = p^{2f}$):

(4.1) $\hat{U} = \{ \hat{\rho}^{2t} \hat{\varepsilon}^j : t = 0,1, \dots , (q-1)/2 ; j = 0,1 \} \cup$
$\cup \{ \hat{\rho}^{2t+1} \hat{\varepsilon}^j \hat{\tau}^f : t = 0,1, \dots , (q-1)/2 ; j = 0,1 \}$
when $\chi(\hat{\tau}^f) = 1$;

(4.2) $\hat{U} = \{ \hat{\rho}^{2t} \hat{\varepsilon}^j : t = 0,1, \dots , (q-1)/2 ; j = 0,1 \} \cup$
$\cup \{ \hat{\rho}^{2t} \hat{\varepsilon}^j \hat{\tau}^f : t = 0,1, \dots , (q-1)/2 ; j = 0,1 \}$
when $\chi(\hat{\tau}^f) = -1$;

(in both cases we have written the subgroup $\hat{U} \cap PSL(2,p^{2f})$ first) .

In case (4.2) the subgroup $\hat{U}$ is not even transitive on $PG(1,q) = PG(V)$. As a matter of fact $\rho^{2t} \varepsilon^j(1) = \rho^{2t} \varepsilon^j \tau^f(1) = \zeta^{2t}$; as in section 3 denote by x and y the points of $PG(1,q)$ deriving from the elements 1 and ζ^{-1} of $V = GF(q^2)$ respectively ; the point y has no representative in V of the form ζ^{2t} , otherwise we would have $\zeta^{2t+1} \in GF(q)$, i.e. $(\zeta^{2t+1})^{q-1} = 1$, whence $(q+1) \mid (2t+1)$, a contradiction because $q+1$ is even ; therefore $\hat{\rho}^{2t} \hat{\varepsilon}^j(x) \neq y$, $\hat{\rho}^{2t} \hat{\varepsilon}^j \hat{\tau}^f(x) \neq y$, thus proving the assertion . It follows that no regular subgroup R of $\hat{U}$ exists in case (4.2) .

In case (4.1) , assume R is a regular subgroup of $\hat{U}$; then $R = \langle \hat{Z}, \hat{\varphi} \rangle$ for some $\hat{\varphi} \in \hat{U} \setminus PSL(2,q)$. If $\hat{\varphi} = \hat{\rho}^{2s+1} \hat{\tau}^f$ for some index $s \in \{0,1,\dots,(q-1)/2\}$, then, since $\hat{Z} = \langle \hat{\rho}^2 \rangle$, we have that R contains at least the q+1 elements $\hat{\rho}^{2t}$, $\hat{\rho}^{2t+1} \hat{\tau}^f$ as t varies in $\{0,1,\dots,(q-1)/2\}$; from $|R| = q+1$ we conclude that R must consist exactly of these elements, but these elements do not form a subgroup (because, for instance $(\hat{\rho} \hat{\tau}^f)^2 = \hat{\rho}^{p^f+1} \hat{\varepsilon} \notin R$) , a contradiction . A similar argument rules out the possibility $\hat{\varphi} = \hat{\rho}^{2s+1} \hat{\varepsilon} \hat{\tau}^f$ for some index $s \in \{0,1,\dots,(q-1)/2\}$, and we are through .

REFERENCES

[1] Barlotti, A. and Strambach, K., k-transitive permutation groups and k-planes, Math. Z. 185 (1984) 465-485 .

[2] Bonisoli, A. and Quattrocchi, P., Sharply k-transitive permutation sets : a survey and some new results, to appear .

[3] Dickson, L.E., Linear groups with an exposition of the Galois field theory (Teubner, Leipzig, 1901) .

[4] Huppert, B., Scharf dreifach transitive Permutationsgruppen, Arch. Math. (Basel) 13 (1962) 61-72 .

[5] Huppert, B., Endliche Gruppen I (Springer, Berlin et al., 1967) .

[6] Huppert, B. and Blackburn N., Finite groups III (Springer, Berlin et al., 1982) .

[7] Korchmaros, G., A combinatorial characterization of the dihedral subgroups of order $2(p^r+1)$ of $PGL(2,p^r)$, Geom. Ded. 9 (1980) 381-384

[8] O'Nan, M.E., Sharply 2-transitive sets of permutations, in: Aschbacher, M. et al. (eds.), Proceedings of the Rutgers Group Theory Year, 1983-1984 (Cambridge Univ. Press., Cambridge et al., 1984) pp. 63-67 .

[9] Quattrocchi, P., Over the non-existence of sharply 3-transitive permutation sets containing sharply 2-transitive permutation subsets, to appear in Geom. Ded. .

[10] Valentini, R.C. and Madan, M.L., A Hauptsatz of L.E. Dickson and Artin--Schreier extensions, J. Reine Angew. Math. 318 (1980) 156-177 .

[11] Wefelscheid, H., Zur Nichtexistenz scharf 2-transitiver Permutationsmengen in scharf 3-fach transitiven Gruppen, Boll. Un. Mat. Ital. (6) 4-A (1985) 105-109 .

[12] Wielandt, H., Finite Permutation Groups (Academic Press, New York et al., 1964) .

Annals of Discrete Mathematics 37 (1988) 87–94
© Elsevier Science Publishers B.V. (North-Holland)

HYPEROVALS IN DESARGUESIAN PLANES OF EVEN ORDER

William CHEROWITZO

Department of Mathematics
University of Colorado at Denver
1100 14th Street, Campus Box 170
Denver, CO 80202

In this survey we shall present all the known hyperovals of the title, including the most recently discovered such hyperovals. The latest examples were found by computer searches and some of the results of these searches in the Desarguesian plane of order 32 will also be given. A number of open problems are posed.

In a finite projective plane of order n, an <u>oval</u> is a set of $n + 1$ points, no three of which are collinear. In a coordinatized Desarguesian plane, a <u>conic</u> is the set of points whose coordinates satisfy a non-degenerate quadratic equation. While every conic is easily seen to be an oval, the converse, proved in 1955 by B. Segre [7] for Desarguesian planes of odd order, is a rather surprising result. Segre's result does not extend to Desarguesian planes of even order and the classification problem for ovals in these planes seems to be very complex.

In the even order case, every oval can be uniquely extended to a set of $n + 2$ points, no three of which are collinear. These sets are called <u>hyperovals</u>. To examine the hyperovals in Desarguesian planes of even order, we introduce projective coordinates in the standard way defined over the Galois field $\mathbf{F} = \mathbf{GF}(2^h)$. By the Fundamental Theorem of Projective Geometry every hyperoval is equivalent to one containing the "Fundamental Quadrangle"; i.e., the points with coordinates (1,0,0), (0,1,0), (0,0,1), and (1,1,1). By assuming that all hyperovals under consideration contain the fundamental quadrangle, we see that they are distinguished solely by their affine points. If Ω is a hyperoval, then it is natural to define:

$y = f(x)$ iff $(x,y,1)$ is a point of Ω.

f is a function since the inclusion of (0,1,0) in Ω guarantees that no two affine points have the same first coordinate and furthermore, f is a permutation of the elements of $\mathbf{F}$ since the inclusion of (1,0,0) in Ω guarantees that no two affine points have the same second coordinate. If $q = 2^h$, by applying LaGrange's Interpolation Formula, it can be shown (see Hirschfeld [3]) that f can be expressed uniquely as a polynomial of degree $\leq q - 2$ over the field **F**. Those permutation polynomials which arise from hyperovals in this way will be called <u>o-polynomials</u>.

A few general remarks can be made about o-polynomials. If f is an o-polynomial, then $f(0) = 0$ and $f(1) = 1$, since the hyperoval contains the fundamental quadrangle. This implies that an o-polynomial has no constant term and the sum of the coefficients is 1. A deeper result concerning the structure of an o-polynomial is:

Result 1 (Segre and Bartocci [10]): <u>If $q > 2$, then the coefficient of each term of odd power in an o-polynomial is zero</u>.

This is an easy consequence of:

Theorem 1 (Hirschfeld [3], Thm. 8.4.2): <u>If $q > 2$, a permutation polynomial f with $f(0) = 0$ and $f(1) = 1$ is an o-polynomial if, and only if, for each element $s \in \mathbf{F}$, $G(x,s) = [f(x+s) + f(s)]/x$, is a permutation polynomial over **F** with $G(0,s) = 0$</u>.

The group of projectivities, **H**, which leaves the fundamental quadrangle invariant, fixes the set of hyperovals which contain this quadrangle. Thus, this group of order 24 leaves the set of o-polynomials invariant. Direct calculation produces

Result 2: <u>If f(x) is an o-polynomial then the following list of images of (x,f(x),1) under the action of group **H** implicitly define o-polynomials equivalent to f(x)</u>:

(1) (x,1,f(x))
(2) (f(x),x,1)
(3) (f(x),1,x)
(4) (1,x,f(x))
(5) (1,f(x),x)
(6) (x+1,f(x)+1,1)
(7) (f(x)+1,x+1,1)
(8) (x+f(x),f(x)+1,1)
(9) (f(x)+1,x+f(x),1)
(10) (x+f(x),x+1,x)
(11) (x+1,x+f(x),x)
(12) (1,f(x)+1,x+1)
(13) (1,x+1,f(x)+1)
(14) (f(x),f(x)+1,x+f(x))
(15) (f(x),x+f(x),f(x)+1)
(16) (x,x+f(x),x+1)
(17) (x,x+1,x+f(x))
(18) (f(x)+1,f(x),x+f(x))
(19) (x+1,1,f(x)+1)
(20) (f(x)+1,f(x),x+f(x))
(21) (x+f(x),f(x),f(x)+1)
(22) (x+1,x,x+f(x))
(23) (x+f(x),x,x+1)

In particular, the following are o-polynomials equivalent to $f(x)$:

(2) $g(x) = f^{-1}(x)$

(5) $g(x) = xf(1/x), \quad g(0) = 0$

(6) $g(x) = f(x+1) + 1$

(11) $g(x) = (x+1)f(1/(x+1)) + 1, \quad g(1) = 1$

(12) $g(x) = x + xf(1+(1/x)), \quad g(0) = 0$

(16) $g(x) = x + (x+1)f(x/(x+1)), \quad g(1) = 1$

as well as each of the above with f replaced by f^{-1}.

Each automorphism of the field **F** induces a collineation of the plane which leaves the fundamental quadrangle fixed pointwise, so they also leave the set of o-polynomials invariant. If $f(x)$ is an o-polynomial, let $f(x,\alpha)$ be the polynomial obtained from $f(x)$ by replacing each coefficient with its image under the automorphism α. We then have:

Result 3: If $f(x)$ is an o-polynomial, then $f(x,\alpha)$ is an o-polynomial equivalent to $f(x)$ under the automorphism α of **F**.

We now consider the known examples of o-polynomials. Until recently, almost all of the examples were monomials, so we shall look at this special case first.

Let $\mathbf{F} = \mathbf{GF}(q)$, with $q = 2^h$ and we will assume that $h > 1$ to avoid the trivial, but exceptional, case of the Fano plane of order 2. Since $f(1) = 1$, any monomial o-polynomial is of the form $f(x) = x^k$. Let us define $E(h)$ as

$$E(h) = \{k \mid x^k \text{ is an o-polynomial over } \mathbf{GF}(2^h)\}.$$

As has been observed by a number of authors (Payne [5], Hirschfeld [3], and Glynn [1]):

Theorem 2: If $k \in E(h)$, then $1/k$, $1-k$, $1/(1-k)$, $k/(k-1)$, and $(k-1)/k \in E(h)$, where these numbers are taken modulo $q-1$. These six o-polynomials give equivalent hyperovals.

The proof consists of observing that the first five maps of Result 2 preserve monomials.

$2 \in E(h)$ for all h. These are the hyperovals that correspond to the conics. The hyperovals that are equivalent to these are called regular.

$2^i \in E(h)$ if, and only if, $(i,h) = 1$. These hyperovals were determined by Segre [8] in 1957. They provided the first examples of irregular ovals (when $i > 1$). They are called Translation Hyperovals since they admit as an automorphism group a group of translations which is transitive on their affine points. Payne [5] has shown that these are the only o-polynomials for which the function f is additive.

$6 \in E(h)$ for h odd. Discovered in 1962 by Segre [9], but see also the 1971 treatment by Segre and Bartocci [10].

$\sigma + \lambda$ and $3\sigma + 4 \in E(h)$ for h odd, where $\lambda^4 \equiv \sigma^2 \equiv 2 \bmod (q-1)$. These two families were discovered by Glynn [1] in 1982. Glynn developed an easily implemented algorithm (see below) for determining when k is in $E(h)$. Using this algorithm he checked all values of h up to and including 19. The values for k which were not already known fell into these two families, which he then proved would always give o-polynomials. Thus, we have:

Result 4 (Glynn [1]): The sets $E(h)$ are completely determined for $h \leq 19$.

Glynn defines a partial ordering $\precsim$ on the set of integers $\{n \mid 0 \leq n \leq q-1\}$ by:

$$\text{if } a = \sum_{i=0}^{h-1} a_i 2^i \text{ and } b = \sum_{i=0}^{h-1} b_i 2^i, \text{ then } a \precsim b \text{ iff } a_i \leq b_i \text{ for all } i.$$

His algorithm can then be expressed as:

Theorem 3 (Glynn [1]): $k \in E(h)$ if, and only if, $d \not\precsim dk$ for all $d \in \{1,2,\ldots,q-2\}$ where dk is reduced modulo q-1 to lie in the set $\{0,1,\ldots,q-1\}$ (that is, 0 is reduced to 0 but all other multiples of q-1 are reduced to q-1).

It should be pointed out that for small values of q there will be some collapsing of the types of hyperovals given by $E(h)$. Thus, with $q = 8$, x^2, x^4, and x^6 are all equivalent. However, with q sufficiently large, the types listed above are all distinct (non-equivalent).

Besides the o-polynomials of $E(h)$, the only other example was the irregular hyperoval discovered by Sce and Lunelli [4] in 1958 and shown to be the only irregular hyperoval in PG(2,16) by Hall [2] in 1975. It may be given by:

$$f(x) = x^{12} + x^{10} + \eta^{11}x^8 + x^6 + \eta^2x^4 + \eta^9x^2,$$

where η is a primitive element of **GF**(16) satisfying $\eta^4 = \eta + 1$.

In 1985 Payne [6], in an investigation of a new family of generalized quadrangles, discovered a new family of o-polynomials for odd h. These are given by

$$f(x) = x^{1/6} + x^{3/6} + x^{5/6} ,$$

where the exponents are taken modulo q-1. The automorphism group of these ovals is determined in [11] by Thas, Payne, and Gevaert and consists of the field automorphisms together with the map (5) of Result 2.

A computer investigation carried out by the author has uncovered some other o-polynomials. We have that

$$f(x) = x^{\sigma} + x^{\sigma+2} + x^{3\sigma+4}, \quad \text{where} \quad \sigma^2 \equiv 2 \bmod (q-1),$$

is an o-polynomial for $h = 3, 5, 7,$ and 9, and conjecture that this is true for all odd h. These o-polynomials are not fixed by any of the maps of Result 2 and the possibility remains that their complete automorphism group is just the group of field automorphisms.

A search was made for all o-polynomials over **GF**(32) with coefficients in **GF**(2). This class of o-polynomials was singled out for a number of reasons although the chief consideration was the reduction of search time. By Result 3, o-polynomials with coefficients in **GF**(2) are fixed by all the field automorphisms, and this class is invariant under the action of group **H**. Since $f(1) = 1$, the o-polynomials in this class must have an odd number of terms.

The completed search produced 60 o-polynomials. Under the action of **H** these reduced to 5 inequivalent o-polynomials, which are, in fact, inequivalent under the full collineation group. Interestingly, each of the irregular hyperovals had a representative among the trinomials in the list. Table 1 lists the 8 trinomial o-polynomials for **GF**(32) in an abbreviated form, i.e., only the exponents of the trinomial are given. Also given in the table are certain equivalences, where $\mathcal{D}(k)$ refers to the hyperoval whose o-polynomial is $f(x) = x^k$, C is the new hyperoval mentioned above, and P is the hyperoval recently discovered by Payne.

TABLE 1

Trinomial o-polynomials over **GF**(32) with Coefficients in **GF**(2)

2	4	6	$\approx D(6)$
2	8	10	$\approx D(\sigma)$
4	22	24	$\approx$ C
6	16	26	- P
8	10	28	- C
8	16	24	$\approx D(\sigma + \lambda)$
22	24	30	$\approx D(4)$
26	28	30	$\approx D(6)$

A number of remarks should be made about this table.

1. It is interesting to note that all of the exponents which appear in these trinomials are also in $E(5) = \{2,4,6,8,10,16,22,24,26,28,30\}$. This does not hold for the o-polynomials with more than three terms.

2. There is a correspondence exhibited in the table. More precisely:

Result 5: <u>$\alpha x^a + \beta x^b + \gamma x^c$ is an o-polynomial if, and only if, $\alpha x^{q-a} + \beta x^{q-b} + \gamma x^{q-c}$ is an o-polynomial, and these two o-polynomials are equivalent.</u>

Proof: Apply map (5) of Result 2.

While this result is stated for trinomials, it clearly holds for any o-polynomial.

3. The equivalences given in the table are consequences of either Result 5 or the following:

Result 6: <u>If $k = 2^i + 2^j$ is in $E(h)$, then $x^{2^i} + x^{2^j} + x^k$ is an o-polynomial which is equivalent to $D(k)$. In particular, if h is odd, there are trinomials equivalent to $D(4)$, $D(6)$, $D(\sigma)$, and $D(\sigma + \lambda)$.</u>

Proof: Let $f(x) = x^k$. By applying map (6) of Result 2, we have that f(x) is equivalent to $g(x) = (x + 1)^k + 1 = (x^{2^i} + 1)(x^{2^j} + 1) + 1 = x^{2^i} + x^{2^j} + x^k$. If h is odd, then $\lambda^4 \equiv \sigma^2 \equiv 2 \bmod (q-1)$ implies that $\sigma = 2^{(h+1)/2}$ and $\lambda = 2^m$, where $m = (h+1)/4$ if $h \equiv 3 \bmod 4$ or $m = (3h+1)/4$ if $h \equiv 1 \bmod 4$. So, 6 and $\sigma + \lambda$ are in $E(h)$ and are clearly the sum of two powers of 2. We also have that 4 and σ are in $E(h)$ by Segre's result [8]; thus by Theorem 2 we have that $(4-1)/4 = 3/4$ and $\sigma/(\sigma-1) = \sigma(\sigma+1) = \sigma^2 + \sigma \equiv 2 + \sigma$ are in $E(h)$ and are equivalent to 4 and σ, respectively. Finally, since $3/4 \equiv 2^{h-1} + 2^{h-2} \bmod (q-1)$, we see that 3/4 and $2 + \sigma$ are also sums of two powers of 2.

4. As with the $E(h)$, we see a collapsing of types due to the smallness of the order. In this plane we have $D(4) \approx D(\sigma) \approx D(\sigma+\lambda)$, and so, there are corresponding spurious equivalences amongst the trinomials.

A complete search for trinomials was also made with $h = 7$. Besides the C and P hyperovals, all others were consequences of Results 5 and 6. In this case however, no collapsing of known types takes place.

A final search for trinomials was made for $h = 9$; however in this case, the search was restricted to trinomials with exponents in $E(9)$. The results were similar to those of the case $h = 7$.

In conclusion, there are a number of open questions which arise from this work that we would like to pose.

1. Is $f(x) = x^{\sigma} + x^{\sigma+2} + x^{3\sigma+4}$ an o-polynomial for all odd h?

2. When the above f(x) is an o-polynomial, what is the automorphism group of the hyperoval?

3. If $x^a + x^b + x^c$ is an o-polynomial, must a, b, and $c \in E(h)$?

4. How is the set of o-polynomials with coefficients in **GF**(2) related to the set of all o-polynomials?

REFERENCES

[1] D. Glynn, Two new sequences of ovals in finite Desarguesian planes of even order, Combinatorial Mathematics X (LNM No. 1036), Springer-Verlag, Berlin, 1983, 217-229.

[2] M. Hall, Jr., Ovals in the Desarguesian plane of order 16, Ann. Mat. Pura App., (4), 102 (1975), 159-176.

[3] J. W. P. Hirschfeld, Projective Geometries over Finite Fields, Clarendon Press, Oxford, 1979.

[4] L. Lunelli and M. Sce, K-archi completi nei piani proietivi desarguesiani di rango 8 e 16, Centro Calcoli Numerici, Politecnico di Milano, 1958.

[5] S. E. Payne, A complete determination of translation ovoids in finite Desarguesian planes, Atti Accad. Naz. Lincei Rend., (8), 51 (1971), 328-331.

[6] S. E. Payne, A new infinite family of generalized quadrangles, Congressus Numerantium, 49 (1985), 115-128.

[7] B. Segre, Ovals in a finite projective plane, Can. J. Math., 7 (1955), 414-416.

[8] B. Segre, Sui k-archi nei piani finiti di caratteristica 2, Revue de Math. Pures Appl., 2 (1957), 289-300.

[9] B. Segre, Ovali e curve σ nei piani di Galois di caratteristica due, Atti Accad. Naz. Lincei Rend., (8), 32 (1962), 785-790.

[10] B. Segre and U. Bartocci, Ovali ed alte curve nei piani di Galois di caratteristica due, Acta Arith., 18 (1971), 423-449.

[11] J. A. Thas, S. E. Payne, and H. Gevaert, A family of ovals with few collineations, preprint.

Annals of Discrete Mathematics 37 (1988) 95–106

CIRCULAR BLOCK DESIGNS FROM PLANAR NEAR-RINGS

James R. Clay
University of Arizona
Tucson, AZ 85721

I. INTRODUCTION

Planar near-rings whose additive group is $(C,+)$, the additive group of complex numbers, have provided direction and motivation for constructing finite planar near-rings with interesting geometric interpretations [1,2]. In particular, balanced incomplete block designs (BIBD) have resulted in two distinct, but related ways [3, 4]. We again take inspiration from a planar near-ring on $(C,+)$ to obtain a third way of constructing BIBD's, and in this case, the blocks of some of the designs have properties analogous to those of circles in the euclidean plane.

Consider the field of complex numbers $(C,+,\cdot)$. For $a,\ b \in C$, define $a * b = a^{(1)} \cdot b$ where $a = (a_1, a_2)$ and $a^{(1)} = a_1$ if $a_1 \neq 0$, and $a^{(1)} = a_2$ otherwise. Then $(C,+,*)$ is a planar near-ring, and if $a \neq 0$, then $C*a$ is the line through 0 and a, so $C*a+b$ is the line through b parallel to $C*a$. So all the lines of the plane coincide with $\mathbf{B} = \{C*a+b \mid a,b \in C,\ a \neq 0\}$. For a finite planar near-ring $(N,+,\cdot)$, if we take $\mathbf{B} = \{Na+b \mid a,b \in N,\ a \neq 0\}$, then sometimes $(N, \mathbf{B}, \in)$ is a BIBD [4,5].

Again, consider $(C,+,\cdot)$. Define $*$ by $a*b = |a| \cdot b$ for $a, b \in C$. Then $(C,+,*)$ is a planar near-ring and $C*a$ is the ray from 0 passing through a, for $a \in C,\ a \neq 0$. Also $C*a+b$ is a translate of $C*a$ by b. So $C*\{a,-a\}$ is the line through 0 and a (and $-a$), and $C*\{a,-a\}+b$ is the line through b parallel to $C*\{a,-a\}$. All the lines in the plane coincide with $\mathbf{B} = \{C*\{a,-a\}+b \mid a,b \in C,\ a \neq 0\}$. Again, for a finite planar near-ring $(N,+,\cdot)$, if we take $\mathbf{B} = \{N\{a,-a\}+b \mid a,b \in N,\ a \neq 0\}$, then $(N,\mathbf{B},\in)$ is sometimes a BIBD [3].

Now, again with $(C,+,\cdot)$, define $*$ by $0*b = 0$ for all $b \in C$, and $a*b = (a/|a|) \cdot b$ if $a \neq 0$. Then $(C,+,*)$ is a planar near-ring, and if $a \neq 0$, then $C*a$ is the circle of radius $|a|$ together with its center 0, and $C*a+b$ is the circle of radius $|a|$ together with its center b. Hence $\mathbf{B} = \{C*a+b \mid a,b \in C,\ a \neq 0\}$ consists of all circles together with their centers. If we let $C^* = C \setminus \{0\}$, then $C^* * a + b$ is the circle of center b and radius $|a|$.

In this paper, we take a finite integral planar near-ring $(N,+,\cdot)$ and let $\mathbf{B}^* = \{N^*a + b \mid a,b \in N,\ a \neq 0\}$, where $N^* = N \setminus \{0\}$. We shall see that $(N,\mathbf{B}^*,\in)$ is always a BIBD, and that sometimes, the blocks $\mathbf{B}^*$ are *circular*, i.e. three distinct points of N belong to at most one block or "circle" of $\mathbf{B}^*$. A "best" upper bound for the number of points on a circle $B \in \mathbf{B}^*$ is given. Using a method of Ferrero, one can easily construct all the planar near-rings on any $(Z_p,+)$, the group of integers modulo a prime p [5]. A computer program was developed to determine all the "circular" near-rings $(Z_p,+,*)$ for primes p, $13 \leq p < 1000$, and the results are tabulated within. A few examples will be examined in greater detail to illustrate several "circular" properties.

II. BACKGROUND

A *near-ring* is a triple $(N, +, \cdot)$ so that (i) $(N, +)$ is a group; (ii) $(N, \cdot)$ is a semi-group; and (iii) $a \cdot (b + c) = (a \cdot b) + (a \cdot c)$ for all $a, b, c \in N$. For any near-ring N define $=_m$ by $a =_m b$ if and only if $ax = bx$ for each $x \in N$. Then $=_m$ defines an equivalence relation on N and on $N^* = N \setminus \{0\}$. Let $N/=_m$ and $N^*/=_m$ denote the equivalence classes of $=_m$ on N and N^*, respectively. If $|N/=_m| \geq 3$ and each equation $ax = bx + c$, $a \neq_m b$, has a unique solution for $x \in N$, then $(N, +, \cdot)$ is a *planar* near-ring. (Here $|X|$ denotes the cardinality of a set X.) For further information on planar near-rings see [5]. The construction method of G. Ferrero [4,5] will be used. We summarize this method here, and emphasize that all finite planar near-rings are constructed in this way.

To begin with, we need a finite group $(N, +)$ with a subgroup of automorphisms Φ, $1 \neq \Phi < \mathrm{Aut}N$, where $\phi \in \Phi \setminus \{1\}$ and $\phi(x) = x$ imply $x = 0$. So, each element of $\Phi \setminus \{1\}$ has only $0 \in N$ as a fixed point. For each orbit $\Phi(a) = \{\phi(a) \mid \phi \in \Phi\}$, $a \in N^* = N \setminus \{0\}$, select a representative $1_a \in \Phi(a)$, so if $b \in N^*$, there is a unique $1_a \in N^*$ so that $b \in \Phi(1_a)$ and there is a unique $\phi_b \in \Phi$ so that $\phi_b(1_a) = b$. Define $\cdot$ on N by $0 \cdot x = 0$ for each $x \in N$, and $b \cdot x = \phi_b(x)$, for $b \in N^*$. Then $(N, +, \cdot)$ is a planar near-ring. In fact, $(N, +, \cdot)$ is an *integral* planar near-ring; i.e. one with $a =_m 0$ if and only if $a = 0$. Planar near-rings that are not integral are constructed by selecting some, but not all, of the orbits $\Phi(a)$ to be included in $[0]_{=_m} = \{a \in N \mid a =_m 0\}$.

For any such planar near-ring constructed by Ferrero's method —and we emphasize that all planar near-rings can be constructed by this method— the sets Na, $N\{a, -a\}$, and N^*a are exactly the sets $\Phi(a) \cup \{0\}$, $\Phi(a) \cup \{0\} \cup \Phi(-a)$, and $\Phi(a)$, respectively. Hence, these sets depend only upon a and Φ. For this reason, we can and do assume that all our near-rings in the remainder of this paper are finite, planar, and integral.

In this paper, for a finite planar near-ring $(N, +, \cdot)$, we refer to the sets N^*a, $a \neq 0$, as *basic blocks*. The set of all *blocks* will be $\mathbf{B}^* = \{N^*a + b \mid a, b \in N,\ a \neq 0\}$. One milestone will be to show that $(N, \mathbf{B}^*, \in)$ is a BIBD, so there must be an integer r, where $[x] = r$ if $x \in N$, where $[x]$ denotes the number of blocks $B \in \mathbf{B}^*$ with $x \in B$. (Similarly, $[x_1, \cdots, x_t]$ denotes the number of blocks $B \in \mathbf{B}^*$ containing $\{x_1, \cdots, x_t\}$.) Also, there will be integers k and λ so that $|B| = k$ if $B \in \mathbf{B}^*$, and $[x, y] = \lambda$ if $x \neq y$ and $x, y \in N$. So the following proposition will prove to be valuable. Note that $0x = 0$ in a planar near-ring [2].

Proposition 1. *For a finite integral planar near-ring $(N, +, \cdot)$, we have $N^*a + b = N^*c + d$ if and only if $b = d$ and $Na = Nc$.*

Proof. If $b = d$ and $Na = Nc$, then $N^*a = N^*c$ and $N^*a + b = N^*c + d$. Conversely, suppose $N^*a + b = N^*c + d$. Then $N^*a = N^*c + (d - b)$. If this forces $d - b = 0$, then we get $b = d$ and $N^*a = N^*c$, hence $b = d$ and $Na = Nc$.

So, suppose $N^*a = N^*b + c$ where $a, b, c \in N^*$. For $u \in N^*$, $N^*a = N^*b + uc$. Let $u_1 = 1_c, u_2, \cdots, u_k$ be representatives of $N^*/=_m$. Since $k \geq 2$, we get $N^*b + c = N^*b + u_2c$ and $N^*b = N^*b + (u_2c - c)$ with $u_2c - c \neq 0$. Let $t = u_2c - c$. Then from $N^*b = N^*b + t$, we get $N^*b = N^*b + t = N^*b + 2t = \cdots = N^*b + (h-1)t$, where h is the order of the subgroup $< t >$ of $(N, +)$ generated by t.

This makes $N^*b = N^*b + < t >$, a union of cosets of $< t >$, hence $k = |N^*b| = wh$ for some integer w. If $v = |N|$, then $k|(v-1)$, so $h|(v-1)$. Certainly $h|v$ since $h = |<t>|$. Thus h divides $(v, v-1) = 1$, forcing $h = 1$ and $t = 0$.

Note: In the above proof, we have used the fact that $|N^*/=_m| = |N^*a| = k$, if $a \in N^*$. This can be easily proven, or the reader is referred to [5].

Theorem 2. *Let* $(N, +, \cdot)$ *be a finite integral planar near-ring. Let*

$$\mathbf{B}^* = \{N^*a + b \mid a, b \in N,\ a \neq 0\}.$$

Then $(N, \mathbf{B}^*, \in)$ *is a BIBD with parameters* $v = |N|$, $k = |N^*/=_m|$, $b = v(v-1)/k$, $r = v-1$, *and* $\lambda = k-1$.

Proof. Certainly $v = |N|$ and $k = |N^*/=_m|$. From proposition 1, $b = v(v-1)/k$. Each of the $(v-1)/k$ blocks N^*a, $a \in N^*$, has k elements, so for a fixed $x \in N$, there are exactly k elements $y_1, \cdots, y_k$ so that $x \in N^*a + y_i$, $1 \leq i \leq k$. Hence, x belongs to exactly $v-1$ blocks of $\mathbf{B}^*$, and so $r = v-1$. There remains to show that $\lambda = k-1$.

Now $x, y \in N^*a + t$ if and only if $0, y - x \in N^*a + (t - x)$. So, the goal is to show that $0, z$ belong to exactly $k-1$ blocks for $z \neq 0$. Let $u_1, \cdots, u_k$ be representatives of the equivalence classes $N^*/=_m$, and let $u_1 = 1_e$ for some $e \in N^*$. So, the only blocks containing 0 are the $N^*a - u_ia$, $a \in N^*$.

For $1 < i \leq k$, $1_ex = u_ix + z$ has a unique solution for x; call it a_i and note that $a_i \neq 0$ since $z \neq 0$. Then $1_ea_i = u_ia_i + z$, so $-u_ia_i + 1_ea_i = z$ and $u_i(-a_i) + a_i = z$. This puts $z \in N^*(-a_i) + a_i$. Note that $0 \in N^*(-a_i) + a_i$ also, so 0 and z belong to $N^*(-a_2) + a_2, \cdots, N^*(-a_k) + a_k$.

If it should be that $i \neq j$ but yet $a_i = a_j$, then from $u_i(-a_i) + a_i = z = u_j(-a_j) + a_j$ we get $u_ia_i = u_ja_j = u_ja_i$. This makes the equation $u_ix = u_jx + 0$ have the solutions $a_i \neq 0$ and 0, contrary to uniqueness.

We now have $k-1$ distinct blocks containing both 0 and z. We must show that this is all of them. If $0, z \in N^*a + b$, then $N^*a + b = N^*a - u_ia$ for some i, $1 \leq i \leq k$, since these are the only blocks containing 0. So $z = u_ja - u_ia$ for some u_j, $i \neq j$. From this we get $-u_ja + z = -u_ia$, $u_j(-a) + z = u_i(-a)$, or $u_i(-a) = u_j(-a) + z$. Now there is a $u_i' \in N$ so that $(u_i'u_i)y = y$ for all $y \in N$ [2], so $1_e[u_i(-a)] = (u_ju_i')[u_i(-a)] + z$ and $u_ju_i' =_m u_l$ for some l, $1 < l \leq k$. Hence $1_ex = u_lx + z$ has the unique solution $a_l = u_i(-a) = -(u_ia)$. So $0, z \in N^*a + b = N^*a - u_ia = N^*(u_ia) + u_i(-a) = N^*(-a_l) + a_l$. This shows that this $N^*a + b$ containing 0 and z must be one of the $k-1$ blocks already accounted for. So $\lambda = k-1$.

Definition. A BIBD $(N, \mathbf{B}^*, \in)$ is *circular* if for each three distinct points $x, y, z \in N$, there is at most one $B \in \mathbf{B}^*$ such that $x, y, z \in B$.

Obviously, a BIBD is circular if $k = 2, 3$, but we will not be concerned with these in this work. The following theorem puts an upper bound on k in terms of v if the design is to be circular. We shall give examples to show that this is the best upper bound attainable in general.

Theorem 3. *Let* $(N, +, \cdot)$ *be a finite integral planar near-ring. For*

$$\mathbf{B}^* = \{N^*a + b \mid a, b \in N,\ a \neq 0\},$$

suppose that $(N,\mathbf{B}^*,\in)$ *is a circular BIBD. Then* $k \le (3+\sqrt{4v-7}\,)/2$, *where* $v = |N|$.

Proof. Each of the $v(v-1)/k$ circles has $\binom{k}{3} = k(k-1)(k-2)/6$ triples that belong to a circle. There are $\binom{v}{3} = v(v-1)(v-2)/6$ distinct triples. Since $(N,\mathbf{B}^*,\in)$ is circular,

$$\frac{v(v-1)}{k}\cdot\frac{k(k-1)(k-2)}{6} > \frac{v(v-1)(v-2)}{6}$$

is false. Hence $(k-1)(k-2) \le v-2$. Thus $k \le (3+\sqrt{4v-7}\,)/2$.

III. TABLE OF CIRCULAR BIBD

The following table has entries for nearly every prime p, $13 \le p < 1000$. For each of these primes p, if $1 < d$ and d divides $p-1$, then the multiplicative subgroup of Z_p^* of order d defines a subgroup Φ_p of $\mathrm{Aut}Z_p^+$ suitable for constructing a planar near-ring on $(Z_p,+)$, the integers modulo p. For such a near-ring $(Z_p,+,\cdot)$, let $\mathbf{B}_d^* = \{Z_p^* a + b \mid a, b \in Z_p,\ a \ne 0\}$. Then $(Z_p,\mathbf{B}_d^*,\in)$ is a BIBD by theorem 2. If $(Z_p,\mathbf{B}_d^*,\in)$ is to be circular, then $k \le u(p) = [(3+\sqrt{4p-7}\,)/2]$ by theorem 3. If $d = 2$ or 3, then $(Z_p,\mathbf{B}_d^*,\in)$ is certainly circular. So, in the following table, for a prime p, $13 \le p < 1000$, there is associated with it a generator for Z_p^*, and all the divisors d, $3 < d \le u(p)$, for which the computer program asserts that $(Z_p,\mathbf{B}_d^*,\in)$ is circular. Also, for each such d, there is associated a generator g_d of the subgroup Φ_d' of Z_p of order d. If there is no entry for such a prime, then there is no appropriate circular $(Z_p,\mathbf{B}_d^*,\in)$; e.g. $p = 23$, 47 and 587.

A typical entry is the cell for the prime $p = 337$.

337	4	6	7	8	12
10	148	129	8	85	72

It has the structure

p	$d_1\ d_2\ \cdots\ d_k$
g_{p-1}	$g_{d_1}\ g_{d_2}\cdots g_{d_k}$

where p is a prime, g_{p-1} is a generator of Z_p^*, each d_i divides $p-1$, each g_{d_i} is a generator of a subgroup Φ_{d_i}' of Z_p^* of order d_i, and the designs defined by $d_1,\cdots,d_k$ are circular, $3 < d_i \le u(p)$. So this table provides 379 nontrivial examples of circular BIBD.

Computation time was decreased tremendously by using

Proposition 4. *Let* $(N,+,\cdot)$ *be a finite integral planar near-ring. Consider the BIBD* $(N,\mathbf{B}^*,\in)$. *For any three distinct points* $x,y,z \in N$, *we have* $[x,y,z] = [0,y-x,z-x]$.

Proof. Let $N^*a_1+b_1,\cdots,N^*a_m+b_m$ be the distinct blocks containing x, y, and z. Then $N^*a_1+(b_1-x),\cdots,N^*a_m+(b_m-x)$ are m distinct blocks containing $0, y-x$ and $z-x$.

If $N^*a + b$ is an additional block containing $0, y - x$, and $z - x$, then $N^*a + (b + x)$ is an additional block containing x, y and z.

So, in checking a BIBD $(N, \mathbf{B}^*, \in)$ to see if it is circular, one only needs to look at those blocks containing 0 and be assured that no pair $x, y \in N^*$, $x \neq y$, belong to more than one of these blocks containing 0.

Notice that the table gives examples of circles of size k, $4 \leq k \leq 18$. The computer indicated that $(Z_{1217}, \mathbf{B}^*_{19}, \in)$ and $(Z_{1801}, \mathbf{B}^*_{20}, \in)$ are also circular.

<table>
<tr><td>13| 4
2| 5</td><td>17| 4
3| 4</td><td>29| 4
2| 12</td></tr>
<tr><td>31| 5 6
3| 2 6</td><td>37| 4 6
2| 6 11</td><td>41| 4 5
6| 9 10</td></tr>
<tr><td>43| 6
3| 7</td><td>53| 4
2| 23</td><td>61| 4 5 6
2| 11 9 14</td></tr>
<tr><td>67| 6
2| 30</td><td>71| 5 7
7| 5 20</td><td>73| 4 6 8
5| 27 9 10</td></tr>
<tr><td>79| 6
3| 24</td><td>89| 4 8
3| 34 12</td><td>97| 4 6 8
5| 22 36 33</td></tr>
<tr><td>101| 4 5
2| 10 36</td><td>103| 6
5| 47</td><td>109| 4 6
6| 33 46</td></tr>
<tr><td>113| 4 7 8
3| 15 16 18</td><td>127| 6 7
3| 20 2</td><td>131| 5 10
2| 53 42</td></tr>
<tr><td>137| 4 8
3| 37 10</td><td>139| 6
2| 43</td><td>149| 4
2| 44</td></tr>
<tr><td>151| 5 6 10
6| 8 33 87</td><td>157| 4 6
5| 28 13</td><td>163| 6 9
2| 59 38</td></tr>
<tr><td>173| 4
2| 80</td><td>181| 4 5 6 9 10
2| 19 42 49 39 46</td><td>191| 5 10
19| 39 7</td></tr>
<tr><td>193| 4 6 8
5| 81 85 9</td><td>197| 4 7
2| 14 36</td><td>199| 6 9
3| 93 43</td></tr>
<tr><td>211| 5 6 7 10
2| 55 15 58 23</td><td>223| 6
3| 40</td><td>229| 4 6 12
6|107 95 18</td></tr>
<tr><td>233| 4 8
3| 89 12</td><td>239| 7
7| 10</td><td>241| 4 5 6 8 10 12
7| 64 87 16 8 36 4</td></tr>
<tr><td>251| 5 10
6| 20 32</td><td>257| 4 8
3| 16 4</td><td>269| 4
2| 82</td></tr>
<tr><td>271| 5 6 10
6| 10 29 27</td><td>277| 4 6 12
5| 60 117 35</td><td>281| 4 5 7 8 10
3| 53 86 59 60 49</td></tr>
<tr><td>283| 6
3| 45</td><td>293| 4
2|138</td><td>307| 6 9
5| 18 46</td></tr>
<tr><td>311| 5 10
17| 6 305</td><td>313| 4 6 8 12
10| 25 99 5 29</td><td>317| 4
2|114</td></tr>
<tr><td>331| 5 6 10 11
3| 64 32 8 74</td><td>337| 4 6 7 8 12
10|148 129 8 25 72</td><td>349| 4 6 12
2|136 123 24</td></tr>
<tr><td>353| 4 8
3| 42 70</td><td>367| 6
6| 84</td><td>373| 4 6 12
2|104 89 69</td></tr>
<tr><td>379| 6 7 9
2| 52 86 84</td><td>389| 4
2|115</td><td>379| 4 6 9 12
5| 63 35 14 157</td></tr>
<tr><td>401| 4 5 8 10
3| 20 39 45 29</td><td>409| 4 6 8 12
21|143 54 31 49</td><td>419| 11
2|334</td></tr>
</table>

421	4 5 6 7 10 12 14	431	5 10	433	4 6 8 9 12
2	29 252 21 33 44 159 36	7	95 25	5	179 199 354 150 64
439	6	443	13	449	4 7 8 14
15	172	2	35	3	67 18 92 5
457	4 6 8 12	461	4 5 10	463	6 7 11 14
13	109 134 170 18	2	48 88 93	3	22 34 15 51
487	6 9	491	5 7 10	499	6
3	233 41	2	101 138 110	7	140
509	4	521	4 5 8 10	523	6 9
2	208	3	235 25 43 5	2	61 19
541	4 5 6 9 10 12	547	6 7 13	557	4
2	52 48 130 15 313 216	2	41 9 46	2	118
569	4 8	571	5 6 10	577	4 6 8 9 12
3	86 76	3	106 110 90	5	24 214 152 287 57
593	4 8	601	4 5 6 8 10 12 15	607	6
3	77 59	7	125 32 25 59 169 5 18	3	211
613	4 6 9 12	617	4 7 8 11 14	619	6
2	35 66 160 142	3	194 142 139 31 62	2	253
631	5 6 7 9 10 14 18	641	4 5 8 10	643	6
3	288 44 21 32 403 30 138	3	154 354 256 79	11	466
653	4	659	7	661	4 5 6 10 11 12 15
2	149	2	12	2	106 197 297 190 9 246 12
673	4 6 7 8 12 14 16	677	4 13	691	5 6 10 15
5	58 256 117 64 16 23 8	2	26 40	3	89 243 371 113
701	4 5 7 10	709	4 6 12	727	6 11
2	135 89 19 63	2	96 228 91	5	282 46
733	4 6 12	739	6 9	743	7 14
6	353 308 113	3	321 197	5	111 151
751	5 6 10	757	4 6 7 9 12 14	761	4 5 8 10
3	80 73 182	2	87 28 59 3 78 127	6	39 67 62 77
769	4 6 8 12 16	773	4	787	6
11	62 361 40 19 27	2	317	2	380
797	4	809	4 8	811	5 6 9 10 15
2	215	3	318 44	3	212 681 796 311 276
821	4	823	6	827	7 14
2	295	3	175	2	124 20
829	4 6 9 12 18	853	4 6 12	857	4 8
2	246 126 5 77 191	2	333 221 98	3	207 188
859	6 11	877	4 6 12	881	4 5 8 10 11
2	261 13	2	151 283 240	3	387 268 177 137 32
883	6 7 9 14 18	907	6	911	5 7 10 13 14
2	339 71 135 134 242	2	385	17	19 49 429 30 7
919	6 9 18	929	4 8 16	937	4 6 8 9 12 13 18
7	53 440 95	3	324 18 40	5	196 323 14 72 333 36 13
941	4 5 10	947	11	953	4 7 8 14
2	97 349 185	2	133	3	442 431 156 79
967	6 7 14	971	5 10	977	4 8 16
5	143 97 175	6	341 168	3	252 227 52
991	5 6 9 10 11 15 18	997	4 6 12		
6	160 114 18 166 42 71 55	7	161 305 91		

IV. EXAMPLES

The reader is encouraged to construct examples for himself, to discover some interesting properties, to make some nice conjectures, and then to prove some nice theorems. To illustrate some of the geometric properties of some circular BIBD's, let us consider $p = 13$ and $p = 31$. We shall let $(N, +, \cdot)$ denote the near-ring under consideration. Taking guidance from $(C, +, *)$, a "circle" $N^*a + b$ is said to have *center* b and *radius* a.

Now, for $p = 13$ and $k = 4$, the circles with center 0 are $N^*1 = \{8, 12, 5, 1\}$, $N^*2 = \{3, 11, 10, 2\}$, and $N^*4 = \{6, 9, 7, 4\}$. The circles of radius 1 and tangent to N^*1 are $N^*1 + 2, N^*1 + 10, N^*1 + 11$, and $N^*1 + 3$. The centers of these circles form the circle N^*2 of radius 2 and center 0, just as in the euclidean plane.

Consider now the circles of radius 1 tangent to the circle N^*2 of radius 2 and center 0. Just as in the plane, these circles fall into natural groupings.

I	II
$N^*1 + 1 = \{9, 0, 6, 2\}$	$N^*1 + 7 = \{2, 6, 12, 8\}$
$N^*1 + 5 = \{0, 4, 10, 6\}$	$N^*1 + 4 = \{12, 3, 9, 5\}$
$N^*1 + 12 = \{7, 11, 4, 0\}$	$N^*1 + 9 = \{4, 8, 1, 10\}$
$N^*1 + 8 = \{3, 7, 0, 9\}$	$N^*1 + 6 = \{1, 5, 11, 7\}$

The centers of those in grouping I form N^*1, as in the euclidean plane. Also, the centers of those in II form N^*4, a circle with center 0. Those in I all pass through 0, as in the plane, and each intersects exactly two of the others.

If one takes the four points on $x^2 + y^2 = 4$ corresponding to $0, \pi/2, \pi$, and $3\pi/2$ radians, and, at each of these four points, places a circle of radius 1 passing through 0 and tangent to $x^2 + y^2 = 4$, then one can see that each of these four circles intersects exactly two of the others, and that the points of intersection lie on $x^2 + y^2 = 2$, a circle with center 0 and radius $\sqrt{2}$. Analogously, the four circles of grouping I intersect at points of N^*4, a circle with center 0 and radius 4. But note that $4^2 \equiv 3 \; mod \; 13$ and $N^*3 = N^*2$, a circle of radius 2 and center 0. That is, one can think of N^*4 as having properties analogous to that of $x^2 + y^2 = 2$.

Each circle in grouping II intersects exactly two of the others. Their points of intersection form N^*8. On each of these four circles, there is exactly one point that does not belong to N^*2 nor to one of the other three; e.g. from $N^*1+9 = \{4, 8, 1, 10\}$, $10 \in N^*2$, $8 \in N^*1+7$, and $1 \in N^*1 + 6$, leaving 4. These four "extra" points constitute N^*4.

The grouping II provides a nice relationship among the 13 points of $N = Z_{13}$. Put 0 at the origin. On the circle $x^2 + y^2 = 4$, at 0, $\pi/2$, π, and $3\pi/2$ radians, put the points 3, 2, 10, and 11, respectively, of N corresponding to N^*2. At 3, 2, 10, and 11, put congruent circles corresponding to $N^*1 + 4$, $N^*1 + 7$, $N^*1 + 9$, and $N^*1 + 6$, respectively, but with large enough radius so that each is tangent to exactly two of the others, as well as to $x^2 + y^2 = 4$. These four circles have points of tangency which we identify with 12, 8, 1, and 5, respectively. Circumscribe the entire picture now with a circle, which will be tangent to each of the four circles at 9, 6, 4, and 7, respectively. Notice now that all 13 points of N are illustrated, and that the circumscribing circle corresponds to N^*4.

Consider now the case where $p = 31$ and $k = 6$. The circles with center 0 are

$$N^*1 = \{26, 25, 30, 5, 6, 1\}, \quad N^*2 = \{21, 19, 29, 10, 12, 2\},$$

$$N^*4 = \{11, 7, 27, 20, 24, 4\}, \quad N^*8 = \{22, 14, 23, 9, 17, 8\},$$

$$\textit{and} \quad N^*16 = \{13, 28, 15, 18, 3, 16\}.$$

First, let us note that the circles of radius 1 tangent to N^*1 are exactly the $N^*1 + t$, where $t \in N^*2$. This is as it is for $p = 13$ and for the euclidean plane. We shall see in theorem 6 that this is no accident.

Again, it will be interesting to look at the circles $N^*1 + t$ tangent to N^*2. As in the $p = 13$ case, there are two groupings.

I	II
$N^*1 + 1 = \{27, 26, 0, 6, 7, 2\}$	$N^*1 + 8 = \{3, 2, 7, 13, 14, 9\}$
$N^*1 + 6 = \{1, 0, 5, 11, 12, 7\}$	$N^*1 + 22 = \{17, 16, 21, 27, 28, 23\}$
$N^*1 + 5 = \{0, 30, 4, 10, 11, 6\}$	$N^*1 + 14 = \{9, 8, 13, 19, 20, 15\}$
$N^*1 + 30 = \{25, 24, 29, 4, 5, 0\}$	$N^*1 + 23 = \{18, 17, 22, 28, 19, 24\}$
$N^*1 + 25 = \{20, 19, 24, 30, 0, 26\}$	$N^*1 + 9 = \{4, 3, 8, 14, 15, 10\}$
$N^*1 + 26 = \{21, 20, 25, 0, 1, 27\}$	$N^*1 + 17 = \{12, 11, 16, 22, 23, 18\}$

The centers of those in I form N^*1 and each circle of I passes through 0. The centers of those in II form N^*8.

Consider $x^2 + y^2 = 4$ in the plane. Mark the six points corresponding to 0, $\pi/3$, $2\pi/3$, π, $4\pi/3$, and $5\pi/3$ radians, and label them 2, 21, 19, 29, 10, and 12, respectively. At each of these six points, place a circle of radius 1 passing through 0. These will correspond to the six circles in I. Notice that adjacent circles in the plane intersect at the origin and on a common circle with center 0. Label these six points with the appropriate points from N^*4. For example, at the point labeled 2 there is the tangent circle corresponding to $N^*1 + 1$. At 21 is the circle corresponding to $N^*1 + 26$. These adjacent circles intersect at 0 and 27. Since $27 \in N^*4$, label this with 27, since 0 will be the label of the common point at the origin.

Also notice that every pair of every other circles intersect at a point other than the origin. This is true in the plane as well as for our circles from I with the arrangement forced by our labeling above. For example, $N^*1 + 1$ and $N^*1 + 25$ is such a pair corresponding to the points 2 and 19. This pair intersects at 0 and 26, so label the corresponding point with 26. In the plane, these points are on a common circle $x^2 + y^2 = 1$, and in our case here, they constitute the points of the circle N^*1.

The relationship of the circles in II is more complicated. Start again with the six points labeled on $x^2 + y^2 = 4$. This time, however, one must place a circle of radius r, $1 < r < 2$,, tangent to $x^2 + y^2 = 4$ at the six points corresponding to 2, 21, 19, 29, 10, and 12, and do so in such a way as to have these six circles lie inside $x^2 + y^2 = 4$. Again, these six circles will correspond to the six circles of II.

To see the desired relationship, it may help to do the above in two steps. First place $N^*1 + 23$, $N^*1 + 17$, and $N^*1 + 22$ at 29, 12, and 21, respectively. Any two of these intersect at two points. Label the "outer" points with 22, 23, and 17, and the "inner" points with 18, 16, and 28. Next, place N^*1+8, N^*1+14, and N^*1+9 at 2, 19, and 10, respectively. Any two of these also intersect at two points. Label the "outer" points with 9, 14, and 8, and the "inner" points with 15, 3, 13.

Note that the "outer" points form N^*8 and that they appear on the same circle in the plane and that this circle has center at the origin. Similarly, the "inner" points form N^*16 and they appear on a circle whose center is the origin.

Each of these six circles from II have five of their six points labeled. We want to find a place to put these missing points. At each of the six points of N^*2 there is a diameter and a circle corresponding to one of the N^*1+t from II. This circle is tangent to $x^2+y^2=4$ and intersects the corresponding diameter at a point. Label this point with the missing point of the N^*1+t. Note that these missing points constitute N^*4, and that in the diagram, they are all on the same circle $x^2+y^2=(r-1)^2$.

Theorem 3 gives an upper bound for the number of points on a circle of a circular design. In general, this is the best that can be attained. However, from our table, it would seem that a much better upper bound could be given for the $(Z_p, \mathbf{B}_d^*, \in)$.

To see that $u(v)$ gives the best upper bound, the following lemma is useful.

Lemma 5. *If $n>1$, then $u(n^2)=[(3+\sqrt{4n^2-7}\,)/2]=n+1$.*

Proof. First show that $n-1<\sqrt{n^2-7/4}<n$. Then $n+1/2<(3+\sqrt{4n^2-7}\,)/2<n+3/2$. This forces $u(n^2)\in\{n,n+1\}$. If $u(n^2)=n$, then $n\le(3+\sqrt{4n^2-7}\,)/2<n+1$, and this forces $n<2$.

Take any field $(F,+,\cdot)$ of p^2 elements, p a prime. Then $p+1$ divides p^2-1 and $u(p^2)=p+1$. Let Φ' be the subgroup of F^* of order $p+1$. Then Ferrero's method yields an integral planar near-ring $(F,+,*)$ and consequently a BIBD $(F,\mathbf{B}_{p+1}^*,\in)$. Our computer program indicates that these are circular for $p\in\{3,5,7,11\}$.

The examples discussed in this section motivated the following

Theorem 6. *Suppose $(N,+,\cdot)$ is an integral planar near-ring of odd order with $(N,+)$ abelian. Suppose that $N^*a=N^*(-a)$ and that $(N,\mathbf{B}^*,\in)$ is circular. Then*

$$\{t\in N \mid N^*a+t \text{ is tangent to } N^*a\}=N^*(2a).$$

Proof. For $ua\in N^*a\cap(N^*a+t)$, there is a $va\in N^*a$ such that $ua=va+t$. This means $t=wa+ua$ for some $wa\in N^*a$. Hence, $t\in N^*a+N^*a$. Conversely, if $t\in N^*a+N^*a$, then there are w_1a, $w_2a\in N^*a$ so that $t=w_1a+w_2a=w_2a+w_1a$. Each $w_i(-a)=-(w_ia)$ is in N^*a also, hence w_1a, $w_2a\in N^*a\cap(N^*a+t)$.

Since $(N,\mathbf{B}^*,\in)$ is circular, we cannot have $0\ne t=w_1a+w_2a=w_1'a+w_2'a$ with each w_ia, $w_i'a\in N^*a$, and $\{w_1a,\ w_2a\}\ne\{w_1'a,\ w_2'a\}$, for otherwise we would have three or more points in common with two distinct circles.

If $0\ne t=w_1a+w_2a$ and $w_1a\ne w_2a$, then $\{w_1a,\ w_2a\}=N^*a\cap(N^*a+t)$. If $0\ne t=wa+wa=w(a+a)=w(2a)\in N^*(2a)$, then $N^*a\cap(N^*a+t)=\{wa\}$ and N^*a+t is tangent to N^*a. Conversely, if $\{wa\}=N^*a\cap(N^*a+t)$, then $t=w_1a+w_2a$, and if $w_1a\ne w_2a$, then we would have $\{w_1 a,\ w_2 a\} = N^* a \cap (N^* a + t)$. So if $\{wa\} = N^* a \cap (N^* a + t)$, then $t = wa + wa = w(2a) \in N^*(2a)$.

V. THE COMPUTER PROGRAM

The computer program used to generate the data in the table was written in BASIC, and was compiled for the Zenith Z-100 Desktop Computer. BASIC is usually an "interpreter" language. The program, left in "interpreter" form, would generate the same data, but the time required would be greatly increased. The program used here was "compiled" thus greatly diminishing the computation time required. Even so, computation time was considerable. There is absolutely nothing about the algorithm that makes BASIC a preferred language. One could execute the algorithm equally well with various languages; e.g. FORTRAN.

The program was designed to test the "circular" property only for $(Z_p, \mathbf{B}_d^*, \in)$, where p is a prime. The program asks the user for a prime p. Upon supplying a suitable p, the program then generates the suitable divisors d of $p-1$, $3 < d \leq u(p)$, and the "first" generator of Z_p^*, g_{p-1}. In an ordered fashion, the program considers each $(Z_p, \mathbf{B}_d^*, \in)$, $3 < d \leq u(p)$. It computes g_d and inquires if you want to check $(Z_p, \mathbf{B}_d^*, \in)$ for circularity. A negative answer results in either considering a new $(Z_p, \mathbf{B}_d^*, \in)$, or requesting another prime value for p.

In testing a particular $(Z_p, \mathbf{B}_d^*, \in)$ for circularity, proposition 4 was used. In particular, for each nonzero value $x \in Z_p$, all the blocks containing 0 and x were calculated. Then, systematically, for each $y \in Z_p \setminus \{0, x\}$, these blocks were searched, and the number of those containing $0, x$, and y were counted. If this number ever exceeded 1, then the design $(Z_p, \mathbf{B}_d^*, \in)$ was rejected for being circular. Failing to be rejected, the program indicated that the particular $(Z_p, \mathbf{B}_d^*, \in)$ "seems to be circular."

In making tables or generating data of this sort, the author feels it best to be somewhat conservative in one's use of language. Hence, the words "seems to be circular" above, and the words "computer program asserts" in the first paragraph of section III were used. In fact, the original table of the author's included six additional $(Z_p, \mathbf{B}_d^*, \in)$. These six, for

$$(p, d) \in \{(271, 9), (449, 16), (541, 18), (701, 14), (859, 13), (919, 17)\},$$

were rejected by a totally independent program written by Matthew Modisett. His program was based on a theorem of his which is very different than proposition 4, and other than this, the author knows nothing about his program. The case for $(p, d) = (271, 9)$ was checked carefully by the author, and he found it not to be circular. With the correction of a parameter in his program, output of the author's program now agrees with that of Modisett's. It is remarkable that for all the primes p, $13 \leq p < 1000$, and all the corresponding divisors of $p-1$, $3 < d \leq u(p)$, that the two completely different programs agreed except for these six cases.

The author wishes to recognize the valuable contribution of Matthew Modisett in writing his program and checking independently all the data in the table. Because of the program's independence and its distinct foundation, the author is now very confident of the data in the table.

REFERENCES

[1] M. ANSHEL & J. R. CLAY, Planar algebraic systems: some geometric interpretations, J. Algebra **10** (1968), 166–173.

[2] J. R. CLAY, Some algebraic and geometric aspects of planarity, Atti del Convegno di Geometria Combinatoria e sue Applicazione, Univ. degli Studi, Perugia (1971), 163–172.

[3] J. R. CLAY, More balanced incomplete block designs from Frobenius groups, Discrete Math. **59** (1986), 229–234.

[4] G. FERRERO, Stems planari e BIB-disegni, Riv. Mat. Univ. Parma (2) **11** (1970), 79-96.

[5] G. BETSCH & J. R. CLAY, Block designs from Frobenius groups and planar near-rings, Proc. Conf. Finite Groups (Park City, Utah), Academic Press, 1976, 473–502.

Annals of Discrete Mathematics 37 (1988) 107–116

EXTENDING THE CONCEPT OF DECOMPOSABILITY FOR TRIPLE SYSTEMS

C. J. Colbourn*, University of Waterloo

E. Mendelsohn*, University of Toronto

A. Rosa*, McMaster University

A triple system of order v and index λ is MPT-decomposable if it contains λ pairwise disjoint maximum packings of triples (MPT) of order v, and it is MPT-free if it contains no MPT of order v. We determine completely the spectrum for MPT-decomposable, as well as for MPT-indecomposable triple systems, and obtain partial results concerning the spectrum of MPT-free triple systems.

1. INTRODUCTION

A triple system TS(v;λ) is a pair (V,B) where V is a v-set of elements and B is a collection of 3-subsets of V called triples such that each 2-subset of V is contained in exactly λ triples. The parameters λ and v are called the index, and the order of the triple system, respectively. Repeated triples (i.e. triples identical as subsets of V) are permitted. When $\lambda = 1$, we have a Steiner triple system (STS) and when $\lambda = 2$, we have a twofold triple system (TTS). It is well known that an STS of order v exists if and only if $v \equiv 1$ or 3 (mod 6), and a TTS of order v exists if and only if $v \equiv 0$ or 1 (mod 3).

When $v \equiv 1$ or 3 (mod 6), one has the following analogue of STSs. A maximum packing of triples of order v (MPT(v)) is a pair (V,B) where V is a v-set, and B is a collection of 3-subsets of V, called triples, such that (i) each 2-subset of V is contained in at most one triple, (ii) no triple can be adjoined to B without violating (i), and (iii) if C is any collection of 3-subsets satisfying (i) and (ii), then $|B| \geq |C|$.

The leave of an MPT (V,B) is the graph (V,E) where E contains as edges all pairs that are not contained in any triple of B. It is well known that the

* Research supported by NSERC Grants No. A0579, A7861, and A7268, respectively.

leave of an MPT(v) is a 1-factor if $v \equiv 0$ or 2 (mod 6), is a 1^*-factor (i.e. a graph whose one component is a 3-edge-star and the other components are single edges) if $v \equiv 4$ (mod 6), and is a quadrangle if $v \equiv 5$ (mod 6) (and, of course, is empty if $v \equiv 1$ or 3 (mod 6)).

If (V,B) is a TS(v;λ) such that $B = B_1 \cup B_2$ where (V,B_1), (V,B_2) is a $TS(v;\lambda_1)$ and $TS(v,\lambda_2)$, respectively, with $1 \leq \lambda_1, \lambda_2 < \lambda$ then (V,B) is said to be decomposable. Otherwise, it is indecomposable.

In some cases, arithmetic conditions automatically ensure indecomposability. Consider, for instance, triple systems TS(v;2) of index 2 (i.e. TTSs). Any TTS of order $v \equiv 0$ or 4 (mod 6) is indecomposable simply because an STS of this order cannot exist. Viewing now a decomposable TTS either as one whose set of triples is partitionable into two STSs, or as one whose set of triples contains an STS, leads to the following (at least) two generalizations:

A TS(v;λ) is MPT-decomposable if its set of triples contains λ pairwise disjoint MPT(v)'s. Otherwise, it is MPT-indecomposable.

A TS(v;λ) is MPT-free if its set of triples contains no MPT(v).

(Note that since repeated triples are allowed, two disjoint MPTs can contain two triples which are identical as subsets of V.)

Another possible definition would involve analogues of MPTs for higher λ (in fact, this would be the "true" analogue of the decomposability concept as defined above), or (at least in the case of $\lambda = 2$, and possibly also $\lambda = 3$) a partitioning of the set of triples of a TS(v;λ) into a maximum packing (=MPT) and a minimum covering, etc. We do not consider any of these in this paper.

We stress again that throughout this paper we allow our triple systems to contain repeated triples.

2. TWOFOLD TRIPLE SYSTEMS

Theorem 2.1. (a) An MPT-decomposable TTS(v) exists if and only if $v \equiv 1$ or 3 (mod 6) or $v = 4$.

(b) An MPT-indecomposable TTS(v) exists if and only if $v \equiv 0$ or 1 (mod 3), $v \neq 3, 4$, or 7.

Proof. For $v \equiv 1$ or $3 \pmod 6$, MPT-decomposability is the same as decomposability. Let now $v \equiv 0$ or $4 \pmod 6$, and assume (V,B) to be an MPT-decomposable TTS(v). Let (V,B_1), (V,B_2) be the two (disjoint) MPT(v)'s contained in it. If $v \equiv 0 \pmod 6$, the leave of (V,B_i), $i = 1, 2$, is a 1-factor F_i. But then the union $F_1 \cup F_2$ must be decomposable into triangles which is absurd. If $v \equiv 4 \pmod 6$, then the leave of (V,B) is a 1^*-factor G_i (one component is a 3-edge-star G_i', the other components are single edges). If $v \geq 10$, delete from the union $G_1 \cup G_2$ all vertices of the stars G_i'. The remaining graph $\bar{G}$ must contain at least one edge; this edge cannot be contained in any triangle of $\bar{G}$, again a contradiction. Observing now that a TTS(4) is MPT-decomposable completes the proof of (a), and noting that an indecomposable TTS(v) exists for all $v \equiv 1$ or $3 \pmod 6$ except when $v = 3$ or 7 [4] yields the proof of (b). ∎

Theorem 2.2. An MPT-free TTS(v) exists if and only if $v \equiv 0$ or $1 \pmod 3$, $v > 7$.

Proof. Case 1. $v \equiv 1$ or $3 \pmod 6$; here a TTS(v) is MPT-free if and only if it is indecomposable. The latter exists if and only if $v > 7$ [4].

Case 2. $v \equiv 0$ or $4 \pmod 6$. Neither the unique TTS(4) nor the unique TTS(6) are MPT-free, so assume $v \geq 10$. If $v \equiv 0 \pmod 4$, $v = 4g$, consider a group divisible design with g groups of size 4, blocks of size 3 and $\lambda = 2$. Such a GDD exists whenever $g \equiv 0$ or $1 \pmod 3$ [3]. Put on each group a TTS(4). This results in a TTS(4g), and since $g \geq 3$, this TTS is MPT-free. If $v \equiv 2 \pmod 4$, $v = 4g+2$, consider a GDD with g-1 groups of size 4 and a unique group of size 6, blocks of size 3 and $\lambda = 2$. Such a GDD exists whenever $g \equiv 1$ or $2 \pmod 3$, $g \geq 4$ [6]. Put on each group a TTS(4) or a TTS(6), respectively. This results in a TTS(4g+2); clearly, this TTS cannot contain an MPT(v). This leaves only the case $v = 10$ to consider. However, any TTS(10) with three sub-TTS(4)'s intersecting in the same element is MPT-free. ∎

3. THREEFOLD TRIPLE SYSTEMS

Theorem 3.1. An MPT-decomposable TS(v;3) exists if and only if $v \equiv 1 \pmod 2$.

Proof. The necessity is obvious. Since for $v \equiv 1$ or $3 \pmod 6$ we can take three copies of an STS(v) to obtain a TS(v;3), it remains only to consider the case $v \equiv 5 \pmod 6$. Consider three MPT(v)'s on the same set V whose leaves are the quadrangles (a,b,c,d), (a,b,d,c), and (a,c,b,d), respectively. The union of the sets of triples of the three MPTs, together with a TTS(4) on $\{a,b,c,d\}$ yields an MPT-decomposable TS(v;3). ■

Before we determine the spectrum for MPT-indecomposable TS(v;3)'s, we need several lemmas.

Lemma 3.2. There exists an MPT-indecomposable TS(11;3).

Proof. Let $V = A \cup X$ where $A = \{a_1,\dots,a_5\}$, $|X| = 6$. Let G be the cubic multigraph with 6 vertices given in Fig. 1. Note that G does not have a 1-factorization.

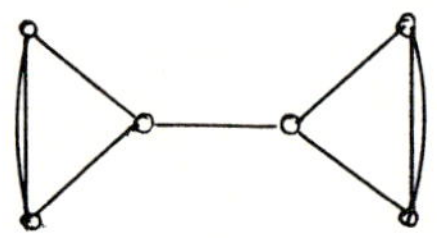

Fig. 1.

Let $H = \{G_1, G_2, G_3, G_4, G_5\}$ be a decomposition of $3K_6$ on X into 5 cubic factors, with $G_1 \simeq G$. Let

$$B = B_1 \cup \bigcup_{i=1}^{5} \{\{a_i, x_i, y_i\}: \ x_i, y_i \in X, \ \{x_i, y_i\} \in E(G_i)\}$$

where (A, B_1) is (the unique) TS(5;3). Then, clearly, (V,B) is a TS(11;3). Assume (V,B) is MPT-decomposable. This means that it contains three disjoint MPT(11)'s with sets of triples, say, C_1, C_2, C_3. Since (V,B) contains a sub-TS(5;3), this forces the vertices of the leaves of these three MPTs to be in A which in turn means that all pairs $\{x,y\}$ with x, y $\in$ X are contained in a triple of each of the MPTs. Thus

$$F_j = \{\{x,y\}: \ \{a_1,x,y\} \in C_j\} \quad j = 1, 2, 3$$

is a 1-factor of G_1, and F_1, F_2, F_3 is a 1-factorization of G_1, a contradiction. Thus (V,B) is MPT-indecomposable. ■

Lemma 3.3. There exists an MPT-indecomposable TS(17;3) with a sub-TS(5;3).

Proof. Let $V = A \cup Z_{12}$ where $A = \{a_1,\dots,a_5\}$. Let G be the graph with V(G) $= Z_{12}$, $E(G) = \{\{x,y\}: |x-y| = \pm 1, \pm 2 \text{ or } 6\}$. Let G_1 be the cubic factor of G given in Fig. 2, and let G_2, G_3 be obtained from G_1 by adding 1 and 2 modulo 12, respectively, to the vertex labels of G_1.

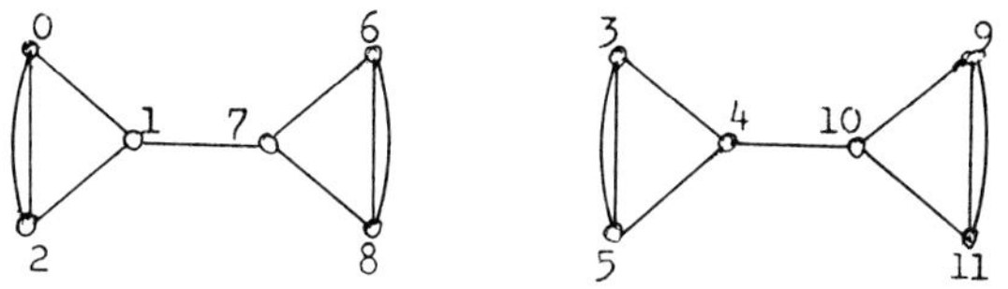

Fig. 2.

The graph $G\setminus(G_1\cup G_2\cup G_3)$ is easily seen to be 1-factorable; combine the six 1-factors in any manner to produce two further cubic factors G_4, G_5 of G. Take now $B = 3C\cup\bigcup_{i=1}^{5}\{\{a_i,x_i,y_i\}: x_i,y_i \in Z_{12}, \{x_i,y_i\} \in E(G_i)\}$ where $C = \{\{x,x+3,x+7\}: x \in Z_{12}\}$. Then (V,B) is an MPT-indecomposable TS(17;3) with a sub-TS(5;3). The verification is similar to that in Lemma 3.2, and we leave out the details. ∎

<u>Lemma 3.4.</u> Let $v \equiv 5 \pmod 6$. If there exists an MPT-indecomposable TS(v;3) then there exists an MPT-indecomposable TS(2v+1;3).

<u>Proof.</u> Use the standard $v \longrightarrow 2v + 1$ construction for STSs using a 1-factorization of K_{v+1} (see, e.g., [5]), with STS(v) replaced by (an MPT-indecomposable) TS(v,3), and each 1-factor of the 1-factorization "triplicated". ∎

Let us remark that an alternative "nonrecursive" construction (patterned after the construction of Lemma 3.2) would be to replace the STS(v) by a (not necessarily MPT-indecomposable) TS(v,3), and the 1-factorization by a 3-factorization, with at least one cubic factor of this 3-factorization not 1-factorable.

<u>Lemma 3.5.</u> Let $v \equiv 5 \pmod 6$. If there exists an MPT-indecomposable TS(v;3) then there exists an MPT-indecomposable TS(4v-3;3).

<u>Proof.</u> Consider a GDD with 4 groups G_1, G_2, G_3, G_4, $|G_i| = v-1 \equiv 4 \pmod 6$, and block size 3. Such a GDD exists by [3]. Let $V = \{\infty\}\cup\bigcup_{i=1}^{4}G_i$. Put on G_1 $\{\infty\}$ a copy of an MPT-indecomposable TS(v,3), and on $G_i\cup\{\infty\}$, i=2, 3, 4, a copy of any TS(v,3). Triplicate all blocks of the GDD. The result is an MPT-indecomposable TS(4v-3;3). ∎

Lemma 3.6. Let $v \equiv 5 \pmod 6$. If there exists an MPT-indecomposable TS(v;3) with a sub-TS(5;3) then there exists an MPT-indecomposable TS(4v-15;3).

Proof. Consider again a GDD with 4 groups G_i, i=1, 2, 3, 4, and block size 3, but let this time $|G_i| = v-5 \equiv 0 \pmod 6$. The existence of such a GDD follows from [3]. Let $A = \{a_1,\ldots,a_5\}$, $V = A \cup \bigcup_{i=1}^{4} G_i$. Put on $G_1 \cup A$ a copy of an MPT-indecomposable TS(v;3) with a sub-TS(5;3) on A, and put on $G_i \cup A$, i=2, 3, 4, a copy of any TS(v,3) with (the same) sub-TS(5;3) on A. Triplicate all blocks of the GDD. The result is an MPT-indecomposable TS(4v-15;3). ∎

Theorem 3.7. An MPT-indecomposable TS(v;3) exists if and only if $v \equiv 1 \pmod 2$, $v \geq 7$.

Proof. If $v \equiv 1$ or $3 \pmod 6$, a TS(v;3) is MPT-indecomposable if and only if it is indecomposable. Indecomposable TS(v,3) are known to exist if and only if $v \geq 7$ [4]. Clearly, the unique TS(5,3) in not MPT-indecomposable. Let now $v \equiv 5 \pmod 6$, $v \geq 11$. An MPT-indecomposable TS(11;3) and TS(17;3) exist by Lemmas 3.2 and 3.3. Note that they both contain a sub- TS(5;3). Assume now that $v \geq 23$, and for all $u < v$ ($u \equiv 5 \pmod 6$, $u \geq 11$) there exists an MPT-indecomposable TS(u;13) with a sub-TS(5;3). If $v \equiv 11 \pmod{12}$, let $u = (v-1)/2$. Then $u \equiv 5 \pmod 6$, $u \geq 11$, and by our hypothesis, there exists an MPT-indecomposable TS(u;3). By Lemma 3.4, there exists an MPT-indecomposable TS(v;3). If $v \equiv 17 \pmod{24}$, let $u = (v+3)/4$. Then $u \equiv 5 \pmod 6$, $u \geq 11$, and there exists an MPT-indecomposable TS(u;3). By Lemma 3.5, there exists an MPT-indecomposable TS(v;3). Finally, if $v \equiv 5 \pmod{24}$, let $u = (v+15)/4$. Then $u \equiv 5 \pmod 6$, $u \geq 11$, and there exists an MPT-indecomposable TS(u;3) with a sub-TS(5;3). By Lemma 3.6, there exists an MPT-indecomposable TS(v;3). Note that in all cases the TS(v;3) contains a sub-TS(5;3). ∎

If $v \equiv 1$ or $3 \pmod 6$, an MPT-free TS(v;3) exists if and only if $v > 7$. This follows easily from [4]. If $v \equiv 5 \pmod 6$, however, the spectrum for MPT-free TS(v;3)'s remains undetermined.

4. THE CASE $\lambda = 4$

By Theorem 2.1(a), there exists no MPT-decomposable TS(v;2) for $v \equiv 0$ or $4 \pmod 6$. However, if $\lambda = 4$, we have the following:

Theorem 4.1. There exists an MPT-decomposable TS(v,4) for every $v \equiv 0$ or 4 (mod 6).

Proof. Take four MPT(v)'s on the same set V in such a way that (i) if $v \equiv 0$ (mod 6), the union of their leaves, each being a 1-factor, yields a graph with v/6 components, each component isomorphic to the cocktail-party graph K_6-L, and (ii) if $v \equiv 4$ (mod 6), the union of their leaves, each being a 1^*-factor, yields a graph with (v+2)/6 components, one component being $2K_4$, and the other components being K_6-L.

In either case, each such component can be decomposed into 4 triangles, which, when adjoined to the triples of the four MPT(v)'s, yield the set of triples of an (MPT-decomposable) TS(v;4). ∎

Theorem 2.2 provides us also with MPT-free TS(v;4)'s for all $v \equiv 0$ or 1 (mod 3), $v > 7$. Let us remark that the results of [1] yield MPT-free TS(v;4) without repeated triples for all $v \equiv 1$ or 3 (mod 6), $v \geq 13$.

5. THE CASE $\lambda = 6$

Theorem 5.1. An MPT-decomposable TS(v;6) exists for all $v \geq 3$.

Proof. For v odd the statement follows trivially from Theorem 2.1 and 3.1. Let now v be even. Distinguish two cases:

Case 1. $v \equiv 0$ or 2 (mod 6). Let $V = Z_v$. Let G = (V,E) be the cyclic graph with $E = \{\{x,y\}: |x-y| = \pm 1, \pm 2, \pm 3\}$. By [8], G has a 1-factorization. Let $F = \{F_1,\ldots,F_6\}$ be such a 1-factorization. Take six MPT(v)'s on V whose leaves are $F_1,\ldots,F_6$. Then add triples $\{\{x,x+1,x+3\}: x \in Z_v\}$. The result is an MPT-decomposable TS(v;6).

Case 2. $v \equiv 4$ (mod 6). For v = 4 this follows from Theorem 2.1(a). Let $v \geq 10$. Let $V = Z_{v-4} \cup \{a,b,c,d\}$. Consider the graph G = (V,E) where

$$E = \{\{x,y\}: x, y \in Z_{v-4}, |x-y| = \pm 1, \pm 2, \pm 3\} \cup E(H)$$

where H is the multigraph in Fig. 3 (the labels indicate the multiplicity of an edge).

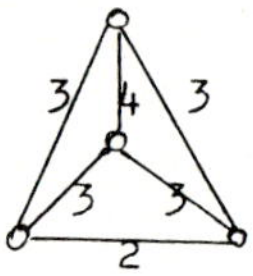

Fig. 3

Clearly, the graph G can be decomposed into six 1^*-factors. The rest is as in Case 1. ∎

Lemma 5.2. There exists an MPT-indecomposable TS(14;6).

Proof. Let $V = A \cup Z_8$, where $A = \{a_1, \ldots, a_6\}$. Let $F_1 = (Z_8, E)$ be the 6-regular graph on Z_8 given in Fig. 4. Clearly, F_1 is not 1-factorable.

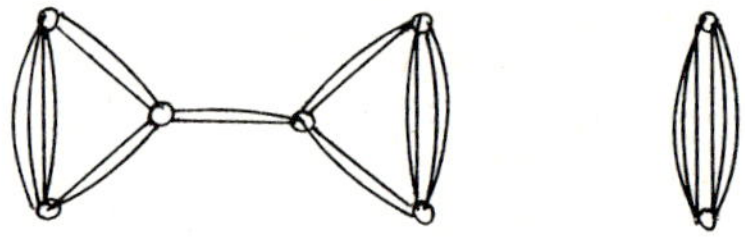

Fig. 4. The graph F_1.

Let $C = \{\{x, x+1, x+3\}: x \in Z_8\}$. Let $F = \{F_1, \ldots, F_6\}$ be a factorization of $6K_8-C$ into 6-regular subgraphs (which exists by Petersen's theorem). Form the set of triples $B = D \cup C \cup \bigcup_{i=1}^{6} \{\{a_i, x_i, y_i\}: \{x_i, y_i\} \in F_i\}$ where (A,D) is a TS(6;6). Then (V,B) is a TS(14;6).

This TS(14;6) is MPT-indecomposable since F_1 is not 1-factorable. Indeed, any MPT(14) has a leave which is a 1-factor, say, f. This 1-factor f must intersect A in a 1-factor (i.e. 3 disjoint edges), otherwise this would contradict the structure of an MPT(6). This implies that f intersects also Z_8 in a 1-factor (i.e. 4 disjoint edges). Assuming MPT-indecomposability would now imply that F_1 is 1-factorable.

Lemma 5.3. There exists an MPT-indecomposable TS(20;6) and TS(26;6).

Proof. Similar to that of Lemma 5.2 and is therefore omitted. ∎

Lemma 5.4. For any $v \equiv 0$ or 2 (mod 6), $v \geq 30$, there exists an MPT-indecomposable TS(v;6).

Proof. By Stern's theorem [7], a TS(v;6) can be embedded in a TS(w;6) if and only if $w \geq 2v+1$. Taking now an MPT-indecomposable TS(14,6) from Lemma 5.2 and noting that for $v \equiv 0$ or 2 (mod 6) an MPT(v) has as its leave a 1-factor, and that embedding in this case preserves MPT-indecomposability completes the proof. ∎

Theorem 5.5. An MPT-indecomposable TS(v;6) exists if and only if $v \geq 6$, $v \neq 8$.

Proof. For v odd, the statement follows, e.g., from Theorem 3.7 (duplicate all blocks). For v even, $v \equiv 0$ or 4 (mod 6), the statement follows from Theorem 2.1(b) (triplicate all blocks). Finally, for $v \equiv 2$ (mod 6), the statement follows from Lemmas 5.2-5.4 and the observation that the (unique) TS(8;6) is MPT-decomposable. ∎

Let us remark that the spectrum of MPT-free TS(v;6)'s remains undetermined.

6. CONCLUSION AND OPEN PROBLEMS

Apart from the open problems concerning the spectrum of MPT-free TS(v;λ)'s that were mentioned in previous sections, several further questions can be asked. For instance, most of our constructions introduce repeated triples. Can one obtain analogues of the results in Sections 2-5 if no repeated triples are allowed?

Another question deals with the complexity of deciding whether a TS(v;λ) is MPT-decomposable. When $\lambda = 2$, deciding MPT-decomposability can be done in polynomial time, as it amounts to deciding decomposability for TTS(v)'s with $v \equiv 1$ or 3 (mod 6). The latter reduces to determining whether the block intersection graph is bipartite (cf. [2], [4]). However, even the complexity of deciding whether a TTS(v) is MPT-free is not known at present. Similar questions may be asked for higher λ.

Finally, given a TS(v;λ), what is the minimum number of triples in a partial triple system (of index 1) contained in the TS(v;λ) (as a function of v)?

REFERENCES

[1] C.J. Colbourn, A. Rosa, Indecomposable triple system with $\lambda = 4$, Studia Sci. Math. Hungar. (to appear)

[2] M.J. Colbourn, Algorithmic aspects of combinatorial designs: a survey, Ann. Discrete Math. 26(1985), 67-136.

[3] H. Hanani, Balanced incomplete block designs and related designs, Discrete Math. 11(1975), 255-369.

[4] E.S. Kramer, Indecomposable triple systems, Discrete Math. 8(1974), 173-180.

[5] A. Rosa, Algebraic properties of designs and recursive constructions, Proc. Conf. Algebr. Aspects Combinatorics (Toronto 1975), Congr. Numer. 13, pp. 183-200.

[6] A. Rosa, D. Hoffman, The number of repeated blocks in twofold triple systems, J. Combinat. Theory (A) 41(1986), 61-88.

[7] G. Stern, Tripelsysteme mit Untersystemen, Arch. Math. 33(1979), 204-208.

[8] G. Stern, H. Lenz, Steiner triple systems with given subspaces; another proof of the Doyen-Wilson theorem, Boll. Un. Mat. Ital. Ser. A (5) 17(1980), 109-114.

Annals of Discrete Mathematics 37 (1988) 117–136

TRANSLATION PARTIAL GEOMETRIES

F. De Clerck - H. Gevaert - J.A. Thas

Seminar of Geometry and Combinatorics, State University of Ghent, Krijgslaan 281, B-9000 GENT (Belgium)

In this paper we introduce a theory of translation partial geometries. These geometries have an automorphism group G acting regularly on the points. An example of such a translation partial geometry is $T_2^*(K)$, embedded in AG(3,q), G being the group of translations of AG(3,q). This theory is closely related to translation generalized quadrangles and translation planes.

1. INTRODUCTION

1.1. Definition

A (finite) partial geometry S = (P,B,I) is an incidence structure with a symmetric incidence relation satisfying the following axioms.

(i) Each point is incident with t+1 lines ($t \geq 1$) and two distinct points are incident with at most one line.

(ii) Each line is incident with s+1 points ($s \geq 1$) and two distinct lines are incident with at most one point.

(iii) If x is a point and L is a line, such that x $\not\mathrm{I}$ L, then there are exactly $\alpha(\alpha \geq 1)$ points $x_1, x_2, \ldots, x_\alpha$ and α lines $L_1, L_2, \ldots, L_\alpha$, such that $x \mathrm{I} L_i \mathrm{I} x_i \mathrm{I} L$, $i = 1,2,\ldots,\alpha$.

The numbers t,s and α are called the parameters of the partial geometry. The integer α is sometimes called the intersection number. This incidence structure was introduced by R.C. Bose in 1963 [4].

Given two (not necessarily distinct) points x,y of S, we write $x \sim y$ and say x and y are collinear, provided that there is some line L for which x I L I y. And $x \not\sim y$ means that x and y are not collinear. Dually for $L,M \in B$, we write $L \sim M$ or $L \not\sim M$ according as L and M are concurrent or nonconcurrent, respectively . The line (resp. point) which is incident with distinct collinear points x,y (resp. distinct concurrent lines L,M) is denoted by xy (resp., LM or $L \cap M$).

For $x \in P$, put $x^\perp = \{y \in P \| y \sim x\}$. Note that $x \in x^\perp$. The trace of a pair $\{x,y\}$ of distinct points is defined to be the set $x^\perp \cap y^\perp$ and is denoted $\{x,y\}^\perp$. We have $|\{x,y\}^\perp| = s+1+t(\alpha-1)$ or $(t+1)\alpha$ according as $x \sim y$ or $x \not\sim y$. More generally, if $A \subset P$, then $A^\perp = \cap\{x^\perp \| x \in A\}$. For $x \neq y$, the span of the pair $\{x,y\}$ is $\{x,y\}^{\perp\perp} = \{u \in P \| u \in z^\perp \ \forall z \in \{x,y\}^\perp\}$. These definitions can be dualized.

1.2. Remarks

a. If $S = (P,B,I)$ is a partial geometry with parameters t,s and α, then $\bar{S} = (\bar{P},\bar{B},\bar{I})$, $\bar{P} = B$, $\bar{B} = P$, $\bar{I} = I$, is a partial geometry with parameters $\bar{t} = s$, $\bar{s} = t$ and $\bar{\alpha} = \alpha$.

b. The number v of points of a partial geometry equals $v = (s+1)(st+\alpha)/\alpha$ while the number b of lines equals $b = (t+1)(st+\alpha)/\alpha$ [4].

c. The point graph of a partial geometry is strongly regular with parameters $v, k = (t+1)s$, $\lambda = s-1+t(\alpha-1)$, $\mu = (t+1)\alpha$.

1.3. The four classes of partial geometries

(I) The generalized quadrangles : $\alpha = 1$ [16], [25].

(II) The 2-$(v,s+1,1)$ designs : $\alpha = s+1$; and their duals : $\alpha = t+1$.

(III) The nets of order $s+1$ and degree $t+1$: $\alpha = t$; and their duals : $\alpha = s$.

(IV) The "proper" partial geometries : $1 < \alpha < \min(s,t)$.

1.4. Some examples of partial geometries

We will give here only two examples of partial geometries. They both are constructed using maximal arcs in a projective plane. For other examples we refer to [8], [13], [19], [20], [23], [26].

1.4.1. Some remarks on maximal arcs in projective planes

A maximal arc K of degree d in a finite projective plane of order n (not necessarily desarguesian) is a non-empty set of $nd-n+d$ points of the plane such that any line of the plane intersects K in 0 or d points. The maximal arcs of degree n of a plane of order n are the sets obtained by deleting a line of the plane. If K is a proper subset of the plane, one can easily prove that d has to divide n. R.H.F. Denniston [12] has proved that this condition is sufficient in the case of any desarguesian plane of order 2^h. J.A. Thas [20], [22] has constructed other families of maximal arcs in desarguesian planes of order 2^h and in translation planes of order 2^h. On the other hand, there are no maximal arcs of degree 3 in $PG(2,3^h)$ $(h>1)$ [6], [21]. In fact, no example of a maximal arc of degree d, with $d < n$, in a projective plane of odd order n is known.

1.4.2. The partial geometry $S(K)$ [19], [20], [27]

Let K be a maximal arc of degree d, $1<d<n$, in a projective plane π of order n. We define the incidence structure $S(K) = (P,B,I)$. The points of $S(K)$ are the points of π which are not contained in K. The lines of $S(K)$ are the lines of π which are incident with d points of K. The incidence is that of π. It is readily proved that $S(K)$ is a partial geometry with parameters $t = n-n/d$, $s = n-d$, $\alpha = n-n/d-d+1$.

1.4.3. The partial geometry $T_2^*(K)$ [19]

Let K be a maximal arc of degree d, $d \neq 1$, in the projective plane $PG(2,q)$

over GF(q) ($q = p^h$, p a prime). We define an incidence structure $T_2^*(K) = (P,B,I)$ as follows. Let PG(2,q) be embedded as a plane H in PG(3,q). The points of $T_2^*(K)$ are the points of PG(3,q)-H. The lines of $T_2^*(K)$ are the lines of PG(3,q) which are not contained in H and which meet K (necessarily in a unique point). The incidence is that of PG(3,q). Again it is readily proved that $T_2^*(K)$ is a partial geometry with parameters $t = qd-q+d-1$, $s = q-1$, and $\alpha = d-1$.

1.4.4. Remarks

As there exist maximal arcs of degree 2^m in $PG(2,2^h)$, $0 < m < h$, there exists a partial geometry S(K) with parameters $s = 2^h-2^m$, $t = 2^h-2^{h-m}$, $\alpha = (2^m-1)(2^{h-m}-1)$. This is a generalized quadrangle iff $s = t = 2$ ($v = b = 15$). Note that a generalized quadrangle with these parameters is unique [16]. The partial geometry $T_2^*(K)$ arising from an arc K of degree 2^m, $m \neq 0$, in $PG(2,2^h)$ has parameters $s = 2^h-1$, $t = (2^h+1)(2^m-1)$, $\alpha = 2^m-1$. This is a generalized quadrangle iff $m = 1$, i.e. iff K is a complete oval. These generalized quadrangles corresponding with complete ovals were first discovered by R.W . Ahrens and G. Szekeres [1] and independently by M. Hall Jr. [14].

1.5. Spreads in partial geometries

Let V be a set of lines of a partial geometry S with parameters t,s,α ($\alpha \neq t+1$), such that any two lines of V are nonconcurrent. It is immediately clear that $|V| \leq (st+\alpha)/\alpha$. If equality holds we call V a spread of S, otherwise V is called a partial spread.

A spread in S(K) has q+1 elements while a spread in $T_2^*(K)$ has q^2 elements. All the lines of S(K) (resp. $T_2^*(K)$) intersecting the maximal arc in a fixed point form a spread in S(K) (resp. $T_2^*(K)$). These spreads are called linear. For more details we refer to [7], [8] and to [16] for the case of the generalized quadrangles $T_2^*(O)$.

2. α-REGULARITY IN PARTIAL GEOMETRIES

2.1. Theorem

Let L and M be two nonconcurrent lines in a partial geometry S with parameters $t,s,\alpha(\alpha \neq t+1)$ then $\{L,M\}^{\perp\perp}$ is a partial spread.

Proof. Suppose $N \in \{L,M\}^{\perp\perp}$ such that L I x I N I y I M. Let z be a point such that z I L, z $\not{I}$ N. The α lines incident with z and concurrent with M are all concurrent with N. This implies however, that z is incident with $\alpha+1$ lines concurrent with N, a contradiction.

Suppose $N \in \{L,M\}^{\perp\perp}$ such that $N \not\sim M$ and N I x I L. Let z be a point of L, $z \neq x$. The α lines incident with z and concurrent with M are all concurrent with N, which yields again that z is incident with $\alpha+1$ lines concurrent with

N, a contradiction.
Let N and N' be two distinct lines of $\{L,M\}^{\perp\perp}$ which are not concurrent with L and M. Then $\{L,M\}^{\perp\perp} = \{L,N\}^{\perp\perp}$ and N' is nonconcurrent with N.

2.2. Corollaries

1. If L and M are nonconcurrent lines in a partial geometry S with parameters t,s,α ($\alpha \neq t+1$) then $|\{L,M\}^{\perp\perp}| \leq s+1$.
2. If $N,N' \in \{L,M\}^{\perp\perp}$, $N \neq N'$, then $\{N,N'\}^{\perp\perp} = \{L,M\}^{\perp\perp}$.

2.3. Definitions

A pair {L,M} of nonconcurrent lines of a partial geometry S with parameters t,s,α ($\alpha \neq t+1$) is called α-regular iff $|\{L,M\}^{\perp\perp}| = s+1$. If $\alpha = 1$, {L,M} is called regular [16]. A set $\mathcal{L}$ of pairwise nonconcurrent lines of S is called a normal set iff every pair $\{L,M\} \subset \mathcal{L}$ is α-regular and $\{L,M\}^{\perp\perp} \subset \mathcal{L}$. Let V be a normal spread, then the incidence structure $\pi(V)$ with point set V, lines $\{L,M\}^{\perp\perp}$ ($L,M \in V$) and natural incidence is obviously a 2-($(st+\alpha)/\alpha$,s+1,1) design. Hence $\alpha|t$ and $(s+1)|t(t-\alpha)/\alpha^2$. If $t = \alpha(s+2)$ then $\pi(V)$ is an affine plane of order s+1.

2.4. Theorem

If a partial geometry S with parameters t,s,$\alpha < t$ has an α-regular pair of lines, then $(s+1)\alpha \leq t+1$.
<u>Proof</u> . Suppose {L,M} is an α-regular pair. Let B' be the set of lines of S which are concurrent with at least one line of $\{L,M\}^{\perp\perp}$. An easy counting yields $|B'| = (s+1)^2(t-\alpha)+(s+1)(\alpha+1)$, and $|B'| \leq b$ implies $(s+1)\alpha \leq t+1$.

2.5. Remarks

(a) If the partial geometry S has an α-regular pair of points, then $(t+1)\alpha \leq s+1$. Hence if $\alpha \neq 1$, $\alpha < s$ and $\alpha < t$, it cannot have also an α-regular pair of lines. For regularity in generalized quadrangles we refer to [16].
(b) Up to now there is no proper partial geometry known with $s+1 = \alpha(t+1)$ or $t+1 = \alpha(s+1)$.

2.6. Lemma

Suppose {L,M} is an α-regular pair of lines in a partial geometry S = (P,B,I) with parameters t,s,α ($\alpha \neq t+1$). Then the substructure $\mathcal{N}(L,M)$ with line set $\{L,M\}^{\perp} \cup \{L,M\}^{\perp\perp}$, and point set the set of all points of S incident with these lines is a net of order s+1 and degree $\alpha+1$.
<u>Proof</u>. As {L,M} is α-regular, $\{L,M\}^{\perp\perp} = \{N,N'\}^{\perp\perp}$ and $\{L,M\}^{\perp} = \{N,N'\}^{\perp}$ for any $\{N,N'\} \subset \{L,M\}^{\perp\perp}$. Let x and y be two points of $\mathcal{N}(L,M)$ and suppose x I L' I y, $L' \in B \setminus \{L,M\}^{\perp\perp}$. Let N (resp., N') be the line of $\{L,M\}^{\perp\perp}$ incident with x

(resp., y), then $L' \in \{N,N'\}^{\perp}$. Hence $L' \in \{L,M\}^{\perp}$. This implies that for any point-line pair (x,A) of $N(L,M)$, $x \not\mathrm{I}\, A$, there exist exactly α lines A_i of $N(L,M)$ and α points a_i $(i=1,\ldots,\alpha)$ of $N(L,M)$ such that $x \,\mathrm{I}\, A_i \,\mathrm{I}\, a_i \,\mathrm{I}\, A$. As every point of $N(L,M)$ is incident with $\alpha+1$ lines of $N(L,M)$, this implies that $N(L,M)$ is a net of order s+1 and degree $\alpha+1$.

Corollary

Let $\{L,M\}$ be an α-regular pair of the partial geometry S, then any pair $\{N,N'\}$ of nonconcurrent lines in $N(L,M)$ is also α-regular and $N(L,M) = N(N,N')$.

3. TRANSLATION PARTIAL GEOMETRIES

3.1. Theorem

Let S = (P,B,I) be a partial geometry with parameters t,s,α ($\alpha \neq t+1$). Suppose that the automorphism group G of S is an abelian group acting regularly on the points of S.

(i) If $|G_L| > 1$, $L \in B$, then the line orbit L^G of L is a spread of S.

(ii) If $t = \alpha(s+2)$, $\alpha \neq s$, then there exists at least one line L, such that $|G_L| > 1$.

<u>Proof</u>.

(i) Let h be an element of G such that $L^h = L$. As G is abelian, $L = L^{ghg^{-1}}$, hence $L^g = (L^g)^h$. This implies that $G_L = G_{L^g}$ $\forall g \in G$. Suppose $h \neq 1$, and suppose g is an element of G such that $L^g \neq L$ and $L \,\mathrm{I}\, p \,\mathrm{I}\, L^g$. As h fixes L^g, we have $p^h = p$, hence h = 1, a contradiction. As $|L^G| = |G|/|G_L| \geq |G|/(s+1) = (st+\alpha)/\alpha$, L^G is a spread of S and $|G_L| = s+1$.

(ii) Suppose that $t = \alpha(s+2)$ and that $G_L = \{1\}$ for every line L of S. Then $|L^G| = |G| = (s+1)(st+\alpha)/\alpha$ $\forall L \in B$, which divides $b = (t+1)(st+\alpha)/\alpha$. Hence $s+1 \mid \alpha(s+2)+1$ which implies that $s+1 \mid \alpha+1$, i.e. $\alpha = s$, a contradiction.

3.2. Definition

Let S = (P,B,I) be a partial geometry with parameters t,s,α and let G be an abelian automorphism group of S.

S is called a translation partial geometry with translation group G, provided

(1) G acts regularly on the points of S.

(2) $t = \alpha(s+2)$

(3) every line orbit of G is a normal spread.

The elements of G are called translations of the partial geometry.

Remark

The definition of translation generalized quadrangle given in [16] is not equivalent to the above definition for $\alpha = 1$, although they are closely related

[16]. For the definition and theorems on translation nets we refer to [17].

Example

The partial geometry $T_2^*(K)$ is a translation partial geometry having as translation group G the group of all translations of AG(3,q) = PG(3,q)-H (H being the plane of K). The line orbits of G are the (normal) linear spreads of $T_2^*(K)$.

3.3. Geometric partitions and group coset geometries

3.3.1. Definition

Let G be an abelian group of order $(s+1)^3$, $s \geq 1$. A set T of t+1 ($t \geq 1$) subgroups A_i, $i \in J = \{0,1,\ldots,t\}$ of order s+1 is called a geometric partition of G iff $t = \alpha(s+2)$ and

(1) for any pair $\{i,j\} \subset J$, there exists a subset V(i,j) of J, $|V(i,j)| = \alpha+1$, such that $A_iA_j = A_kA_l$ $\forall k,l \in V(i,j)$ with $k \neq l$;

(2) $A_iA_j \cap A_m = \{1\}$ $\forall m \in J-V(i,j)$.

The corresponding group coset geometry is the incidence structure S(G,T) = (P,B,I) with P = G, with B the set of cosets A_ig, $g \in G$, $i \in J$, and with the natural incidence relation. For a related construction of generalized quadrangles using group cosets we refer to [16].

3.3.2. Lemma

If $A_iA_j = A_kA_l$, with $i,j,k,l \in J$, $i \neq j$, $k \neq l$, then $k,l \in V(i,j)$.
Proof. Now $A_k \subset A_iA_j$, so $A_iA_j \cap A_k = A_k$. By (2) we have $k \in V(i,j)$. Analogously $l \in V(i,j)$.

Corollary

For all $i \neq j$ we have $i,j \in V(i,j)$.

3.3.3. Lemma

If $A_iA_j = A_kA_l$, with $i,j,k,l \in J$, $i \neq j$, $k \neq l$, then $V(i,j) = V(k,l)$.
Proof. Let $V(i,j) = \{i_0,i_1,\ldots,i_\alpha\}$. Then $A_{i_m}A_{i_n} = A_iA_j = A_kA_l$, $m \neq n$. By lemma 3.3.2. we have $i_m,i_n \in V(k,l)$, and so $V(k,l) = \{i_0,i_1,\ldots,i_\alpha\} = V(i,j)$.

3.3.4. Lemma

For any pair $\{i,i'\} \subset J$, $A_i \cap A_{i'} = \{1\}$.
Proof. Let $j \notin V(i,i')$, $j \in J$. Then by lemma 3.3.2. we have $A_iA_j \neq A_iA_{i'}$, and so $i' \notin V(i,j)$. By (2) we have $A_iA_j \cap A_{i'} = \{1\}$. Consequently $A_i \cap A_{i'} = \{1\}$.

3.3.5. Theorem

The group coset geometry S(G,T) is a translation partial geometry with parameters t,s,α and translation group isomorphic to G.

Proof. From 3.3.1. and 3.3.4. it is immediately clear that the axioms (i) and (ii) for a partial geometry are satisfied. We only have to prove axiom (iii). Let A_i, A_j and A_k be three different groups from T. Suppose $z \in (A_iA_j - A_i) \cap (A_iA_k - A_i)$, so $z = a_ia_k = a_i'a_j$, $a_i, a_i' \in A_i$, $a_j \in A_j$, $a_k \in A_k$. Hence $a_k = a_i^{-1}a_i'a_j \in A_k \cap A_iA_j$. If $k \notin V(i,j)$ then $a_k = 1$ and $z \in A_i$, a contradiction. Hence $k \in V(i,j)$ and $A_iA_j - A_i = A_iA_k - A_i$. This implies that for any $i \in J$, there exists a subset $W(i)$ of $s+2$ indices of J ($i \notin W(i)$), such that the union of A_i and the $s+2$ sets $A_iA_j - A_i$, $j \in W(i)$, is a partition $\Delta(i)$ of G. Let x be an element of G and let A_i be a group of T, $x \notin A_i$. Then $x \text{ I } A_kx \text{ I } a_i \text{ I } A_i$ iff $A_kx = A_ka_i$ with $a_i \in A_i$, i.e. iff $x \in A_kA_i - A_i$. There is exactly one element l of W(i) for which $x \in A_lA_i - A_i$. Now it is clear that $x \in A_kA_i - A_i$ iff $A_kA_i = A_lA_i$, iff $k \in V(l,i) - \{i\}$. Hence x is incident with α such lines A_kx. Next let A_iy, $y \notin A_i$ be a line of S(G,T) such that $x \not\text{I}\, A_iy$. Then $xy^{-1} \not\text{I}\, A_i$ and by the preceding argument xy^{-1} is incident with α lines A_kxy^{-1} which are concurrent with A_i. Hence x is incident with α lines A_kx concurrent with A_iy.

Let θ_a be the automorphism of S(G,T) defined by $g^{\theta_a} = ga$, $\forall g \in G$, and $(A_ig)^{\theta_a} = A_iga$ for all lines A_ig of S(G,T). Let G* be the group of all these automorphisms θ_a, $a \in G$. Then G* acts regular on the points of S(G,T) and moreover $\theta : G \to G^*$, $a \to \theta_a$ is an isomorphism. As $S_i = \{A_ig \parallel g \in G\}$ is the line orbit of A_i with respect to G*, S_i is a spread of S(G,T). We only have to prove that S_i is a normal spread for all $i \in J$. First of all we will prove that $\{A_i, A_ix\}$, $x \notin A_i$ is α-regular. There exists an index $j \in W(i)$ such that $A_ix \subset A_iA_j - A_i$. Let $k \in V(i,j) - \{i,j\}$. Then $A_k \subset A_iA_j$. The s+1 cosets A_ia_j, $a_j \in A_j$, are pairwise disjoint. Hence there exists an element h such that $A_k \cap A_ix = \{h\}$. For any $g \in A_i$, $A_kg \subset A_kA_i = A_iA_j$, and hence there exists an y such that $A_kg \cap A_ix = \{y\}$. By the preceding arguments, $\{A_i, A_ix\}^{\perp} = \{A_kg \parallel g \in A_i,\ k \in V(i,j) - \{i\}\}$. Moreover $\{A_i, A_ix\}^{\perp\perp} = \{A_ih \parallel h \in A_j\}$ which is by definition of a geometric partition equal to the set $\{A_ih' \parallel h' \in A_k\}$ for any $k \in V(i,j) - \{i\}$. Note that $\{A_i, A_ix\}^{\perp\perp} \subset S_i$. Next, let A_ix, A_iy be two distinct lines of S_i. As $\{A_i, A_iyx^{-1}\}$ is α-regular and $\{A_i, A_iyx^{-1}\}^{\perp\perp} \subset S_i$, it is immediately clear that $\{A_ix, A_iy\}$ is α-regular and $\{A_ix, A_iy\}^{\perp\perp} \subset S_i$. Hence any S_i is a normal spread and S(G,T) is a translation partial geometry with translation group isomorphic to G.

3.3.6. Theorem

For every translation partial geometry $S = (P,B,I)$ with translation group G, there exists a geometric partition T of G such that $S \cong S(G,T)$.

Proof. Let o be a fixed point of S. Let ω be the bijection from P to G defined by $p^{\omega} = g$ iff $o^g = p$. Let $L_0, L_1, \ldots, L_t$ be the t+1 lines of S incident with o. As the line orbit of L_i with respect to G is a spread S_i, the points of L_i will

be mapped by ω on the elements of a subgroup A_i of G, $i \in J = \{0,1,\dots,t\}$. Moreover if $i \neq j$, $A_i \cap A_j = \{1\}$.
Let T be the set $\{A_i \| i \in J\}$. The points on a line L_i^g, $g \in G$, of the spread S_i will be mapped by ω onto the points of the coset $A_i g$ of A_i. Let L_i and L_j, $i \neq j$, be two lines incident with o. Let p be a point on L_j, $p \neq o$. As S is a translation partial geometry, there exists a unique element $g \in G$ such that $o^g = p$. Then p I L_i^g. Moreover o is incident with α lines $L_j, L_{i_1}, L_{i_2}, \dots, L_{i_{\alpha-1}}$, which are concurrent with L_i^g. Put $V(i,j) = \{i,j,i_1,i_2,\dots,i_{\alpha-1}\}$.
Using the map $\omega : p^\omega = g \in A_j$ and $A_i g \cap A_j = \{g\}$. Let h be an arbitrary element of A_i. Then, as G is an abelian group, $A_i g \cap A_j h = \{gh\}$. The same reasoning goes through, replacing g by any $g' \in A_k$, $k \in V(i,j)-\{i\}$. As the pair $\{L_i, L_i^g\}$ is α-regular and S_i is a normal spread, the set $\{A_i g' \| g' \in A_k\}$ is a parallel class of lines in the net $N(L_i, L_i^g)$, for any $k \in V(i,j)-\{i\}$. This implies that $A_i A_k = A_i A_{k'}$, $\forall k,k' \in V(i,j)-\{i\}$. As $N(L_i,L_i^g) = N(L_k,L_k^h)$, $g \in A_j - A_i$, $k \in V(i,j)-\{i\}$, $h \in A_i$, $h \neq 1$; we have $A_i A_k = A_{k'} A_{k''}$ for any three different elements k, k', k" of $V(i,j)-\{i\}$. Finally, let l be an element of $J-V(i,j)$ and suppose that $A_i A_j \cap A_l \neq \{1\}$. Then there exists a group element $h_l \in A_l$ such that $h_l = a_i^{(k)} h_k$, $k \in V(i,j)-\{i\}$, $a_i^{(k)} \in A_i$, $h_k \in A_k$, $h_k \neq 1$, $a_i^{(k)} \neq 1$. This implies that $A_i h_l = A_i h_k$ $\forall k \in V(i,j)-\{i\}$, hence the line L_l as well as the α lines L_k, $k \in V(i,j)-\{i\}$ are incident with o and concurrent with the line $L_i^{h_l}$, a contradiction. Hence T is a geometric partition and $S \simeq S(G,T)$.

Corollary

If S is a translation partial geometry with translation group G, then the net $N(L_i,L_i^g)$ is a translation net of order s+1 and degree α+1. Moreover, each translation of this net has an order dividing s+1 [17]. The translation group of this net is a group $A_i A_j$ of order $(s+1)^2$, $A_j \in T$ such that $N(L_i,L_i^g) = N(L_j,L_j^h)$, $h \in A_i$, $h \neq 1$.

3.3.7. Theorem

Let G be the translation group of a translation partial geometry S, then the order of each translation of S is a divisor of s+1.

Proof. Consider for $A_i \in T$, as in theorem 3.3.5., the partition $\Delta(i)$ of G, with the corresponding subset W(i) of s+2 indices of J.
Let g be a translation of S. If $g \in A_i$, then there is nothing to prove. So let there exist an element j of W(i), such that $g \in A_i A_j - A_i$, and consequently g is a translation of the translation net $N(L_i,L_i^h)$, $h \in A_j$, $h \neq 1$. So the order of g divides s+1.

3.3.8. Theorem

Let S be a translation partial geometry with translation group G. Let π_i be the affine plane with as points the lines of a normal line orbit S_i and with lines the sets $\{L_i, L_i^g\}^{\perp\perp} \subset S_i$, $g \notin A_i$. Then π_i is a translation plane of order s+1 with elementary abelian translation group isomorphic to G/A_i.

Proof. Let $T = \{A_0, A_1, \ldots, A_t\}$ be the geometric partition of G. The points of π_i are the cosets of A_i, while the lines of π_i can be identified with the cosets of A_iA_j, $j \in J-\{i\}$.

Any translation h of S, induces a permutation h_i of the points of π_i. Any line A_iA_jg is mapped onto a parallel line A_iA_jgh of π_i, so h_i is a dilatation of π_i. By 3.3.7. the order of h_i divides s+1, and so h_i is a translation of π_i.

If G_i is the group of all translations of the affine plane π_i, then ϕ_i : $G \to G_i$, $h \to h_i$ is a homomorphism of G into G_i. The kernel $Ker(\phi_i)$ of ϕ_i exists of all translations of S fixing all lines of S_i, so $|Ker\ (\phi_i)| \leq s+1$. On the other hand, $|\phi_i(G)| \leq (s+1)^2$, and since $|G| = (s+1)^3$ there results $|Ker(\phi_i)| = s+1$, $|\phi_i(G)| = (s+1)^2$ and $\phi_i(G) = G_i$. Hence $Ker(\phi_i) = A_i$ and the affine plane π_i is a translation plane of order s+1 and elementary abelian translation group $G_i = G/A_i$.

4. THE KERNEL OF A TRANSLATION PARTIAL GEOMETRY

4.1. Definition

Consider a translation partial geometry S with translation group G and the corresponding group coset geometry S(G,T) with geometric partition $T = \{A_0, A_1, \ldots, A_t\}$ of G.

The kernel K of S(G,T) is by definition the set of all endomorphisms α of the group G with $A_i^\alpha \subseteq A_i$, $i \in J$. With the usual addition and multiplication (K,+,.) is a ring.

4.2. Theorem

(K,+,.) is a field. Moreover, if $\alpha \in K^\circ = K-\{0\}$, with 0 the null endomorphism of G, then $A_i^\alpha = A_i$.

Proof. We prove that every $\alpha \in K^\circ$ is an automorphism of G.

Suppose that $g^\alpha = 1$, $g \in A_i$, $g \neq 1$ and $i \in J$. For any $j \in J-\{i\}$ there is at least one $k \in J-V(i,j)$. If g' is an arbitrary element of A_j, $g' \neq 1$, then since $A_iA_j \cap A_k = \{1\}$, we have $gg' \notin A_k$. Moreover, there exists an element l of the subset W(k) of s+2 indices of J, $l \neq i$, $l \neq j$, such that $gg' = hh'$ with $h \in A_k$ and $h' \in A_l$. Consequently $g'^\alpha = h^\alpha h'^\alpha \in A_j \cap A_kA_l = \{1\}$ and so $A_j^\alpha = \{1\}$. Hence, for every $j \in J-\{i\}$, we have $A_j^\alpha = \{1\}$. Choose $j, j', j'' \in J-\{i\}$

such that $j'' \notin V(j,j')$. Then $G = A_jA_{j'}A_{j''}$. Hence $G^\alpha = A_j^\alpha A_{j'}^\alpha A_{j''}^\alpha = \{1\}$. There follows that $\mathrm{Ker}(\alpha) = G$. We conclude that α is the null endomorphism of G, a contradiction. Next, suppose that $g^\alpha = 1$ with $g \notin \bigcup_{i=0}^{t} A_i$. There exist elements $i,i',i'' \in J$ such that $G = A_iA_{i'}A_{i''}$, hence g can be written as $g = h_ih_{i'}h_{i''}$ with $h_i \in A_i$, $h_{i'} \in A_{i'}$, $h_{i''} \in A_{i''}$. Then $h_i^\alpha h_{i'}^\alpha h_{i''}^\alpha = 1$. Since $A_i \cap A_{i'}A_{i''} = \{1\}$ there follows that $h_i^\alpha = 1$. Analogously $h_{i'}^\alpha = h_{i''}^\alpha = 1$. Again a contradiction by the preceding paragraph. So any $\alpha \in K^\circ$ is an automorphism of G.
Now we prove that $A_i^\alpha = A_i$ for any $\alpha \in K^\circ = K - \{0\}$. Suppose $A_i^\alpha \subset A_i$, then there exist two different elements g and h of A_i such that $g^\alpha = h^\alpha$, or $(gh^{-1})^\alpha = 1$, in view of the foregoing a contradiction.
Let $\alpha \in K^\circ = K - \{0\}$. Then the automorphism α^{-1} of G satisfies $A_i^{\alpha^{-1}} = A_i$ for all $i \in J$. Consequently $\alpha^{-1} \in K$. Now the theorem is completely proved.

4.3. Remark

The multiplicative group K° of the kernel is isomorphic to a subgroup of the automorphism group H of the translation partial geometry $S(G,T)$ which fixes the identity 1 of G, all lines $A_i \in T$ through 1, and all spreads S_i, $i \in J$. Consider for this the following natural monomorphism ϕ of K° into H

$$\phi : K^\circ \to H,\ \alpha \to \alpha^*$$

with $g^{\alpha^*} = g^\alpha$, $\forall g \in G$, and $(A_ig)^{\alpha^*} = A_ig^\alpha$, $\forall i \in J$, $\forall g \in G$.
Using notations of theorem 3.3.5., $G^* = \{\theta_a \,\|\, a \in G\}$. If $\gamma^* \in H$, then γ^* induces a permutation γ of G with $A_i^\gamma = A_i$. We can prove that γ belongs to K°, i.e. $\phi(\gamma) = \gamma^*$, iff $\gamma^{*-1}G^*\gamma^* = G^*$. Indeed, if $\gamma^{*-1}G^*\gamma^* = G^*$ then $\gamma^{*-1}\theta_a\gamma^* = \theta_{(a^{\gamma^*})}$ since $1^{\gamma^{*-1}\theta_a\gamma^*} = a^{\gamma^*}$. If h and h' are two elements of G, then $(hh')^\gamma =$
$(hh')^{\gamma^*} = 1^{\theta_{(hh')^{\gamma^*}}} = 1^{\gamma^{*-1}\theta_h\gamma^*\gamma^{*-1}\theta_{h'}\gamma^*} = 1^{\theta_{h^{\gamma^*}}\theta_{h'^{\gamma^*}}} = 1^{\theta_{h^{\gamma^*}h'^{\gamma^*}}} = h^{\gamma^*}h'^{\gamma^*}$
$= h^\gamma h'^\gamma$. So γ is an endomorphism of G. Conversely, if $\gamma \in K^\circ$, then for any element h of G we have $h^{\gamma^{*-1}\theta_a\gamma^*} = (h^{\gamma^{*-1}}a)^{\gamma^*} = (h^{\gamma^{-1}}a)^\gamma = ha^\gamma = ha^{\gamma^*}$ so $\gamma^{*-1}\theta_a\gamma^* = \theta_{(a^{\gamma^*})}$. Hence $\gamma^{*-1}G^*\gamma^* = G^*$.

4.4. The vector space (G,F)

If K is the kernel of the translation partial geometry $S(G,T)$, then for any subfield F of K there exists a vector space (G,F), with the group G as the additive group of (G,F) and with the scalar multiplication defined by $g\alpha = g^\alpha$ $g \in G$, $\alpha \in F$.
The groups A_i of the geometric partition T are subspaces over F of (G,F), because for every element α of F we have $A_i\alpha = A_i^\alpha \subset A_i$. All groups A_i have the same order $s+1$, so that $[A_i:F] = n$ is independent of i. As there exist three distinct elements i,j and k of J, such that $G = A_iA_jA_k$, with $A_j \cap A_k = \{1\}$ and $A_i \cap A_jA_k = \{1\}$, we have $[G:F] = [A_i:F]+[A_j:F]+[A_k:F] = 3n$.

5. α-GEOMETRIC (n-1)-SPREADS OF PG(3n-1,q)

5.1. Definitions

A partial (n-1)-spread K in a projective space PG(3n-1,q), is a set of pairwise disjoint (n-1)-dimensional subspaces of PG(3n-1,q). A partial (n-1)-spread K is called geometric iff for every two distinct elements X and X' of K, the elements of K which have a point in common with $\langle X,X'\rangle$ are completely contained in $\langle X,X'\rangle$. A partial (n-1)-spread K is α-geometric iff it is geometric and for every two distinct elements X and X', the projective (2n-1)-dimensional space $\langle X,X'\rangle$ contains exactly $\alpha+1$ elements of K [3].

Consider an α-geometric partial (n-1)-spread K of PG(3n-1,q), with $|K| = m$. Let X be an element of K and let r be the number of (2n-1)-dimensional subspaces of PG(3n-1,q) through X which contain $\alpha+1$ elements of K. By counting in two ways the number of ordered pairs (L,X') with $X' \in K$, $X' \neq X$ and $\langle X,X'\rangle = L$ we find $r.\alpha = m-1$. Moreover r must be less than or equal to q^n+1. So $\alpha \mid m-1$ and $m \leq \alpha q^n+\alpha+1$.

5.2. Theorem

An α-geometric partial (n-1)-spread K of PG(3n-1,q) has $|K| = m = \alpha q^n+\alpha+1$ iff every n-dimensional projective subspace of PG(3n-1,q) through a given element of K intersects exactly α elements of K in exactly one point.

<u>Proof</u>. Suppose that $m = \alpha q^n+\alpha+1$ and $K = \{PG^{(i)}(n-1,q) \parallel i = 1,2,\ldots,m\}$. By the definition of α-geometric partial spread, through any $PG^{(i)}(n-1,q)$ we find exactly $(m-1)/\alpha = q^n+1$ (2n-1)-dimensional projective subspaces $PG^{(j)}(2n-1,q)$, $j \in \{1,2,\ldots,q^n+1\}$ containing an element of $K-\{PG^{(i)}(n-1,q)\}$ and these q^n+1 spaces have mutually only $PG^{(i)}(n-1,q)$ in common. Consider now the residual structure R in $PG^{(i)}(n-1,q)$ w.r.t. PG(3n-1,q). So, R is a (2n-1)-dimensional projective space, and with the q^n+1 projective spaces $PG^{(j)}(2n-1,q)$ of PG(3n-1,q) correspond exactly q^n+1 pairwise disjoint (n-1)-dimensional projective subspaces of R. In other words, we find a (n-1)-spread of R. With an n-dimensional projective subspace N of PG(3n-1,q) through $PG^{(i)}(n-1,q)$ corresponds a point of R. Consequently, there exists exactly one space $PG^{(k)}(2n-1,q)$, $k \in \{1,2,\ldots,q^n+1\}$ of PG(3n-1,q) through N. Hence each of the α elements of K distinct from $PG^{(i)}(n-1,q)$ contained in $PG^{(k)}(2n-1,q)$ intersect N in exactly one point.

Conversely, we suppose that every n-dimensional projective space through $PG^{(i)}(n-1,q) \in K$ intersects exactly α elements of K in exactly one point. We count in two ways the number of ordered pairs (N,k) with N an n-dimensional projective space through $PG^{(i)}(n-1,q)$, such that N has exactly one point in common with $PG^{(k)}(n-1,q)$, $k \in \{1,2,\ldots,m\}-\{i\}$. There are $\alpha(q^{2n}-1)/(q-1) =$

$(m-1)(q^n-1)/(q-1)$ such pairs, hence $m = \alpha(q^n+1)+1$.

5.3. Remark

A partial $(n-1)$-spread K of a projective space PG(d,q) such that every n-dimensional subspace of PG(d,q) containing an element of K intersects exactly α elements of K in exactly one point is defined in [3] to be an α-uniform partial spread.
If K is an α-geometric and α-uniform partial $(n-1)$-spread of PG(3n-1,q), then K is a particular SPG-regulus [24]. It is a SPG-regulus with no tangent spaces.
If $\alpha = 1$, then K is a (q^n+2)-set of PG(3n-1,q) and then q is even [18].

5.4. Definition

We define W to be the set of $(2n-1)$-dimensional projective subspaces of PG(3n-1,q) which contain exactly $\alpha+1$ elements of K. An element of W will be called an intersecting block of K.

5.5. Theorem

If K is an α-geometric and α-uniform partial $(n-1)$-spread of PG(3n-1,q) with $\alpha \leq q^n-1$, then the number of intersecting blocks of K through a point of PG(3n-1,q) not contained in an element of K, is equal to $q^n+1-(q^n/(\alpha+1))$.
Hence $\alpha+1$ is a divisor of q^n.
Proof. We construct a design $D = (K,W,\subset)$, with the elements of K as points, the blocks are the elements of W and incidence is the set-theoretic inclusion.
So D is a $2-(q^n\alpha+\alpha+1,\alpha+1,1)$ design, hence $|W| = (q^n\alpha+\alpha+1)(q^n+1)/(\alpha+1) = a$.
If $\alpha+1 = q^n+1$, then D is a $2-(q^{2n}+q^n+1,q^n+1,1)$ design i.e. a projective plane of order q^n.
In the case $\alpha+1 = q^n$, D is an affine plane of order q^n.
Let $\alpha+1 \leq q^n$. Call V the set of points of PG(3n-1,q) which are not contained in an element of K. Then $d = |V| = (q^n-1)(q^n+1)(q^n-\alpha)/(q-1)$.
Let x_i be an element of V and let t_i be the number of elements of W through x_i. First of all we count the number of ordered pairs (x_i,H), $x_i \in V$, $x_i \in H$ and $H \in W$, and we obtain :

$$\sum_i t_i = a(q^n-1)(q^n-\alpha)/(q-1).$$

Next we count in different ways the ordered triples (x_i,H,H'), $x_i \in V$, $x_i \in H' \cap H$ with $H,H' \in W$ and $H \neq H'$, and we obtain :

$$\sum_i t_i(t_i-1) = a(a-1-q^n(\alpha+1))(q^n-1)/(q-1).$$

Hence $\sum_i t_i^2 = a(q^n-1)m(q^n-\alpha)/((q-1)(\alpha+1))$ and as $d\sum_i (t_i-\bar{t})^2 = d\sum_i t_i^2-(\sum_i t_i)^2 = 0$, we get that t_i is a constant, $t_i = \bar{t} = q^n+1-(q^n/(\alpha+1))$. This proves the theorem.

Remark that for n = 1 we obtain the classical result for maximal arcs in projective planes [6], [15].

5.6. The partial geometry $S_{3n-1}(K)$

Throughout this section we use the notations and terminology of theorem 5.5. Let $K = \{X_1, X_2, \ldots, X_m\}$ with $m = q^n\alpha+\alpha+1$.
Consider a (2n-1)-dimensional projective subspace S of PG(3n-1,q), skew to a certain X_i, $i \in \{1,2,\ldots,m\}$. Let S_i be the set of intersections of S with q^n+1 intersecting blocks of W through X_i. It is trivial that S_i is a (n-1)-spread of S and we will refer to S_i as the spread of S arising from the projection of K onto S from X_i [5]. We say that a (2n-1)-dimensional projective subspace S of PG(3n-1,q) is projection-stable w.r.t. K iff for every two elements X_i and X_j of K, both skew to S, there holds $S_i = S_j$.
We now define a relation R on the point-set V = PG(3n-1,q)-K : xRy iff $\{H \parallel H \in W \text{ and } x \in H\} = \{H \parallel H \in W \text{ and } y \in H\}$. It is clear that R is an equivalence relation and that every equivalence class C_x determined by $x \in V$ is a projective subspace of PG(3n-1,q) with dimension less than or equal to n-1. Suppose the following axiom holds.

Axiom (W)

Every intersecting block of K is projection-stable w.r.t. K. □
Then we show that every equivalence class of R is an (n-1)-dimensional projective subspace of PG(3n-1,q). Consider two distinct intersecting blocks H and H' of K through a point x of V. Then, $N = H \cap H'$ is an element of the (n-1)-spread of H, obtained by projection of K onto H. Let X_i be an element of K disjoint from H and H'. By axiom (W), N must be an element of the spread of H, obtained by projection of K onto H from X_i, and so the intersecting block of K through x and X_i intersects H in N. Hence the equivalence class C_x determined by x must be N.
Let C be the set of equivalence classes of R. By the previous results, there follows that $S^* = C \cup K$ is a (n-1)-spread of PG(3n-1,q). Moreover, through every element of C there are exactly $q^n+1-(q^n/(\alpha+1))$ intersecting blocks of K. The incidence structure $S_{3n-1}(K) = (C,W,\subset)$ with as points the elements of C, as lines the intersecting blocks of K and as incidence the set-theoretic inclusion is a partial geometry with parameters $s = q^n-\alpha-1$, $t = q^n-(q^n/(\alpha+1))$ and intersection number $q^n-\alpha-(q^n/(\alpha+1))$. For this, we remark that by axiom (W) every two distinct intersecting blocks of K intersect each other in an element of the spread S^* of PG(3n-1,q).
If n = 1, then this partial geometry is S(K), the geometry introduced by J.A. Thas [19] and W.D. Wallis [27].

6. THE PARTIAL GEOMETRY $T^*_{3n-1}(K)$

6.1. Definition

Let K be an α-geometric and α-uniform partial (n-1)-spread of PG(3n-1,q). Now we embed $H^* = PG(3n-1,q)$ as a hyperplane in $P^* = PG(3n,q)$. We construct a point-line geometry $T^*_{3n-1}(K)$ as follows : the points of $T^*_{3n-1}(K)$ are the points of P^*-H^*, the lines of $T^*_{3n-1}(K)$ are the n-dimensional subspaces of P^*, not contained in H^* and intersecting H^* in an element of K, and incidence is that of P^*. We refer to the paper [24] of J.A. Thas for the proof that $T^*_{3n-1}(K)$ is a partial geometry with parameters $s = q^n-1$, $t = \alpha(s+2)$ and α.

6.2. $T^*_{3n-1}(K)$ and the translation partial geometries S(G,T)

6.2.1. Theorem

$T^*_{3n-1}(K)$ is a translation partial geometry with translation group G isomorphic to the group of all translations of the affine space $AG(3n,q) = P^*-H^*$. Moreover, $F = GF(q)$ is a subfield of the kernel of $T^*_{3n-1}(K)$.

Proof. First notice that $t = \alpha(s+2)$.

The group of all translations of the affine space $AG(3n,q) = P^*-H^*$ induces a point-regular automorphism group G of $T^*_{3n-1}(K)$. If X_i is an element of K and if L is a line of $T^*_{3n-1}(K)$ which contains X_i, then the line orbit LG of L is the spread S_i of $T^*_{3n-1}(K)$ consisting of all lines of $T^*_{3n-1}(K)$ which contain X_i. Now we prove that every spread S_i is a normal spread of $T^*_{3n-1}(K)$. Let L_1 and L_2 be two distinct lines of $T^*_{3n-1}(K)$, which contain X_i. The (n+1)-dimensional projective subspace of P^* joining L_1 and L_2, intersects H^* in an n-dimensional projective space N. This projective space N is contained in an intersecting block PG(2n-1,q) of K. Denote the elements of K contained in PG(2n-1,q) by $X_i, X_{i_1}, X_{i_2}, \ldots, X_{i_\alpha}$. Let PG(2n,q) be the projective space joining PG(2n-1,q), L_1 and L_2. Then $\{L_1,L_2\}^{\perp\perp}$ is the set of the q^n lines of $T^*_{3n-1}(K)$ contained in PG(2n,q) and intersecting H^* in X_i. In fact, the partial subgeometry of $T^*_{3n-1}(K)$ with as points, the points of $PG(2n,q)-H^*$ and as lines, the lines of $T^*_{3n-1}(K)$ contained in PG(2n,q), is the net $\mathcal{N}(L_1,L_2)$ of order q^n and degree $\alpha+1$. The lines of $T^*_{3n-1}(K)$ contained in PG(2n,q) and intersecting H^* in a fixed element X_j of K, $j \in \{i,i_1,i_2,\ldots,i_\alpha\}$, define a parallel class of lines of $\mathcal{N}(L_1,L_2)$. So each spread S_i of $T^*_{3n-1}(K)$ is normal.

The group of all homologies of P^* with center $o \notin H^*$ and axis H^* induces a subgroup of the automorphism group of $T^*_{3n-1}(K)$ which fixes o, fixes all lines of $T^*_{3n-1}(K)$ through o, and fixes the spreads S_i. Since this group of dilatations of AG(3n,q) with center $o \notin H^*$ normalizes the group of translations of AG(3n,q), by remark 4.3., we have that it is isomorphic to a subgroup of the multiplicative group of the kernel $T^*_{3n-1}(K)$.

Let us introduce an affine coordinate system for $AG(3n,q)$ such that o is the point $(0,0,\dots,0)$. The translation of $AG(3n,q)$ which maps $(0,0,\dots,0)$ onto $(t_1,t_2,\dots,t_{3n})$ will be denoted by $\tau(t_1,t_2,\dots,t_{3n})$. We define η_k, $k \in GF(q)$, to be the following mapping on the translations of $AG(3n,q)$: $(\tau(t_1,t_2,\dots,t_{3n}))^{\eta_k} = \tau(t_1k,t_2k,\dots,t_{3n}k)$. By the preceding paragraph η_k belongs to the kernel of $T^*_{3n-1}(K)$. Moreover if $k = 0 \in GF(q)$, then η_k is the null endomorphism of the translation group of $AG(3n,q)$. Clearly $\eta_{kk'} = \eta_k\eta_{k'}$ and $\eta_{k+k'} = \eta_k+\eta_{k'}$, with $k,k' \in GF(q)$. This completely proves that $GF(q)$ is isomorphic to a subfield of the kernel of $T^*_{3n-1}(K)$.

6.2.2. Theorem

Let S be a translation partial geometry with translation group G and parameters $s,t = \alpha(s+2)$ and α. If $F = GF(q)$ is a subfield of the kernel of S, then S is isomorphic to a partial geometry $T^*_{3n-1}(K)$ with $[G:F] = 3n$.

Proof. Consider the translation partial geometry $S(G,T)$ defined by S, with $T = \{A_i \| i \in J\}$ the geometric partition of G. The vector space (A_i,F) is a subspace of (G,F) (cfr. 4.4.), so $|A_i| = s+1 = q^n$, $n \geq 1$, $i \in J$. With the 3n-dimensional vector space (G,F) corresponds a 3n-dimensional affine space $AG(3n,q)$. Let $PG(3n,q)$ be the projective closure of $AG(3n,q)$, and let $H^* = PG(3n,q)-AG(3n,q)$. With the q^{2n} cosets A_ig of A_i, $g \in G$, correspond the q^{2n} n-dimensional projective subspaces of $PG(3n,q)$ which intersect H^* in a fixed (n-1)-dimensional projective space $PG^{(i)}(n-1,q)$.

Let $K = \{PG^{(i)}(n-1,q) \| i \in J\}$. We prove that K is an α-geometric and α-uniform (n-1)-spread of H^*. Let A_i and A_j be two distinct elements of T. Since $A_i \cap A_j = \{1\}$ we have that $PG^{(i)}(n-1,q) \cap PG^{(j)}(n-1,q) = \phi$. Also $|K| = t+1 = \alpha q^n+\alpha+1$. By condition (1) of 3.3.1., there exists a subset $V(i,j)$ of $\alpha+1$ elements of J, such that the n-dimensional subspaces A_k of $AG(3n,q)$, $k \in V(i,j)$, are subspaces of the 2n-dimensional subspace of $AG(3n,q)$ joining A_i and A_j. By 3.3.2. $\langle PG^{(i)}(n-1,q), PG^{(j)}(n-1,q)\rangle$ contains exactly $\alpha+1$ elements of K and by condition (2) of 3.3.1., i.e. $A_iA_j \cap A_m = \{1\}$, $\forall m \in J-V(i,j)$ these $\alpha+1$ elements of K are the only elements of K which have at least one point in common with $\langle PG^{(i)}(n-1,q), PG^{(j)}(n-1,q)\rangle$.

This essentially proves that S is isomorphic to a partial geometry $T^*_{3n-1}(K)$.

6.3. A characterization of $T^*_2(K)$

Theorem

If the α-geometric and α-uniform partial (n-1)-spread K of H^* is embeddable in an (n-1)-spread S^* of H^*, such that S^* induces an (n-1)-spread in each intersecting block of K, then $T^*_{3n-1}(K)$ is isomorphic to a $T^*_2(K')$.

Proof. Through every point x of H^*, not contained in an element of K, there are exactly $q^n+1-(q^n/(\alpha+1)) = (t+1)/(\alpha+1)$ intersecting blocks of K (theorem 5.5.). If X is the element of S* containing x, then by assumption all these intersecting blocks through x have mutually X in common. In fact, axiom (W) is satisfied for K.

Since every two distinct intersecting lines of $T^*_{3n-1}(K)$ define in a natural way a net of order s+1 and degree $\alpha+1$, and the same fact is true for two distinct lines of $T^*_{3n-1}(K)$ belonging to a same line orbit of G (such lines are called parallel in [10]), there follows that $T^*_{3n-1}(K)$ is a net-inducible partial geometry, in the sense of [10].

Let (x,y,z) be a triad of points of $T^*_{3n-1}(K)$, i.e. $x \not\sim y \not\sim z \not\sim x$. We consider now the number β of nets of order s+1 and degree $\alpha+1$ through x, y and z. Suppose first of all that x,y and z are incident with a line L of P*. This line intersects an element X of S*, $X \notin K$. Every net through x,y and z must contain X. So there are $\beta = (t+1)/(\alpha+1)$ nets through x,y and z.

Next, let us suppose that x,y and z span a plane γ of P*. Let $L = \gamma \cap H^*$. If L is completely contained in an element X of S*, $X \notin K$, then there arises again the previous case. If L is not completely contained in an element of S* then according to L is contained in an intersecting block of K or not, $\beta = 1$ or $\beta = 0$.

So β is 0,1 or $(t+1)/(\alpha+1)$. Now by a theorem of [10] , $T^*_{3n-1}(K)$ is isomorphic to a $T^*_2(K')$.

Corollaries

1. If for an α-geometric and α-uniform partial (n-1)-spread of PG(3n-1,q) axiom (W) holds, then the partial (n-1)-spread is embeddable in a regular (n-1)-spread [11] S* of PG(3n-1,q).

Proof. Let K be an α-geometric and α-uniform partial (n-1)-spread of PG(3n-1,q), $n > 1$, satisfying axiom (W). As in 5.6. we can construct an (n-1)-spread S* of PG(3n-1,q) which contains K. Clearly K satisfies the conditions of the previous theorem, so $T^*_{3n-1}(K)$ is isomorphic to a $T^*_2(K')$.

Before proving that S* is geometric, let us introduce some notations concerning $T^*_{3n-1}(K) = (P,B,I)$ and $T^*_2(K') = (P',B',I')$. Remind that K' is a maximal arc of degree $\alpha+1$ of $PG(2,q^n)$, and that $PG(2,q^n)$ is embedded in a $PG(3,q^n)$ such that the points of $AG(3,q^n) = PG(3,q^n)-PG(2,q^n)$ are exactly the points of $T^*_2(K')$. Let A (resp. A') be the set of n-dimensional subspaces of AG(3n,q) (resp. lines of $AG(3,q^n)$) intersecting PG(3n-1,q) (resp. $PG(2,q^n)$) in an element of S* not in K (resp. in a point not in K'). Finally Γ is the isomorphism of $T^*_{3n-1}(K)$ onto $T^*_2(K')$.

Let Z be an element of A, intersecting PG(3n-1,q) in an element N of S*. Consider an intersecting block M of K through N. The lines of $T^*_{3n-1}(K)$ contained in the 2n-dimensional subspace <M,Z> of PG(3n,q) will be mapped by Γ onto lines of $T_2^*(K')$. If $\alpha+1 > 2$, then it is clear that these lines lie in an affine plane π of $AG(3,q^n)$. Now let $\alpha = 1$, and assume that the mentioned $2q^n$ lines of $T_2^*(K')$ do not lie in a common affine plane. Then the lines are generators of a hyperbolic quadric of $PG(3,q^n)$ and intersect $PG(2,q^n)$ in $2q^n$ distinct points. Since these $2q^n$ points belong to a complete oval, we have a contradiction. Hence also for $\alpha = 1$ the $2q^n$ lines of $T_2^*(K')$ lie in common affine plane π of $AG(3,q^n)$. So the q^n points of $T_{3n-1}^*(K)$ contained in Z correspond under Γ with q^n points of $T_2^*(K')$ contained in π. If M' is another intersecting block of K through N, then in the same way we get an affine plane π', $\pi' \neq \pi$. Since $|\pi \cap \pi'| = q^n$, the q^n points of $T_{3n-1}^*(K)$ contained in Z will be mapped by Γ onto the q^n points of $T_2^*(K')$ in $\pi \cap \pi'$. Thanks to the previous result, the incidence structure $D = (P^*,B^*,I^*)$ with $P^* = P$, $B^* = B \cup A$, and $p\ I^*\ L$ iff $p \in L$, is a 3-dimensional affine space of order q^n. Hence, D possesses a parallelism among the blocks. We will prove now that two distinct blocks L and L' of B^* are parallel iff they determine at infinity the same element of S^*. Let N resp. N' be the element of S^* at infinity of L resp. L'. Consider an affine plane π (resp. π', $\pi \neq \pi'$) of D through L such that π (resp. π') is parallel to L' and such that the (2n-1)-dimensional subspace of $PG(3n-1,q)$ M (resp. M') at infinity of π (resp. π') is an intersecting block of K. Since L' is parallel to π, we have $M \cap N' \neq \phi$, so N' is contained in M. Analogously N' is contained in M'. So N' = N.

Consider two elements Z and Z' of A having just one point of $T_{3n-1}^*(K)$ in common. Let N resp. N' be the element of S^* at infinity of Z resp. Z'. Then Z and Z' determine an affine plane π of D. Let Y be a line of that plane and suppose that the element $N'' \in S^*$ at infinity of Y is not completely contained in $\langle N,N' \rangle$. Then there exists a line Y' of D having N" as element at infinity and intersecting Z' but not Z. This implies that Y' and Z are parallel in the affine plane π, and so N = N", a contradiction. Hence the q^n+1 points at infinity of π correspond to q^n+1 elements of S^* contained in $\langle N,N' \rangle$. In other words, S^* is geometric , i.e. regular.

2. Any partial geometry $S_{3n-1}(K)$ is isomorphic to a $S_2(K') = S(K')$, where K' is a maximal arc of a Desarguesian plane.

Proof. Immediately from Corollary 1.

REFERENCES

[1] R.W. Ahrens, G. Szekeres, On a combinatorial generalization of 27 lines associated with a cubic surface. J. Austr. Math. Soc. 10 (1969) 485-492.

[2] L. Berardi and F. Eugeni, On generalized partial geometries constructed by uniform spreads, preprint.

[3] A. Beutelspacher, Partial spreads in finite projective spaces and partial designs, Math. Z. 145 (1975) 211-229.

[4] R.C. Bose, Strongly regular graphs, partial geometries and partial balanced designs. Pacific J. Math. 13 (1963) 389-419.

[5] L.R.A. Casse, J.A. Thas and P.R. Wild, (q^n+1)-sets of PG(3n-1,q), Generalized Quadrangles and Laguerre Planes. Simon Stevin 59 (1985) 19-40.

[6] A. Cossu, Su alcune proprietà dei {k;n}-archi di un piano proiettivo sopra un corpo finito. Rend. Mat. e Appl. 20 (1961) 271-277.

[7] F. De Clerck, Een kombinatorische studie van de eindige partiële meetkunden. Dissertation Ph.D., Univ. of Ghent (1978).

[8] F. De Clerck, R.H. Dye, J.A. Thas, An infinite class of partial geometries associated with the hyperbolic quadric in PG(4n-1,q). European J. of Comb. 1 (1980) 323-326.

[9] F. De Clerck, Substructures of partial geometries. Seminario di Geometrie, Quaderno 5 (1984) Università degli Studi de L'Aquila, 1-16.

[10] F. De Clerck, M. De Soete and H. Gevaert, A characterization of the partial geometry $T_2^*(K)$, to appear in European J. of Comb.

[11] P. Dembowski, Finite geometries, Springer-Verlag, Berlin, Heidelberg New York (1981).

[12] R.H.F. Denniston, Some maximal arcs in finite projective planes, J. of Comb. Th. 6 (1969) 317-319.

[13] W. Haemers, A new partial geometry constructed from the Hoffman-Singleton graph. In : Finite geometries and designs (ed. P.J. Cameron, J.W.P. Hirschfeld, D.R. Hughes), London Math. Soc., Lect. Notes Series, 49 (1981) 119-127.

[14] M. Hall Jr., Affine generalized quadrilaterals. Studies in Pure Math. (ed. L. Mirsky), Academic Press (1971) 113-116.

[15] J.W.P. Hirschfeld, Projective Geometries over Finite Fields. Clarendon Press - Oxford (1979).

[16] S.E. Payne, J.A. Thas, Finite generalized quadrangles. Research Notes in mathematics 110, Pitman Adv. Publ. Progr. (1984).

[17] A.P. Sprague, Translation nets, Mitt. aus dem math. Sem. Giessen, 157 (1982) 46-68.

[18] J.A. Thas, The m-dimensional projective space $S_m(M_n(GF(q)))$ over the total matrix algebra $M_n(GF(q))$ of the nxn-matrices with elements in the Galois field GF(q), Rend. di mat. (6) 4 (1971) 459-532.

[19] J.A. Thas, Construction of partial geometries. Simon Stevin 46 (1973) 95-98.

[20] J.A. Thas, Construction of maximal arcs and partial geometries. Geom. Dedicata 3 (1974) 61-64.

[21] J.A. Thas, Some results concerning {(q+1)(n-1);n}-arcs and {(q+1)(n-1)+1;n}-arcs in finite projective planes of order q. J. Comb. Th. 19 (1975) 228-232.

[22] J.A. Thas, Construction of maximal arcs and dual ovals in translation planes. European J. of Comb. 1 (1980) 189-192.

[23] J.A. Thas, Some results on quadrics and a new class of partial geometries. Simon Stevin 55 (1981) 129-139.

[24] J.A. Thas, Semi-partial geometries and spreads of classical polar spaces J. Comb. Th. A35 (1983) 58-66.

[25] J. Tits, Sur la trialité et certains groupes qui s'endéduisent. Publ. Math. I.H.E.S. Paris 2 (1959) 14-60.

[26] J.H. Van Lint, A. Schrijver, Construction of strongly regular graphs, two-weight codes and partial geometries by finite fields. Combinatorica 1 (1984) 63-73.

[27] W.D. Wallis, Configurations arising from maximal arcs. J. Comb. Th. A15 (1973) 115-119.

Annals of Discrete Mathematics 37 (1988) 137–146

ON ADMISSIBLE SETS WITH TWO INTERSECTION NUMBERS IN A PROJECTIVE PLANE

Marialuisa J. de RESMINI

Dipartimento di Matematica
Università di Roma "La Sapienza"
I-00185 Rome, Italy

A set K of points in a finite projective plane π has two intersection numbers, m and n, if $|K \cap \ell| = m$ or n for any line ℓ in π. When $k=|K|$, m and n satisfy all necessary arithmetic conditions for K to exist, K is an admissible set and (k;m,n) an admissible triple. An infinite class of admissible triples is exhibited in planes of order $q=a^{2s+1}$, $(2s+1,3) = 1$, $a \geq 2$, and a general approach is provided to find all admissible triples which, in particular, recovers the known results in square order planes.

1. INTRODUCTION

Let π be a projective plane of order q, where q is any integer not ruled out by Bruck-Ryser theorem. A set of points in π, say K, has two intersection numbers m and n if $|K \cap \ell| = m$ or n for any line ℓ in π. K is said to be of type (m,n) and (k;m,n), where $k = |K|$, is an admissible triple whenever all necessary conditions are satisfied for K to exist [10,13]. Under these assumptions, K is an admissible set. Infinite families of admissible triples are known in square order planes and sets were constructed with those parameters [2,3,4,5,6,7,8,9]. Furthermore, under additional assumptions, sets of type (m,n), $m \geq 2$, are just possible in square order planes [12]. On the other hand, only a few sporadic admissible triples are known in non-square order planes [1,8,12], but no set with those parameters has been constructed.

Here an infinite family of admissible triples is shown to exist in planes of order $q = a^{2s+1}$, $(2s+1,3) = 1$, $a \geq 2$ an integer. Moreover, a general procedure is given which enables to find admissible triples in any plane and the construction of some of them is provided with details.

Obviously, whenever there is an admissible triple (k;m,n), both the complement and the dual sets of such an admissible set provide other admissible triples. The latter ones will always be understood. Recall that the dual sets of a set of type (m,n) are defined as the sets of m-secants and n-secants, respectively, in the dual plane [13].

2. PRELIMINARIES

Since a set of type (0,n) is a maximal arc and sets of type (1,n) were completely characterized, from the arithmetical point of view, in [12], we are mainly concerned with intersection numbers m and n satisfying $2 \leq m < n \leq q-1$. It is well known [10,13] that for a set of type (m,n) to exist n-m must be a proper divisor of q and its size k must be a root of

$$x^2-x((n+m)(q+1)-q)+mn(q^2+q+1) = 0. \tag{2.1}$$

Therefore, for K to exist, the number of points of π, q^2+q+1, must split. This trivial observation suggests to look for divisors, in particular for prime divisors, of q^2+q+1.
When $q = a^2$, a any integer ≥ 2, the natural splitting $q^2+q+1 = (a^2+a+1)(a^2-a+1)$ leads to the following admissible triples

$$(k;m,n) = (m(a^2+a+1);\ m,\ m+a), \tag{2.2}$$

$$(k;m,n) = ((m+a)(a^2-a+1);\ m,\ m+a). \tag{2.3}$$

Sets with parameters (2.2) were constructed, for any m, both in the Desarguesian [3] and in the Hughes planes [6,8]. On the other hand, to the author's best knowledge, not many sets are known with parameters (2.3); namely, an infinite family in Desarguesian planes whose order is a fourth power [5], an infinite family in planes of order a^2 containing a subplane of order a-1 (e.g. the non-Desarguessian planes of order nine) [4]; two sporadic examples corresponding to the triples (35;2,5) in the Desarguesian plane of order nine [7], and (78;2,6) in the Desarguesian plane of order 16 [9]. Some constructions of sets with parameters (2.3) were proposed in [3].
In sect. 3 the admissible triples coming from the splitting of q^2+q+1 when $q = a^{2s+1}$, $(2s+1,3) = 1$, will be discussed.
The next result is well known to number theorists (e.g. see [11], thm. 88).

LEMMA 2.4. Assume p is a prime divisor of q^2+q+1, q any positive integer, in particular the order of a plane. Then either $p = 3$ or $p \equiv 1\ (6)$.

PROOF. Of course, $3|q^2+q+1$ iff $q \equiv 1\ (3)$. Thus, assume $p > 3$, p a prime, and $p|q^2+q+1$. Reduce mod. p, $q \equiv q_p\ (p)$. Then $p|q_p^2+q_p+1$, i.e. q_p is a root of $x^2+x+1 = 0$ over GF(p). Recall that $x^2+x+1 = 0$ has two distinct roots in GF(p) iff -3 is a square in GF(p) iff $(p-1,3) = 3$ [10]. Since $(p-1,3) = 3$ implies $p \equiv 1\ (3)$, the statement is proved. □

COROLLARY 2.5. Assume q is the order of a projective plane and $p > 3$ a prime. Then $p|q^2+q+1$ iff $q \equiv q_p$ or $q_p^2\ (p)$, where q_p and q_p^2 are the two roots of $x^2+x+1=$ $=0$ over GF(p). □

Notice that q_p is the smallest order for a plane whose number of points is divided by p; in particular, this number of points is p or 3p.

Obviously, 3^2 is never a divisor of q^2+q+1. On the other hand, Lemma 2.4 and Cor. 2.5, taking into account the previous remark, allow to prove (by calculations) the following result.

COROLLARY 2.6. Suppose $p > 3$ is a prime and $p^2 | q^2+q+1$. Then one of the following holds.

(i) $q = py+q_p$ and $(2q_p+1)y+1 \equiv 0$ or $-2(p)$ according to $q_p^2+q_p+1 = p$ or $3p$;

(ii) $q = py+q_p^2$ and $(2q_p+1)y+2q_p \equiv 0$ or $-4q_p(p)$ according to $q_p^2+q_p+1 = p$ or $3p$.

Conversely, if either (i) or (ii) holds, then $p^2|q^2+q+1$. □

Of course, cor. 2.6 can be generalized requiring that p^s, $s \geq 3$, divides the number of points of a plane.
Notice that, whenever $p^2|q^2+q+1$, either $3|q^2+q+1$ or another non-squared prime, which may be the same p, is a divisor of q^2+q+1.
Combining lemma 2.4 and cor. 2.5, by the same argument that gives cor. 2.6, allows to find the expression of q under the requirement $p \cdot p'|q^2+q+1$, p and p' distinct primes.
Finally, again by lemma 2.4 and cor. 2.5, all (prime) divisors of q^2+q+1 have one of the following forms:

$$3(3y^2+3y+1), \qquad q_p = 3y+1$$

$$9y^2+3y+1, \qquad q_p = 3y$$

$$9y^2+15y+7, \qquad q_p = 3y+2.$$

4. ADMISSIBLE TRIPLES IN PLANES OF ORDER $q = a^{2s+1}$, $(2s+1,3) = 1$

Suppose π has order $q = a^{2s+1}$, $a \geq 2$ any integer and $(2s+1,3) = 1$. Under these assumptions, it is well known that a^2+a+1 divides q^2+q+1 [10,11].

PROPOSITION 3.1. Let π be as above. Then admissible triples (k;m,n) are provided by

$$m = \frac{(a^s-1)(a^{s+1}-1)}{a^2+a+1}, \qquad n = m+a^s,$$

and either $k = mn(a^2+a+1)$ or $\bar{k} = (a^{4s+2}+a^{2s+1}+1)/(a^2+a+1)$.

PROOF. Since $a^2+a+1|q^2+q+1$, we look for sets having sizes $k=mn(a^2+a+1)$ and $\bar{k} = (a^{4s+2}+a^{2s+1}+1)/(a^2+a+1)$, respectively. Since $n-m|q$, assume $n-m = a^s$. Recall that k and $\bar{k}$ must be roots of eq. (2.1). Thus, $k+\bar{k} = (2m+a^s)(a^{2s+1}+1)-a^{2s+1}$. Whence, with the considered values for k and $\bar{k}$ and $n-m = a^s$,

$$m^2(a^2+a+1)-m(2a^{2s+1}+2-a^s(a^2+a+1))$$
$$+(a^{4s+2}+a^{2s+1}+1)/(a^2+a+1)-a^s(a^{2s+1}+1)+a^{2s+1} = 0,$$

whose discriminant is $\Delta = a^{2s}(a^2-a-1)^2$. Therefore, either $m=(a^{2s+1}+1-a^{s+2})/(a^2+a+1)$ which is never an integer, or $m = (a^s-1)(a^{s+1}-1)/(a^2+a+1)$ which is always an integer. Indeed, when $s \equiv 0$ or 3 (6), $a^3-1|a^s-1$ as $3|s$; when $s \equiv 2$ or 5 (6), then $a^3-1|a^{s+1}-1$ as $3|s+1$. The assumption $(2s+1,3) = 1$ rules out $s \equiv 1$ or 4 (6). □

Notice that for the admissible triples $(k^*;m^*,n^*)$ provided by the dual sets of the set in prop. 3.1 $n^*-m^* = a^{s+1}$.

REMARKS. For no triple in prop. 3.1 $m = 2$. On the other hand, $m = 3$ iff $a = 2$ and $s = 2$, i.e. $q = 32$. In this case, $n = 7$, $k = 147$ and $\bar{k} = 151$. The next smallest values for m are $m = 15$ $(a = 2, s = 3)$ and $m = 16$ $(a = 3, s = 2)$. The admissible triples provided by such smallest values for a and s were found by computer in [1] with no hint of their being the first ones in an infinite family.

The existence of sets with the parameters in prop. 3.1 is an open problem. However, a non-existence observation is the following one. Assume π is the Desarguesian plane of order 32. Then the orbit of a point under σ^7, σ the Singer cycle of π, has length 151. On the other hand, such an orbit has more than two intersection numbers.

4. ADMISSIBLE TRIPLES IN PLANES OF ORDER $q \equiv 1$ (3)

Recall that $3|q^2+q+1$ iff $q \equiv 1$ (3); hence, $q = 3y+1$. If $q^2+q+1 = 3p$, p a prime, then the only possible sets in π of type (m,n) have sizes

$$(4.1) \qquad k = 3mn \quad \text{and} \quad \bar{k} = p = 3y^2+3y+1.$$

PROPOSITION 4.2. For sets of sizes (4.1) to exist in π, q must be a square.

PROOF. Since $n-m|q$, set $n-m = a$; so that $a|3y+1$ implies $3y+1 = a\gamma$. Assume π contains sets of sizes (4.1). Taking into account eq. (2.1),

$$(4.3) \qquad (n+m)(q+1)-q = 3mn+3y^2+3y+1.$$

Since $n-m = a$ and $q = a\gamma = 3y+1$, (4.3) becomes

$$(4.4) \qquad 9m^2-3m(2a\gamma+2-3a)+a^2\gamma^2-3a^2\gamma+4a\gamma-3a+1 = 0,$$

and its discriminant

$$\Delta/9 = a(9a-8\gamma)$$

must be a square.

Next, recall that the n-secants of K form a set of type (m^*,n^*) and size k^*= no. of n-secants in the dual plane, where m^* and n^* are the numbers of n-secants on a point off K and on K, respectively [13]. Thus, $m^*=m(3m+3a-a\gamma-1)/a$, $n^* = m(3m+3a-a\gamma-1)/a+\gamma = m^*+\gamma$. Obviously, $nk^* = 3mnn^*$, i.e. $k^* = 3mn^*$. Writing eq. (2.1) with m^* and n^* instead of m and n and calling k^* and $\bar{k}^*$ its roots, $k^*\bar{k}^* = 3m^*n^*p$. Set $\beta = (3m+3a-a\gamma-1)/a$, then $m^* = m\beta$, $n^* = m\beta+\gamma$, $k^*\bar{k}^* = 3mn^*\beta p$, implying $\bar{k}^* = \beta p$. Since $\bar{K}^*$ is a proper subset of π, $\beta < 3$. Assume $\beta = 2$. Then $m = y-(a-2)/3$. Hence, $a \equiv 2$ (3). Write $a = 3b+2$, so that $m = y-b$, $n = y+2b+2$. Consequently, $m = (2\gamma-1)/3+b\gamma-b$, which implies $2\gamma-1 \equiv 0$ (3), i.e. $\gamma = 3c+2$. Therefore, $m = 3bc+2c+b+1$. Since m must satisfy (4.4), substituting, $(b-c)(3b+2) = 0$, whence $b = c$. Hence, $q = a\gamma = (3b+2)^2$. Solving eq. (4.4) with $a = \gamma$,

$$(4.5) \qquad m = (a^2+1-a)/3 \quad \text{and} \quad m = (a-1)^2/3,$$

and just the former is an integer when $a = 3b+2$.
Next, assume $\beta = 1$, which implies $m = (a\gamma-2a+1)/3$. Since m must satisfy eq. (4.4), substituting, $2a(\gamma-a) = 0$. Hence, $\gamma = a$. On the other hand, $\beta = 1$ implies $m = y-2(a-1)/3$, so that $a \equiv 1$ (3), i.e. $q = (3b+1)^2$ and m takes the second possible value in (4.5). □

REMARK. In the proof to prop. 4.2 no use has been made of $(q^2+q+1)/3$ being a prime. When this is the case, the sets with sizes (4.1) are the only possible ones of type (m,n) and m is given by (4.5), whereas $n = m+a$. On the other hand, whenever $3|q^2+q+1$ one can look for sets of sizes $k = 3mn$ and $\bar{k} = (q^2+q+1)/3$ and the argument that proves prop. 4.2 proves the next more general result too.

PROPOSITION 4.6. Assume $q \equiv 1$ (3). Then for π to contain sets of type (m,n) with sizes $k = 3mn$ and $\bar{k} = (q^2+q+1)/3$, q must be a square. Moreover, $q=(3b+1)^2$ implies $m = 3b^2$, $n = 3b^2+3b+1$, and $q = (3b+2)^2$ implies $m = 3b^2+3b+1$, $n=3b^2+6b+3$. □

Finally, we observe that, whenever π is a cyclic plane of order $q \equiv 1$ (3), prop. 4.6 recovers a result in [2]. Examples of sets with the parameters in prop. 4.6 can be found in [5].

5. SETS IN PLANES OF ORDER q SUCH THAT $p|q^2+q+1$, $p > 3$ A FIXED PRIME

Recall (cor. 2.5) that $p|q^2+q+1$ iff $q \equiv y$ or y^2(p), where y and y^2 are the two roots of $x^2+x+1 = 0$ over GF(p). To generalize the results in sect. 4, we look for sets of type (m,n) and sizes

$$(5.1) \qquad k = mnp \quad \text{and} \quad \bar{k} = (q^2+q+1)/p.$$

Therefore, since $n-m|q$, writing $n-m = a$ and $q = a\gamma$, by eq. (2.1),

$$(n+m)(q+1)-q = mnp+(q^2+q+1)/p,$$

which implies

$$(5.2) \qquad m^2p^2-mp(2a\gamma+2-ap)+a^2\gamma^2+a\gamma+1+a\gamma p-ap-a^2\gamma p = 0,$$

whose discriminant is

$$(5.3) \qquad \Delta/p^2 = a(ap^2+4\gamma-4\gamma p)$$

and must be a square. In case $a = \gamma$, $\Delta/p^2 = (ap-2a)^2$, so that either

$$(5.4) \qquad m = (a^2-a+1)/p \quad \text{or} \quad m = (a^2+a+1)/p-a.$$

Taking into account that $q = a^2 \equiv y$ or $y^2(p)$ and $(p-y)^2 \equiv y^2(p)$, the first value for m in (5.4) holds when $a^2 = (pw+p-y)^2$; namely, $m = (w+1)(p(w+1)-2y-1)$ provided that $y^2+y+1 = p$, whereas $m = (w+1)(p(w+1)-2y-1)+3$ in case $y^2+y+1=3p$. On the other hand, the second value for m in (5.4) holds when $q=a^2=(pw+y)^2$, so that $m = (w-1)(pw+y)+w(y+1)$ if $y^2+y+1 = p$, and $m = (w-1)(pw+y)+w(y+1)+3$ if $y^2+y+1 = 3p$. Of course, $n = m+a$ and k, $\bar{k}$ can be computed from (5.1).
Notice that if $a = \gamma$, i.e. $q = a^2$, then p is a prime divisor of one of the two natural factors of $q^2+q+1 = (a^2+a+1)(a^2-a+1)$. (See [5] for examples.)
Next, suppose $a \neq \gamma$. The n-secants of K form a set of type (m^*,n^*) and size k^* in the dual plane. Since $m^*(n^*)$ is the number of n-secants on a point off (on) K,

$$(5.5) \qquad m^* = m(pm+pa-a\gamma-1)/a,$$

which implies $a|pm-1$, as $(m,a) = 1$ by eq. (5.2), and $n^* = m^*+\gamma$. Furthermore, $nk^* = kn^* = pmnn^*$ implies $k^* = pmn^*$ and, by eq. (2.1), $k^*\bar{k}^* = m^*n^*p((q^2+a+1)/p$. Set

$$(5.6) \qquad \beta = (pm+pa-a\gamma-1)/a,$$

so that $m^* = m\beta$ and $n^* = m\beta+\gamma$. Consequently, $\bar{k}^* = \beta(q^2+q+1)/p$.
If $\beta = 1$, then $m = (a\gamma+a+1)/p-a$. Substituting in (5.2), $(a-\gamma)(p+1) = 0$ which implies again $a = \gamma$ and $q = a^2$. Hence, assume $\beta > 1$. Therefore, $k^*=pmn^* = pm(m\beta+\gamma)$, $\bar{k}^* = \beta(q^2+q+1)/p$ and, by eq. (2.1), $k^*+\bar{k}^* = (m^*+n^*)(q+1)-q$, whence

$$pm(m\beta+\gamma)+\beta(q^2+q+1)/p = (2m\beta+\gamma)(a\gamma+1)-a\gamma,$$

so that

$$\beta = \frac{p(a\gamma^2+\gamma-a\gamma-pm\gamma)}{p^2m^2-2mpa\gamma-2mp+a^2\gamma^2+a\gamma+1} . \tag{5.7}$$

The two values for β given by (5.6) and (5.7) must be equal. Thus, writing $(pm-1)/a = b$.

$$p\gamma(\gamma-b) = (b-\gamma)(ab^2+a\gamma^2-\gamma-2a\gamma b)+pa(b-\gamma)^2. \tag{5.8}$$

Therefore, either $b = \gamma$, implying $m = (a\gamma+1)/p$, a contradiction, or

$$ab^2-b(2a\gamma-ap)+a\gamma^2-\gamma-pa\gamma+p\gamma = 0, \tag{5.9}$$

whose discriminant is $\Delta = a^2p^2+4a\gamma-4a\gamma p$, and must be a square. From (5.9)

$$b = \frac{2a\gamma-ap\pm(a^2p^2+4a\gamma-4a\gamma p)^{1/2}}{2a} \tag{5.10}$$

follows. Since b must be an integer, $a|(a^2p^2+4a\gamma-4a\gamma p)^{1/2}$, impying $a|4\gamma(p-1)$. Set

$$4\gamma(p-1) = as. \tag{5.11}$$

Then $\Delta = a^2(p^2-s)$. Consequently, p^2-s must be a square; write $p^2-s = w^2$. Moreover, from (5.10) $b = \gamma-(p\pm w)/2$ follows, so that $w = 2t+1$. Taking into account that $b = (pm-1)/a$, either

$$m = (1+a\gamma-at-a(p+1)/2)/p, \text{ or} \tag{5.12}$$

$$m = (1+a\gamma+at-a(p-1)/2)/p, \tag{5.13}$$

and conditions can be found on t in order to m be an integer.
On the other hand, whenever q is known and we fix a, γ is known too. Since p is known, the equality

$$4\gamma(p-1) = a(p^2-(2t+1)^2) \tag{5.14}$$

yields the possible values for t. Notice that $w = 2t+1 = p-2$ gives $a = \gamma$. If this is the case, then s is minimum.
The just proved results can be used to construct, starting from any prime p, all the orders q of planes such that $p|q^2+q+1$ and the plane may contain sets of

type (m,n) and sizes (5.1), with m given by (5.12) and (5.13). In sect. 6 examples are given of such an application. We observe that all sporadic admissible triples found in [1] are recovered by the here described techniques.

6. APPLICATIONS

In this section it is shown how the results in sect. 5 can be applied by constructing the admissible triples (pmn;m,n) in planes of order q such that $p|q^2+q+1$ for p = 7,13,19.

(i) p = 7.

Recall (sect. 2) that $7|q^2+q+1$ iff $q \equiv 2$ or 4(7). By the argument in sect. 5, p = 7 implies w = 2t+1 with t = 0, 1, 2 and t = 2, i.e. w = p-2, corresponds to the square order plane case which will not be dealt with.

Assume t = 0; then (5.12) and (5.13) become

$$(6.1) \qquad m = (1+a\gamma-4a)/7 \quad \text{and} \quad m = (1+a\gamma-3a)/7,$$

respectively. Moreover, (5.14) yields $\gamma = 2a$. Therefore, $q = a\gamma = 2a^2 \equiv 2$ or 4(7) which gives the forms for q. Taking into account (6.1) and that m must be an integer, the possible solutions are:

$$a = 7h+6,\ q = 2a^2,\ m = 14h^2+20h+7;$$

$$a = 7h+3,\ q = 2a^2,\ m = 14h^2+8h+1 \text{ (and } m = 1 \text{ iff } q = 18 \text{ [8,12]);}$$

$$a = 7h+1,\ q = 2a^2,\ m = 14h^2+h;$$

$$a = 7h+4,\ q = 2a^2,\ m = 14h^2+19h+3.$$

Of course, n = m+a and k = 7mn in all cases. Notice that since p = 7, all planes of order $q = 2^{2c+1}$ with (2c+1,3) = 1 come in these classes (sect. 3). Next, assume t = 1, i.e. w = 3. Thus, (5.12) and (5.13) yield

$$(6.2) \qquad m = (1+a\gamma-5a)/7 \quad \text{and} \quad m = (1+a\gamma-2a)/7.$$

By (5.14), $3\gamma = 5a$. Consequently, $\gamma = 5d$, a = 3d, impying $q = 15d^2$. Thus, all planes in this class are unknown. To compute solutions, take into account that $q = 15d^2 \equiv 2$ or 4(7) implies $d^2 \equiv 2$ or 4(7). Hence, $m = (1+15d^2-15d)/7$ is an integer when $d \equiv 3$ or 5(7), whereas $m = (1+15d^2-6d)/7$ is an integer when $d \equiv 2$ or 4(7). Obviously, n = m+3d, k = 7mn.

(ii) $p = 13$.

Then $q = a\gamma \equiv 3$ or $9\ (13)$, $w = 2t+1$ and $t = 0, 1, 2, 3, 4$, omitting the square order planes occuring when $t = 5$.

When $t = 0$, by (5.14), $2\gamma = 7a$ so that $a = 2d$, $\gamma = 7d$ and $q = 14d^2$, with $d^2 \equiv 3$ or $9\ (13)$. Hence, $m = d(d-1)+(d^2-d+1)/13$ whenever $d \equiv 4$ or $10\ (13)$, and $m = d(d-1)+(d^2+d+1)/13$ when $d \equiv 3$ or $9\ (13)$. In both cases, $n = m+2d$, $k = 13mn$. By the same argument, $t = 1$ implies $a = 3d$, $\gamma = 10d$, $q = 30d^2$ with $d \equiv 2,5,8,11\ (13)$, and using (5.12) and (5.13) the values for m can be computed. Similarly, $t = 2$ yields $\gamma = 3a$ and $q = 3a^2$ with $a \equiv 1, 4, 9, 12\ (13)$. In particular, $a = 9$ yields $q = 3^5$, $m = 16$, $n = 25$, $k = 13 \cdot 16 \cdot 25$ which is the set in sect. 3. The values for m are computed as before.

Finally, $t = 3$ gives $a = 2d$, $\gamma = 5d$, $q = 10d^2$ and $d \equiv 5, 6, 7, 8\ (13)$; $t = 4$ yields $a = 6d$, $\gamma = 11d$, $q = 66d^2$ with $d \equiv 3, 4, 9, 10\ (13)$.

(iii) $p = 19$.

If this is the case, $q \equiv 7$ or $11\ (19)$, $w = 2t+1$ and $t = 0, 1, \ldots, 8$. For the first values for t the solutions are given by

$$t = 0,\ \gamma = 5a,\ q = 5a^2,\ a \equiv 3, 5, 14, 16\ (19);$$

$$t = 1,\ a = 9d,\ \gamma = 44d,\ q = 9 \cdot 44d^2,\ d \equiv 2, 3, 16, 17\ (19);$$

$$t = 2,\ a = 24d,\ \gamma = 121d,\ q = 24 \cdot 121d^2,\ d \equiv 2, 3, 16, 17\ (19);$$

$$t = 3,\ a = 3d,\ \gamma = 13d,\ q = 39d^2,\ d \equiv 7, 8, 11, 12\ (19);$$

$$t = 4,\ a = 9d,\ \gamma = 35d,\ q = 9 \cdot 35d^2,\ d \equiv 1, 7, 12, 18\ (19).$$

Notice that the described method might provide some value for q which is ruled out by Bruck-Ryser theorem.

7. CONCLUDING REMARKS

The arguments used to get some admissible triples for sets of type (m,n) rely on some possible splittings of $mn(q^2+q+1)$. Therefore, other admissible triples can be found by similar techniques looking at other splittings and distributing, in suitable ways, between k and $\bar{k}$ all prime divisors of m, n, and q^2+q+1. This provides a general approach to find all admissible triples which is the first step in the construction of sets of type (m,n). Furthermore, one can try to construct some unknown plane under the assumption it contains a set with two intersection numbers.

REFERENCES

[1] Antonelli, S., Sugli archi di tipo (m,n) in un piano grafico π, Relazione no. 23, Istituto Matematico, Università di Napoli, (1973).

[2] Brouwer, A.E., A series of separable designs with applications to pairwise orthogonal Latin squares, Europ. J. Comb. 1 (1980), 39-41.

[3] de Finis, M., On k-sets of type (m,n) in projective planes of square order, in: "Finite Geometries and Designs", LMS Lecture Note Series 49 (1981), 98-103.

[4] de Resmini, M.J., On k-sets of type (m,n) in a Steiner system S(2,ℓ,v), in: "Finite Geometries and Designs", LMS Lecture Note Series 49(1981), 104-113.

[5] de Resmini, M.J., An infinite family of type (m,n) sets in $PG(2,q^2)$,q a square, J. Geometry 20 (1983), 36-43.

[6] de Resmini, M.J., On 2-blocking sets in projective planes, Ars Combinatoria, 20B, Dec. 1985, 59-69.

[7] de Resmini, M.J., A 35-set of type (2,5) in PG(2,9), J. Comb. Th. A, in print.

[8] de Resmini, M.J., On 3-blocking sets in projective planes, in print.

[9] de Resmini, M.J., and Migliori, G., A 78-set of type (2,6) in PG(2,16), Ars Combinatoria, 22 (1986).

[10] Hirschfeld, J.W.P., Projective Geometry over Finite Fields, Clarendon Press, Oxford, 1979.

[11] Nagell, T., Introduction to Number Theory, Wiley, New York, 1951.

[12] Tallini, G., Some new results on sets of type (m,n) in projective planes, in print.

[13] Tallini Scafati, M., {k,n}-archi di un piano grafico finito, con particolare riguardo a quelli con due caratteri, Rend. Accad. Naz. Lincei, 40 (1966), 812-818, 1020-1025.

Annals of Discrete Mathematics 37 (1988) 147–156

COMMUTATIVE FINITE A-HYPERGROUPS OF LENGTH TWO

Mario DE SALVO*
Via Palermo, 836 - Scala Ritiro (ME), Italy

1. INTRODUCTION

A *hypergroupoid* $H = \langle H, \circ \rangle$ is a non-empty set H equipped with a *hyperoration* $\circ$ (i.e. a mapping whose domain is H X H and whose range is the set of non-empty subsets of H).

If $A \subseteq H$, $B \subseteq H$, then $A \circ B$ denotes $\cup\{a \circ b / a \in A,\ b \in B\}$.

A hypergroupoid H is called *hypergroup* if the hyperoperation satisfies the following axioms:

(i) $(x \circ y) \circ z = x \circ (y \circ z)$, $\forall (x,y,z) \in H^3$;

(ii) $x \circ H = H \circ x = H$, $\forall x \in H$.

We say that a hypergroupoid H has *length* n, if $\forall (x,y) \in H^2$, $|x \circ y| = n$.

In [6], T. Vougiouklis introduced a new class of hypergroups, constructed from ordinary groups as follows:

let $G = \langle G, \cdot \rangle$ be a group and $A \subset G$, $A \neq \emptyset$; we shall call *A-hypergroup* the hypergroup $G_A = \langle G, \overset{A}{*} \rangle$, equipped with the following hyperoperation

$$\overset{A}{*} : (x,y) \longrightarrow x \cdot A \cdot y.$$

Since $\forall (x,y) \in G \times G$, $|x \overset{A}{*} y| = |A|$, these hypergroups have length $|A|$.

Throughout the paper, we consider A-hypergroups of length 2, constructed from abelian groups. These hypergroups are studied and their number is determined when G is cyclic or $|G| < 12$.

2. NOTATIONS AND RECALLS

We recall some definitions and notations typical of hypergroups.

Let A be a non-empty subset of a hypergroup H; A is *complete* if $\forall n \in N^*$ and $\forall (x_1, \ldots, x_n) \in H^n$ such that $\prod_{i=1}^{n} x_i \cap A \neq \emptyset$ then $\prod_{i=1}^{n} x_i \subseteq A$.

We denote by $C_H(A)$ the *complete closure* of A, i.e. the intersection of all subsets of H which are complete and contain A.

Let β_H^* denote the transitive closure of the binary relation β_H, defined in the following way:

$x \beta_H y$ iff there exist $n \in N^*$ and $(z_1, \ldots, z_n) \in H^n$ such that $\{x,y\} \subseteq \prod_{i=1}^{n} z_i$.

---

* Research supported by the G.N.S.A.G.A. of the C.N.R. (Italy).

It is known that the relation $x\mathcal{C}_H y$ iff $x\in\mathcal{C}_H(\{y\})$ is an equivalence and moreover $\mathcal{C}_H = \beta^*_H$ ([5]).

Let ϕ: $H \longrightarrow H/\beta^*_H$ denote the canonical projection. If H is a hypergroup, then H/β^*_H is a group. Hence, we define the *core* of H, $\omega(H)$, as the kernel of ϕ.

H is said to be a *n-hypergroup* iff $\forall x \in H$, $|\mathcal{C}_H(x)| = n$.

H is said to be *complete* iff $\forall(x,y) \in H^2$, $x \circ y = \mathcal{C}_H(x \circ y)$.

Let $H = \langle H, \circ \rangle$, $H' = \langle H', \square \rangle$ be two hypergroups and f be a mapping of H into H'; f is said to be a *homomorphism* (*good homomorphism*) if $f(x \circ y) \subseteq f(x) \square f(y)$ ($f(x \circ y) = f(x) \square f(y)$, respectively) $\forall(x,y) \in H \times H$.

We say that a homomorphism f is an *isomorphism* iff f is bijective and good.

If $f : H \longrightarrow H'$ is a homomorphism (isomorphism) then $\forall x \in H$, $f(\mathcal{C}_H(x)) \subset \mathcal{C}_{H'}(f(x))$ ($f(\mathcal{C}_H(x)) = \mathcal{C}_{H'}(f(x))$, respectively).

H is said to be *regular* iff $E(H) = \{e \in H / \forall x \in H, x \in e \circ x \cap x \circ e\} \neq \emptyset$ and $\forall x \in H$, $i(x) = \{x' \in H / \exists e \in E(H)$ such that $e \in x \circ x' \cap x' \circ x\} \neq \emptyset$ ($[1]_1$).

3. A-HYPERGROUPS CONSTRUCTED FROM ABELIAN GROUPS

In the following let G be an abelian group. We shall use the additive notation; so, + will denote the operation in the group, 0 will be the unit element and -a the inverse element of a. Clearly, if G is abelian, then G_A is commutative. It is known (see [6]) that G_A is regular and $E(G_A) = -A = \{-a / a \in A\}$.

REMARK 3.1. The hyperproducts of two elements in G_A may be defined as follows:

$\forall x \in G$, $0 \overset{A}{*} x = A + x$;

$\forall (a,b) \in G \times G$, $a \overset{A}{*} b = 0 \overset{A}{*} (a+b)$.

We begin with the following result.

PROPOSITION 3.1. G_A *is a complete hypergroup iff* G *is partitioned by the family of cosets of* A.

PROOF. If $\{A + x\}_{x\in G}$ is a partition of G, then by REMARK 3.1, the hyperproducts of two elements are pair-wise disjoint sets, thus G_A is complete. Similarly, we may show the converse.

As an immediate consequence, we deduce

COROLLARY 3.1. *If* G *is a finite group and* G_A *is complete, then* $|A| \mid |G|$.

We can characterize the complete commutative A-hypergroups of length 2, with the

PROPOSITION 3.2. *If* $A = \{a,b\}$, *then* G_A *is complete iff* $2a = 2b$.

PROOF. Suppose $2a = 2b$. We show that G is partitioned by the family $\{A+g\}_{g\in G}$. In fact,
$\forall (g,g')\in G \times G$, $g \neq g'$, such that $\{A + g\}\cap\{A + g'\} \neq \emptyset$, there are two possibilities: (i) $a + g = b + g'$;
(ii) $b + g = a + g'$.

In the first case, $a+g = b+g' \Rightarrow 2a+g = a+b+g' \Rightarrow 2b+g = a+b+g' \Rightarrow b+g = a+g'$ and then $A+g = A+g'$. The other case brings the same result. By PROPOSITION 3.1, G_A is complete. Conversely, let G_A be complete. From PROPOSITION 3.1, $\{A+g\}_{g\in G}$ is a partition of G. Consider the cosets $A+a = \{2a, a+b\}$, $A+b = \{a+b, 2b\}$.

Since $\{A + a\}\cap\{A + b\} \neq \emptyset$, it follows that $2a = 2b$. Observe that if G is a finite group, then, by COROLLARY 3.1, $|G|$ must be even.

REMARK 3.2. $\forall n \in N^*-\{1\}$, $\forall (x_1, \ldots, x_n)\in G^n$, $\exists g\in G$ such that $\overset{A}{*}\prod_{i=1}^{n} x_i = (n-1)A+g$.

In fact we have $\overset{A}{*}\prod_{i=1}^{n} x_i = x_1 + A + x_2 + \ldots + x_{n-1} + A + x_n) =$ (by commutativity) $= (n-1)A + (x_1 + \ldots + x_n)$.

In the rest of the paper, consider A-hypergroups of length 2. Put $A = \{a,b\}$. It is easy to see that $\forall n\in N^*$, we obtain

$$nA = \{\mu a + \nu b / 0 \leq \mu \leq n,\ 0 \leq \nu \leq n,\ \mu + \nu = n\}.$$

The number of distinct elements of nA depends on the order of $(a-b)$. In fact, there is the following result, where $O(a-b)$ denotes the order of $(a-b)$, and $n < \infty$.

PROPOSITION 3.3. *If* $O(a-b) = n$, *then* $|mA| = m+1$ $\forall m < n$, *and* $|kA| = n$ $\forall k \geq N$.

PROOF. If $O(a-b) = n$, then $na = nb$, and so $|nA| \leq n$. Now, we prove that for every $\{x,y\}\subset nA - \{na\}$, such that $x = ua + vb$, $y = ra + sb$, $u \neq r$, we have $x \neq y$. If $u > r$ then $u-r = s-v = k < n$. Now, assume $x = y$. Thus $ka = kb$, and consequently $k(a-b) = 0$, a contradiction. Therefore, $|nA| = n$. The rest may be proved, using analogous methods.

We next require some lemmas.

LEMMA 3.1. *If* $O(a-b) = n$ *and* $nA = (n-1)A$, *then* $mA = nA$ $\forall m > n$.

PROOF. Let $m = n + q$. The result may be obtained by induction on q.

LEMMA 3.2. *If* $O(a-b) = n$ *and* $nA \neq (n-1)A$, *then* $nA \cap (n-1)A = \emptyset$.

PROOF. From PROPOSITION 3.3., we have $|nA| = |(n-1)A| = n$. If $x\in nA\cap(n-1)A$, then there are integers u,v,r,s such that $x = ua+vb = ra+sb$ and $1 \leq u \leq n$, $0 \leq v \leq n-1$, $0 \leq r \leq n-1$, $0 \leq s \leq n-1$, $n = u+v = r+s+1$. If $u=r$, then $v = s+1$ and so $ra+(s-1)b = ra+sb$, hence $b = 0$ and $nA = (n-1)A$. If $u > r$, then set $u = r+q$. We have $r+q+v = r+s+1$, hence $v = s+1-q$ and $ra+qa+sb+(1-q)b = ra+sb$, that is

$qa = (q-1)b$. Note that $1 \leq q \leq n$, since $q = u-r$. If $q = 1$ then $a = 0$ and $nA = (n-1)A$. If $q = n$ then $na = (n-1)b$, hence, since $na = nb$, we have $b = 0$ and $nA = (n-1)A$. Let $q \neq 1$ and $q \neq n$. We have:

(α) $qa = (q-1)b$.

Set $t = n-q$ and add to (α) all the elements $\rho a + \nu b$ such that $\rho+\nu = t$. After $(t+1)$ additions, we have:

(1) $na = ta+(q-1)b$

(2) $(n-1)a+b = (t-1)a+qb$

....

(t+1) $qa+tb = (n-1)b$.

From (1), since $na = nb$, we have:

(γ) $(n-q+1)b = (n-q)a$.

Now, add to (γ) all the elements $\mu a + \delta b$ such that $\mu+\delta = q-1$ and $n-q+1+\mu<n$. After $(q-1)$ additions, obtain:

(1)' $(n-q+1)b+(q-1)a = (n-1)a$

(2)' $(n-q+2)b+(q-2)a = (n-2)a+b$

....

(q-1)' $(n-1)b+a = (n-q+1)a+(q-2)b$.

From (1), (2), ..., (t+1), (1)', (2)', ..., (q-1)', $nA = (n-1)A$.

LEMMA 3.3. *If* $O(a-b) = n$, *then* $(mA+g)\cap(mA+g') \neq \emptyset \longrightarrow mA+g = mA+g'$, *for every* $(g,g')\in G^2$, $g \neq g'$, *and for every* $m \geq n-1$.

PROOF. Let $m = n+d$ ($d\in N\cup\{-1\}$). We have $\forall g\in G$, $mA + g = \{ ma+g, (m-1)a+b+g, \ldots, (d+2)a+(n-2)b+g, (d+1)a+(n-1)b+g\}$. Suppose $\alpha\in(mA+g)\cap(mA+g')$. There are integers r,s,v,w such that $\alpha = ra+sb+g = va+wb+g'$, $r+s = v+w = m$, $r\in\{d+1, d+2, \ldots, m\}\ni v$, $s\in\{0, 1, \ldots, n-1\}\ni w$. Let $r > v$ and $r = v+\varepsilon$. Then $s = m-r = m-v-\varepsilon = w-\varepsilon$, hence $va+\varepsilon a+wb-\varepsilon b+g = va+wb+g'$, that is:

(Σ) $\varepsilon a+g = \varepsilon b+g'$.

Therefore $ma+g = [\varepsilon+(m-\varepsilon)]a+g = \varepsilon a+(m-\varepsilon)a+g = \varepsilon b+(m-\varepsilon)a+g' = (m-\varepsilon)a+\varepsilon b+g'$. Similarly, using ($\Sigma$), we obtain the following, where $k = n-\varepsilon$:

$(m-1)a+b+g = (m-\varepsilon-1)a+(\varepsilon+1)b+g'$

....

$[m-(k-1)]a+(k-1)b+g = (d+1)a+(n-1)b+g'$

$(m-k)a+kb+g = ma+g'$

....

$[m-k-(\varepsilon-1)]a+[k+(\varepsilon-1)]b+g = (m-\varepsilon+1)a+(\varepsilon-1)b+g'$

Whence, $mA+g = mA+g'$.

Now we come the main theorem of this section.

THEOREM 3.1. *If* $A = \{a,b\}$ *and* $O(a-b) = n$, *then* G_A *is a n-hypergroup.*

PROOF. From PROPOSITION 3.3 we have $\forall h < n-1$, $|hA| = h+1$ and $\forall k \geq n-1$, $|kA| = n$. The following cases are possible:

(i) $nA\cap(n-1)A \neq \emptyset$;

(ii) $nA\cap(n-1)A = \emptyset$.

In the first case, from LEMMAS 3.1, 3.2, we have $mA = (n-1)A$ $m \geq n-1$. So, for every hyperproduct P of k elements with $k \geq n$, there exists $g\in G$ such that $P = (n-1)A+g$. But, from LEMMA 3.3, the family $\mathcal{F} = \{(n-1)A+g\}_{g\in G}$ is a partition of G. Since G_A is regular, for every hyperproduct Q of s elements with $s < n$, $\exists g' \in G$ such that $Q \subset (n-1)A+g'$, whence the sets of $\mathcal{F}$ are the classes mod. $\beta^*_{G_A}$. Thus, since $\forall g \in G$, $|(n-1)A+g| = n$, we deduce that G_A is a n-hypergroup.

In the second case, set $\mathcal{O}(a) = q$. Clearly, $q \neq 1$, since $nA \neq (n-1)A$. We have $(n-1)A = (n-1+q)A$; so, every hyperproduct of k elements with $k \geq n$ is of the form $hA+g$ where $g \in G$ and $n-1 \leq h \leq n-2+q$. But, for regularity, all aforesaid hyperproducts can be considered of the form $(n-2+q)A+g$, $g \in G$. As regards the remaining hyperproducts, each of them is contained in $(n-2+q)A + g'$, for some g'. By LEMMA 3.3, we can say that G is partitioned in the classed mod. β^* by the family $\{(n-2+q)A+g\}_{g\in G}$. Since $|(n-2+q)A+g| = n$ $\forall g \in G$, the theorem is proved.

The theorem leads to the following corollary.

COROLLARY 3.1. *If* G *is a finite group and* $A = \{a,b\}$, *then* $|G_{A/\beta^*}| = |G|/\mathcal{O}(a-b)$.

4. FINITE A-HYPERGROUPS FROM CYCLIC GROUPS OR GROUPS OF ORDER LESS THAN 12.

Now, we find (to within an isomorphism) all A-hypergroups of length 2, constructed from finite abelian groups of order t, when the group is cyclic or $t < 12$. In [6] the following statements are proved:

(4.1) *Let* α: $G \to G'$ *be an isomorphism of groups and* A *a non-empty subset of* G. *If* $A' = \alpha(A)$, *then* $G_A \simeq G_{A'}$.

In particular:

(4.2) *If* α *is an automorphism of the group* G, *then* $G_A \simeq G_{\alpha(A)}$ *for every* $A \in P(G)-\{\emptyset\}$.

For later use, we recall, from $[1]_2$:

(4.3) *If* $\mathcal{F}$ *is the family of complete* n*-hypergroups, of order* nh, *then there are (to within an isomorphism) as many hypergroups in* $\mathcal{F}$, *as the pair-wise non-isomorphic groups of order* h.

REMARK 4.1. In order to generate (to within an isomorphism) all A-hypergroups of finite order t, by (4.1), it is sufficient to consider the A-hypergroups constructed from the representatives of the classes of isomorphism of groups of order t.

We begin with an useful lemma;

LEMMA 4.1. *If* $A = \{a,b\}$, $A' = \{c,d\}$ *are subsets of a finite group* G *such that* $G_A \simeq G_{A'}$, *then* $\mathcal{O}(a-b) = \mathcal{O}(c-d)$.

PROOF. If $O(a-b) \neq O(c-d)$, then from COROLLARY 3.1, it follows that $|G_{A/\beta^*}| \neq |G_{A'/\beta^*}|$. But $G_A \simeq G_{A'}$, implies $G_{A/\beta^*} \simeq G_{A'/\beta^*}$, a contradiction.

Now we come to the important theorem:

THEOREM 4.1. *Let* $A = \{a,b\}$, $A' = \{c,d\}$ *be two subsets of a finite group* G *such that* $O(a-b) = O(c-d)$. *If* $\exists (x,y) \in G \times G$ *such that* $A + x = A' + y$, *then* $G_A \simeq G_{A'}$.

PROOF. Suppose $O(a-b) = O(c-d) = n$. Let $n = 2$. From THEOREM 3.1 and PROPOSITION 3.2, it follows that G_A and $G_{A'}$ are complete, 2-hypergroups. By COROLLARY 3.1, $|G|$ must be an even number; set $|G| = 2k$ and denote $G_A = \{a_1, \dots, a_k, b_1, \dots, b_k\}$, $G_{A'} = \{c_1, \dots, c_k, d_1, \dots, d_k\}$ with $a_i \beta^* b_i$, $c_i \beta^* d_i$ $\forall i \in \{1, 2, \dots, k\}$.

Since β^* is a strongly regular equivalence, we obtain $0 \overset{A}{*} a_i = 0 \overset{A}{*} b_i$ and $0 \overset{A'}{*} c_i = 0 \overset{A'}{*} d_i$ $\forall i \in \{1, 2, \dots, k\}$.

Then we may write: $G_A = \bigcup_{i \in \{1, \dots, k\}} \{A+a_i\}$, $G_{A'} = \bigcup_{i \in \{1, \dots, k\}} \{A'+c_i\}$ where $A + a_i = A + a_j$ iff $i = j$; and analogously $A' + c_i = A' + c_j$ iff $i = j$. $\forall a_i \in \{a_1, \dots, a_k\}$, $\exists w \in G$ such that $a_i = w + x$, and thus $A + a_i = A + x + w = A' + y + w$, whence $\exists c_\rho \in \{c_1, \dots, c_k\}$ such that $A' + c_\rho = A' + y + w$, and c_ρ is clearly unique. In the same way, $\forall c_i \in \{c_1, \dots, c_k\}$ there exists an unique element $a_\mu \in \{a_1, \dots, a_k\}$ such that $A' + c_i = A + a_\mu$. So, we may define a bijection $\alpha_1 : G_A \to G_{A'}$ in the following way:
$\forall i \in \{1, \dots, k\}$ $\alpha_1(a_i) = c_j$ and $\alpha_1(b_i) = d_j$ if $A + a_i = A' + c_j$. Later, we shall show that α_1 is an isomorphism.

Now, let $n > 2$. Then G_A and $G_{A'}$ are n-hypergroups. We have $G_A = \bigcup_{x \in G} \{A + x\}$ and $G_{A'} = \bigcup_{x \in G} \{A' + x\}$. Consider $A + z \subset G_A$. Then $\exists t \in G$ such that $z = x + t$, whence $A + z = A + x + t = A' + y + t \subset G_{A'}$. Note that $y + t = z'$ is the only element such that $A + z = A' + z'$; in fact, if there exists $w \neq z'$ such that $A + z = A' + w$, then it follows that $A' + z' = A' + w$, that is $\{c + z', d + z'\} = \{c + w, d + w\}$, whence $c + z' = d + w$ and $d + z' = c + w$. Hence $z' = d + w - c$, $d + d + w - c = c + w$, and so, $2d = 2c$, a contradiction, since $O(c-d) > 2$.

In a similar way, $\forall \{A' + v\} \subset G_{A'}$ there exists an only s such that $A + s = A' + v$. Therefore, we may define a bijection $\alpha_2 : G_A \to G_{A'}$ such that $\forall x \in G_A$, $\alpha_2(x) = y$ iff $A + x = A' + y$. Later on, the bijections α_1 and α_2 will be denoted by the same letter α.

Now, we prove that α_1 and α_2 satisfy the following property:

(Δ) $\forall (x,y) \in G \times G$, $\alpha(x + y) = \alpha(x) + y = x + \alpha(y)$.

If $\alpha(x) = x'$ and $\alpha(y) = y'$, then we have $A + x = A' + x'$ and $A + y = A' + y'$, whence $A + x + y = A' + x' + y = A' + x + y'$.

So, $\alpha(x + y) = x' + y = \alpha(x) + y$ and $\alpha(x + y) = x + y' = x + \alpha(y)$. At last, using ($\Delta$), we prove that α_1 and α_2 are isomorphisms:
we have $\forall (z,v) \in G \times G$ with $\alpha(z) = w$, $\alpha(v) = u$,
$\alpha(z) \overset{A'}{*} \alpha(v) = A' + \alpha(z) + \alpha(v) = A' + w + u = A + z + u$ and $\alpha(z \overset{A}{*} v) = \alpha(A + v + z)$; let $\gamma \in \alpha(A + v + z)$, then $\exists p \in A$ such that $\gamma = \alpha(p + v + z) = p + z + \alpha(v) = p + z + u \in A + z + u$; let $\mu \in A + z + u$, then $\exists p \in A$ such that $\mu = p + z + u = p + z + \alpha(v) = \alpha(p + z + v) \in \alpha(A + z + v)$. This completes the proof.

As a consequence of the preceding theorem, we obtain the

COROLLARY 4.1. *If* G *is a finite group of order* m *and* $|A| = 2$, *then (to within an* *isomorphism) the number of* A-*hypergroups, constructed from* G, *is strictly less than* m.

PROOF. From THEOREM 4.1, it is sufficient to consider the A-hypergroups with $A = \{0,x\}$ and $x \in G - \{0\}$. In fact, for each $\{a,b\} \subset G - \{0\}$, $\exists (x,y) \in G \times G$ such that $\{a,b\} = \{0,x\} + y$, where $x = b - a$ and $y = a$.

We require some lemmas.

LEMMA 4.2. *Assume that* $G \simeq Z_m$; *let* $\mathcal{G} = \{a \in G / O(a) = m\}$ *and let* $I(d) = \{x \in G / O(x) = d\}$, $d \neq 1$. *Then* $I(d) = (m/d)\mathcal{G} = \{(m/d)a / a \in \mathcal{G}\}$.

PROOF. The case $d = m$ is trivial. Let $x \in (m/d)\mathcal{G}$, $m \neq d$; then $\exists a \in \mathcal{G}$ such that $x = (m/d)a$, whence $dx = ma = 0$ and $O(x) \mid d$. Suppose that $O(x) = p \neq d$; then $\exists t$ such that $d = pt$ and $px = p(m/d)a = (d/t)(m/d)a = (m/t)a = 0$, a contradiction since $O(a) = m > (m/t)$. Therefore $x \in I(d)$. Conversely, if $x \in I(d)$, then x generates the only subgroup of order d. So, $\exists a \in \mathcal{G}$ such that $x = (m/d)a$. This completes the proof.

LEMMA 4.3. *Let* $G \simeq Z_m$ *and denote* $A_i = \{0, x_i\}$ $\forall x_i \in G - \{0\}$. *If* $O(x_i) = O(x_j) = d$ *then* $G_{A_i} \simeq G_{A_j}$.

PROOF. From LEMMA 4.2, it follows that $\exists (a_i, a_j) \in G \times G$ such that $x_i = (m/d)a_i$ and $x_j = (m/d)a_j$. If α is the automorphism of G such that $\alpha(a_i) = a_j$, then we have $\alpha(x_i) = \alpha((m/d)a_i) = (m/d)\alpha(a_i) = (m/d)a_j = x_j$ and so, $\alpha(A_i) = A_j$. Therefore, using (4.2), the lemma is proved.

By means of LEMMAS 4.1, 4.2, 4.3, and COROLLARY 4.1, we have the following theorem

THEOREM 4.2. *If* G *is a finite cyclic group of order* m, *then there are (to within an isomorphism) as many* A-*hypergroups of length 2, constructed from* G, *as divisors of* m, *except* 1.

As an immediate consequence, we state:

COROLLARY 4.2. *If* G *is a finite group of prime order, then there exists (to within an isomorphism) exactly one* A-*hypergroup of length 2, constructed from* G.

Now, we are going to find the number of A-hypergroups of length 2, constructed from abelian groups of order t less than 12. The cases when t is prime or $t \in \{6,10\}$ lead respectively to COROLLARY 4.2 and THEOREM 4.2. The cases $t = 4$, $t = 8$, $t = 9$, remain to be studied.

We begin with a lemma, where mZ_2 denotes the direct sum $Z_2 \oplus Z_2 \oplus \dots \oplus Z_2$, with m summands.

LEMMA 4.4. *If* $G \simeq mZ_2$, *then there exists (to within an isomorphism) exactly one A-hypergroup of length 2, constructed from* G, G_A. *Moreover,* G_A *is a complete hypergroup and* $G_{A/\beta^*} \simeq (m-1)Z_2$.

PROOF. Clearly, $\forall x \in G - \{0\}$, $O(x) = 2$; therefore, for every $A \in P(G)$ such that $|A| = 2$, G_A is a complete 2-hypergroup.

By COROLLARY 4.1, it is sufficient to consider the A-hypergroups of the form G_{A_i} with $A_i = \{0,i\}$ and $i \in G - \{0\}$. For each $j \in G - \{0\}$, denote by $\bar{j}$ the class mod. β^* of which j is the representative.

If $A = \{0,i\}$, then we have $\bar{j} = A + j = \{j, i+j\}$, whence $2\bar{j} = A = \bar{0}$; therefore, taking into account the COROLLARY 3.1, we have that $G_{A/\beta^*} \simeq (m-1)Z_2$. Lastly, by COROLLARY 2.1 of $[4]_2$, the statement of the lemma.

In the rest, $\forall n \in N^*$, we shall denote $Z_n = \{0, 1, \ldots, n-1\}$. Now, let $|G| = 4$. There are two cases:

(i) $G \simeq Z_4$;

(ii) $G \simeq Z_2 \oplus Z_2$.

If $G \simeq Z_4$, then, using THEOREM 4.2, there are two A-hypergroups G_A, $G_{A'}$, where we may choose $A = \{0,1\}$, $A' = \{0,2\}$. Note that, as a consequence of PROPOSITION 3.2, $G_{A'}$ is complete.

In the case (ii), by (3.3) and LEMMA 4.4, we obtain hypergroups which are isomorphic to $G_{A'}$.

Now, let $|G| = 8$. By commutativity, there are three cases:

(I) $G \simeq Z_8$;

(II) $G \simeq Z_2 \oplus Z_2 \oplus Z_2$;

(III) $G \simeq Z_2 \oplus Z_4$.

(I) By THEOREM 4.2, we obtain three A-hypergroups: G_{A_1}, G_{A_2}, G_{A_3}, where, according to LEMMA 4.3 and COROLLARY 4.1, we may suppose

$$A_1 = \{0,1\},\ A_2 = \{0,2\},\ A_3 = \{0,4\}.$$

Note that $G_{A_1} = \omega(G_{A_1})$, G_{A_2} is a 4-hypergroup and G_{A_3} is a complete 2-hypergroup. Moreover, it is easy to see that $G_{A_3/\beta^*} \simeq Z_4$.

(II) By LEMMA 4.4, this case leads to an only A-hypergroup, $G_{\bar{A}}$. $G_{\bar{A}}$ is complete and $G_{\bar{A}/\beta^*} \simeq Z_2 \oplus Z_2$, whence $G_{\bar{A}} \not\simeq G_{A_3}$.

(III) Let $G = \{a_0, a_1, \ldots, a_7\}$ where $\forall i \in Z_4$, $a_i = (0,i)$ and $a_{i+4} = (1,i)$. $\forall a_i \in G - \{a_0\}$ $O(a_i) \in \{2,4\}$. If we consider the sets $\{a_0, a_k\}$ with $O(a_k) = 2$, then we obtain hypergroups, each of which, by (4.3), is isomorphic to $G_{\bar{A}}$ or to G_{A_3}. Therefore, using COROLLARY 4.1, the two following hypergroups remain to be considered:

$$G_{C_1},\ G_{C_2},\ \text{where } C_1 = \{a_0, a_1\},\ C_2 = \{a_0, a_5\}.$$

But the bijection $\begin{pmatrix} a_0 & a_1 & a_2 & a_3 & a_4 & a_5 & a_6 & a_7 \\ a_0 & a_5 & a_2 & a_7 & a_4 & a_1 & a_6 & a_3 \end{pmatrix}$ is as isomorphism and so,

$G_{C_1} \simeq G_{C_2}$. Since G_{C_1} and G_{A_2} are 4-hypergroups, they could be isomorphic. We have $C_1 = \{a_0, a_1\}$, $A_2 = \{0,2\}$ $E(G_{C_1}) = \{a_0, a_3\}$, $E(G_{A_2}) = \{0,6\}$, $\omega(G_{C_1}) = \{a_0, a_1, a_2, a_3\}$, $\omega(G_{A_2}) = \{0,2,4,6\}$. If there exists an isomorphism $\delta: G_{C_1} \to G_{A_2}$ then $\delta(E(G_{C_1})) = E(G_{A_2})$, whence we have two possibilities:

(j) $\delta(a_0) = 0$, $\delta(a_3) = 6$;

(jj) $\delta(a_0) = 6, \quad \delta(a_3) = 0.$

In the first case, $\delta(a_0 \overset{C_1}{*} a_0) = \delta(\{a_0, a_1\}) = \{0, \delta(a_1)\}$ and $\delta(a_0) \overset{A_2}{*} \delta(a_0) = 0 \overset{A_2}{*} 0 = \{0,2\}$, whence $\delta(a_1) = 2$ and since $\delta(\omega(G_{C_1})) = \omega(G_{A_2})$, it follows that $\delta(a_2) = 4$. Then $\delta(a_4 \overset{C_1}{*} a_4) = \delta(\{a_0, a_1\}) = \{0,2\}$. But $\delta(a_4) \overset{A_2}{*} \delta(a_4) = \{0,2\}$ iff $\delta(a_4) \in \{0,4\} = \delta(\{a_0, a_2\})$, which is a contradiction. In a similar way, also the second case leads to a contradiction, hence we may say that there exist (to within an isomorphism) exactly 5 commutative A-hypergroups of order 8 and lenght 2: G_{A_1}, G_{A_2}, G_{A_3}, $G_{\bar{A}}$, G_{C_1}.

Finally, we examine the case $|G| = 9$. We have:

(1) $G \cong Z_9$ or

(2) $G \simeq Z_3 \oplus Z_3$.

(1) Reasoning as in the preceding cases, we obtain two hypergroups G_{S_1}, G_{S_2} where we may choose $S_1 = \{0,1\}$ and $S_2 = \{0,3\}$.

(2) Let $G = \{x_0, x_1, \ldots, x_8\}$, where $\forall i \in Z_3$, $x_i = (0,i)$, $x_{i+3} = (1,i)$, $x_{i+6} = (2,i)$. We have $\forall i \neq 0$, $O(x_i) = 3$. For COROLLARY 3.1, it is sufficient to consider two hypergroups: G_T, G_V with $T = \{x_0, x_1\}$, $V = \{x_0, x_5\}$. It is easy to see that $\omega(G_T) = \{x_0, x_1, x_2\}$, $C_{G_T}(x_3) = \{x_3, x_4, x_5\}$, $C_{G_T}(x_6) = \{x_6, x_7, x_8\}$ and $\omega(G_V) = \{x_0, x_5, x_7\}$, $C_{G_V}(x_1) = \{x_1, x_3, x_8\}$, $C_{G_V}(x_2) = \{x_2, x_4, x_6\}$. If there exists an isomorphism $f: G_T \to G_V$, then $\forall x \in G$, $f(C_{G_T}(x)) = C_{G_V}(f(x))$ and $f(\omega(G_T)) = \omega(G_V)$. Therefore, we must have $f(\{x_3, x_4, x_5\}) = \{x_1, x_3, x_8\}$ or $f(\{x_3, x_4, x_5\}) = \{x_2, x_4, x_6\}$. But, in every case, we obtain $f(x_3 \overset{T}{*} x_4) \neq f(x_3) \overset{V}{*} f(x_4)$. Thus, G_T is not isomorphic to G_V. Since G_{S_2} is a 3-hypergroup, it could be isomorphic to G_T or to G_V.

If $g : G_{S_2} \to G_T$ is an isomorphism, then, reasoning as before, we always find $g(1 \overset{S_2}{*} 1) \neq g(1) \overset{T}{*} g(1)$ or $g(2 \overset{S_2}{*} 2) \neq g(2) \overset{T}{*} g(2)$.

So, G_{S_2} is not isomorphic to G_T, and in a similar way, it is possible to see that G_{S_2} is not isomorphic to G_V.

In the following table, we list all different commutative A-hypergroups of length 2, of each order from 2 to 11:

ORDER	2	3	4	5	6	7	8	9	10	11
NUMBER	1	1	2	1	3	1	5	4	3	1

REFERENCES

[1] P. CORSINI: $[\cdot]_1$ *Ipergruppi semiregolari e regolari*, Rend. Sem. Mat. Univ. Politec. Torino, vol. 40, (1982), 35-46; $[\cdot]_2$ *Recenti risultati in teoria degli ipergruppi*, B.U.M.I., 2-A, (1983), 133-138; $[\cdot]_3$ *Prolegogomeni alla teoria degli ipergruppi*, Quaderni dell'Istituto di Matematica, Informatica e Sistemistica dell'Università di Udine (1986).

[2] P. CORSINI and G. ROMEO, *Hypergroups completes et T-groupoides*, Atti Convegno su Sistemi Binari e loro Applicazioni, Taormina Italy (1978), 129-146.

[3] F. DE MARIA, *Gruppi selezioni di ipergruppi*, Atti Sem. Mat. Fis. Univ. Mo-

dena, XXX, (1981), 76-82.

[4] M. DE SALVO: $[\cdot]_1$ *Su le potenze ad esponente intero in un ipergruppo e gli r-ipergruppi*, Riv. Mat. Univ. Parma (4) 11 (1985), 409-421; $[\cdot]_2$ *Nuovi risultati sugli* (H,G)- *ipergruppi*, Atti Sem. Mat. Fis. Univ. Modena; XXXIV, (1985-86), 1-14; $[\cdot]_3$ *Ipergruppi finiti di lunghezza costante*, Rend. Ist. Lombardo Accad. Sci. Lett. Sez. A, 120 (1986), 41-56.

[5] M. KOSKAS, *Groupoides, demi-hypergroupes et hypergroupes*, J. Math. Pures et Appl. 49 (1970), 155-192.

[6] T. VOUGIOUKLIS, *Generalization of* P- *hypergroups*, to appear in Rend. Circolo Mat. di Palermo.

Annals of Discrete Mathematics 37 (1988) 157–168

ON SETS OF FIXED PARITY IN STEINER SYSTEMS

Franco EUGENI and Stefano INNAMORATI

Istituto di Scienze dell'Universita` di Chieti (Italia)

A set H of points in a Steiner system S(t,k,v) is called of even (odd) type if $B\cap H$ is even (odd) for any block B. In this paper we give conditions for the non-existence of such sets and examples.

INTRODUCTION

It is well known that fixed parity sets play an important role in coding theory. Recently, in two papers G.Tallini (cf. [15]) and V. Tonchev (cf. [16]) constructed codes by using fixed parity sets of Steiner systems or their duals

In this paper we begin a systematic study of fixed parity sets with particular regard to the case of blocking sets.

Let S be a finite non-empty set and denote by $\mathscr{B}$ a family of k-subsets of S, put $|S|=v$. The elements of S are called points and those of $\mathscr{B}$ are called blocks. The pair $(S,\mathscr{B})$ is called a Steiner system S(t,k,v), if every t-subset of S is exactly contained in one block of $\mathscr{B}$ and $2\leqslant t<k<v$.

If $(S,\mathscr{B})$ is an S(t,k,v) Steiner system the pair $(S_x,\mathscr{B}_x)$, where $S_x:=S-\{x\}$, and $\mathscr{B}_x:=\{B-\{x\},\ x\in B\ ,\ B\in\mathscr{B}\}$, is a Steiner system S(t-1,k-1,v-1); it is called the contraction of $(S,\mathscr{B})$ in x.

The number r_s of blocks containing a fixed s-set of S with $0\leqslant s\leqslant t$ is:

$$(1.1)\qquad r_s=\binom{v-s}{t-s}\Big/\binom{k-s}{t-s}$$

Denote by C a c-set in S(t,k,v), and by t_i the number of blocks that are i-secant of C (i=0,...,k). We have

$$(1.2)\qquad \sum_{i=0}^{c}\binom{i}{s}t_i=\binom{c}{s}r_s \qquad s=0,1,\ldots,t\ .$$

1.1 DEFINITION. A point-set H of $(S,\mathscr{B})$ is called an even type set (or odd type set) if H is a proper subset of S which is intersected by every block in an even (or odd , respectively) number of points.
Moreover, H is said to be a fixed parity set if it is either an even type set or an odd type set.

1.2 DEFINITION. A point-set C of $(S,\mathscr{B})$ is called a blocking set if C intersects every block and contains no block.

If a blocking set C is a fixed parity set, we call it a fixed parity blocking set (in particular even blocking set or odd blocking set).

In Section 2 we shall give some non-existence theorems for Steiner systems, containing fixed parity sets and bounds for the cardinality of such sets . The non-existence theorems imply that only a small number of Steiner systems with $v\leqslant 28$ can contain fixed parity sets. In some cases, we shall be able to give a complete classification of the fixed parity sets. Moreover, the case of projective and affine spaces is investigated. They contain fixed parity sets if and only if its order is even. In the affine case the fixed parity sets are even type sets.

In Section 3 we shall deal with the case of even blocking sets in projective and affine spaces. We give a lower bound for the cardinality of even blocking sets. We prove that each affine space AG(r,q), q even, contains even blocking sets, if its order is large enough . The paper suggests many open problems and questions.

2. MAIN RESULTS

We begin with some lemmas which will be useful in the sequel. The following three ones are trivial.

2.1 LEMMA. If H is a fixed parity set in a Steiner system S(t,k,v),then the complement of H is a fixed parity set too. In particular H and its complement have the same parity iff k is even; H and its complement have different parity iff k is odd.

2.2 LEMMA. If v and k have a different parity, then an S(t,k,v) has no fixed parity set.

2.3 LEMMA. An even type set (or odd type set) of a Steiner system has an even number of points (or an odd number of points,respectively) (cf. Tallini [15]).

Now we prove the following

2.4 LEMMA. Let H be a fixed parity set in a Steiner system S(t,k,v). Then

$$|H| \geqslant r_{t-1} + t-2 .$$

PROOF. Suppose that the proper subset H of S is an even type set with cardinality $|H| = h \geqslant 2$, h even.

We claim that $h \geqslant t$. (Assume that $h < t$. Fix t-1 points taking h-1 of them in H and the others t-h in S-H. Each of the r_{t-1} blocks through the t-1 fixed points contains at least another point of H (otherwise a block would intersect H in exactly h-1 points with h even). Then $h = |H| \geqslant h-1+r_{t-1}$, a contradiction).

Suppose that t is odd. Fix t-2 points in H and one point outside H. Each of the r_{t-1} blocks through the fixed points has at least another point in H, so

$$h = |H| \geqslant r_{t-1} +t-2 .$$

Now suppose that t is even. Fix t-1 points in H. Each of the r_{t-1} blocks containing these t-1 points has at least another point in H, so

$$h = |H| \geqslant r_{t-1} +t-1 .$$

Suppose that H is an odd type set with $|H| = h \geqslant 1$, h odd. Then it follows similarly as above $h \geqslant t$. If t is odd, fix t-1 points in H, it follows

$$h = |H| \geqslant r_{t-1} +t-1 .$$

If t is even, fix t-2 points in H, and one point outside H. So the assertion is completely proved.

2.5 LEMMA. If in an S(t,k,v) the number r_{t-1} is even, then there is no fixed parity set.

PROOF. Assume that H is an even type set and t even. Fix t-1 points in H; each of the r_{t-1} blocks containing the t-1 points of H intersects H in an odd number of points different from the t-1 fixed points. So H is odd, a contradiction.

Suppose that t is odd. Fix t-2 points in H and one point outside H. Each of the r_{t-1} blocks containing the t-1 fixed points intersects H in an odd number of other points. So H is odd, a contradiction.

The case of an odd type set can be proved similarly.

2.6 LEMMA. Suppose that a Steiner system S=S(t,k,v), with $k \mid v$, admits a spread, i.e. a family of v/k mutually disjoint blocks. If $v/k \not\equiv r_{t-1} \pmod 2$, then S has no odd type sets.

PROOF. We note that v and k are even necessarily. Assume that H is an odd type set. Each of the v/k blocks of the spread intersects H in an odd number of points, so we have $|H| \equiv v/k \pmod 2$.

Fix t-1 points outside H (cf 2.4). Then each of the r_{t-1} blocks through these t-1 points intersects H in an odd number of points,so $|H| \equiv r_{t-1} \pmod 2$, a contradiction.

2.7 LEMMA. Suppose that a Steiner system S(t,k,v) contains no fixed parity set. Then no extension of it contains a fixed parity set. (An extension of a Steiner system S=S(t,k,v) is a Steiner system S(t+1,k+1,v+1) whose contraction is S).

PROOF. Trivial.

2.8 LEMMA. Let H be a fixed parity set of a Steiner system S(t,k,v). Then H is a fixed parity set in any contraction of the system in a point.

PROOF. Trivial.

2.9 REMARK. In S(2,3,v) the even type sets are exactly the sets of type (0,2). They exist iff $v \equiv 3$ or 7 (mod 12). (cf. de Resmini [8] and Lenz-Zeitler [12]). Clearly the odd type sets are the complements of the even type sets.

2.10 PROPOSITION. A Steiner system S(3,4,v) contains no fixed parity set.

PROOF. Suppose H an even type set with h points.

By (1.2), we have

$$\begin{cases} t_0 + t_2 + t_4 = (v-1)(v-2)(v-3)/24 & (1) \\ 2t_2 + 4t_4 = h(v-1)(v-2)/6 & (2) \\ 2t_2 + 12t_4 = h(h-1)(v-1)/3 & (3) \\ 24t_4 = h(h-1)(h-2) & (4) \end{cases}$$

By (2) and (3) it follows

$$24t_4 = h(v-1)(h-v/2) \geqslant 0 \qquad (5)$$

consequently $h \geqslant v/2$. The complement of H is also of even type, so $v-h \geqslant v/2$, and so h=v/2. By (4) and (5) we have (v-1)(h-v/2) = (h-1)(h-2), which implies h=1 or h=2, a contradiction.

Assume now that H is an odd type set, then necessarily H is of type (1,3). By a well-known result of Tallini (cf. [13]), we have $|H| = v/2$. Hence

$$\begin{cases} t_1 + t_3 = (v-1)(v-2)(v-3)/24 \\ 3t_3 = v(v-1)(v-2)/12 \\ 6t_3 = v(v-1)(v-2)/12 \end{cases}$$

It follows t =0, a contradiction since $v \geqslant 5$.

2.11 PROPOSITION. In an $S(3,5,17)$, i.e. the inversive plane of order 4, there are no fixed parity sets.

PROOF. By (2.1) it is sufficient to prove that there are no even type sets. Suppose that H is an even type set with $|H| = h$. Then $t_0 + t_2 + t_4 = 68$, $t_2 + 2t_4 = 10h$, $2t_2 + 12t_4 = 5h(h-1)$, $24t_4 = h(h-1)(h-2)$. So, we have $h^2 - 18h + 77 = 0$ which implies $h = 7$ or 11, a contradiction by 2.3.

2.12 PROPOSITION. In a Steiner system $S(2,4,28)$ there are no odd type sets.

PROOF. Suppose H an odd type h-set, then H is of type $(1,3)$ necessarily. The (1.2) become:

$$t_1 + t_3 = 63 \quad , \quad t_1 + 3t_3 = 9h \quad , \quad 6t_3 = h\,(h-1);$$

which give $h^2 - 28h + 189 = 0$. Since this equation has no integer solution, the assertion is proved.

Now we consider the list of admissible parameters of a Steiner system with $v \leqslant 28$, contained in 7 (cf. Appendix 4).

According to the previous Lemmas fixed parity sets may exist only in the following Steiner systems:

	$S(2,3,7)$	$S(2,3,19)$	$S(4,7,23)$	
(2.1)	$S(2,3,15)$	$S(2,5,21)$	$S(5,8,24)$	$S(2,4,28)$
	$S(2,4,16)$	$S(3,6,22)$	$S(2,3,27)$	

Moreover, when $29 \leqslant v \leqslant 50$ the admissible parameters of a Steiner system containing fixed parity sets are the followings: $(t,k,v)=(2,3,31)$, $(2,6,36)$, $(3,7,37)$, $(2,3,39)$, $(2,4,40)$, $(3,5,41)$, $(4,6,42)$, $(2,3,43)$, $(2,7,43)$, $(5,7,43)$, $(6,8,44)$), $(2,5,45)$, $(7,9,45)$, $(2,6,46)$, $(3,6,46)$, $(2,10,46)$, $(8,10,46)$, $(9,11,47)$, $(10,12,48)$, $(11,13,49)$, $(12,14,50)$.

Now we shall characterize all fixed parity sets in $S(t,k,v)$'s with $v \leqslant 28$, except $S(2,4,28)$.

Note that in (2.1) the fixed parity sets of $S(2,3,v)$ are characterized by 2.9.

Now we deal with $S(2,4,16)$ and $S(2,5,21)$. The fixed parity sets in $PG(3,4)$ have been characterized by Hirschfeld and Hubaut in [11]. Consequently the even type sets in $S(2,5,21) \cong PG(2,4)$ are characterized too. We give a new characterization of this case, from which the case $S(2,4,16)$ will be deduced.

2.13 PROPOSITION. The even type sets in $PG(2,4)$ are the following sets:
(1) the hyperovals;
(2) the symmetric differences of two lines;
(3) the complements of the union of an hyperoval and an external line;
(4) the complements of Baer (or Fano) subplanes or the hermitian arcs;
(5) the complements of a line.
The odd type sets are the complements of sets (1),...,(5).

PROOF. Let H be an even type 2h-set. By (1.2) we have $2t_0 = (h-6)(h-7)$, $2t_2 = 2h(8-h)$, $2t_4 = h(h-3)$. Then $3 \leqslant h \leqslant 8$.
If $h=3$, then H is of type $(0,2)$, hence an hyperoval. Suppose $h=4$. Then H is the symmetric difference of two lines. In fact the common point of the two 4-secant lines is outside H (otherwise H would have at least two 1-secant lines). Suppose $h=5$, then $t_0=1$, $t_2=15$, $t_4=5$.
The complement of H is an odd type set with 11 points and $t_1=5$, $t_3=15$, $t_5=1$.
The 6-set outside the 5-secant line is of type $(0,2)$ and hence an hyperoval.
If $h=6$ or $h=7$, then H is of type $(2,4)$. It follows that H is the complement of

a Baer subplane or of a Hermitian arc (cf. [10], [11]).
Suppose h=8, then H is the complement of a line, since H is of type (0,4) with $t_0=1$.

Now we can characterize the even type sets of $S(2,4,16) \cong AG(2,4)$. Note that AG(2,4) has no odd type sets by 2.6

2.14 DEFINITION. An elliptic hyperoval in $\underline{A}$ = AG(2,q), q even, is a (q+2)-arc K of the projective extension $\underline{P}$=PG(2,q) of $\underline{A}$ with the property that the complement of $\underline{A}$ in $\underline{P}$ is external to K.

2.15 PROPOSITION. In AG(2,4) the even type sets are
(1) the elliptic hyperovals;
(2) the unions of two parallel lines;
(3) the complements of elliptic hyperovals.

PROOF. Denote by H an even type 2h-set. Then, by (1.2), $2t_0=(h-5)(h-8)$, $2t_2 = 2h(8-h)$, $2t_4=h(h-3)$. So, $3 \leqslant h \leqslant 5$.
If h=3, then H is a 6-set of type (0,2), i.e. a 6-arc. The complement is exactly the case h=5.
Suppose h=4, then H is a 8-set of type (0,2,4) with $t_0=2$, $t_2=16$, $t_4=2$, hence the union of two parallel lines.

Now we deal with the case S(3,6,22).

2.16 PROPOSITION. Denote by H an even type set in S(3,6,22). Then H is one of the following sets
(1) a block;
(2) three blocks through two fixed points;
(3) the union of two disjoint blocks;
(4) the complement of a set considered above.

PROOF. Since H is an even type 2h-set, we have

$$t_0+t_2+t_4+t_6=77 \quad , \quad t_2+2t_4+3t_6=21h,$$

$$2t_4+6t_6=h(5h-13) \quad , \quad 6t_6=h(h-4)(h-5), \qquad \text{then } h \geqslant 3.$$

Suppose h=3, then H is a 6-set with $t_6 = 1$, hence a block.
In the case h=4, H is a 8-set with $t_0=7$, $t_2=56$, $t_4=14$, $t_6=0$. It follows that the complement $\overline{H}$ of H is a 14-set with $t_0=0$, $t_2=14$, $t_4=56$, $t_6=7$. Denote by B_1, B_2, B_3, three 6-secant blocks of $\overline{H}$. We claim that these three blocks have two common points. In fact suppose $|B_1 \cap B_3| = 0$, $|B_1 \cap B_2| = |B_2 \cap B_3| = 2$ and $\{x,y\} = B_2-(B_1 \cup B_3)$. A new 6-secant block of $\overline{H}$ contains x and y necessarily. So there are at most two of such blocks and $t_6 \leqslant 5$, a contradiction. So, $\overline{H}$ is the union of three blocks through two fixed points. Suppose h=5, then H is a 10-set with $t_6=0$, $t_4=30$, $t_2=45$, $t_0=2$, and the complement $\overline{H}$ is a 12-set with $t_6=2$. Denote by B_1 and B_2 the two 6-secant blocks of $\overline{H}$. Now we prove that $B_1 \cap B_2 = \emptyset$. Suppose $|B_1 \cap B_2| = 2$ and $\{p,q\} = \overline{H}-(B_1 \cup B_2)$ and $\{x,y\} = B_1 \cap B_2$. The block through x,y,p contains q. So there is a 3-secant block of $\overline{H}$, through x disjoint to the set $\{y,p,q\}$, a contradiction. So, $\overline{H}$ is the union of two disjoint blocks.

2.17 REMARK. S(3,6,22) contains also odd type sets. More precisely, L.Berardi proved in [2] that the odd type blocking sets, and then the odd type sets, necessarily, are the 7-sets of type (1,3) (called Fano sets), the 11-sets of type (1,3,5) which are the symmetric difference of a Fano set with a 1-secant block, and their complements.
Now we give the characterization of the fixed parity sets in S(4,7,23).

2.18 PROPOSITION. Denote by H and odd type set of S(4,7,23), then H is one of the following sets.
(1) a block;
(2) an odd blocking set;
(3) the union of three blocks through three fixed points.
The complements are, of course, the even type sets of the system.

PROOF. By (1.2) we have

$$\begin{cases} t_1 + t_3 + t_5 + t_7 = 253, & t_1 + 3t_3 + 5t_5 + 7t_7 = 77h, \\ 6t_3 + 20t_5 + 42t_7 = 21h(h-1), & 6t_3 + 60t_5 + 210t_7 = 5h(h-1)(h-2), \\ \qquad 120t_5 + 840t = h(h-1)(h-2)(h-3). & \end{cases}$$

It follows:

$$h^4 - 56\,h^3 + 1106\,h^2 - 9136\,h + 26565 = 0; \quad \text{hence} \quad h = 7,11,15,23\ .$$

Suppose h=7, then H is a block because $t_7 = 1$.
In the case h=11,H is a 11-set with $t_7 = 0$;so H is the irreducible blocking set of odd type constructed by L.Berardi in [3]. Now suppose h=15. Then H is a 15-set of type (3,5,7) because $t_1 = 0$. We claim that H is the union of three blocks through three fixed points. In fact let x be a point of H. If we consider the contraction of S(4,7,23) in x, we obtain a Steiner system S(3,6,22) in which $H-\{x\}$ is a 14-set of type (2,4,6). Then $H - \{x\}$ is the union of three blocks through two fixed points of the contraction S(3,6,22) by 2.16. So, the assertion follows.

Now we deal with the case S(5,8,24).

2.19 PROPOSITION. Denote by H an even type set of S(5,8,24). Then H is one of the following sets
(1) a block;
(2) an even blocking set;
(3) the complement of a block.

PROOF. By (1.2) we have,being h the cardinality of H:

$$\begin{cases} t_0 + t_2 + t_4 + t_6 + t_8 = 759 \\ 2t_2 + 4t_4 + 6t_6 + 8t_8 = 253\ h \\ 2t_2 + 12t_4 + +30t_6 + 56t_8 = 77h(h-1) \\ 24\ t_4 + 120\ t_6 + 336\ t_8 = 21\ h(h-1)(h-2) \\ 24\ t_4 + 360\ t_6 + 1680\ t_8 = 5h(h-1)(h-2)(h-3) \\ \qquad 720\ t_6 + 6720\ t_8 = h(h-1)(h-2)(h-3)(h-4)\ . \end{cases}$$

It follows:

$$h^5 - 60\,h^4 + 1280\,h^3 - 11520\,h^2 + 36864\,h = 0, \quad \text{hence } h = 0,8,12,16,24\ .$$

Suppose h=8, then H is a 8-set with $t_8 = 1$, a block. The case h=12 implies $t_0 = t_8 = 0$, then H is a blocking set, the even reducible blocking set of type (2,4,6) considered by Jonnson (cf.[4]). Suppose h=16, then H is a 16-set with t = 1, the complement of a block.

2.20 REMARK. In S(5,8,24) the odd type sets are blocking sets. Their characterization can be found in [4] .

Now we deal with the case of projective and affine spaces.

2.21 PROPOSITION. Let Σ be a projective or affine space of dimension r and order q. Then Σ contains fixed parity sets iff q is even. Moreover Σ contains odd type sets iff Σ is projective.

PROOF. Suppose Σ a projective space with dimension $r \geq 2$. Without loss of generality, we may suppose that r is even, (otherwise, we consider a hyperplane of Σ). If q is odd, then $r_1 = q^{r-1}+...+ q+1$ is even. So, by 2.5, Σ contains no fixed parity set. Suppose now that Σ is projective. If q is even, a hyperplane is an odd type set and the complement of a hyperplane is an even type set.

Now suppose that Σ is an affine space.

If q is even, then $v/k = q^{r-1}$ is even and $r = q^{r-1}+...+q+1$ is odd. So, by 2.6 Σ contains no odd type sets. Finally, we note that the union of two parallel hyperplanes in Σ is an even type set.

We conclude this Section by the following example.

2.22 EXAMPLE. We recall that a $\{k;n\}$-arc in PG(2,q) is a set K such that any line intersects it in at most n points. As well known we have $|K| \leq (n-1)q+n$. If equality holds, K is called a maximal arc and in this case K is a set of type (0,n). So, when q and n are even, a maximal arc is an even type set and its complement is an odd type set. Moreover, if K has at least one external line L, we call K an elliptic arc. An elliptic maximal arc in AG(2,q) of type (0,n) is an even type set K. Since its complement $\bar{K}$ is of type (q-n,q), it is an even type set too.

We recall that if K is a maximal $\{(n-1)q+n;\ n\}$-arc then the dual of the complement is a maximal $\{q(q+1-n)/n\ ;\ q/n\}$-arc. So, if q/n is even, this is an even type set too.

Now we consider in $PG(2,2^h)$ the family of conics F_λ which have, with respect to a system of homogeneus coordinates, the following equations:

$$x_0^2 + b\,x_0x_1 + x_1^2 + \lambda x_2^2 = 0$$

where $\lambda \in GF(q)$, and the equation $t^2 + bt + 1 = 0$ is irreducible in GF(q). We define F_λ as the double-line of equation $x_2^2 = 0$. So the family F_λ with $\lambda \in GF(q) \cup \{\infty\}$ is a partition of the plane formed by the point F_0 i.e. (0,0,1), the line F_∞ and the q-1 conics F_λ, $\lambda \in GF(q) - \{0\}$. F_0 is the knot of each of them.

Fix a subgroup H of GF(q) with $n=|H|$ and $n|q$. Then the family F_λ where H forms a $\{k;n\}$-arc, which is a maximal arc called a Denniston arc (cf. [9]). This maximal arc K is the union of the n-1 conics F_λ with $\lambda \in H - \{0\}$ and the point (0,0,1) which is the knot of each of them.

Note that if q/n is even, we can add to the $\{k;n\}$-arc K of Denniston an even number of lines through a point $x \notin K$. If we delete the point x, we get an even type set which we call an even type set derived from Denniston arc.

Moreover if one of the lines through the point x is fixed as the line at infinity and we fix an even number of parallel lines in the direction of x , then we have an even type set derived from Denniston's arc of the affine plane AG(2,q).

As an example we consider PG(2,8), where we have two possibilities.

(a) H=GF(2). In this case K is a 10-arc. There are 4 lines through an external point x. So we fix two lines L, M containing x and consider the 26-set $K \cup (L-\{x\}) \cup (M - \{x\})$, that is of type (0,2,4,8).

(b) H=GF(4). Then we have three conics and the common knot, which is a 28-set of type (0,4). The complement is a 45-set of type (5,9). The dual of the complement is the set considered in (a). In the affine plane, obtained by

fixing a line outside K, we have a 28-set of type (0,4), the complement being a 36-set of type (4,8).
We can add to K two lines and in this case we have a 44-set of type (0,4,6,8). We can consider the complement, which is an odd type set and the two associated affine sets.

Other examples of fixed parity sets and blocking sets can be found in [1] .

3. FIXED PARITY BLOCKING SETS IN AFFINE SPACES OF LARGE ORDER

In this Section we prove some bounds for the cardinality of blocking sets with a fixed parity. Many examples are given, in particular in every affine space of even order q.

We recall the following.

3.1 DEFINITION. Let Σ be a projective or affine space of dimension r and order q. A blocking set C in Σ is said to be of level ℓ if each line contains at least ℓ points of C and at least ℓ points outside C.

Now we prove

3.2 LEMMA. Let Σ be an affine or projective plane of even order $q>4$. Denote by C an even blocking set in Σ . Then

$$|C| \geqslant 2q + 6 . \tag{3.1}$$

PROOF. Denote by t_i the number of lines in Σ which intersect C in exactly i points ("i-secants") and by c the number of points of C. Suppose that every line is a 2-secant. Then

$$2t_2 = c(c-1) = c(q+1) = 2(q^2+ q + a)$$

where a = 0, 1 according whether Σ is affine or projective.
It follows $c = q + 2 = 2q + 2a/(q+1)$ which implies $q\leqslant 2$, a contradiction. So there is at least a 2j-secant with $j \geqslant 2$, and we have $c\geqslant 2q + 2i$ with $i \geqslant 2$. So, $c \geqslant 2q+4$ with equality only if C is of type (2,4). In this case we have

$$t_2 + t_4= q^2+ q + a \quad ; \quad 2t_2 + 4t_4 = (q+1)(2q+4) \ , \quad 2t_2+ 12t_4= (2q+4)(2q+3).$$

So, $4t_2 = 2q^2+ 4q$, $4t_4 = q^2 + 4q + 4$ and therefore $q^2- 4q + 4(a-1) = 0$

which implies $q\leqslant 5$, a contradiction. Hence, $i\geqslant 3$ and the assertion is proved.

3.3 COROLLARY. Let Σ be a projective or affine space of dimension r and even order $q>4$. Denote by C an even blocking set in Σ . Then

$$|C| \geqslant 2\Theta_{r-1} + 4 \quad , \Theta_{r-1}:=q^{r-1}+\ldots+q+1 . \tag{3.2}$$

PROOF. We may assume that C has minimal cardinality. Fix a 2-secant line L of C. (This line exists since C is minimal). Each of the Θ_{r-2} planes through L intersects C in at least 2q+4 points, so

$$|C|\geqslant 2 +\Theta_{r-2}(2q+4) = 2\Theta_{r-1}+ 4 .$$

3.4 COROLLARY. Let $\Sigma= AG(r,q)$ an affine space of dimension r and even order $q>4$. Denote by C an even blocking set in Σ . Then

$$2\Theta_{r-1}+ 4 \leqslant |C| \leqslant q^r-2\Theta_{r-1}-4 . \tag{3.3}$$

3.5 COROLLARY. Let $\Sigma=PG(r,q)$ a projective space of even order $q>4$. Denote by

C an odd blocking set. Then

(3.4) $$|C| \leq q^r - \Theta_{r-1} - 4 .$$

3.6 REMARK. In PG(r,2) and AG(r,2) there are no blocking sets ($r \geq 2$). In AG(2,4) there is only one blocking set which has a 3-secant line and a 2-secant line; therefore AG(2,4) contains no fixed parity blocking set. AG(r,4) contains no blocking sets for $r \geq 3$. In PG(2,4) there exist only two odd blocking sets namely the Baer subplanes and the Hermitian arcs, the complements being the even blocking sets.

3.7 LEMMA. In PG(r,4), $r \geq 3$, there are no fixed parity blocking set.

PROOF. Without loss of generality, we may assume that $r = 3$. Suppose that there exists an even blocking set in PG(3,4) having c points. We have $t_2 + t_4 = 357$, $2t_2 + 4t_4 = 21c$, $2t_2 + 12t_4 = c(c-1)$. This yields an equation in c without a real solution.

3.8 LEMMA. In a projective plane π_q (desarguesian or not) there are fixed parity blocking sets iff q is even and $q \geq 4$.

PROOF. If q is odd there are no fixed parity sets. If $q = 2$ there are no blocking sets. Thus the lemma is a consequence of the following examples.

3.9 EXAMPLE. (Triangle H without vertices). Let L_1, L_2, L_3 be three non-confluent lines in π_q. Let H consist of the points of L_1, L_2 and L_3 except their points of intersection. Then the set H is a (3q-3)-set of type (1,3,q-1), i.e. an odd type blocking set when q is even. The complement is an even type blocking set in π_q.

3.10 EXAMPLE. (Projective triade in a desarguesian plane). If q is even, put:

$$G_0 := \{ c : x^2 + x + c = 0 \quad \text{has a solution in } GF(q) \} .$$

The set G_0 is called the set of elements of category zero. The structure (G , +) is a group with q/2 elements.

In PG(2,q), q even, the set of points: (0,1,a) , (1,0,b), (1,1,m) , (0,0,1) where $a,b,m \in G$ is a blocking set of type (1,3,(q+2)/2) having (3q+2)/2 elements. This set is called a projective triade (cf. [10]). The complement of a triade is an even type blocking set.

3.11 EXAMPLE. In PG(2,q), q even, we fix an oval $\mathcal{O}$, a point $x \in \mathcal{O}$, the tangent T of $\mathcal{O}$ at x and the knot k of $\mathcal{O}$. The set

$$\mathcal{O} \cup T - \{x,k\}$$

is a blocking (2q-1)-set of type (1,3,q-1). This set is the symmetric difference of the (q+2)-arc $\mathcal{O} \cup \{k\}$ and the line T.

3.12 EXAMPLE. Suppose q even and square. Then a Baer subplane, the union of an odd number ($< q-\sqrt{q}+1$) of Baer subplanes and an Hermitian arc are odd blocking sets of π_q. In particular the union of 2s, with $2 \leq 2s \leq q-\sqrt{q}+1$, Baer subplanes is a blocking $2s(q-\sqrt{q}+1)$-set of type

$$(2s, 2s+\sqrt{q}, \ldots, 2s+t\sqrt{q}, \ldots, 2s+2s\sqrt{q})$$

with $1 \leq t \leq 2s$.

In the next part of this Section we prove that if q is large enough, in AG(r,q) there exist even blocking sets.

3.13 LEMMA. Let $h \geq 3$ be an integer. Then $\underline{A} = AG(2,2^h)$ contains an even blocking set of level 2^{h-2}.

PROOF. Let δ, δ' be two distinct directions in $AG(2,2^h)$. Denote by $\mathscr{L}$ the union of 2^{h-2} lines having direction δ and by $\mathscr{M}$ the union of 2^{h-1} lines in direction δ'. Put $H=\mathscr{L}\cap\mathscr{M}$. Denote by C the following symmetric difference

$$C := \mathscr{L} \triangle \mathscr{M} \quad .$$

We will prove that C is an even blocking set of level 2^{h-2}. Each line of direction δ or δ' intersects C in an even number of points, respectively.
If a line L, having a direction $\neq \delta, \delta'$, is a 0-secant, then L intersects C in exactly $2^{h-2}+2^{h-1}< 2^h$ points. If L is i-secant H, then L intersects $\mathscr{L}$ in $(2^{h-1}-i)$ points and $\mathscr{M}$ in $(2^{h-2}-i)$ points. So, in this case, L intersects C in $2^{h-2}+2^{h-1}-2i$ points where $1 \leq i \leq 2^{h-2}$. Then C is a blocking set of level 2^{h-2}. Moreover, since and are even type sets, their symmetric difference has the same property too (cf. [15]).

3.14 LEMMA. Suppose that in $\Sigma = AG(r,q)$ there is an even type blocking set C of level 2^m, $m \geq 3$. Suppose, moreover that q is even and $q \geq 2^{m+1}$. Then $AG(r+1,q)$ contains an even type blocking set of level $\ell = 2^{m-2}$.

PROOF. Let H_i (i = 1,...,) be ℓ parallel hyperplanes in Σ. Fix an even type bloking set C of level 2^m in H_1. Fix in Σ a direction δ which is not parallel to H_1. The family of lines M_x parallel to δ through the points $x \in C_1$ intersects H_i in a set C_i which is an even blocking set of level 2^m in H_i. We define C as the set of points which lie either in some C_i or outside H_i and outside the lines projecting C_i.

We claim that C is an even type blocking set of level ℓ in Σ. Denote by $\mathscr{L}$ the union of lines parallel to δ and containing points of H_1-C_1. Let L be a line parallel to δ. Then
1) $L \cap C_1 \neq \emptyset$ implies that L is ℓ-secant of C and $(q-\ell)$-secant of Σ-C. (Note that $(q-\ell) > \ell$).
2) $L \cap C_1 = \emptyset$ implies that L is $(q-\ell)$-secant of C and ℓ-secant of Σ-C.

Suppose now that L is not parallel to δ. The plane α through L and parallel to δ intersects H_i in a line L_i and the line set $\mathscr{L}$ in t lines with $2^m \leq t \leq q - 2^m$. Put

$$\mathscr{L}' = \bigcup_{i-1}^{\ell} L_i \qquad \text{and} \qquad \mathscr{M} = \bigcup_{i-1}^{\ell} M_i \quad .$$

Since $C \cap \alpha = \mathscr{L}' \triangle \mathscr{M}$, by 3.13, our Lemma is proved.

The following theorem follows by a repeated application of Lemma 3.13.

3.15 THEOREM. Fix an integer $r \geq 2$ and an even prime power $q \geq 2^{2r-1}$. Then $AG(r,q)$ contains an even type blocking set.

ACKNOWLEDGEMENT. The authors wish to thank Prof. L. Berardi and Prof. A. Beutelspacher for their helpfull comments.

REFERENCES

[1] V. Abatangelo, B.Larato, Ovali e {k,n}-archi in piani ciclici di ordine pari. To appear.

[2] L.Berardi, Blocking sets in the large Mathieu designs, I:the case S(3,6,22). In these Proceedings.

[3] L.Berardi, Blocking sets in the large Mathieu design , II: the case S(4,7,23). Preprint 1986.

[4] L.Berardi, F.Eugeni, Blocking sets in big Mathieu systems, III: the case S(5,8,24). Preprint 1986.

[5] A.Beutelspacher, Einfuhrung in die endliche Geometrie II. Projektive Raume (B.Istitut, Mannheim, Wien, Zurich, 1983).

[6] A. Beutelspacher, F. Eugeni, Sui blocking sets di dato indice con particolare riguardo all'indice tre. Boll. UMI 4-A (1985), 441-450.

[7] P.J. Cameron, J.H.Van Lint, Graph, codes and designs. London Math. Soc. 43 (1980), Cambridge University Press.

[8] M.J.de Resmini, On k-sets of type (m,n) in a Steiner system S(2,k,v). London Math. Soc., 49 (1981), 104-113.

[9] R.H.F. Denniston, Some maximal arcs in finite projective planes. J. Combinatorial Theory, 6 (1969), 317-319.

[10] J.W.P. Hirschfeld, Projective Geometries over finite fields. (Claredon Press, Oxford, 1979).

[11] J.W.P. Hirschfeld, X.Hubaut, Sets of even type in PG(3,4), alias the binary (85,24) projective geometry code. J. Combinatorial Theory, A 29 (1980), 101-112.

[12] H.Lenz, H.Zeitler, Arcs and ovals in Steiner triple system. Combin.Theory Proc. Lectures Notes 969, Springer Verlag 1982, 229-250.

[13] G.Tallini, Spazi combinatori e sistemi di Steiner. Riv. Mat. Univ. Parma 4 (1979), 221-248.

[14] G.Tallini, k-insiemi e blocking sets in PG(r,q) ed in AG(r,q). Quaderno n. 1, Ist. Mat. Appl. L'Aquila, 1982.

[15] G.Tallini, Spazi parziali di rette e codici correttori. To appear.

[16] V.D. Tonchev, A characterization of designs related to the Witt System S(5,8,24), Math. Z. 191 (1986), 225-230.

F.EUGENI

Istituto di Scienze
Facoltà di Architettura

Università di Chieti (ITALIA)

S.INNAMORATI

Via Colli Innamorati

Pescara(ITALIA)

Annals of Discrete Mathematics 37 (1988) 169–176

BLOCKING SETS OF INDEX TWO

Franco EUGENI and Erika MAYER

ABSTRACT. The index of a blocking set C of a t-(v,k,λ) design is the minimal number i(C) of blocks whose union contains C. It is well-known that i(C) ≥ 3 for projective and affine planes. In this paper we will show that i(C) ≥ 3 for all Steiner systems with t = 2. We also prove that a blocking set of index 2 exists in an inversive plane I(q) if and only if q = 3. This is followed by a survey of those Steiner systems S(t,k,v) which may admit blocking sets of index 2 and examples of such sets. Furthermore, we give a direct proof of a result due to Drake which says that no 2-(v,3,λ) design with v>4 admits a blocking set. We also show that the number of points of a blocking set in an S(v,4,λ) with v>5 is v/2.

1. INTRODUCTION

A t-(v,k,λ) *design* **D** is a pair (S,**B**), where S is a set of elements called *points* and the set **B** consists of k-subsets of S called *blocks* such that through any t points there pass exactly λ blocks of **B**. To exclude trivial cases we suppose that $2 \le t < k < v$. If **B** consists of all k-sets of S, then **D** is called a *complete design*. A t-(v,k,1) design is called a *Steiner system* and is also denoted by S(t,k,v) while a 2-(v,k,λ) design is called a *block design*.

A t-(v,k,λ) design is an s-(v,k,r_s) design for all $0 \le s \le t$, where

$$(1.1) \qquad r_s = \binom{v-s}{t-s}\lambda / \binom{k-s}{t-s}.$$

If we set $b = r_0$, the number of blocks, and $r = r_1$, the number of blocks through a point, then $vr = bk$.

A set C of points is called a *blocking set* of **D** if it meets every block in at least one point but not in all points. Obviously, the complement of a blocking set is a blocking set. Blocking sets have mainly been studied in projective and affine spaces, whereas in Steiner systems only a few results have been obtained so far.

For every integer $0 \le h \le k$ we define the *character* x_h of a set C containing c points to be the number of blocks which meet C in exactly h points. Counting in two ways the number of ordered pairs consisting of s points of C and a block through them we obtain the following equations:

$$(1.2) \qquad \sum_{h=0}^{k} \binom{h}{s} x_h = \binom{c}{s} r_s \qquad \textit{for } s=0,\dots,t.$$

If a t-(v,k,λ) design contains a blocking set of index i, then we can obtain an upper bound for the number of blocks in terms of i, k and r. Let D be the union of the i blocks which contain the blocking set, and suppose that D has exactly d points. Each of these d points is incident with r blocks each of which contains exactly $1 \le h \le k$ points of D. As each block intersects D non-trivially we have

$$b = dr - \sum_{h=2}^{k} x_h (h-1) \le ikr - i(k-1).$$

Since $vr = bk$ we obtain the following proposition for the number of points.

1.1 Proposition. The number of points of a t-(v,k,λ) design which admits a blocking set of index i satisfies

$$v = dk - \frac{k}{r}\sum_{h=2}^{k} x_h(h-1) \le ik\,(k - \frac{k-1}{r}) < ik^2. \quad \square$$

The number of points of an S(t,k,v) satisfies the well-known Fisher-inequality

(1.3) $v \ge (k-t+2)(k-t+1)+t-1$,

while the number of points of a blocking set C of such a system lies between the following bounds (see [2] or [12])

(1.4) $r_{t-1} < |C| < v-r_{t-1}$.

2. BLOCK DESIGNS

In this section we show that no Steiner system S(2,k,v) contains a blocking set of index 2 . Furthermore, we give a direct proof for Drake's result that a 2-(v,3,λ) design with more than 4 points does not contain a blocking set.

2.1 Theorem. No Steiner system S(2,k,v) admits a blocking set of index 2 .

Proof. Suppose there exists a blocking set C of index 2 in a Steiner system S(2,k,v) . The two blocks B and B' which contain C have at most 1 point in common. If they have a common point x , say, then x lies on at least one block which does not contain a further point of $B \cup B'$. So x is in C .

Since C does not contain a block, there exist in either case a point y on $B \setminus B'$ and a point z on $B' \setminus B$ which are not in C . So the unique block through y and z is not blocked by C . This contradiction proves the theorem. □

Drake [8] obtained the following result as a corollary of theorems on blocking sets in block designs. The case $\lambda = 1$ had been eliminated before by Beutelspacher [3] (section 7.3, exercise 4) and Tallini [12] .

2.2 Theorem. The only 2-(v,3,λ) design which admits a blocking set is the complete design with 4 points. Its blocking sets are exactly the sets of 2 points which necessarily have index 1 .

Proof. Clearly, any 2 points of the complete design on 4 points form a blocking set. Now suppose that a 2-(v,3,λ) design **D** with v>4 admits a blocking set C . Since every block meets C in at least 1 point and in at most 2 points, we obtain the following equations for the characters of C :

$$\begin{aligned} x_1 + x_2 &= v(v-1)\lambda/6 \\ x_1 + 2x_2 &= c(v-1)\lambda/2 \\ x_2 &= c(c-1)\lambda/2 \end{aligned} \qquad (2.1)$$

This gives the quadratic equation $3c^2-3cv+v^2-v=0$ whose discriminant $3v(4-v)$ is negative for v>4 . This is the required contradiction. □

In his paper Drake made the assumption $k+2 \le v$. But any 2-(v,v−1,λ) design is complete, as can easily be seen by using (1.1) , and admits a blocking set.

2.3 Proposition. There exists a blocking set in a complete 2-(v,k,λ) design **D** if and only if v<2k−1 . In this case every c-set with v−k<c<k is a blocking set.

Proof. Let C be a set of order c . Since **D** is complete , every k-set is a block. Hence C does not contain a block if and only if c<k ; it has no exterior blocks if and only if v−k<c. So C is a blocking set if and only if v−k<c<k . Combining the 2 inequalities gives v−k<k+1 or v<2k+1 . This proves the proposition. □

By Drake's result, the contraction of a 3-(v,4,λ) design **D** with v>5 does not admit a blocking set. The design **D** , however, might contain a blocking set; an example is the inversive plane I(3) (see Section 3). The number of points of such a blocking set is

$v/2$. This was shown by Tallini [12] for the case $\lambda = 1$.

2.4 Theorem. If C is a blocking set of a 3-$(v,4,\lambda)$ design **D**, then
(i) **D** is the complete 3-(5,4,2) design, or
(ii) C consists of $v/2$ points and $x_1 = x_3 = v(v-2)(v-4)\lambda/48$, $x_2 = v(v-2)\lambda/8$.

Proof: We obtain the following system of equations for the characters of a blocking set C with c points:

$$
(2.2)\qquad
\begin{aligned}
x_1+x_2+x_3&=b\\
x_1+2x_2+3x_3&=cr\\
x_2+3x_3&=c_2r_2\\
x_3&=c_3\lambda,
\end{aligned}
$$

where $c_2 = \binom{c}{2}$ *and* $c_3 = \binom{c}{3}$.

Subtracting the first from the second and the fourth from the third equation yields

$$
\begin{aligned}
x_2+2x_3&=cr-b\\
x_2+2x_3&=c_2r_2-c_3\lambda .
\end{aligned}
$$

Hence $cr-b = c_2r_2-c_3\lambda$. Substituting the respective values for r, r_2, c_2 and c_3 gives

$$\frac{c(v-1)(v-2)}{6}\lambda - \frac{v(v-1)(v-2)}{24}\lambda = \frac{c(c-1)(v-2)}{4}\lambda - \frac{c(c-1)(c-2)}{6}\lambda$$

or $2(c-v/2)(2c^2-2cv+v^2-3v+2) = 0$. Hence $c = v/2$ or $(2c^2-2cv+v^2-3v+2) = 0$. The second equation has no solution for $v>5$. If $v = 5$, then **D** is the complete 3-(5,4,2) design in which every 2-set and every 3-set is a blocking set.
If $c = v/2$, then $x_3 = c_3\lambda$. Now a block intersects C in exactly 1 point if and only if it contains exactly 3 points of the complement of C. Since this complement is also a blocking set on $v/2$ points, we have $x_1 = x_3$. It follows from the first equation of (2.2) that $x_2 = b-2x_3 = v(v-2)\lambda/8$. □

3. INVERSIVE PLANES

In this section we examine the inversive planes I(q) which are the Steiner systems $S(3,q+1,q^2+1)$. We shall prove the following theorem in a series of steps.

3.1 Theorem. An inversive plane I(q) admits a blocking set of index 2 if and only if $q = 3$. The blocking sets of I(3) are precisely the sets $B \cup B' \setminus \{x\}$, where B and B' are two blocks intersecting in x and a further point, and where the symmetric difference $B \Delta B'$ is not a block. □

We begin by considering the case $q = 3$. The existence of blocking sets of index 2 in I(3) has already been looked at in [2].

3.2 Lemma. Let B and B' be two blocks of I(3) which intersect each other in exactly 2 points. Then the symmetric difference $B\Delta B'$ is a block if and only if the complement of $B \cup B'$ is a block.

Proof. Let D be the complement of $B \cup B'$, then D consists of 4 points. We obtain the following equations for the characters of $B \cup B'$:

$$
(3.1)\qquad
\begin{aligned}
x_0+x_1+x_2+x_3+x_4&=30\\
x_1+2x_2+3x_3+4x_4&=72\\
x_2+3x_3+6x_4&=60\\
x_3+4x_4&=20 .
\end{aligned}
$$

Assume that D is a block. Then it is the only block which does not contain a point of $B \cup B'$. Hence $x_0 = 1$ and $x_1 = 0$, $x_2 = 18$, $x_3 = 8$ and $x_4 = 3$. This means that $B \cup B'$

contains a block E different from B and B'. Since $t=3$, E does not contain a point of $B \cap B'$. Thus E must consist of the 4 points of $B \Delta B'$.

Conversely, let $B \Delta B'$ be a block of I(3). Then $x_4 = 3$ and (3.1) yields $x_0 = 1$. This implies that D is a block. □

3.3 Theorem. A set C in I(3) is a blocking set if and only if $C = B \cup B' \setminus \{x\}$ for 2 blocks B and B' whose symmetric difference is not a block and where x is one of the 2 points of $B \cap B'$.

Proof. Let C be a blocking set of I(3). By Theorem 2.4, both C and its complement C^* consist of 5 points. Since C does not contain any block, there exist exactly 5 choose 3 equals 10 blocks which contain 3 points of C and 1 of C^*. Now let x be any point of C^*. Then x lies on at most 2 of these 10 blocks, since they cannot have x and more than 1 of the 5 points of C in common. As there are 10 blocks and 5 points, each point of C^* lies on exactly 2 blocks which contain 3 points of C. Call these blocks B and B', then $C = B \cup B' \setminus \{x\}$. Since C is a blocking set contained in $B \cup B'$, the complement of $B \cup B'$ and thus $B \Delta B'$ cannot be blocks.

Conversely, let B and B' be 2 distinct blocks which intersect in the 2 points x and y. Let $C = B \cup B' \setminus \{x\}$. Since $B \Delta B'$ is not a block, C does not contain a block. Moreover, by Lemma 3.2, a block E has at most 3 points in common with the complement D of $B \cup B'$. Let E contain exactly 3 points of D. If x was one of the three points, then E would be one of the 4 blocks through x and y which cover the points of I(3). This contradiction shows that x is not in E and that C intersects E in a point. If E contains at most 2 points of D, then C intersects E in at least one point. This proves the theorem. □

By searching through the table of the unique I(3) which is listed in [10] we see that most but not all blocks which intersect in exactly 2 points give rise to blocking sets. Note that not every 5-set is a blocking set; the set $\{P_1, \ldots, P_5\}$, for instance, is not one.

3.4 Corollary. The inversive plane I(3) contains blocking sets of index 2. □

To show that no I(q) with $q > 3$ admits a blocking set of index 2, we need the following lemma and remark.

3.5 Lemma. Let B and B' be 2 blocks of I(q). For a point x in $B \setminus B'$ let ψ be the number of blocks which contain x but no further point of $B \cup B'$. Then the following hold:

(i) If $B \cap B' \neq \emptyset$, then
$\psi = (q-2)/2$ for q even, and
$\psi \geq (q-3)/2$ for q odd.

(ii) If $B \cap B' = \emptyset$, then
$\psi = (q-4)/2$ for q even, and
$\psi \geq (q-5)/2$ for q odd.

Proof. The contraction of I(q) in the point x is an affine plane **A** in which B' forms an oval. So ψ is just the number of exterior lines of B' which are parallel to the line $B^* = B \setminus \{x\}$ and not equal to B^*. Let **P** be the projective closure of **A** with line at infinity L.

If q is even, then all tangents contain the same point y (see [3] or [9]). The point y is not on L, since L is an exterior line. Every point $\neq y$ lies on a unique tangent to B'. It follows that $\psi = q-(1+q/2)$ if $B \cap B' \neq \emptyset$, and $\psi = q-1-(1+q/2)$ if $B \cap B' = \emptyset$.

If q is odd, then every point of **P**, which is not in B', lies on 0 or 2 tangents to B'. It follows that $\psi \geq q-1-(2-(q-1)/2)$ if the two blocks B and B' have no point in common and $\psi \geq q-(2-(q-1)/2)$ otherwise. □

3.6 Remark. Every inversive plane with $q = 2^h$ is ovoidal by a well-known theorem of Dembowski [5]. For $q = 5$ there exists a unique inversive plane which is ovoidal; this was proved by Yi Chen [4]. □

We shall now complete the proof of the theorem.

Proof of Theorem 3.1. The case $q=3$ was already dealt with. Let $q>3$ and suppose that there exists a blocking set C which is contained in 2 blocks B and B'.

Case 1: $B\cap B' \neq \emptyset$. By Lemma 3.5(i), every point of the symmetric difference of B and B' lies on a block which does not contain a further point of $B\cup B'$. Hence $B\Delta B' \subseteq C$. If B and B' have exactly one point y in common, then the contraction in y is an affine plane in which $B\backslash\{y\}$ and $B'\backslash\{y\}$ are parallel. Hence there exist $q-2\geq 1$ blocks each of which intersects $B\cup B'$ just in y. Hence $y\in C$ and $B\cup B'\subseteq C$, a contradiction. Now assume that $B\cap B' = \{x,y\}$. If neither x nor y is in C, then none of the remaining $r_2-2 \geq 1$ blocks through x and y is blocked by C. So C contains exactly one of the two points, say x. Hence $|C|=2q-1$. Let x'_4 be the number of blocks other than B or B' which intersect C in exactly 4 points. For $q\geq 5$ we have of course $x_4 = x'_4$, since $|B\cap C| = |B'\cap C| = q$. We obtain the following equations for the characters of C.

$$\begin{aligned}
x_1+x_2+x_3+x'_4 \qquad\qquad &+2=b=q(q^2+1)\\
x_1+2x_2+3x_3+4x'_4 \qquad\quad &+2q=(2q-1)(q+1)q\\
x_2+3x_3+6x'_4 \qquad &+q(q-1)=(2q-1)(2q-2)(q+1)/2\\
x_3+4x'_4&+q(q-1)(q-2)/3=(2q-1)(2q-2)(2q-3)/6\ ,
\end{aligned}$$

This has the solution $x_1 = q^3-4q^2+8q-5$, $x_2 = -q^3+7q^2-10q+4$, $x_3 = q^3-3q^2+3q-1$, $x_4' = 0$. As $x_2<0$ for $q\geqq 6$ it remains to exclude the cases $q = 4$ and $q = 5$. Now choose three points p, q, u on $B\backslash B'$ and a point w on $B'\backslash B$. The blocks E through p, q, w and F through p, u, w are distinct and, since $x_4' = 0$, neither of them contains a further point of $B\cup B'$. In the contraction in w the line $B'\backslash\{w\}$ is parallel to both $E\backslash\{w\}$ and $F\backslash\{w\}$. But this is impossible since E and F intersect in p. This is the required contradiction.

Case 2: $B\cap B'=\emptyset$. It follows from part (ii) of Lemma 3.5 that $q<6$. It remains to exclude the cases $q=5$ and $q=4$.

By Remark 3.6, an inversive plane with $q=4$ or $q=5$ is ovoidal. Hence there exists an elliptic quadric Q whose sections are exactly the blocks of the system. So there exist 2 planes α and β which intersect Q in the disjoint conics B and B' and which are tangent to an external line of Q. This line is contained in 2 planes which are tangent to Q. Furthermore, there is at least a third plane through this line which intersects Q in a block external to the 2 blocks B and B'. This completes the proof of Theorem 3.1. □

4. BLOCKING SETS IN STEINER SYSTEMS

In this section we present some results about blocking sets of index 2 in arbitrary Steiner systems $S(t,k,v)$. We begin with the following examples, a detailed discussion of which can be found in [1].

4.1 Example. The Steiner system $S(3,4,8)$ is the unique design $AG_2(3,2)$ of planes in a 3-dimensional affine space over $GF(2)$. Its blocking sets are the sets of 4 independent points. These blocking sets have index 2.

4.2 Example. In $S(3,6,22)$ the blocking sets of index 2 are precisely the sets $B\cup B'\backslash\{x,y\}$, where B, B' are disjoint blocks and $x\in B$, $y\in B'$.

4.3 Example. The blocking sets of index 2 in $S(4,7,23)$ are precisely the sets $B\cup B'\backslash\{x\}$, where B, B' are blocks with $B\cap B'=\{x\}$.

4.4 Example. In $S(5,8,24)$ the blocking sets of index 2 are precisely the sets $B\cup B'\backslash\{x,y\}$ and $B\cup B'\backslash\{x\}$, where B and B' are blocks with $B\cap B'=\{x,y\}$.

4.5 Example. In $S(5,6,12)$ the blocking sets of index 2 are the 6-sets different from a block.

We now give a classification of Steiner systems which admit blocking sets of index 2.

4.6 Theorem. If a Steiner system $S(t,k,v)$ contains a blocking set C of index 2, then the system is one of the following:

(a) $S(t,k,v) = S(3,4,8)$ and C is a set of 4 independent points,

(b) the inversive plane $I(3)$ and C is as stated in Theorem 3.1,

(c) the extension $S(5,6,12)$ of $I(3)$ and C is any 6-set different from a block,

(d) $S(t,k,v)$ is one of $S(3,14,170)$ or $S(4,15,171)$,

(e) the big Mathieu systems $S(3,6,22)$, $S(4,7,23)$ and $S(5,8,24)$, where the blocking sets are as described in Examples 4.2 and 4.4.

(f) the systems with $t \geq 6$ and where $v = (k-t+1)(k-t+2+j)+t-1$, $3 \leq j \leq t-2$,

(g) the systems $S(t,k,v)$ with $v = (k-t+1)(k-t+2+j)+t-1$, where $h \leq j \leq t+s-3$ and $h = \max(3,t-1)$, $1 < s \leq k-2$.

Proof. By Theorem 2.1 we may assume $t \geq 3$. We now proceed in several steps.

Step 1. Let C be a blocking set contained in the union of 2 blocks, then $|C| \leq k+s \leq 2k-2$ for some s with $0 \leq s \leq k-2$. Using (1.1), (1.2) and (1.3), we obtain

(4.1) $v = (k-t+1)(k-t+2)+t-1+j(k-t+1)$, where $0 \leq j \leq t+s-3$ and $0 \leq s \leq k-2$ and $j(j-1) \equiv 0 \mod (k-t+2)$.

Step 2. If $j = 0$ $(s = k-2)$, then $S(t,k,v)$ is $S(3,4,8)$, $S(3,6,22)$, $S(4,7,23)$ or $S(5,8,24)$: By (4.1) we have $v = (k-t+1)(k-t+2)+t-1$. So, $r_{t-3} = A(k,t) + 12/(k-t+3)$ for an appropriate integer $A(k,t)$. Since $k \geq t+1$, it follows that $k-t+3 \in \{4,6,12\}$. We are left with the cases $S(t,t+1,t+5)$, $S(t,t+3,t+19)$ and $S(t,t+9,t+109)$. A system $S(t,t+1,t+5)$, whose $(t-3)$-rd contraction is $S(3,4,8)$ can only exist for $t \leq 3$. A system $S(t,t+3,t+19)$ exists if and only if $t \leq 5$. Finally note that any $S(t,t+9,t+109)$ is an extension of a projective plane of order 10. The non-existence of such an extension was proved in [11].

Step 3. For $j = 1$ the system is $I(3)$ or one of $S(4,5,11)$, $S(4,15,171)$, $S(5,6,12)$: By (4.1) we have $v = (k-t+1)(k-t+3)+t-1$. The contraction of such an $S(t,k,v)$ in $t-3$ points gives the system $S(3,q+1,q^2+1)$ where $q = k-t+2$. The assertion follows, since $S(4,5,11)$ does not contain a blocking set [2].

Step 4. Let $2 \leq j \leq t-2$. We note that $r_{t-1} \equiv 0 \mod (k-t+2)$ if and only if $j(j-1) \equiv 0 \mod (k-t+2)$. For $j = 2$ we have $k-t+2 = 1$ or 2. Hence $k = t-1$ or $k = t$, a contradiction. So $j \geq 3$ and $j(j-1) \neq 0$. Now $k-t+2 \leq j(j-1) \leq (t-2)(t-3)$ implies $k \leq (t-2)^2$. For $t = 3$ or $t = 4$ we have $k \leq 1$ or $k \leq 4$, a contradiction.
Suppose that $t = 5$, then $k \leq 9$. But the parameters of the systems which we obtain from (4.1) are not admissible: since $j = 3$ we have $6 \equiv 0 \mod (k-3)$ and $k-3 \in \{3,6\}$. This yields the two systems $S(5,6,16)$ and $S(5,9,49)$, the first of which does not exist [6]. The existence of an $S(5,9,49)$ implies the one of an $S(3,7,47)$. But the parameters $(3,7,47)$ contradict (1.1).
Consequently we have $t \geq 6$ and the systems described in (f).

Step 5. For $j > t-2$ we have $t-2 < j \leq t+s-3$. Hence $s > 1$ and there remain the Steiner systems described in (g). □

We note that the existence of systems $S(3,14,170)$ and $S(4,15,171)$ is undecided.

4.7 Corollary. Let B be a block in an $S(t,k,v)$ with $t \leq 5$, and let x, y be 2 points not in B. If C is a blocking set with $C \subset B \cup \{x,y\}$, then
(a) $S(t,k,v) = S(3,4,8)$,
(b) $S(t,k,v) = S(3,14,170)$ or $S(4,15,171)$, or
(c) $S(t,k,v) = S(5,6,12)$.

Proof. Since $C \subset B \cup \{x,y\}$ we have $|C| \leq k+1$ (as $s \leq 1$). This excludes case (g) of Theorem 4.6; case (f) is impossible as $t \leq 5$. Case (e) does not occur, since those blocking sets are not of the required type. This proves the assertion. □

In the following theorems arising from Theorem 4.6 we give some "classes" of Steiner systems $S(t,k,v)$ which (independent of the question of their existence) can admit blocking sets of index 2. To find these systems we used the Doyen-Rosa-table for Steiner systems with $v \le 100$ [7].

4.8 Theorem. Let the Steiner system $S(3,k,v)$ admit a blocking set of index 2.
(a) If $v \le (k-2)(k+3)+2$, then
$(3,k,v) = (3,4,8)$ or $(3,14,170)$, or
$S(3,k,v)$ is the inversive plane $I(3)$ of order 3,
(b) If $v \le (k-2)(k+6)+2$, then we have the systems described in (a), or
$S(3,k,v) = S(3,4,20)$, $S(3,4,22)$, or
$(3,k,v) = (3,6,42)$, $(3,6,46)$ or $(3,21,477)$.

Proof. (a) By Theorem 4.6 (g) and with $t=3$ we have $v=(k-2)(k-1+j)+2$ and $3 \le j \le k-2$. Since we assumed $v \le (k-2)(k+3)+2$ we obtain $k-1+j \le k+3$ and $j \le 4$. If $j=3$, then $(k-1)$ divides 6. Hence $k=7$ and the system has the parameters $(3,7,47)$, which is arithmetically impossible.
If $j=4$, then $(k-1)$ divides 12, so $S(3,k,v) = S(3,7,52)$ or $S(3,13,178)$. But both possibilities lead to a contradiction.

(b) Combining $v=(k-2)(k-1+j)+2$ $(3 \le j \le k-2)$ and $v \le (k-2)(k+6)+2$ gives $j \le 7$.
If $j=3,4$, then we have the systems described in (a).
If $j=5$, then $(k-1)$ divides 20. Thus $k=5$, 6, 11 or 21 and $v=(k-2)(k+4)+2$. The values $k=5$, 11 are arithmetically impossible. The existence of Steiner systems $S(3,6,42)$, $S(3,6,46)$ and $S(3,21,477)$ is undecided.
The cases $j=6$ and $j=7$ are treated in the same way. □

4.9 Theorem. Let C be a blocking set contained in the union of 2 blocks of an $S(4,k,v)$ with $v \le (k-3)(k+3)+3$. Then the only possible cases are
(a) $(k,v) = (6,27)$ or $(7,23)$,
(b) $(k,v) = (5,17)$, $(7,43)$ or $(15,171)$.

Proof. Using the same arguments as in the proof of Theorem 4.8 we obtain $j \le 5$. The rest follows from Theorem 4.6. □

4.10 Theorem. Let C be a blocking set contained in the union of 2 blocks of an $S(5,k,v)$ with $v \le (k-4)(k+1)+4$. Then we have the following systems:
(a) $S(5,6,12)$, $S(5,7,28)$, $S(5,8,24)$,
(b) $S(5,6,18)$.

Proof. Use Theorem 4.6. □

4.11 Theorem. Let C be a blocking set of index 2 of an $S(t,k,v)$ with $t \ge 6$ and $v \le (k-t+1)(k+1)+t-1$. Then $v=(k-t+1)(k-t+2+j)+t-1$, where $3 \le j \le t-1$.

Proof. By Theorem 4.6 (f), (g). □

4.12 Corollary. There is no known Steiner system $S(6,k,v)$ for $v \le (k-5)(k+1)+5$ which contains a blocking set of index 2.

Proof. Theorem 4.11 yields the Steiner systems $S(6,7,19)$, $S(6,8,29)$, $S(6,9,45)$, $S(6,14,140)$, $S(6,16,181)$, $S(6,24,480)$ whose existence is undecided. The parameters $(k,v) = (7,17)$, $(8,32)$, $(10,50)$, $(10,55)$ are arithmetically impossible. □

4.13 Corollary. There is no known Steiner system $S(7,k,v)$ with $v \le (k-6)(k+1)+6$ which contains a blocking set of index 2.

Proof. By using Theorem 4.11 and taking into account that systems with parameters $(k,v) = (8,20)$, $(8,24)$, $(9,30)$, $(10,46)$, $(15,141)$, $(15,150)$, $(17,182)$, $(20,300)$, $(25,481)$, $(35,1050)$ are not known to exist. □

REFERENCES

[1] L. Berardi and F. Eugeni, Blocking sets in big Mathieu systems I, II, III (to appear).

[2] L. Berardi, F. Eugeni and O. Ferri, Sui blocking sets nei sistemi di Steiner, Boll.Un.Mat.Ital., Serie VI, Vol.III-D, N.1 (1984).

[3] A. Beutelspacher, *Einführung in die endliche Geometrie I, II*, Bibliographisches Institut, Mannheim/Wien/Zürich, 1982.

[4] Y. Chen, The Steiner system S(3,6,26), J. Geometry 2 (1972), 7-28.

[5] P. Dembowski, *Finite Geometries*, Springer-Verlag, Berlin Heidelberg New York, 1968.

[6] M.J. de Resmini, On blocking sets in symmetric BIBD's with $\lambda = 2$, J. Geometry 18 (1982), 194-198.

[7] J. Doyen and A. Rosa, An updated bibliography and survey of Steiner systems, Ann. Discr. Math. 7 (1980), 317-349.

[8] Q.A. Drake, Blocking sets in block designs, J. Comb. Theory (A) 40 (1985), 459-462.

[9] D.R. Hughes and F.C. Piper, *Projective Planes*, Springer-Verlag, Berlin Heidelberg New York, 1973.

[10] F. Karteszi, *Introduction to Finite Geometries*, North-Holland, 1976.

[11] C.W.H. Lam, L. Thiel, S. Swiercz and J. McKay, The non-existence of ovals in a projective plane of order 10, Discr. Math. 45 (1983), 319-321.

[12] G. Tallini, Blocking sets nei sistemi di Steiner e d-blocking sets in PG(r,q) e in AG(r,q), Sem.Geom.Comb.Fac.Ing.Univ.L'Aquila 3, 1982.

Franco Eugeni
Dipartimento di Matematica Applicata
Università dell'Aquila
I-67100 L'Aquila
Italy

Erika Mayer
Fachbereich Mathematik
Universtät Mainz
Saarstr.
D-6500 Mainz
West Germany

Annals of Discrete Mathematics 37 (1988) 177–180
Elsevier Science Publishers B.V. (North-Holland)

A SHORT PROOF THAT ORDERED LINEAR SPACES ARE LOCALLY PROJECTIVE

Rolfdieter FRANK

Mathematisches Seminar der Universität Hamburg
Bundesstraße 55
2000 Hamburg 13, West Germany

1. INTRODUCTION

Pasch [4] proves that geometries satisfying certain axioms of incidence and of order, which are true for example in the geometry induced on a convex open subset of the point set of euclidean space, can be imbedded into a 3-dimensional projective space. One of his axioms is that two planes never have exactly one point in common. Schur [6] showed that without this axiom the geometries still can be embedded into a projective space, but now the dimension may be greater than three. This result can also be found in [7, pp.5-23], [1, pp.94-116], [8], or [2, pp.59-66]. An essential part of Schur's proof is to show that the geometries are locally projective, i.e. the lines and planes through any given point are the points and lines of a projective space if incidence is defined by inclusion. By [9] and [3], the fact that the geometries are locally projective implies their embeddability, except for the 3-dimensional case, where the validity of the Bundle Theorem is needed. See [4, pp.33-34] for a proof of the Bundle Theorem. In this note, we give a short proof that the geometries investigated by Schur are locally projective. Our key lemma (see (3) below) is Teorema V of [5], but appearently Peano did not see its consequences.

2. DEFINITIONS AND THEOREM

A *linear space* is a set P whose elements are called *points* together with a set L of subsets of P, called *lines*, such that any two points X, Y $\in$ P are contained in exactly one line XY and every line contains at least two points. A subset S $\subset$ P is *closed* if XY $\subset$ S for any two points X, Y $\in$ S. The *closure* of a set T $\subset$ P is the intersection of all closed sets containing T. A *plane* is the closure ABC of three points A, B, C not on a line. A linear space (P, L) is *ordered* if it contains four points not on a plane, and if a ternary relation "*between*" is defined on P, which satisfies:

(i) B between A, C implies that A, B, C are on a line.

(ii) On every line l $\in$ L one can define a total order "<" such that l has neither maximal nor minimal elements and that for all A, B, C $\in$ l B is between A, C if and only if A < B < C or C < B < A.

(iii) (Axiom of Pasch, see [4, p.20]) Let A, B, C be three points not on a line. If a line l in the plane ABC with A, B, C $\notin$ l contains a point D between A, B, then it contains a point E between B, C or between C, A.

Though Schur postulates weaker axioms, his geometries turn out to be ordered linear spaces. In the next section, we will give a short proof of the following result due to Schur (see [6, p.269] or [7, p.13]):

THEOREM. *Every ordered linear space is locally projective.*

3. PROOF OF THE THEOREM

Let (P, L) be an ordered linear space. Instead of "the Axiom of Pasch applied to the triangle ABC" we write "Pasch (ABC)".

(1) *For any two points A, B ∈ P there exists a point X between A, B.*

Proof. Choose a point C not on AB. By (ii), we can find points D ∈ BC and E ∈ AD such that C is between B, D and A is between D, E. Now Pasch (ABD) implies that CE contains a point X between A, B.

(2) *If A, B, C are noncollinear points in a plane $\varepsilon \subset P$, then $\varepsilon = ABC$.*

Proof. Let $\varepsilon = UVW$ and $X \in \varepsilon \setminus UV$. Since ε is closed, we have $UVX \subset \varepsilon$. Choose a point T between U, V. If $W \notin TX$, then by Pasch (UVW), TX contains a point S between U, W or between V, W. Hence $W \in TX \subset UVX$ or $W \in SU \cup SV \subset UVX$. This implies the other inclusion $\varepsilon \subset UVX$. Applying what we have just proved, we obtain $\varepsilon = UVW = UVA = UBA = ABC$, since we may assume $A \notin UV$ and $U \notin AB$.

(3) *Let $\varepsilon, \delta \subset P$ be planes with $\varepsilon \cap \delta = l \in L$. Then any plane containing a line $m \subset \varepsilon$ with $m \neq l$ and a point $A \in \delta \setminus l$ intersects δ in a line.*

Proof. We may assume $l \cap m = \emptyset$, since the existence of a point $H \in l \cap m$ implies $mA \cap \delta = AH$. Choose points $B \in m$, $C \in AB$, $D \in l$, $E \in AD$ such that B is between A, C and A is between D, E. By Pasch (ACE), BD contains a point F between C, E, and by Pasch (DEF), B is between D, F. By Pasch (DFG) and (2), m contains a point S between F, G, and by Pasch (EFG), CS contains a point T between E, G. Since $T \in \delta \setminus \{A\}$, the plane containing m and A intersects δ in the line AT.

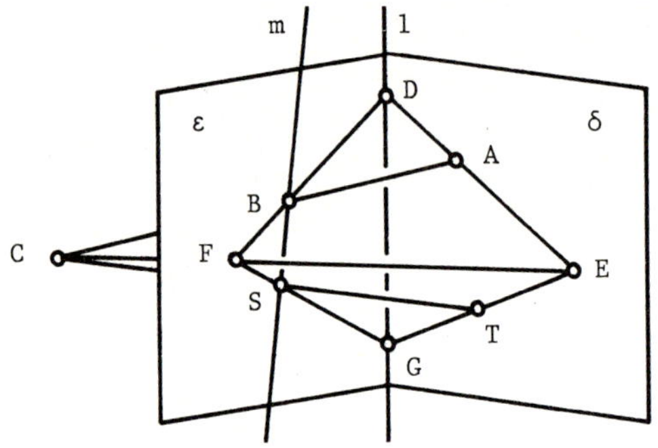

(4) *Let XA, XB, XC be three lines not in a plane, and let XD be a third line in XAC and XE a third line in XBC. Then the planes XAB and XDE intersect in a line.*

Proof. (see [9, p.603]) Let $\varepsilon := XCA$ and $\delta := XCB$. Then $m := AD \subset \varepsilon$ and $B \in \delta \setminus XC$, and hence by (3), $\gamma := ADB$ and δ intersect in a line BQ. Now $XE \subset \delta$, $D \in \gamma \setminus BQ$, and (3) imply that $\beta := XED$ and γ intersect in a line DR. Finally β and ABX intersect in a line by (3), since $\gamma \cap \beta = DR$, $AB \subset \gamma$, and $X \in \beta \setminus DR$.

Since (4) is the Veblen-Young-Axiom for the geometry of lines and planes through the point X, the proof of the Theorem is complete.

REFERENCES

[1] Baker, H.F., Principles of Geometry, Vol.1 (Cambridge, 1922)
[2] Hotje, H., Angeordnete geometrische Strukturen, lecture notes (Hannover, 1985)
[3] Kahn, J., Locally projective-planar lattices which satisfy the Bundle Theorem, Math.Z. 175 (1980) 219-247.
[4] Pasch, M., Vorlesungen über neuere Geometrie, 2nd ed. (Berlin, 1926)
[5] Peano, G., Sui fondamenti della Geometria, Rivista di Matematica 4 (1894) 51-90.
[6] Schur, F., Ueber die Grundlagen der Geometrie, Math.Ann. 55 (1901) 265-292.
[7] Schur, F., Grundlagen der Geometrie (Leipzig, 1909)
[8] Sörensen, K., Projektive Einbettung angeordneter Räume, Beiträge zur Geometrie und Algebra 15 (Technische Universität München, 1986) 8-35.
[9] Wyler, O., Incidence geometry, Duke Math.J. 20 (1953) 601-610.

Annals of Discrete Mathematics 37 (1988) 181–188

Midpoints and Midlines in a Finite Hyperbolic Plane

Cyril W.L. Garner
Carleton University
Ottawa, Ontario
Canada K1S 5B6

1. Introduction

Let **P** be a finite projective plane of arbitrary odd order n, and let π be a regular polarity of **P**: that is, a polarity for which there exists an integer $s = s(\pi)$ such that every line containing two or more absolute points of π contains s+1 absolute points [**8**, p. 247]. Baer [**1**] has shown that the absolute points form an oval when n is odd and nonsquare, and Segre [**10**] has shown that every oval in a Desarguesian projective plane is a conic. This implies s = 1, and just as in the real projective plane, there are two disjoint classes of nonabsolute points:

O = {outer points, or points having 2 absolute lines},
I = {inner points, or points having 0 absolute lines};

and two disjoint classes of nonabsolute lines:

o = {outer lines, or lines having 0 absolute points},
i = {inner lines, or lines having 2 absolute points}.

Clearly $\mathbf{I}\pi = \mathbf{o}$ and $\mathbf{i}\pi = \mathbf{O}$.

In analogy with the real projective plane, we might expect the incidence structure HA(n) whose points are **I** and lines are **i**, with incidence as given in **P**, to be a finite hyperbolic plane, since intersecting, parallel, and ultraparallel lines can be defined in the obvious way. Baer [**1**] has shown that this is not true - a simpler, but less general argument [**5**, p. 316] shows that the incidence structure consisting of **I** and **i** involves "parallel" points, that is, points which do not determine a common line.

In previous papers ([**6**] and [**7**]) we have investigated the geometry of this incidence structure HA(n) defined over a Desarguesian projective plane of order $n \equiv 3 \pmod 4$. Concepts such as line reflections, point reflections and their products, circles, horocycles and hypercycles, analogous to those in the classical hyperbolic plane, were introduced, and it is noteworthy that parallel points do not cause a great divergence from the classical situation.

In this paper we study the concepts of "midpoint" and "midline" of two points, and find the existence of parallel points is intimately connected with the existence or non-existence of midpoints and midlines.

The words "point" and "line" will always refer to elements of HA(n) unless the adjective "projective", "outer" or "absolute" is explicitly inserted. In the figures, points and lines are represented by solid dots and lines as usual. However, outer points and lines appear as open dots and dashed lines, while absolute points are marked by an x.

2. Definitions

If HA(n) is defined over a desarguesian projective plane **P** of non-square order $n \equiv 3 \pmod 4$, then **P** is pappian [**3**, p. 160] and also fanonian, i.e. the diagonal points of a quadrangle are not collinear [**9**, pp. 190 - 194]. Thus **P** satisfies the axioms of projective geometry as enunciated by Coxeter [**2**, p. 25], and in particular we can exploit the properties of harmonic conjugates and involutory homologies.

Definition 1: In **P**, $\sigma_{A,a}$ denotes the involutory homology with centre A, axis $a = A\pi$. Since only one of A, a is an inner element, $\sigma_{A,a}$ is called a *point reflection* σ_A or *line reflection* σ_a according as A or a belongs to HA(n).

Many properties of point- and line-reflections are derived in [**5**], and a list of all possible products of reflections is given in Table 1 [**6**, p. 493]. As usual, two lines are said to be perpendicular with respect to π if each passes through the pole of the other. Thus lines x and y are perpendicular if $X \in y$ and $Y \in x$. The points X and Y are, of course, called conjugate points. This notation is used for any points and lines in **P**, whether inner, outer or absolute.

Definition 2: Let A, B be any two points of HA(n). If there exists a point O so that $\sigma_O(A) = B$, then O is called *a midpoint of A and B*.

Definition 3: Let A, B be any two points of HA(n). If there exists a line o so that $\sigma_o(A) = B$, then o is called *a midline of A and B*.

Clearly a midpoint of A and B must lie on the projective line joining A and B, and a midline must be perpendicular to that projective line. In the following, we show that any two non-parallel points have a unique midpoint and a unique midline line, as in the classical hyperbolic plane. Two parallel points, however, have either two midpoints or two midlines, but it is impossible for them to have both a midpoint and a midline.

3. Midpoints and Midlines

Theorem 1. Let A and B be two points of HA(n) $(n \geq 3)$ having a common line t. Then A and B have a unique midpoint.

Proof: Let a_1, b_1 be the unique lines through A, B respectively which are perpendicular to t [**6**, p. 486 (3)]. Let t, a_1, b_1 intersect the absolute $\mathcal{C}$ in the three pairs of absolute points: M, N; M_1, N_1; M_2, N_2 respectively as shown in figure 1.

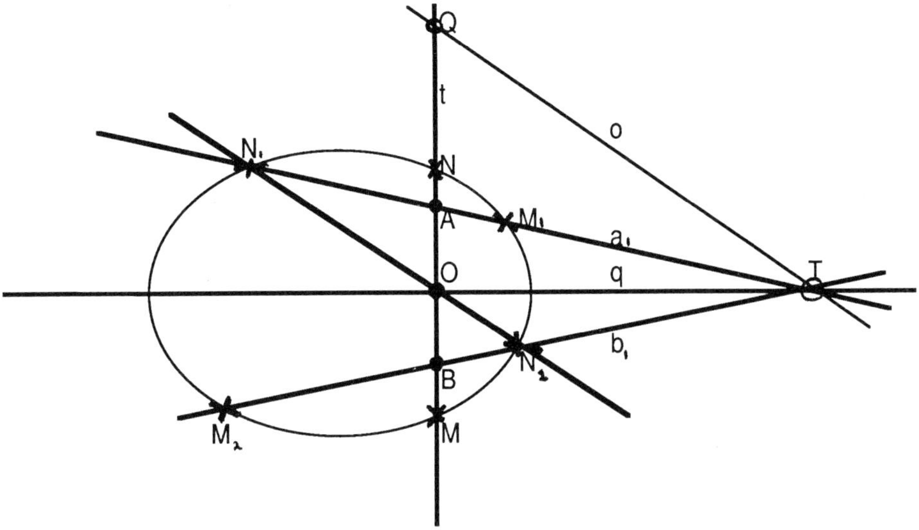

Figure 1a. O is inner.

Since a_1, b_1 are perpendicular to t, they intersect in the outer point T [**5**, p. 322, Result 8]. In **P**, let $t \cap N_1N_2 = O$. Since $O \in t$, $T \in o$; let $t \cap o = Q$.

We show first that the involutory homology $\sigma_{O,o}$ interchanges A and B; clearly it maps $\mathcal{C}$ onto itself [**5**, Result 3, p. 321] and so interchanges N_1 and N_2 (and incidentally, M_1 and M_2). We use the standard procedure to find the image of A under an homology [**2**, 6.22, p. 53]: since $AN_1 \cap o = t$, the image of A is $OA \cap LN_2$, which is B.

We now distinguish two cases:

1) O is a point so that $\sigma_{O,o} = \sigma_O$ and so O is a midpoint of A and B. Since o is an outer line, $o \cap t = Q$ is an outer point, being the intersection of a line with a perpendicular outer line [**5**, lemma 1, p. 325], and so its polar q = OT is an inner line. Thus $\sigma_{Q,q}$ is a line reflection which also interchanges A and B: for since TOQ is a self-polar triangle, $\sigma_{Q,q} = \sigma_{O,o} \cdot \sigma_{T,t}$ [**3**, 3.1.7, p. 120], or $\sigma_Q = \sigma_O \cdot \sigma_t$; we have $\sigma_q(A) = B$.

2) O is an outer point so that $\sigma_{O,o}$ is a line reflection interchanging A and B. Since o is a line, $Q = o \cap t$ is a point, being the intersection of two perpendicular lines. As above, $\sigma_{Q,q} = \sigma_{O,o} \cdot \sigma_{T,t}$ or $\sigma_Q = \sigma_o \cdot \sigma_t$ and $\sigma_Q(A) = B$. Thus Q is a midpoint of A and B.

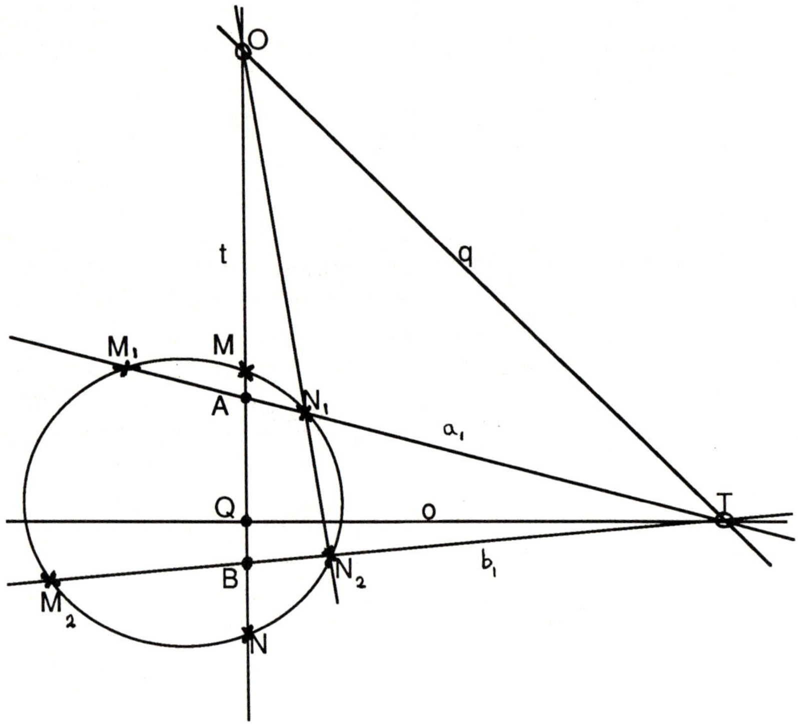

Figure 1b. O is outer

To show that the midpoint is unique in either case, suppose there exist two points X, Y with $\sigma_X(A) = \sigma_Y(A) = B$. Necessarily, X and Y are both incident with t, and so σ_X, σ_Y both interchange M and N as well as A and B [**5**, result 3, p.

321]. The product $\sigma_Y \cdot \sigma_X$ then fixes four points A, B, M, N on t so that $\sigma_Y \cdot \sigma_X$ induces on t a projectivity which, by the fundamental theorem in a pappian plane, is the identity [**2**, p. 34]. $\sigma_Y . \sigma_X$ is thus a central collineation with axis t; a dual argument shows that T is its centre. Since t is a line, $\sigma_Y . \sigma_X = \sigma_t$. But then $X \in y$ and $Y \in x$ [**3**, 3.1.7, p. 120] and so the perpendicular through X to t, namely y, is a line. This forces Y to be an outer point which is the desired contradiction.

Corollary: Any two points of HA(n) ($n \geq 3$) having a common line have a unique midline.
Proof: In case 1), the midline is q, and in case 2), it is o.

The situation for parallel points is quite different, as the following two theorems show. For the proofs, it is necessary to distinguish between conjugate parallel points and non-conjugate parallel points, although the results are the same in both cases.

Theorem 2: Let A and B be two conjugate parallel points in HA(n) ($n > 3$). Then either they have precisely two midpoints or they have precisely two midlines.

Proof: Let A, B be conjugate points on the outer line t. The self-polar triangle ABT has points as vertices and outer lines as sides. Since $q \equiv 3 \pmod 4$, ABT is a "canonical triangle" [**4**, p. 364], and so there exists a *unique* pair of (not necessarily inner) points O and Q on t which are harmonic with respect to A and B and are also conjugate with respect to π; that is, **H**(OQ,AB) and so o = QT, q = OT. Thus both $\sigma_{O,o}$ and $\sigma_{Q,q}$ interchange A and B. See Fig. 2.

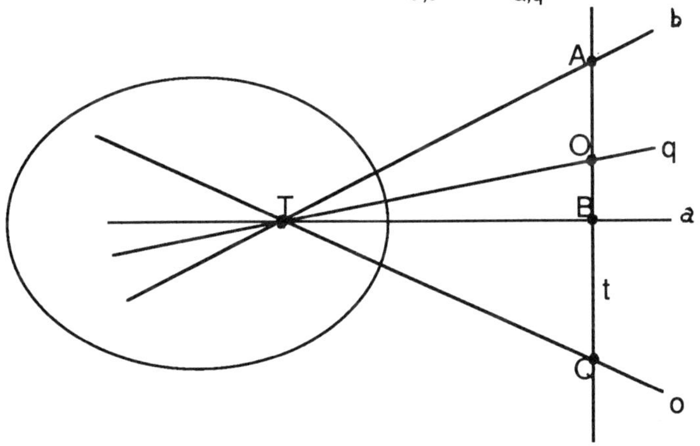

Figure 2. O and Q are midpoints.

Since -1 is not a square in GF(q) with $q \equiv 3 \pmod 4$, O and Q are either both inner or both outer [**4**, p. 367 or **5**, lemma 1, p. 325]. Thus if both O and Q are points, they are midpoints of A and B. But if O and Q are outer points, there is no midpoint for A and B but two midlines, o and q. Thus by the uniqueness of the points O and Q mentioned in the theorem, A and B have either two midpoints or two midlines, but not both.

Theorem 3: Let A and B be two non-conjugate parallel points in HA(n) (n > 3). Then either they have precisely two midpoints or they have precisely two midlines.

Proof: Let A and B be non-conjugate points on the outer line t. In Coxeter's terminology [**2**, exercise 6, p. 101], A and B are *accessible* points, and since in PG(2,n) all inner points are mutually accessible [**2**, exercise 7, p. 129], there exists at least one pair of (not necessarily inner) points O and Q which are conjugate with respect to π, and **H**(AB,OQ). Since t is outer, O and Q are either both inner or outer points. Since O is conjugate to T and Q, o = TQ and q = TO. We distinguish two cases:

1). If O (and so Q) is a point, then $\sigma_O(A) = A'$ implies **H**(AA',OQ). But **H**(AB,OQ) and so B = A'. Thus A and B have a midpoint O. Similarly, Q is a midpoint.

2). If O (and so Q) is an outer point, then o and q are lines and $\sigma_o(A) = \sigma_q(A) = B$. Thus o and q are midlines.

We must show that A and B cannot have more than two midpoints or midlines and cannot have a midpoint and a midline simultaneously. Suppose that on t there exist O, Q = $t \cap o$ and R, S = $t \cap r$ such that **H**(OQ,AB) and **H**(RS,AB) where O, Q, R, S are not assumed to be either inner or outer, and O, Q $\neq$ R, S. As before, O and Q will be together inner or outer and the same holds for R and S. Thus we are assuming that A and B have two pairs of midpoints or two pairs of midlines or one of each which are not conjugate: $\sigma_{O,o}(A) = \sigma_{R,r}(A) = B$.

Now $\sigma_{O,o}$ induces on t the involution (OO)(QQ) in which A,B is a pair of mates. Similarly, A,B is a pair of mates in the involution (RR)(SS) induced on t by $\sigma_{R,r}$. The product of these two involutions is a projectivity which has A and B as fixed points [**2**, p. 46], i.e. the hyperbolic involution (AA)(BB) is induced on t by the product $\sigma_{O,o} \cdot \sigma_{R,r}$.

Clearly the product $\sigma_{O,o} \cdot \sigma_{R,r}$ fixes T and t. The only other fixed points must be on t [**3**, 3.1.7, p. 120] . But since $\sigma_{O,o} \cdot \sigma_{R,r}$ is an automorphism of $\mathcal{C}$, it is a motion of HA(n) and so must appear in Table 1 of [**6**]. Since a point reflection is the only motion admitting more than one fixed inner point, we must have $\sigma_{O,o} \cdot \sigma_{R,r} = \sigma_L$. But this happens if and only if O and R are conjugate points [**3**, 3.1.7, p. 120] which forces R to be Q. This contradiction completes the theorem.

Examples: In HA(7), there are only 4 points on any outer line t. Thus given two points A and B, the other two points on t are their conjugate points A' and B', and the midlines of A and B are A'T and B'T. There are no midpoints.

In HA(11), there are 6 points on an outer line t. Given two points A and B, and their conjugates A', B', there remain two further points on t, say C and its conjugate C'. Both σ_C and $\sigma_{C'}$ interchange A and B, so that A and B have two midpoints and no midlines.

References

1. Baer, R., Polarities in finite projective planes. *Bull. Am. Math. Soc.* **51** (1946), 77-93.
2. Coxeter, H.S.M., *Projective Geometry* . 2nd Edition, University of Toronto Press 1974.
3. Dembowski, P., *Finite Geometries.* Springer-Verlag 1968.
4. Edge, W.L., Conics and orthogonal projectivities in a finite plane. *Can. J. Math.* **8** (1956), 362 - 382.
5. Garner, C.W.L., A finite analogue of the classical hyperbolic plane and Hjelmslev groups. *Geometriae Dedicata* **7** (1978), 315-331.
6. Garner, C.W.L., Motions in a finite hyperbolic plane. *The Geometric Vein, The Coxeter Festschrift,* Springer-Verlag 1982, 485 - 493.
7. Garner, C.W.L., Circles, horocycles and hypercycles in a finite hyperbolic plane (to appear).
8. Hughes, D.R. and Piper, F.C., *Projective Planes.* Springer-Verlag 1973.
9. Pickert, G., *Projektive Ebenen* . 2nd Edition, Springer-Verlag 1975.
10. Segre, B., Ovals in finite projective planes. *Can. J. Math* . **7** (1955), 414-416.

Supported by grant A809 from NSERC of Canada.

Annals of Discrete Mathematics 37 (1988) 189–194

HALL-RYSER TYPE THEOREMS FOR RELATIVE DIFFERENCE SETS

Dina GHINELLI SMIT

Dipartimento di Matematica, Università di Roma "La Sapienza"
I-00185 Roma, Italy

In this paper we give applications of the nonexistence theorems proved in [4] (see also [3]), to point-divisible and divisible designs with a Singer group. Using known results on the links between these designs and Relative Difference Sets, we also obtain nonexistence theorems for *Relative Difference Sets*. The results in this paper generalize the famous Hall-Ryser theorem for symmetric designs with a Singer group.

1. INTRODUCTION

Let Γ be a group of order $v=cm$ and let U be a subgroup of Γ of order c. A k-subset Δ of Γ is called a *Relative Difference Set* with parameters c,m,k,λ' and λ (relative to U) or, briefly, a (c,m,k,λ',λ)-RDS, provided that:

(a) the differences $d-d'$ ($d,d'\in\Delta$, $d\neq d'$) contain each element of U (except zero) precisely λ' times, and each element of $\Gamma-U$ exactly λ times.

We call a (c,m,k,λ',λ)-RDS *nonsingular*, if it satisfies

(b) $\qquad k\neq\lambda'$, and $k^2\neq v\lambda$.

Counting the differences arising from Δ, we obtain

(1) $$k(k-1) = c(m-1)\lambda+(c-1)\lambda'.$$

Hence, if we set $n=k-\lambda'+c(\lambda'-\lambda)$, then

(2) $$n = k-\lambda'+c(\lambda'-\lambda) = k^2-v\lambda,$$

and $n>0$ for nonsingular Relative Difference Sets.

The above definition of Relative Difference Set is essentially in Jungnickel [7]; As pointed out by Jungnickel in [7], Elliot and Butson [2] have considered the special case of a normal subgroup U and of $\lambda'=0$. This had earlier been introduced for cyclic groups by Butson [1]. If $c=1$, then $U=\{0\}$ and we have the well known notion of (v,k,λ)-Difference Set Δ in the group Γ (see Hughes and Piper [6]). A necessary condition for the existence of a (v,k,λ)-Difference Set Δ in a finite group Γ is the following theorem, first proved by Hall and Ryser for cyclic groups (see [5]).

HALL-RYSER THEOREM 1.1 *Assume the existence of a* (v,k,λ)*-Difference Set in a finite group* Γ*, and let* H *be a (normal) subgroup of order* $\mu=v/u$*, then*

(i) $(k-\lambda)^{u-1}$ *is a square,*
(ii) the equation

(3) $$(k-\lambda)\, x^2+ (-1)^{u(u-1)/2}u\, y^2= z^2,$$

has a nontrivial solution in integers (x,y,z) *(i.e. a solution other than* $(0,0,0)$*).*

In this paper we prove the following nonexistence theorems for Relative Difference Sets.

THEOREM 1.2 *Let* Γ *be a group of order* $v=cm$ *and let* U *be a subgroup of* Γ, *with order* c. *Assume the existence of a nonsingular* (c,m,k,λ',λ)*-RDS (relative to* U*) and let* H *be a normal subgroup of* Γ *such that* $U\cap H=\{0\}$. *Then* $u=m/|H|$ *is an integer and the following conditions must hold*

(i) $(k-\lambda')^{u(c-1)}n^{u-1}$ *is a square,*
(ii) the diophantine equations

$$(4)\qquad (k-\lambda')\,x^2+(-1)^{u(c-1)\{u(c-1)+1\}/2}\,c^u\,y^2=z^2,$$

and

$$(5)\qquad n\,x^2+(-1)^{u(u-1)/2}\,u\,y^2=z^2,$$

are equivalent (i.e. (4) *has a nontrivial solution in integers if and only if* (5) *has such a solution).*

If we make the hypothesis of normality for the subgroup U, we get the following stronger result.

THEOREM 1.3 *Let* Γ *be a group of order* $v=cm$, *with a normal subgroup* U *of order* c. *Assume the existence of a nonsingular* (c,m,k,λ',λ)*-RDS (relative to* U*) and let* H *be any subgroup of* Γ *such that* $U\cap H=\{0\}$. *Then the following conditions must hold with* $u=m/|H|$,

(I) $(k-\lambda')^{u(c-1)}$ *and* n^{u-1} *are both squares,*

(II) *the equations* (4) *and* (5) *both possess a nontrivial solution in integers.*

In particular, if $c=1$, then $U=\{0\}$ is normal and each subgroup H of Γ satisfies $U\cap H=\{0\}$. In this case, (4) has clearly the solution $(0,1,1)$ and $n=k-\lambda$; hence theorem 1.3 gives, for $c=1$ a different proof of theorem 1.1 without the hypothesis that H is normal.

In section 2, we shall see that 1.2 and 1.3 can be formulated, equivalently, in terms of point-divisible and divisible designs with a Singer group. We will prove nonexistence theorems for these designs, so that the proofs of 1.2 and 1.3 will be an immediate consequence.

We observe that, for cyclic groups, Shrikhande [9], obtained, in a different way, the nonexistence condition (II), while the cases $c=2,3$ and $\lambda'=0$ are also in Ryser [8].

2. HALL-RYSER TYPE THEOREMS FOR P-DIVISIBLE AND DIVISIBLE DESIGNS WITH A SINGER GROUP

Let Δ be a nonsingular (c,m,k,λ',λ)-RDS in a group Γ of order $v=cm$, relative to a subgroup U with order c. If $c=1$, then $U=\{0\}$ and the *development* of Δ, namely the incidence structure

$$(6)\qquad \text{dev}\,\Delta=(\Gamma,\{\Delta+g\mid g\in\Gamma\},\varepsilon)$$

is a symmetric (v,k,λ)-design admitting Γ as a Singer group. In general (see Jungnickel [7] th. 2.7), dev Δ will be a *point-divisible design* (shortly: p-divisible) (1) *with* Γ *as a Singer group*; furthermore, all p-divisible designs with a Singer group may be represented in this way. When the subgroup U is normal, then (see Jungnickel [7] 2.10 and 2.11) the dual structure $(\text{dev}\,\Delta)^*$ is also a p-divisible design (i.e. dev Δ is both point and block-divisible). In this case, we say that dev Δ is a *divisible design*. Furthermore, dev Δ will admit Γ as a *normal Singer group* (see [7], section 2).

From 1.3 of [3], it is not difficult to deduce that the main nonexistence results announced by the author in [3] (namely: 4.6 and 5.1 of [3]) can be stated as follows. Let (m,c,k,λ',λ) denote the parameters of a p-divisible design.

THEOREM 2.1 *If a point-divisible design exists with a standard automorphism group G of order* μ, *which fixes* N *points and* Φ *classes (and so has* $v'=\{(v-N)/\mu\}+N$ *orbits on points and* $u=\{(m-\Phi)/\mu\}+\Phi$ *orbits on classes) and if* $n=(k-\lambda')+c(\lambda'-\lambda)$, *then* $n=k^2-v\lambda$ *and*

(i) $(k-\lambda')^{v'-u}\, n^{u-1}$ *is a square,*

(ii) the diophantine equations

(7) $$(k-\lambda')\, x^2 + (-1)^{(v'-u)(v'-u+1)/2}\, c^u\, \mu^{N+\Phi}\, y^2 = z^2,$$

and

(8) $$n\, x^2 + (-1)^{\{u(u+1)/2\}+1}\, c^u\, \lambda\, \mu^{\Phi+1}\, y^2 = z^2,$$

are equivalent.

For further details and for the proof of 2.1, which is simply sketched in [3], see [4]. Here we will only show how theorem 2.1 can be applied to obtain nonexistence conditions for p-divisible designs with a Singer group. These conditions, by the above mentioned results on devΔ, will give theorem 1.2.

THEOREM 2.2 *Let* $D=D(m,c,k,\lambda',\lambda)$, $\lambda'\neq\lambda$, *be a p-divisible design with a Singer group* Γ, *and choose a point* P *of* D. *Let* U *be the subset of all* g *in* Γ *such that* P^g *and* P *have joining number* λ' (2). *If for each normal subgroup* H *of* Γ *such that* $U\cap H$ *is the identity we write*

(9) $$u = m/|H|,$$

then the following conditions must hold:

(i) $(k-\lambda')^{u(c-1)}\, n^{u-1}$ *is a square,*

(ii) the diophantine equations

(10) $$(k-\lambda')\, x^2 + (-1)^{u(c-1)\{u(c-1)+1\}/2}\, c^u\, y^2 = z^2,$$

and

(11) $$n\, x^2 + (-1)^{u(u-1)/2}\, u\, y^2 = z^2,$$

are equivalent.

Proof It is easy to see that the conditions $U\cap H$ is the identity and H is normal, imply that H is a standard automorphism group of D with $N=\Phi=0$; thus, in particular, the order of H must divide m and $u=m/|H|$, $v'=uc$. (3) Thus 2.1 implies that (i) holds and (10) is equivalent to

(12) $$n\, x^2 + (-1)^{\{u(u+1)/2\}+1}\, c^u\, \lambda\, |H|\, y^2 = z^2.$$

If u is *even*, by (i) n is a square; thus (11) and (12) both have a solution (i. e. they are equivalent). To prove the equivalence between (11) and (12) also in the case u *odd*, we observe that if u is odd, then

(13) $$\frac{u(u+1)}{2} + 1 \equiv \frac{u(u+1)}{2} - u \equiv \frac{u(u-1)}{2} \pmod 2,$$

hence (12) (as well as (10)) is equivalent to

(14) $$n\, x^2 + (-1)^{u(u-1)/2}\, c^u\, \lambda\, |H|\, y^2 = z^2.$$

Using the Hilbert p-norm residue symbol (see [3] for properties and notation), (14) has a nontrivial solution in integers if and only if

(15) $\quad (n \,,\, (-1)^{u(u-1)/2}\, c^u\, \lambda\, |H| \,)_p = 1$ for all primes p (including $p=\infty$).

Since u is odd we have, by the properties of the Hilbert symbol

$$(n,c^u\lambda|H|)_p = (n,c\lambda|H|)_p = (n,mc\lambda m/|H|)_p = (n,v\lambda)_p(n,u)_p.$$

Now, $n+v\lambda=k^2$ so that (see [3], 1.10)

$$(n,v\lambda)_p = (-nv\lambda,n+v\lambda)_p = (-nv\lambda,k^2)_p = 1;$$

therefore, for all primes p (including $p=\infty$) we have

$$(n,(-1)^{u(u-1)/2}c^u\lambda|H|)_p = (n,(-1)^{u(u-1)/2}u)_p.$$

This proves that (14) (and thus (12) and (10)) is equivalent to

(11) $\quad n\, x^2 + (-1)^{u(u-1)/2}\, u\, y^2 = z^2,$

also when u is odd, which proves the theorem.

For divisible designs theorem 2.1 and theorem 2.2 can be improved. The idea behind these improvements is simple and is based on the same method which allows us to prove 2.1. The method gives a general nonexistence theorem for tactical decompositions of rational matrices (see [4], chapter 3). To obtain 2.1 we apply this method to the tactical decomposition given by the orbits of an automorphism group. For a divisible design we can also apply the method to the tactical decomposition given by the divisibility classes. This leads us to consider a pseudo-structure, that we call a *quotient symmetric pseudodesign*, which for our purposes behaves as a symmetric $(m,k+\lambda'(c-1),c\lambda)$ design. Using 2.1 on this quotient symmetric pseudodesign, the following improvement of 2.1 can be obtained (see [4], chapter 4, or [3], for further details).

THEOREM 2.3 *If a divisible design* $D = D(m,c,k,\lambda',\lambda)$ *exists with a standard automorphism group G of order* μ, *and* Φ, N, v', u, n *are defined as in* 2.1, *then*

(i) $(k-\lambda')^{v'-u}$ *and* n^{u-1} *are both squares,*

(ii) the diophantine equations (7) *and* (8) *both possess a non trivial solution in integers.*

As an application of 2.3 we have the following result, which improves 2.2 for divisible designs.

THEOREM 2.4 *Let* D *be a divisible design with a Singer group* Γ *and choose a point* P *of* D. *If* U *is the subgroup of all* g *in* Γ *such that* P^g *and* P *have joining number* λ', *then for each normal subgroup* H *of* Γ *such that* $U \cap H$ *is the identity the following conditions must hold, with* $u=m/|H|$.

(I) $(k-\lambda')^{u(c-1)}$ *and* n^{u-1} *are both squares,*

(II) the equations (10) *and* (11) *both possess a nontrivial solution in integers.*

If U *is normal, then (I) and (II) hold for each subgroup* H *(not necessarily normal) of* Γ.

Proof In the proof of 2.2, replace "p-divisible" by "divisible", "2.1" by "2.3" and "(i)" by "(I)". If U is normal, the last part of the statement follows by (3). Only when U is normal, can an alternative proof be given, using the quotient symmetric pseudodesign D/∿. In this case, the quotient group G/U acts as a Singer group on D/∿; hence, applying 2.2 (for c=1) to D/∿ (i.e. the Hall-Ryser theorem 1.1 for symmetric designs with a Singer group), we get the desired result.

Now, let Δ be a nonsingular (c,m,k,λ',λ)-RDS in a group Γ, relative to a *normal* subgroup U with order c. As we mentioned above, dev Δ is a p-divisible design with a *normal* Singer group and thus is divisible (see Jungnickel [7], th. 2.10). Applying 2.4 to dev Δ, we obtain theorem 1.3.

It would be interesting to decide whether or not the hypothesis of normality is essential in theorem 1.3 (and therefore in theorem 2.10 of Jungnickel [7]); the author knows no examples of p-divisible designs (with a Singer group) which are not divisible.

ACKNOWLEDGEMENTS

The author would like to thank the referee for his valuable comments. Many thanks go also to Westfield College London for its hospitality during the time of this research and to the Accademia Nazionale dei Lincei, the Royal Society and the British Council for their assistance during all of her studies in London.

NOTES AND REFERENCES

(1) See [3], 4.2, for the definition. We note that, with Bose and Connor's terminology, this structure is called Symmetric Regular Group Divisible Design.

(2) We note that U is actually a subgroup (see Jungnickel [7], 2.2).

(3) If U is normal, we do not need the hypothesis of normality for H since then each subgroup H of Γ such that U ∩ H is the identity is standard with N=Φ=0.

[1] Butson, A.T., Relations among generalized Hadamard matrices, relative difference sets and maximal length linear recurring sequences, *Can. J. Math.* 15 (1963) 42-48.
[2] Elliot, J.E.H., and Butson, A.T., Relative Difference Sets, *Illinois J. Math.* 10 (1966) 517-531.
[3] Ghinelli Smit, D., Automorphisms and generalized incidence matrices of point-divisible designs, *Ann. Discr. Math.* 18 (1983) 377-400.
[4] Ghinelli Smit, D., Nonexistence theorems for automorphism groups of divisible square designs, *Ph.D. Thesis*, University of London (1983).
[5] Hall, M. Jr., and Ryser, H.J., Cyclic incidence matrices, *Canad. J. Math.* 3 (1951) 495-502.
[6] Hughes, D.R., and Piper, F.C., *Design Theory* (Cambridge University Press, 1985).
[7] Jungnickel, D., On automorphism groups of divisible designs, *Can. J. Math.* XXXIV, 2 (1982), 257-297.
[8] Ryser, H.J., Variants of cyclic difference sets, *Proc. Amer. Math. Soc.* 41, 1 (1973) 45-50.
[9] Shrikhande, S.S., Cyclic solutions of symmetrical group divisible designs, *Calcutta Stat. Ass. Bull.* 5 (1953) 36-39.

Annals of Discrete Mathematics 37 (1988) 195–208

COORDINATIZATION OF GENERALIZED QUADRANGLES

HANSSENS G. - VAN MALDEGHEM H.

A coordinatization method for any thick generalized quadrangle is worked out, using a new algebraic structure, i.e. a quadratic quaternary ring.

1. INTRODUCTION

A *(thick) generalized quadrangle* (GQ) is an incidence structure $\mathbf{S} = (\mathbf{P},\mathbf{L},I)$ with point set $\mathbf{P}$ and line set $\mathbf{L}$, satisfying the following axioms :

(i) each point is incident with $1 + t$ lines ($t \geqslant 2$) and two distinct points are incident with at most one line ;

(ii) each line is incident with $1 + s$ points ($s \geqslant 2$) and two distinct lines are incident with at most one point ;

(iii) if P is a point and L is a line not incident with P, then there is a unique pair $(Q,M) \in \mathbf{P}\times\mathbf{L}$ for which P I M I Q I L.

We say that $\mathbf{S}$ has order (s,t), where $s,t \in \mathbb{N} \cup \{\infty\}$. In view of the point-line duality for GQ, we assume that the dual of a given definition or theorem has also been given implicitly. It is a nice exercise to show that axioms (i) and (ii) can be replaced by :

(i)' each point is incident with at least three lines ;

(ii)' each line is incident with at least three points ;

(iv)' there is a non-incident point-line pair.

Given two points P and Q of $\mathbf{S}$, we write $P \perp Q$ and say that P and Q are collinear provided that there is some line L incident with both. If this is not the case, we write $P \not\perp Q$.

For $P \in \mathbf{P}$, put $P^{\perp} = \{Q \in \mathbf{P} \mid Q \perp P\}$ and note that $P \in P^{\perp}$. If $A \subset \mathbf{P}$, we write $A^{\perp} = \cap\{P^{\perp} \mid P \in A\}$. For distinct points P and Q, $\{P,Q\}^{\perp}$ is called the trace of P and Q and $\{P,Q\}^{\perp\perp}$ the span.

Generalized quadrangles were introduced by J. Tits and appeared first in [1]. They arose as natural objects "succeeding" the projective planes, which could be viewed as generalized triangles.

One of the most powerful concepts in the modern theory of projective planes is that of coordinatization. This is certainly the case in constructing non-classical planes and to determine whether two given projective planes are isomorphic. For details we refer to [2] and [8].

It is surprising that an analogous general coordinatization theory for GQ is not yet available. The known non-classical GQ are constructed using geometrical methods or matrices. Especially in the last case, one gets the impression that there might be an underlying coordinatization without being explicit.

S. Payne [9] worked out a preliminary version of such a theory for a special class of GQ of order (s,s), $s > 1$, namely those having an axis of symmetry. He essentialy uses the coordinatization by a planar ternary ring of an underlying projective plane.

To be useful such a general coordinatization theory for GQ should satisfy some "beauty" conditions.

Firstly it must provide an easy, algebraically more concrete description of the existing GQ, and if possible, also of their automorphism groups [3].

Secondly, the fact of having certain automorphism groups for the GQ should be reflected by "nice" properties of the corresponding algebraic structure [4].

Finally, important geometrical conditions should have a simple algebraic reformulation (see § 3 and [5]).

We mention here also that the proposed method of coordinatization works in the case of generalized n-gons [6] as well. The theory of quadratic quaternary rings should also be the crucial tool in describing an affine building of type $\tilde{C}_2$ in terms of its (infinite) generalized quadrangle at infinity (see also [7] for further references and comments).

2. COORDINATIZATION BY QUADRATIC QUATERNARY RINGS

2.1. Introduction of coordinates

Let **S** be a GQ of order (s,t), $s,t > 1$. Choose an arbitrary point (∞) and an arbitrary line $[\infty]$ incident with it. Let R_1 be a set of cardinality s not containing the symbol ∞, and assign bijectively a coordinate (a), with $a \in R_1$, to every point on $[\infty]$ different from (∞).

Dually, let R_2 be a set of cardinality t not containing the symbol ∞, and give every line on (∞) different from $[\infty]$ a coordinate [k] with $k \in R_2$, such that there is a bijection between the lines on (∞) and R_2.

We pick out of R_1 resp. R_2 two distinguished elements denoted by 0 and 1.

Now we choose a point A not on $[\infty]$ and collinear with (0), and call B the point of [0] collinear with A. Like before we choose a bijection between R_1 and the points of the line (0)A with the only restriction that A corresponds to 0. The point of (0)A corresponding to $a' \in R_1$ will have coordinate $(0,0,a') \in R_1 x R_2 x R_1$. Dually, we give coordinates $[0,0,k'] \in R_2 x R_1 x R_2$ to the lines on B different from [0], with the restriction that BA has coordinate [0,0,0].

We define next the points with two coordinates : a point P collinear with (∞), but not lying on $[\infty]$ has coordinate $(k,a) \in R_2 x R_1$ if and only if P lies on [k] and is collinear with (0,0,a). Dually, lines meeting $[\infty]$ not passing (∞) are given coordinates $[a,k] \in R_1 x R_2$.

Finally, consider a point P not collinear with (∞). By axiom (iii) there is exactly one line on P meeting $[\infty]$. This line must have two coordinates, say [a,l]. On the other hand, P is collinear with exactly one point (0,a') on [0]. Now P is given the coordinate (a,l,a'). Conversely, let (a,l,a') be any element of $R_1 x R_2 x R_1$, then we construct a point P having this element as coordinate. Indeed, given the line [a,l] and the point (0,a') not incident with it, then there is exactly one point collinear with (0,a') and lying on [a,l] by axiom (iii).

The coordinate of a line [k,b,k'] is defined dually. It is easy to check that there arises no ambiguity for the coordinates (0,0,a') and [0,0,k']. To avoid complications, we identify each element with its coordinate.

In this way we have coordinatized every point and line of **S** (see fig.1). Now we must build into our coordinatization system a criterium for determining whether two given elements are incident.

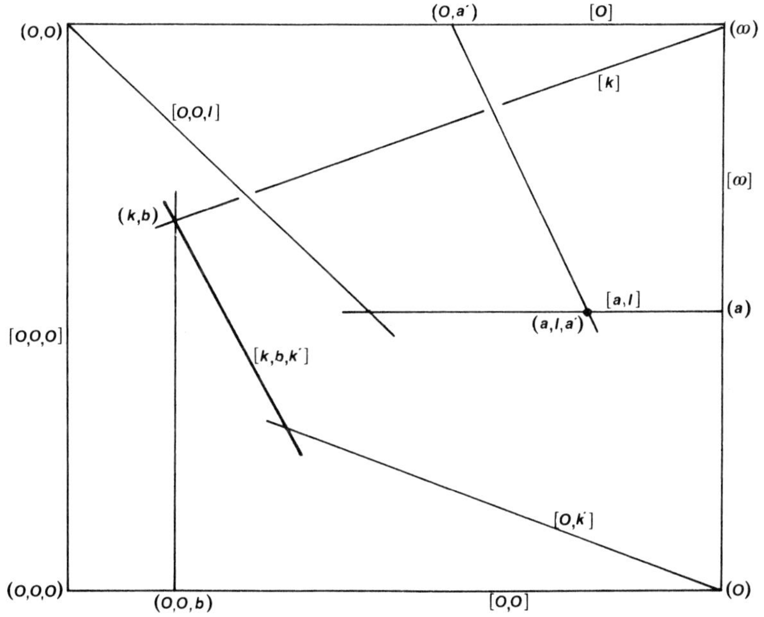

2.2. Quadratic quaternary rings

If a GQ **S** has been coordinatized by the elements of the sets R_1 and R_2 as described in section 2.1, then we use incidences of **S** to define two quaternary operations Q_1 and Q_2 as follows : if $a,a',b \in R_1$ and $k,k',l \in R_2$,

$$Q_1(k,a,l,a') = b \quad \text{if and only if } (k,b) \perp (a,l,a'), \tag{1}$$

$$Q_2(a,k,b,k') = l \quad \text{if and only if } [a,l] \perp [k,b,k'] \tag{2}$$

The operations are uniquely determined, for given the point (a,l,a') and the line $[k]$, there is exactly one point (k,b) on $[k]$ which is collinear with (a,l,a'). Dually, given the line $[k,b,k']$ and the point (a), there is exactly one line $[a,l]$ on (a) meeting $[k,b,k']$.

It is clear that (a,l,a') I $[k,b,k']$ only if (1) and (2) are satisfied. Conversely, suppose that (1) and (2) hold, then (a,l,a') is collinear with $(k,b) \in [k,b,k']$ and $[k,b,k']$ meets $[a,l]$ containing (a,l,a'). Because this meeting point of $[a,l]$ and $[k,b,k']$ cannot be (k,b), it follows that (a,l,a') is incident with $[k,b,k']$. We will call the quadruple (R_1,R_2,Q_1,Q_2) a *coordinatization* of the GQ **S**.

We remark that Q_1 and Q_2 are dual operations.

2.2.1. Theorem

Let **S** be a GQ coordinatized by (R_1,R_2,Q_1,Q_2), then the following properties hold :

(O) $Q_1(k,0,0,a) = a$

$Q_1(0,a,k,a') = a'$.

($\bar{O}$) $Q_2(a,0,0,k) = k$

$Q_2(0,k,a,k') = k'$.

(A) If $a,b \in R_1$ and $k,l \in R_2$, then there is a unique $x \in R_1$ such that $Q_1(k,a,l,x) = b$.

($\bar{A}$) If $a,b \in R_1$ and $k,l \in R_2$, then there is a unique $p \in R_2$ such that $Q_2(a,k,b,p) = l$.

(B) If $a,b \in R_1$ and $k,l,k' \in R_2$ with $k \neq l$, then there is a unique pair $(x,y) \in R_1^2$ such that

$Q_1(k,x,Q_2(x,k,a,k'),y) = a$

$Q_1(l,x,Q_2(x,k,a,k'),y) = b$.

($\bar{B}$) If $a,b,a' \in R_1$ and $k,l \in R_2$ with $a \neq b$, then there is a unique pair $(p,q) \in R_2^2$ such that

$$Q_2(a,p,Q_1(p,a,k,a'),q) = k$$

$$Q_2(b,p,Q_1(p,a,k,a'),q) = l.$$

(C) If $a,a',b \in R_1$ and $k,k',l \in R_2$ then the system of equations in the unknowns x,p,x',p'

$$Q_1(k,x,Q_2(x,k,b,k'),x') = b$$

$$Q_1(p,x,Q_2(x,k,b,k'),x') = Q_1(p,a,l,a')$$

$$Q_2(a,p,Q_1(p,a,l,a'),p') = l$$

$$Q_2(x,p,Q_1(p,a,l,a'),p') = Q_2(x,k,b,k'),$$

has a unique solution $(x,p,x',p') \in R_1 x R_2 x R_1 x R_2$ if $Q_1(k,a,l,a') \neq b$ and $Q_2(a,k,b,k') \neq 1$ and none if one of both equalities holds.

Proof

We have only to show one of each pair of dual properties marked with a same letter. As to property (0), it follows from the coordinatization that (0,0,a) is incident with [k,a,0] and (a,k,a') with [0,a,l] for some $l \in R_2$.

In order to derive the others, it suffices to express the axiom (iii) of the generalized quadrangle **S** with respect to the following point-line pairs :

(A) : (k,b) and [a,l]

(B) : (l,b) and [k,a,k']

(C) : (a,l,a') and [k,b,k'].

We remark that (C) is self-dual.

2.2.2. Definition

Let R_1, (resp.R_2) be a set containing distinguished elements 0 and 1 but not ∞ and Q_1(resp.Q_2) quaternary operations from $R_2 x R_1 x R_2 x R_1$ to R_1 (resp.$R_1 x R_2 x R_1 x R_2$ to R_2). We call the quadruple (R_1,R_2,Q_1,Q_2) satisfying the properties (0), ($\bar{0}$), (A), ($\bar{A}$), (B), ($\bar{B}$), and (C) of theorem 2.2.1 a *quadratic quaternary ring*, which we shall abbreviate to QQR.

We denote the unique solution of $Q_1(k,a,l,x) = b$ by $Q_1^*(a,k,b,l) = x$, and dually $p = Q_2^*(k,a,l,b)$ if p satisfies $Q_2(a,k,b,p) = 1$.

2.2.3. Theorem

If (R_1,R_2,Q_1,Q_2) is a QQR then the structure S defined as follows is a generalized quadrangle. The points of S are elements of $R_1xR_2xR_1$, R_2xR_1 or R_1, denoted by parenthesis, together with (∞) where is a symbol not contained in R_1 or R_2. Lines are represented in square brackets by elements of $R_2xR_1xR_2$, R_1xR_2 or R_2 together with $[\infty]$. Incidence is defined in the following manner :

$$(a,l,a') \text{ is on } [k,b,k'] \Leftrightarrow Q_1(k,a,l,a') = b$$
$$Q_2(a,k,b,k') = l,$$

(a,l,a') is on $[a,l]$,

(k,a) is on $[k]$ and $[k,a,k']$ for all $k' \in R_2$,

(a) is on $[\infty]$ and $[a,k]$ for all $k \in R_2$,

(∞) is on $[\infty]$ and $[k]$ for all $k \in R_2$,

and there are no further incidences.

Proof

It suffices to show that for any point P and any line L not incident with P, there is a unique line on P meeting L and a unique point on L collinear with P.

To begin with, suppose $P = (\infty)$, then L has to be $[k,b]$ or $[k,b,k']$. It is easy to check that for the first case only $(\infty)I[\infty]I(a)I[a,l]$ is possible whereas for the second $(\infty)I[k]I(k,b)I[k,b,k']$.

Next, take $P = (a)$, then the cases where L is $[k]$ or $[b,k]$ with $b \neq a$ are straightforward. So assume $L = [k,b,k']$. A line on (a) has to be of the form $[a,p]$ with $p \in R_2$, or $[\infty]$, ruled out by $[\infty] \not\!I L$, and a point on that line of the form (a,p,x). But if this point has to be incident with $[k,b,k']$ then $p = Q_2(a,k,b,k')$ is known and x is uniquely determined by the equation

$$Q_1(k,a,p,x) = b$$

in view of (A).

Now we suppose $P = (l,a)$. The cases where L is $[\infty]$ or $[k]$ with $k \neq l$ are already proved dually. Let L be $[b,k]$. Then any point on L has the form (b,k,x) or (b') and any line on p the form $[l,a,p]$ or $[l]$. These elements are incident iff $a = Q_1(l,b,k,x)$ and $k = Q_2(b,l,a,p)$, which uniquely determines x and p by (A) and $(\bar{A})$. Let L be of the form $[k,b,k']$ with $b \neq a$ or $k \neq l$.
Consider a chain of elements :

$$(l,a)I[l,a,q]I(x,p,y)I[k,b,k']$$

then we have

$$Q_1(l,x,p,y) = a \qquad (1)$$

$$Q_2(x,l,a,q) = p \qquad (2)$$

$$Q_1(k,x,p,y) = b \qquad (3)$$

$$Q_2(x,k,b,k') = p \qquad (4)$$

Suppose first that $k = l$, then $(k,a)I[k]I(k,b)I[k,b,k']$. On the other hand, (1) and (3) imply that $a = b$, contradicting our assumption. Hence, $[k]$ is the only line on P meeting L and (k,b) is the only point on L collinear with P. Suppose now $k \neq l$, then it follows from (B) that the equations

$$Q_1(l,x,Q_2(x,k,b,k'),y) = a$$

$$Q_1(k,x,Q_2(x,k,b,k'),y) = b$$

with $k \neq l$ uniquely determine x and y. From these p follows by (4) and q by (2) in view of $(\bar{A})$. Hence, in both cases the chain exists and is unique.

Finally, let $P = (a,l,a')$. Considering the dual, we can assume that $L = [k,b,k']$ with $Q_1(k,a,l,a') \neq b$ or $Q_2(a,k,b,k') \neq l$. If $Q_1(k,a,l,a') = b$ then

$$(a,l,a')I[k,b,p]I(k,b)I[k,b,k']$$

where p follows from $Q_2(a,k,b,p) = l$. Dually, if $Q_2(a,k,b,k') = l$, then

$$(a,l,a')I[a,l]I(a,l,x)I[k,b,k']$$

where x follows from $Q_1(k,a,l,x) = b$.

Consider now a chain

$$(a,l,a')I[p,y,p']I(x,q,x')I[k,b,k']$$

then this is equivalent to

$$Q_1(p,a,l,a') = y$$

$$Q_2(a,p,y,p') = l$$

$$Q_1(p,x,q,x') = y$$

$$Q_2(x,p,y,p') = q$$

$$Q_1(k,x,q,x') = b$$

$$Q_2(x,k,b,k') = q$$

and also to

$$Q_1(k,x,Q_2(x,k,b,k'),x') = b$$

$$Q_2(a,p,Q_1(p,a,l,a'),p') = l$$

$$Q_1(p,x,Q_2(a,k,b,k'),x') = Q\ (p,a,l,a')$$

$$Q_2(x,p,Q_1(p,a,l,a'),p') = Q_2(x,k,b,k').$$

If $Q_1(k,a,l,a') = b$ or $Q_2(a,k,b,k') = l$, then this set of equations has no solution, whereas in the opposite case it has exactly one by (C).
This proves the theorem.

2.2.4. Remark

Note that if we coordinatize **S** again in the obvious way, we get a QQR identical to the one we started with.

2.2.5. Definition

If (R_1,R_2,Q_1,Q_2) and (R_1',R_2',Q_1',Q_2') are two quadratic quaternary rings we say that they are isomorphic if there is a bijection α from R_1 onto R_1' and a bijection β from R_2 onto R_2' such that

$$(Q_1(a,k,b,l))^\alpha = Q_1'(a^\alpha,k^\beta,b^\alpha,l^\beta)$$

and

$$(Q_2(k,a,l,b))^\beta = Q_2'(k^\beta,a^\alpha,l^\beta,b^\alpha)$$

for all $a,b \in R_1$ and $k,l \in R_2$.

2.3. Normalization of a QQR

Let S be a GQ coordinatized by a QQR (R_1,R_2,Q_1,Q_2). As we have seen in 2.1 there does not have to be a connection between the bijections from R_1 onto $[\infty]-\{(\infty)\}$ and onto $(0)A-\{(0)\}$.

We now demand that the bijections are chosen such that $(a,0,0)$ is collinear with $(1,a)$.

For the QQR we get :

$$Q_1(1,a,0,0) = a \qquad \text{(N)}$$

We can do the same dually and obtain :

$$Q_2(1,k,0,0) = k \qquad (\overline{\text{N}})$$

If (R_1,R_2,Q_1,Q_2) is a QQR not satisfying (N) and $(\overline{\text{N}})$, then define permutations α of R_1 and β of R_2 by :

$$Q_1(1,a,0,0) = a^\alpha$$

$$Q_2(1,k,0,0) = k^\beta$$

Defining moreover

$$\tilde{Q}_1(k,a,l,a') = [Q_1(k,a,l^\beta,a'^\alpha)]^{\alpha^{-1}}$$

and

$$\tilde{Q}_2(a,k,b,k') = [Q_2(a,k,b^{\alpha},k'^{\beta})]^{\beta^{-1}}$$

it is easy to check that $(R_1,R_2,\tilde{Q}_1,\tilde{Q}_2)$ is a QQR satisfying (N) and $(\bar{N})$.
We call such a QQR *normalized*, and the procedure just described, *normalization*.

2.4. Algebraic properties of quadratic quaternary rings

2.4.1. Definition

Let (R_1,R_2,Q_1,Q_2) be a (normalized) QQR.
We introduce a binary operation of addition into R_1 and R_2, and a "twisted" multiplication as follows : for any $a,b \in R_1$ and $k,l \in R_2$, we define

$$a + b = Q_1(1,a,0,b) \in R_1$$

$$k + l = Q_2(1,k,0,l) \in R_2$$

$$k \,.\, a = Q_1(k,a,0,0) \in R_1$$

$$a \,.\, k = Q_2(a,k,0,0) \in R_2$$

We denote $R_i - \{0\}$ by R_i^*, $i = 1,2$.

2.4.2. Theorem

If (R_1,R_2,Q_1,Q_2) is a normalized QQR, then the following properties hold :

(i) $a + 0 = 0 + a = a$ for all $a \in R_1$

(ii) $k + 0 = 0 + k = k$ for all $k \in R_2$

(iii) $a + x = b$ has a unique solution for any $a,b \in R_1$

(iv) $k + p = l$ has a unique solution for any $k,l \in R_2$

(v) $1 \,.\, a = a$ for all $a \in R_1$

(vi) $1 \,.\, k = k$ for all $k \in R_2$

(vii) $k \,.\, x = a$ has a unique solution for any $a \in R_1$ and $k \in R_2^*$

(viii) $a \,.\, p = k$ has a unique solution for any $a \in R_1^*$ and $k \in R_2$

Proof

We have $a + 0 = Q_1(1,a,0,0) = a$ by (N), and $0 + a = Q_1(1,0,0,a) = a$ by (O). This proves (i), and dually (ii).

If we apply (A) for $k = 1$ and $l = 0$, we obtain (iii) and dually (iv). By (N) and $(\bar{N})$, we get (v) and (vi) respectively.

Now we apply (B) for $k = k' = 0$ and $a = 0$, and obtain a unique pair $(x,y) \in R_1^2$ such that

$$Q_1(0,x,Q_2(x,0,0,0),y) = 0$$

$$Q_1(1,x,Q_2(x,0,0,0),y) = b$$

But $Q_2(x,0,0,0) = 0$ by $(\bar{0})$ and $Q_1(0,x,0,y) = 0$ forces $y = 0$ by (0). Hence,

$$1 \cdot x = b$$

has a unique solution $x \in R_1$ for $1 \neq 0$. Of course, (viii) can be proved dually.

2.5. Case of a finite QQR

Theorem

Let (R_1,R_2,Q_1,Q_2) be a quadruple where R_1 and R_2 are distinguished finite sets containing the symbol 0, and Q_1 resp. Q_2 a quaternary operation from $R_1 x R_2 x R_1 x R_2$ resp. $R_2 x R_1 x R_2 x R_1$ to R_1 resp. R_2. Suppose that a weaker version of the properties (0)-(C) of 2.2.1, namely where "a unique" is replaced by "at most one", holds. Then (R_1,R_2,Q_1,Q_2) is QQR.

Proof

Put $|R_1| = s$ and $|R_2| = t$, and construct an incidence structure as was done in theorem 2.2.3. It is straightforward to check that the following hold :

(i) each point is incident with $1 + t$ lines and two distinct points are incident with at most one line ;

(ii) each line is incident with $1 + S$ points and two distinct lines are incident with at most one point ;

(iii) if P is a point and L is a line not incident with P, then there is at most one pair $(Q,M) \in \mathbf{P} \times \mathbf{L}$ for which P I M I Q I L.

(iv) $|\mathbf{P}| = v = (s+1)(st+1)$

(v) $|\mathbf{L}| = b = (t+1)(st+1)$

Now the number of points collinear with at least one point of a line L, but not lying on L, equals $(s+1)ts$. But this is exactly the number of points not lying on L, proving that in property (iii) we can replace "at most one" by "a unique". It follows that (R_1,R_2,Q_1,Q_2) is a QQR.

3. REGULARITY

3.1. Definition

A pair of points $(p,q), p \neq q$, is called *regular* if $p \perp q$, or if $p \not\perp q$ and for any pair of distinct points $a,b \in \{p,q\}^\perp$, we have that each point of $\{p,q\}^\perp$ is collinear with each point of $\{a,b\}^\perp$. If the GQ **S** is finite with parameters (s,t), then (p,q) is regular iff $|\{p,q\}^{\perp\perp}| = t + 1$ provided $p \not\perp q$. The point p is *regular* if (p,q) is regular for all points $q \neq p$.

3.2. Theorem

Let **S** be a GQ coordinatized by a QQR (R_1,R_2,Q_1,Q_2). Then the point (∞) is regular if and only if Q_1 is independent of the third argument, i.e.

$$Q_1(k,a,l,a') = Q_1(k,a,0,a')$$

for all $a,a' \in R_1$ and $k,l \in R_2$. Dually the line $[\infty]$ is regular if and only if Q_2 is independent of the third argument.

Proof

Suppose (∞) is a regular point. We have $(a),(0,a) \in \{(\infty),(a,m,a')\}^{\perp}$, and $\{(a),(0,a')\}^{\perp} = \{(a,l,a') \mid l \in R_2\} \cup \{(\infty)\}$. Now we express that $(k,b) \perp (a,l,a')$ for some $p \in R_2$, $[k,b,p]I(a,l,a')$, so

$$b = Q_1(k,a,l,a')$$

By the regularity of (∞) this must hold for all $l \in R_2$, so

$$Q_1(k,a,l,a') = b = Q_1(k,a,0,a')$$

for all $a,a' \in R_1$ and $k,l \in R_2$. Conversely, if $Q_1(k,a,0,a') = Q_1(k,a,l,a')$ then

$$(a,l,a') \perp (k,Q_1(k,a,0,a'))$$

hence, (∞) is regular if Q_1 does not depend on the third argument.

3.3. Theorem

Let S be a finite GQ of order s having (∞) as regular point. The incidence structure $\pi_{(\infty)}$ with pointset $(\infty)^{\perp}$, with lineset consisting of spans $\{p,q\}^{\perp\perp}$, where $p,q \in (\infty)^{\perp}$, $p \neq q$, and with the natural incidence, is a projective plane of order s coordinatized by the PTR $T(k,a,a') = Q_1(k,a,0,a')$ (Coordinatization Method of Hall) if R_1 and R_2 are identified by $k = Q_1(k,1,0,0)$ and the QQR is normalized.

Proof

If s = t it is possible to identify R_1 and R_2. We can do that in the following way provided (∞) is regular : given $k \in R_2$, we denote the point on $[k]$, collinear with $(1,0,0)$ by (k,k^{σ}).

This means that $k^{\sigma} = Q_1(k,1,0,0)$, and in particular $0^{\sigma} = 0$ and $1^{\sigma} = 1$.
We show that the map σ from R_2 tot R_1 is injective. In order to derive a contradiction, let for $k \neq 1$, (k,a) and $(1,a)$ be collinear with $(1,0,0)$. Then (∞), $(1,0,0)$, $(0,0,a) \in \{(k,a),(1,a)\}^{\perp}$, and by regularity of (∞), $(0,0,a) \perp (1)$, clearly a contradiction.

The fact that the incidence structure $\pi_{(\infty)}$ is a projective plane is known [9].

The lineset is thus given by the set of spans $\{(a), (0,a')\}^{\perp\perp}$ together with the lines [k], $k \in R_2$ and $[\infty]$. In view of the proof of theorem 3.2, we have :

$$\{(a),(0,a')\}^{\perp\perp} = \{(\infty),(a,0,a')\}^{\perp}$$
$$= \{(k,Q_1(k,a,0,a')) \mid k \in R_2\} \cup \{(a)\}$$

Define a ternary operation on R_1, $T(k^{\sigma},a,a') = Q_1(k,a,0,a')$, and give the span $\{(a),(0,a')\}^{\perp\perp}$ the coordinate [[a,a']], and the point (k,b) the coordinate $((k^{\sigma},b))$. Clearly ((x,y)) is on [[m,k]] if and only if $(x^{\sigma^{-1}},y)$ is on the span $\{(m),(0,k)\}^{\perp\perp}$ and this holds if and only if $y = Q_1(x^{\sigma^{-1}},m,0,k)$, i.e. $y = T(x,m,k)$. Moreover there hold :

(A) $T(x,0,b) = Q_1(x^{\sigma^{-1}},0,0,b) = b$

$T(0,x,b) = Q_1(0,x,0,b) = b$

for all $x,b \in R_1$.

(B) $T(x,1,0) = Q_1(x^{\sigma^{-1}},1,0,0) = x$

$T(1,x,0) = Q_1(1,x,0,0) = x$

for all $x \in R_1$.

This proves the theorem.

4. ACKNOWLEDGEMENT

Both authors are very grateful to Prof. Dr. J. Tits for an interesting conversation on the subject.

BIBLIOGRAPHY

[1] Dembowski P., Finite geometries, Springer Verlag 1968.

[2] Hall M., Projective planes, Trans. Am. Math. Soc. 54, 229-277 (1943)

[3] Hanssens G. and Van Maldeghem H., A new look upon the classical generalized quadrangles (to appear in Ars Combinatoria).

[4] Hanssens G. and Van Maldeghem H., Algebraic properties of quadratic quaternary rings (preprint).

[5] Hanssens G. and Van Maldeghem H., The QQR ofthe known generalized quadrangles (in preparation).

[6] Hanssens G. and Van Maldeghem H., A coordinatization method for generalized n-gons (in preparation).

[7] Hanssens G. and Van Maldeghem H., Incidence geometries related to rank 3 affine buildings (preprint).

[8] Hughes D.R. and Piper F.C., Projective planes, Springerverlag 1973.

[9] Payne S.E. and Thas J.A., Finite generalized quadrangles, Research Notes in Math. 110, 1984.

Address : Seminarie voor Meetkunde en Kombinatoriek
Krijgslaan, 281,
9000 GENT
Belgium

(*) both authors are supported by the NFWO (National Fund for Scientific Research of Belgium).

Annals of Discrete Mathematics 37 (1988) 209–216

CONSTRUCTION OF SOME PLANAR TRANSLATION SPACES

Dedicated to Prof. H.Lenz in occasion of his 70th birthday

Armin HERZER, Fachbereich Mathematik
Joh.Gutenberg-Universität, Saarstr.21
D-6200 Mainz, Germany

Planar translation spaces come from "linear partitions" of groups. Here a construction method in two variations for such linear partitions is given and the related planar translation spaces are considered.

1. INTRODUCTION

In the study of projective embeddings of geometries, linear spaces etc. the locally (generalized) projective semimodular (or specially geometric) lattices play an exceptional role; for they can be embedded only if certain properties are satisfied, see [5],[11], [15]. Therefore one is interested in examples of geometries which are not embeddable.

A series of planar translation spaces containing non-desarguesian planes is constructed in [11] and [12]. These spaces therefore cannot be projectively embedded. Here we generalize this construction, initiating a theory of planar translation spaces.

Planar translation spaces are realized by means of "linear partitions of groups" (definition below). This is similar to the construction of linear translation spaces (Translationsstrukturen, strutture di André) by partitions of groups.

We give two different constructions of such linear partitions and show that there is a wide class of linear translation spaces which can occur as planes of planar translation spaces. So for any linear translation space of this class which cannot be embedded in a desarguesian projective plane, we obtain a planar translation space which is not projectively embeddable. The construction of the underlying groups give abelian and Frobenius groups. In an appendix we show that a non-abelian group permitting a linear partition has trivial center.

For further information about incidence structures and related topics see [9], for groups (in particular Frobenius groups) see [13]. The basic ideas for translation structures (in the concept of "parallel structures") were given in [1] and [3].

2. BASIC DEFINITIONS

A <u>linear space</u> is an incidence structure $\mathbb{L} = (P, G, I)$, where the elements of P are called <u>points</u> and the elements of G are called <u>lines</u> and the following hold:

(i) Every line contains at least two points.
(ii) Through any two points p,q there goes exactly one line (denoted by pq).

Points are called collinear, iff they lie on a common line.

(iii) There are three non-collinear points.

Let G and E be sets of subsets of a set P. We call $\mathsf{P} = (P,G,E)$ a planar space, if $(P,G,\in)$ is a linear space, and for E (the elements of which are called planes) the following hold:

(i) Every plane contains at least three non-collinear points.
(ii) For any two points p,q of a plane E the whole line pq is contained in E.
(iii) Through any three non-collinear points there goes exactly one plane.
(iv) There are at least two planes.

It follows that if two different planes have two points in common, they intersect in a line.

For $p \in P$ let $G(p)$ and $E(p)$ be the set of lines through p and planes through p respectively. Thus $\mathsf{P}(p) := (G(p), E(p), \subset)$ is a linear space. P is called locally projective, if for every $p \in P$ the linear space $\mathsf{P}(p)$ is a projective space. - For any plane E we also may consider the linear subspace $(E, G \cap E, \in)$, where $G \cap E = \{l \in G \mid l \subset E\}$. We shall also use the word "plane" to refer to such a linear subspace.

A dilatation of the incidence structure $\mathsf{I} = (P,B,\in)$ is any permutation α of P respecting B with

$$B \cap B^{\alpha} \neq \emptyset \Rightarrow B = B^{\alpha} \quad \forall B \in B.$$

A dilatation is said to be proper if it is not the identity but fixes some $p \in P$. This point is known to be unique and is called the center of the dilatation. It follows that if Γ is a group of dilatations of I, then for $p \in P$ the stabilizer $\Gamma(p)$ consists of the identity and perhaps proper dilatations of I with center p; in this case we call $\Gamma(p)$ a group of proper dilatations. A translation of I is any dilatation which is not proper. The translations consist of the identity map and all fixed-point-free dilatations. The incidence structure I is called a translation space, if it possesses a point-transitive group of translations. In particular the translation space $\mathsf{L} = (P,G,\in)$ is called linear, if L is also a linear space, and the planar space $\mathsf{P} = (P,G,E)$ is called a planar translation space, if $(P,E,\in)$ is a translation space.

For a group G with identity element 1 and a set $\varphi(G)$ of subgroups of G let $\overline{\varphi}(G)$ consist of all right cosets of the subgroups in $\varphi(G)$:

$$\overline{\varphi}(G) = \{Ug \mid U \in \varphi(G),\ g \in G\}.$$

Recall that a (group-)partition $\pi(G)$ is a set of subgroups of G, such that every element of G different from 1 is contained in exactly one member of $\pi(G)$. We only consider non-trivial partitions: $|\pi(G)| > 1$ and $\{1\} \notin \pi(G)$. As is wellknown (see e.g. [3]),

$$\gamma(G) := (G, \overline{\pi}(G), \in)$$

is a linear translation space, where G is acting as a group of translations by multiplication from the right; and every linear translation space - up to isomorphism - can be represented in this way.

To define the projective closure of a linear translation space - for the concept see also [2], where it appears under the name of "projective parallel structure" - we may assume that the space has the form $\gamma(G)=(G,\overline{\pi}(G),\in)$. For $U\in\pi(G)$ we define the (parallel-)class $[U]$ of $\gamma(G)$ as the set of all right cosets of U in G. Adjoin all $[U]$ for $U\in\pi(G)$ as ideal points and one ideal line, the points of which are exactly all ideal points. A line X has $[U]$ as ideal point, iff $X\in[U]$.

Let V be a (right-)vector space over the field F (the multiplication in F not necessarily commutative). A generalized spread of V is a partition of the abelian group $(V,+)$, consisting of F-subspaces of V. A translation space is F-linear for a field F, if it is isomorphic to $(V,\overline{\pi}(V),\in)$ for an F-vector space V and a generalized spread $\pi(V)$ of V.

The partition $\pi(G)$ of G is called linear, if there is a "companion set" $\varepsilon(G)$ of subgroups of G with the following properties:

(i) $\forall X\in\pi(G)\ \forall Y\in\varepsilon(G): X\cap Y \neq \{1\} \Rightarrow X \subset Y$.

(ii) $\omega(G) := (\pi(G),\varepsilon(G),\subset)$ is a linear space.

PROPOSITION 1. From every linear partition $\pi(G)$ of a group G with companion set $\varepsilon(G)$ we obtain a planar translation space

$$\mathcal{P} = (G,\overline{\pi}(G),\overline{\varepsilon}(G)).$$

For any $g\in G$ the linear space $\mathcal{P}(g)$ is isomorphic to $(\pi(G),\varepsilon(G),\subset)$. Every planar translation space - up to isomorphism - can be represented in this way. In particular $\mathcal{P}$ is locally projective, if and only if $\omega(G)$ is a projective space.

The proof follows the usual lines in this field (see e.g.[1],[3]).

Here it is possible to explain the terminology: Why a partition of a group is called linear, if one may construct a planar space, whereas each partition of a group gives rise to a linear space? The first reason is that a linear partition in itself is a linear space. On the other hand we have this phenomenon already in classical geometry: Considering projectively the sets of 1- and 2-dimensional subspaces of a vector space one gets a linear space, namely a projective space. Considering the same thing from an affine point of view (the vectors beeing the points) one sees lines and planes through the origin in an affine space.

In the sequel we give constructions of linear partitions of groups using vector spaces as well as Frobenius groups. As a definition for the latter, which also works for infinite groups, we prefer the following: A group G is a Frobenius group, if there is a proper normal subgroup N of G and a complement H of N with the following properties:

(i) $G = N \cup \bigcup_{a\in N} H^a$

(ii) $H\cap H^a = \{1\}$, if $a\in N$, $a\neq 1$.

3. THE CONSTRUCTIONS

For a group G and a group Δ of automorphisms of G the semidirect product ΔG is the set Δ×G with multiplication

$$(\alpha,g)(\beta,h) = (\alpha\beta, g^{\beta}h) \qquad \forall \alpha,\beta\in\Delta \quad \forall\, g,h\in G.$$

By the embeddings $G \to \Delta G;\ g \rightsquigarrow (1,g)$ and $\Delta \to \Delta G;\ \alpha \rightsquigarrow (\alpha,1)$ we consider Δ and G as subgroups of ΔG. Then G is a normal subgroup of ΔG and Δ is a complement of G in ΔG. For $U \subset G$ let be

$$\Delta^U = \{\Delta^g \mid g \in U\}.$$

DEFINITION. Let $\pi=\pi(G)$ be a partition of the group G and Δ a group of automorphisms of G.

(i) Δ is π-<u>admissible</u>, iff $\forall X\in\pi\ \forall\alpha\in\Delta$: $X = X^{\alpha}$
(ii) Δ is <u>subtractive</u>, iff $\forall\alpha\in\Delta$: $\alpha\neq 1 \Rightarrow$ the mapping $-\alpha+1\colon G \to G;\ g \rightsquigarrow (g^{\alpha})^{-1}g$ is surjective.

REMARK 1. If Δ is π-admissible, then every $\alpha\in\Delta$, $\alpha\neq 1$, is a fixed point free automorphism; so the mapping $-\alpha+1$ on G is injective. Thus for finite G a π-admissible automorphism group necessarily is subtractive.

Now we are able to give our first construction (inspired somewhat by the construction of the Frobenius groups in [18]):

THEOREM 1. Let $\pi=\pi(G)$ be a partition of the group G and Δ a non-trivial group of automorphisms of G, which is π-admissible and subtractive. Then ΔG is a Frobenius group. If G is abelian, then

$$\pi(\Delta G) := \pi(G) \cup \Delta^G$$

is a linear partition.

<u>Proof</u>: Since Δ is π-admissible and subtractive the mapping $-\alpha+1$ is a bijection on G. From this it follows by wellknown arguments that ΔG is a Frobenius group and $\pi(\Delta G)$ is a group-partition.

From now we suppose G to be abelian. Then since every member U of $\pi(G)$ is fixed by Δ, also U is a normal subgroup of ΔG and we have the semidirect products

$$T(Ug) := \Delta^g U = \{(\alpha, g^{-\alpha+1}u) \mid \alpha\in\Delta,\ u\in U\}, \qquad \forall U\in\pi(G) \quad \forall g\in G.$$

We claim that

$$\varepsilon(\Delta G) := \{G\} \cup \{T(Ug) \mid g\in G,\ U\in\pi(G)\}$$

is a companion set for $\pi(\Delta G)$. For this purpose we give an isomorphism α of the projective closure of $\gamma(G)=(G,\overline{\pi}(G),\in)$ onto $\omega(\Delta G) := (\pi(\Delta G), \varepsilon(\Delta G), \subset)$ defining $g^{\alpha}=\Delta^g$ for any point $g\in G$ and $[U]^{\alpha}=U$ for any ideal point [U] of $\gamma(G)$, $U\in\pi(G)$. Then the line Ug of $\gamma(G)$ is mapped onto T(Ug) and the ideal line of $\gamma(G)$ is mapped onto G.

In other words, for $\omega(\Delta G)$ the line through $U,U'\in\pi(G)$, $U\neq U'$, is G, the line through U and Δ^g is T(Ug), and the line through Δ^g and Δ^h for $g\neq h$ is T(Wh), where $W\in\pi(G)$ is defined by $gh^{-1}\in W$. □□

REMARK 2. In the situation of the theorem we have the following partial converse of the last sentence:

If $\pi(\Delta G)$ is linear and G is a line, then G is abelian.

<u>Proof</u>: For different elements X and Y of $\pi(\Delta G)$ we denote the line through X an Y by $X \vee Y$. For $g,h \in G$, $g \neq h$, and $1 \neq \alpha \in \Delta$ we have

$$(\alpha^{-1}, g^{-\alpha^{-1}+1})(\alpha, h^{-\alpha+1}) = (1, g^{-1} g^{\alpha} h^{-\alpha} h) \in (\Delta^{g} \vee \Delta^{h}) \cap G.$$

So for $U \in \pi(G)$ and $1 \neq u \in U$ we have $U \vee \Delta^{h} = \Delta^{hu} \vee \Delta^{h}$, therefore

$$u^{-1}(h^{-\alpha+1})^{-1} u^{\alpha} h^{-\alpha+1} = (hu)^{-1}(hu)^{\alpha} h^{-\alpha} h \in (U \vee \Delta^{h}) \cap G = U.$$

Since $-\alpha+1$ is surjective, we conclude, that U is a normal subgroup of G. But if $\pi(G)$ consists of normal subgroups, by a theorem of Kontorowitsch ([16], see also [3]) G is abelian.

REMARK 3. For G, $\pi(G)$ and Δ fulfilling the conditions of theorem 1 we have the following geometric interpretation: If $\mathbb{L} = (P,G,I)$ is a linear translation space isomorphic to $\gamma(G) = (G,\pi(G),\in)$, then $\mathbb{L}$ has a group Γ of dilatations isomorphic to ΔG, where the translation group T is isomorphic to G and every point p is the center of a group $\Gamma(p)$ of proper dilatations isomorphic to Δ such that

$$\Gamma = T \cup \bigcup_{p \in P} \Gamma(p).$$

Conversely we obtain

COROLLARY 1. Let $\mathbb{L}$ be a linear translation space with abelian translation group T. If $\mathbb{L}$ possesses a group Γ of dilatations containing proper dilatations such that Γ is the union of T and the groups $\Gamma(p)$ of proper dilatations for all points p of $\mathbb{L}$, then Γ is a Frobenius group and there is a planar translation space $\mathbb{P}$ containing a class of planes isomorphic to $\mathbb{L}$. Moreover $\mathbb{P}$ has a translation group isomorphic to Γ and for any point p of $\mathbb{P}$ the linear space $\mathbb{P}(p)$ is isomorphic to the projective closure of $\mathbb{L}$.

The <u>proof</u> has to use the transfer from the geometric language of Corollary 1 to the conditions of theorem 1 and then to apply the results of its proof and of Proposition 1.

REMARK 4. All the other planes of Cor.1 come from the so-called Frobenius partitions of the groups T(Ug). So in the finite case they can be considered as Frobenius-Jungnickel transversal designs ([14], see also [6]), where the point classes are the cosets of U in T(Ug).

EXAMPLES. 1. For any ring R and R-right-module M let $\pi(M)$ be a partition of (M,+) consisting of R-submodules. Let S be a group of <u>units</u> of R such that for $a \in S$, $a \neq 1$, also $1-a$ is a unit in R. Let $\bar{a}$: $M \to M$; $m \mapsto ma$, and $\Delta = \{\bar{a} \mid a \in S\}$. Then for G = (M,+) and Δ theorem 1 can be applied. For a not necessarily commutative integral domain R with greatest common right-divisor (gcd) and $M = R^{n}$ a minimal partition is $\pi(M) = \{(x_1,\dots,x_n)R \mid \gcd(x_1,\dots,x_n)=1\}$ as constructed in [17].

2. Let R=F be a field. Then for S we can choose any non-trivial subgroup of the multiplicative group of F. So theorem 1 can be applied to all F-right-vectorspaces V with generalized spread $\pi(V)$ for a field F with $|F| > 2$; and Cor.1 applies to all F-linear spaces for such an F, in particular (affine) translation planes with

kernel F.

3. If $\mathbb{L}$ is any finite linear translation space with abelian translation group and proper dilatations, then in fact Cor.1 applies to $\mathbb{L}$ by a theorem of Frobenius (see e.g.[3], [9], [13]).

Now our second construction.

THEOREM 2. Let V be an F-vector space and H a hyperplane (maximal subspace) of V. Let $\pi(H)$ be a generalized spread of H and

$$\pi(V) = \pi(H) \cup \{vF \mid v \in V \setminus H\}.$$

Then the generalized spread $\pi(V)$ is a linear partition of $(V,+)$.

Proof: Let $\varepsilon(V)$ consist of H and all subspaces X of V, which are not contained in H but fulfill $X \cap H \in \pi(H)$. Then it is easy to be seen that $\varepsilon(V)$ is a companion set to $\pi(V)$. In fact again $(\pi(H), \varepsilon(H), \subset)$ is isomorphic to the projective closure of $\gamma(H) = ((H,+), \overline{\pi}(H), \in)$. □□□

This is a generalization of the construction in [8], where $\pi(H)$ is a spread (belonging to a translation plane). This construction also was used in [4], [7], [10]. Since the linear translation spaces $\gamma(H)$ often contain non-desarguesian translation planes, we obtain examples of projectively non-embeddable translation spaces.

COROLLARY 2. For any F-linear translation space $\mathbb{L}$ there exists a planar translation space $\mathbb{P}$ containing a class of planes isomorphic to $\mathbb{L}$. The translation group of $\mathbb{P}$ is abelian and for any point p of $\mathbb{P}$ the linear space $\mathbb{P}(p)$ is isomorphic to the projective closure of $\mathbb{L}$.

Proof analogous to the proof of Cor.1.

REMARK 5. The other planes of $\mathbb{P}$ in Cor.2 can be described in the following way: The points of such a plane are the points of an affine space $\mathbb{A}$ over F. We have two kinds of lines: 1. One parallel class $\mathcal{H}$ of hyperplanes of $\mathbb{A}$ as the "great lines" of $\mathbb{P}$, 2.All lines of $\mathbb{A}$ not contained in any hyperplane of $\mathcal{H}$ as the "normal lines" of $\mathbb{P}$. In the finite case considering the hyperplanes of $\mathcal{H}$ as point-classes we obtain again a classical translation transversal design (for the concept see [6]).

EXAMPLES of F-linear translation spaces (to which, therefore, Cor.2 applies):

1. All (affine) translation planes
2. All finite linear translation spaces with abelian translation group (see [3], "zentrale Translationsstrukturen").

4. FINAL REMARKS

We consider the following rather trivial statement, which nevertheless sometimes is useful:

PROPOSITION 2. Let $\pi(G)$ be a linear partition of G and H a subgroup of G with $H \cap U \neq \{1\}$ $\forall U \in \pi(G)$. Then $\pi(G) \cap H := \{U \cap H \mid U \in \pi(G)\}$ is a linear partition of H. (Namely $\varepsilon(G) \cap H := \{X \cap H \mid X \in \varepsilon(G)\}$ is a companion set.)

EXAMPLE. Let be $V = F^n$ for a field F and $\pi(V)$ a generalized spread which at the same time is a linear partition of $G = (V,+)$, e.g. by the construction of th.2. For a subring R of F consider $H = (R^n,+)$ in a natural way as subgroup of G. Then the conditions of Prop.2 are satisfied.

Which non-abelian groups other than Frobenius groups allow a linear partition? As a recent result we have that in the finite case there are no other examples [19]. For a first step to answer the question in the infinite case we give the following

PROPOSITION 3. If G is a non-abelian translation group of a planar translation space, then G has trivial center.

Proof: We have sets $\pi(G)$, $\varepsilon(G)$ of subgroups of G such that $\pi(G)$ is a partition of G and $(\pi(G), \varepsilon(G), \subset)$ is a linear space. For $G^* = G \setminus \{1\}$ and $a \in G^*$ let (a) be the unique member of $\pi(G)$ containing a; and for $a,b \in G$ with $(a) \neq (b)$ let (a,b) be the unique member of $\varepsilon(G)$ containing a and b. The center of G is denoted by Z and $[a,b] := a^{-1}b^{-1}ab$ is the commutator of a and b.

1. For $(a)\neq(b)$ and $ab \neq ba$ is $Z \subset (a,b)$: W.l.o.g. we may assume $([a,b]) \neq (b)$. Consider $z \in Z$. If $(az)=(a)$ or $(az)=(b)$, then $z \in (a,b)$. Now assume $(a)\neq(az)\neq(b)$. Thus, since $[a,b]=[az,b] \in G^*$, $[a,b] \in (a,b)$, $[az,b] \in (az,b)$, we have

$$az \in (az,b) = ([az,b],b) = ([a,b],b) = (a,b),$$

therefore $z \in (a,b)$. □

2. There is $a,b \in G^*$ with $ab \neq ba$ and $(a)\neq(b)$: Since $Z \neq G$ we may choose $a \in G \setminus Z$. Then (a) and C(a), the centralizer of a in G, are proper subgroups of G and so $(a) \cup C(a) \neq G$. Therefore there is some $b \in G \setminus ((a) \cup C(a))$, and so a,b are as desired. □

3. There is $c \in G \setminus (a,b)$ with $ac \neq ca$: We argue as before, using that (a,b) and C(a) are proper subgroups of G. □

4. $Z \subset (a)$: From 2. and 1. we obtain $Z \subset (a,b)$ and from 3. we have $ac \neq ca$ and $(a) \neq (c)$, and so again from 1. we obtain $Z \subset (a,c)$. Therefore $Z \subset (a,b) \cap (a,c) = (a)$. □

5. $Z = \{1\}$: Changing a and b, by symmetry from 4. we get also $Z \subset (b)$, thus

$$Z \subset (a) \cap (b) = \{1\}.$$ □□

REFERENCES

[1] André, J., Über nicht-Desarguessche Ebenen mit transitiver Translationsgruppe, Math.Z.60(1954)156-186
[2] André, J., Über Parallelstrukturen. I.Grundbegriffe, Math.Z. 76(1961)85-102
[3] André, J., Über Parallelstrukturen. II.Translationsstrukturen, Math.Z.76(1961)155-163
[4] Barlotti, A. and Cofman, J., Finite Sperner Spaces Constructed from Projective and Affine Spaces, Abh.Math.Sem.Univ.Hamburg 40(1974)230-241

[5] Batten, L.M., Locally Generalized Projective Lattices satisfying a Bundle Theorem, Geom.Dedic.22(1987)363-369

[6] Beth, T., Jungnickel, D. and Lenz, H., Design Theory (Bibliographisches Institut Mannheim Wien Zürich 1985)

[7] Bonetti,F. and Lunardon, G., Un esempio di S-spazio di traslazione, Atti Accad. Naz. Sci., Lettere e Arti di Modena, (6) 19(1977)153-157

[8] Bruck, R.H. and Bose, R.C., The Construction of Translation Planes from Projective Spaces, J.Algebra 1(1964)85-102

[9] Dembowski, P., Finite Geometries (Springer New York 1968)

[10] Herzer, A., Konstruktion von Spernerräumen aus einer vorgegebenen projektiven Ebene, Abh.Math.Sem.Univ.Hamburg 46 (1977) 25-54

[11] Herzer, A., Projektiv darstellbare stark planare Geometrien vom Rang 4, Geom.Dedic.5(1976)467-484

[12] Herzer, A., Semimodular locally projective lattices of rank 4 from v.Staudt's point of view, in: Plaumann, P. and Strambach, K.,(edts.), Geometry - von Staudt's Point of View (Reidel, Dordrecht 1981) pp 373-400

[13] Huppert. B., Endliche Gruppen I (Springer, Berlin Heidelberg New York 1967)

[14] Jungnickel, D., Transversal designs associated with Frobenius groups, J.Geometry 17(1981)140-154

[15] Kahn, J., Locally projective-planar lattices which satisfy the bundle theorem, Math.Z.175(1980)219-247

[16] Kontorowitsch, P., Sur les groupes a base de partition, Mat.Sbornik 12(1943)56-70

[17] Permutti, R., Geometria affine su un anello, Atti Acc.Naz. Lincei, Memorie 8, I (1968) 259-287

[18] Schulz, R.-H., Transversal designs and partitions associated with Frobenius groups, Journal r.u.a.Math.355(1985) 153-162

[19] Schulz, R.-H., On a conjecture of Herzer concerning linear partitions, J.Geometry 26 (1986) 186-190

Annals of Discrete Mathematics 37 (1988) 217–224

REGULAR SETS IN GEOMETRIES

J.D. KEY

Department of Mathematics, University of Birmingham,
P.O.Box 363, Birmingham B15 2TT, U.K.

Given a permutation group G acting on a set Ω we look for conditions under which G has a regular orbit in its natural action induced on the power set $P(\Omega)$.

1. INTRODUCTION

Let $S = (P,B,I)$ be an incidence structure with point set P, block set B and automorphism group $G = \mathrm{Aut}\,(S)$. A *regular set* of points for S is a subset Q of P which is such that the set-stabilizer $G_{\{Q\}}$ of Q in G is the identity. Thus no non-trivial automorphism of S fixes the set Q. We ask the general question: when do regular sets exist in structures with non-trivial automorphism groups? In this paper we collect together some results obtained for structures that have automorphism groups doubly transitive on the point set, in particular, the finite projective and affine desarguesian geometries. The results quoted include joint work of the author with J. Siemons and A. Wagner, and recent work of F. Dalla Volta.

Closely linked to the above notion is the analogous concept for permutation groups in general: let G be a group with a permutation representation ϕ on a set Ω. A subset Λ of Ω is a *regular set for* G^{ϕ} if $G^{\phi}{}_{\{\Lambda\}} = 1$. Amongst those groups for which there are no regular sets at all are the symmetric groups S_n for $n \geqslant 3$, and the alternating groups A_n for $n \geqslant 4$. Using essentially counting arguments and the Classification Theorem for Finite Simple Groups, it was shown in [1] that only a finite number of finite primitive groups not containing the alternating group of the same degree have no regular set at all. In Section 2 we quote results that give all those with this property that act as groups of automorphisms on finite desarguesian geometries. In [3] Gluck found all the primitive soluble groups of finite degree that do not have regular sets: his list agrees with that given in Result 7 (Section 2) below, to the extent to which the groups there happen to be soluble.

The counting arguments used in some of the results mentioned above give no information on the size of regular sets when they exist for given classes of groups or incidence structures. If Λ is a regular set for G, where $G \leqslant Sym(\Omega)$, then clearly $\Omega\setminus\Lambda$ is also a regular set for G , so we may assume that $|\Lambda| \leqslant |\Omega|/2$. Then we may ask the following questions:

(i) for a given class of groups or designs, what is the minimum size of a regular set (when such exists) and for what members of the class do no regular sets exist?
(ii) if $G \leqslant Sym(\Omega)$ and G has a regular set of size k where $k < |\Omega|/2$, then does G have a regular set of size m for any m such that $k < m \leqslant |\Omega|/2$? In fact we conjecture that (ii) is generally, but not always, true if G is primitive. In this connection the following is easily obtained:

RESULT 1 ([5], Lemma 2.1)
If $G \leqslant Sym(\Omega)$ is semi-regular on Ω where $|\Omega| = n$, then G has a regular set of size k for every k for which $1 \leqslant k \leqslant n-1$, unless G is a transitive elementary abelian 2-group and k=2 or k = n-2.

The question raised here, of the existence of regular sets for permutation groups, developed in [5] out of more general considerations of the induced action of a permutation group on the power set $P(\Omega)$, and, more specifically, on the set $\Omega^{\{k\}}$ of all subsets of Ω of size k. The existence of a regular k-set for G ensures the existence of an orbit of G on $\Omega^{\{k\}}$ of the maximal possible length, viz $|G|$. In [8] the *k-closure* $G^{\{k\}}$ of G was defined to be the largest subgroup of $Sym(\Omega)$ having the same orbits on $\Omega^{\{k\}}$ as G: this is directly analogous to the definition in Wielandt [9] of closure, where the action of G on k-sequences is considered. A group is *k-closed* if $G = G^{\{k\}}$. In [8] Siemons shows that if G is k-closed then G is also m-closed for any m such that $k \leqslant m \leqslant |\Omega|/2$. The full automorphism group of a t-(v,k,λ) design *D* must be k-closed in its action on points of *D*: for desarguesian geometries the groups will generally be 3-closed. In [5] the 3-closure of $P\Gamma L(d,q)$ for $d \geqslant 3$ and the general existence of regular k-sets was used to prove that any subgroup of $P\Gamma L(d,q)$ is k-closed. Again, by the same counting arguments as mentioned above, and assuming the classification theorem, only a finite number of finite primitive groups not containing the alternating group of the same degree can not be k-closed for any k. We find those acting on geometries, using regular sets (see Results 8,9 in Section 2).
In this partial survey of the situation regarding the existence of regular sets for structures or groups, we present some of the known results, concentrating in particular on joint results of the author with J. Siemons.

2. FINITE DESARGUESIAN GEOMETRIES.

For the finite geometries PG(d,q) and AG(d,q) for $d \geq 2$, $q \geq 3$, the question of the existence of regular sets, and the type of configuration that may be involved, has been examined in Key and Siemons [5]. The only geometries that do not have regular sets here are PG(2,3), PG(2,4), AG(2,3) and AG(2,4). For q=2 the arguments are somewhat different, due to the nature of the geometry; these have been examined by Dalla Volta [2]. Those that do not have regular sets for $d \geq 2$ are PG(2,2), PG(3,2), AG(2,2), AG(3,2), AG(4,2) and AG(5,2). For d=1 the arguments are necessarily more algebraic ; the question is answered in Key, Siemons and Wagner [6], for the more difficult projective case.

We give outline statements of the results and the type of configuration involved: for more precise statements and proofs, the sources quoted above should be consulted. Some attempt has been made to find the smallest size of a regular set, but we have tended towards general arguments that are applicable to most dimensions and fields. Thus for $d \geq 2$, $q \geq 3$ two general types of configurations of regular sets were established, either of which is applicable if the field is large enough with respect to the dimension d (see Results 4 and 5 below). We describe these for the projective case, and give a brief outline of the reasoning involved.

A lemma from [5] that has been crucial in obtaining the smaller type of regular set for the larger fields, and that has proved to be of use in finding regular sets for some other designs and groups is the following:

RESULT 2 ([5], Lemmas 2.2 and 2.3)
Let $G = P\Gamma L(2,q)$, $q=p^n$, p a prime, acting on the points Ω of PG(1,q) where $\Omega = \{\infty\} \cup GF(q)$. Let $\Lambda \subsetneq \Omega$, $|\Lambda| = 3$. For any number d such that $1 \leq d \leq q-3$, for $q > 16$ there exists a set Δ where $\Delta \subsetneq \Omega \setminus \Lambda$ and $|\Delta| = d$ such that $G_{\{\Lambda\},\{\Delta\}} = 1$.

Remark: the proof involves use of Result 1 quoted in Section 1. It involves the observation that for q large enough $G_{\{\Lambda\}}$ has an orbit of length 6n. Note also that Result 2 is also true for some values of $q \leq 16$: see [5].

RESULT 3 ([6], Theorem)
For $q \geq 29$, $P\Gamma L(2,q)$ has a regular 5-set.

Remark: $P\Gamma L(2,q)$ has a regular 5 or 6-set for some values of $q < 29$. The proof here involves careful choice of two points in $GF(q)\setminus\{0,1\}$ to complete the set $\{\infty,0,1\}$ to a regular 5-set: the cross-ratio is involved in this.

RESULT 4 ([5], Lemma 3.2)
For $q > d+1 \geqslant 3$, PG(d,q) has a regular (d+4)-set

Outline of proof : take any line ℓ in PG(d,q) and let Λ be a set of three points on ℓ. Then $P\Gamma L(d+1,q)$ acts as $P\Gamma L(2,q)$ on the line ℓ. If we write $G = P\Gamma L(d+1,q)$, and $H = G^{\ell}{}_{[\ell]}$, then by Result 2, if q is large enough, we can pick (d+1) points Δ on ℓ such that $H_{[\Lambda],[\Delta]} = 1$. Thus take Δ^* to be any simplex in the space with the property that the faces (hyperplanes) of the simplex meet the line ℓ in precisely these (d+1) points Δ. Choose our set Λ^* of d+4 points in PG(d,q) to be the set Λ together with the set of d+1 vertices of Δ^*. Then if $g \in G$ fixes Λ^* it must fix the set Λ (being the only set of three collinear points in Λ^*), and hence it fixes the line ℓ, and also the sets Δ^* and Δ. By our choice then every point of ℓ is fixed. Hence g is linear, and fixes every vertex of a simplex, and is thus the identity.
If q is not large with respect to d, we have the following that holds for all $q \geqslant 3$:

RESULT 5 ([5], Lemma 3.3)
For $d \geqslant 3$, $q \geqslant 3$, PG(d,q) has a regular k-set where $k = 2d+1$ if d is odd and $k = 2(d+1)$ if d is even.

Remark: the regular set Λ involved has the following geometrical configuration: $\Lambda = \Delta \cup \Delta^*$ where Δ is the set of (d+1) vertices of a simplex and Δ^* is a set of d or d+1 points, (respectively if d is odd or even), with the properties:

(i) the points of Δ^* form triples of collinear points centred at a point $P \in \Delta^*$;
(ii) none of the points of Δ^* are on an edge of the simplex Δ;
(iii) none of the points of Δ are collinear with any set of two or more points of Δ^*.

For q=2 the arguments are quite different:

RESULT 6 ([2])
PG(d,2) has regular sets for $d \geqslant 4$.

For affine groups and geometries a similar analysis can be made: see [5] and [2]. Taken together, Results 3 to 6, and the analogous results for affine geometries, give the following theorem for regular sets in primitive permutation groups:

RESULT 7 ([2], [5] and [6])
Let G be a subgroup of $P\Gamma L(d+1,q)$ or $A\Gamma L(d,q)$ for $d \geqslant 1$, $q \geqslant 2$, q a prime power, and suppose that G acts as a primitive permutation group on the points of

the geometry, in the natural way, and that G does not contain the alternating group of the same degree. If G has no regular k-set for any k then G is one of the following:

(i) in projective geometries:

D_{10} (< PΓL(2,4)); PSL(2,5), PGL(2,5); PSL(2,7), PGL(2,7); PGL(2,8), PΓL(2,8); Sym(5) (< PΓL(2,9)), PSL(2,9), PSL(2,9).C_2, M_{10}, PGL(2,9), PΓL(2,9); PGL(2,11); PGL(2,16).C_2, PΓL(2,16); PGL(3,2); PGL(3,3); PΓL(3,4); PGL(4,2).

(ii) in affine geometries:

ASL(1,5), AGL(1,5); AGL(1,7); AΓL(1,8); ASL(1,9).C_2, AΓL(1,9); ASL(2,3), AGL(2,3); AΓL(2,4); AGL(3,2); $(C_2)^4$. Sp(4,2)', $(C_2)^4$. Sp(4,2), $(C_2)^4.A_7$, AGL(4,2); AGL(5,2).

(Here D_{10} denotes the dihedral group of order 10, C_2 the cyclic group of order 2, $(C_2)^4$ the elementary abelian 2-group of order 2^4).

We may also answer the question for k-closure:

RESULT 8 ([2],[5])

Let G be as in the first sentence of Result 7. If G is not k-closed for any k then G is one of the following:

(i) in projective geometries:

PGL(2,5); PGL(2,8), PΓL(2,8); PGL(2,9);

(ii) in affine geometries:

C_5 (< AGL(1,5)), AGL(1,5); AGL(1,8), AΓL(1,8); AGL(1,9); ASL(2,3).

RESULT 9 ([5], Theorem 5.2)

Let G $\leqslant$ Sym(Ω) be transitive of odd degree $|\Omega| = p^d > 9$ (where p is a prime) with an elementary abelian normal subgroup. Then there exists k $\leqslant$ 2(d+1) such that G is k-closed.

As mentioned in Section 1 above, we have not necessarily found the smallest value of k for which regular k-sets exist in finite desarguesian geometries. In Result 3, the value five is smallest for PΓL(2,q) when regular sets do exist. For the affine groups AΓL(1,q), three is the smallest size for a regular set [5]. For the plane we have the following:

RESULT 10

The minimal size of a regular set for PG(2,q) for q $\geqslant$ 5 is k where k is given as follows:

q $\geqslant$ 29 or q $\in$ {19,23}: k=5; 7 $\leqslant$ q $\leqslant$ 27, q $\notin$ {19,23}: k=6; q=5: k=7.

(While for q $\in$ {2,3,4} there is no regular set at all).

Outline of proof: that PG(2,q) cannot have a regular 4-set follows from simple geometric reasoning. For configurations of five points, only the case of a 5-arc may possibly lead to a regular set. If $G = P\Gamma L(3,q)$ and O is an oval in the plane then $G^{O}_{\{O\}} = P\Gamma O(3,q) \cong P\Gamma L(2,q)$ on the oval. Now by Result 3, if $q \geq 29$ there is a regular 5-set for $P\Gamma L(2,q)$. If we choose our set Λ in the plane to be such a 5-set then $G_{\{\Lambda\}}$ is the identity on the oval and as $G_{\{O\}}$ acts faithfully on O, $G_{\{\Lambda\}} = 1$. The other values given are minimal by similar reasoning.
For the affine plane we find that four is the smallest size for a regular set for $q \geq 25$.

3. OTHER DESIGNS AND GROUPS

The results from Section 2 give some application to other designs with doubly transitive automorphism groups:

RESULT 11 ([4])
Let $U(q)$ denote the classical (Hermitian) unital of q^3+1 points, where q is a prime power, $q \geq 3$. Then $U(q)$ has a regular 5-set for all $q \geq 3$.

Outline of proof: $U(q)$ is a 2-$(q^3+1, q+1, 1)$ design with $\mathrm{Aut}(U(q)) = P\Gamma U(3,q) = G$. Since $U(q)$ is embedded in $PG(2,q^2)$, $G \leq P\Gamma L(3,q^2)$. If b is a block of the design then it is part of a non-absolute line ℓ of the plane and $G^b_{\{b\}} \cong P\Gamma L(2,q) < P\Gamma L(2,q^2)$. We use Result 2 to find a point P of ℓ such that for a given set Δ of three points of b, $G^b_{\{\Delta\},P} = 1$. This can be done if $q^2 > 16$, i.e. $q > 4$. Notice that the point P might not be a point of the design, but this has no relevance to our argument. Now we choose our 5-set Λ of points of $U(q)$ as follows: $\Lambda = \Delta \cup \{A,B\}$ where A,B are points of $U(q)$ such that P, A, B are on a line (necessarily also a block) but such that R, the image of ℓ under the unitary polarity , is not on this line. Then if $g \in G$ fixes Λ it must fix Δ and {A,B}, thus Δ, {A,B} and P, and thus every point of ℓ aswell. It follows easily that Λ is a regular 5-set for $U(q)$. For q=3 or 4 the regular 5-sets were found by computation.
Note: regular 6-sets were also constructed for $U(q)$ for all $q > 3$; their nature is as in Result 4.
Result 2 can be applied in a similar way to the Ree unital:

RESULT 12 ([4])
Let $RU(q)$ denote the Ree unital of $q^3 + 1$ points where $q = 3^m$ and m is odd. Then $RU(q)$ has a regular 5-set for all $q \geq 3$.

Gluck [3] examined primitive soluble groups: his results agree with Result 7. He also proves the following:

RESULT 13 ([3])
If $G \leqslant Sym(\Omega)$, $|G|$ odd, then G has a regular k-set for some k.

Recently Pierce [7] considered transitive groups with cyclic point stabilizer that have no regular sets: these are precisely those that occur in Result 7 with this property.

Finally it should be mentioned that immediate counting arguments show that the Mathieu groups M_{11},M_{12}, M_{23} and M_{24} have no regular sets, and, with some computation, M_{10} and M_{22} are also included. However M_{21} = PSL(3,4) does have regular sets, as does M_9 = AGL(1,9).

REFERENCES.

[1] P. J. Cameron, P. M. Neumann and J. Saxl, On groups with no regular orbits on the set of subsets, Arch. Math. 43 (1984) 295-296.

[2] F. Dalla Volta, Regular sets for affine and projective groups over the field of two elements (submitted).

[3] D. Gluck, Trivial set-stabilizers in finite permutation groups, Can. J. Math. Vol.35, No.1 (1983) 59-67.

[4] J.D. Key and N. Rostom, Unitary designs with regular sets of points (submitted).

[5] J.D. Key and J. Siemons, Regular sets and geometric groups, Resultate Math, 11 (1987), 97-116.

[6] J.D. Key, J. Siemons and A. Wagner, Regular sets on the projective line, J. Geometry, 27 (1986), 188-194.

[7] R.S. Pierce, Permutation representations with trivial set-stabilizers, J.Algebra 95 (1985) 88-95.

[8] J. Siemons, On partitions and permutation groups on unordered sets, Arch. Math. 38 (1982) 391-403.

[9] H. Wielandt, Permutation Groups through Invariant Relations and Invariant Functions, (Lecture Notes, Ohio State University, 1969).

Annals of Discrete Mathematics 37 (1988) 225–230

GROUP PRESERVING EXTENSIONS OF SKEW PARABOLA PLANES

Norbert Knarr

Mathematisches Seminar der Universität
Olshausenstr.40
D-2300 Kiel 1
West-Germany

0. INTRODUCTION

In [6] H.Salzmann has constructed a class of 2-dimensional affine planes using a differentiable function $f: \mathbb{R} \to \mathbb{R}$ whose derivative is strictly increasing from $-\infty$ to $+\infty$. We will show that each of these planes can be embedded as a Baer subplane into a 4-dimensional affine plane. The embedding is group preserving, i.e. every collineation of the embedded plane extends to a collineation of the whole plane.

1. PLANAR FUNCTIONS

Let G, H be (additively written) abelian groups. A function $f: G \to H$ is called planar if it has the following property:

(P) For all $d \in G-\{0\}$ the mapping $f_d: G \to H: x \to f(x+d)-f(x)$ is bijective.

The incidence structure $I(G,H;f)$ with point set P and line set $\mathcal{L}$ is constructed as follows:

Define $L_c := \{(c,y) \mid y \in H\}$ for $c \in G$ and $L(a,b) := \{(x, f(x-a)+b) \mid x \in G\}$ for $(a,b) \in G\times H$. Set $P := G\times H$ and $\mathcal{L} := \{L_c \mid c \in G\} \cup \{L(a,b) \mid (a,b) \in G\times H\}$.

Then $I(G,H;f)$ is an affine plane iff f is planar. This is proved for finite G and H in [2: Lemma 12]. A proof which works for arbitrary groups can be found in [1: Satz 1], the assumption made there that $G \simeq H \simeq \mathbb{R}^2$ is inessential.

2. SKEW PARABOLA PLANES

The simplest examples of planar functions are quadratic polynomials over commutative fields of characteristic not 2. The resulting planes are isomorphic to the pappian planes over the respetive fields (so called parabola model of pappian planes). The first nontrivial examples of planar functions were constructed by H.Salzmann [6], actually before the term planar function was invented. He showed that a differentiable function $f: \mathbb{R} \to \mathbb{R}$ whose derivative is strictly increasing from $-\infty$ to $+\infty$ is planar. Later on H.Groh has determined all continuous planar functions $f: \mathbb{R} \to \mathbb{R}$ [3: 3.6 A1 B1]. Their results are summarized in the following

THEOREM: Let $f: \mathbb{R} \to \mathbb{R}$ be continuous. Then f is planar iff it has the following properties:

(i) f (or $-f$) is strictly convex.

(ii) $\lim_{x\to\pm\infty} (f(x)-l(x)) = +\infty$ $(-\infty)$ for every linear function $l: \mathbb{R} \to \mathbb{R}$.

The plane $I(\mathbb{R},\mathbb{R};f)$ or for short $\mathcal{A}_f$ is a 2-dimensional locally compact topological affine plane.

$\mathcal{A}_f$ is desarguesian iff f is a quadratic polynomial.

Denote the projective extension of $\mathcal{A}_f$ by $\mathcal{P}_f$ and its collineation group by Γ_f. If $\mathcal{P}_f$ is nondesarguesian then every collineation of $\mathcal{P}_f$ fixes the ideal line and can therefore be viewed as a collineation of $\mathcal{A}_f$. The dimension of Γ_f is at most 3. If this happens to be the case the graph of f is, up to isomorphism, a skew parabola, i.e. we have $f(x)=x^d$ for $x \geq 0$ and $f(x)=c|x|^d$ for $x<0$ with $0<c\leq 1<d$. The group Γ_f consists of all mappings

$$\lambda: \mathbb{R}^2 \to \mathbb{R}^2: (x,u) \to (ax+r, |a|^d u+s)$$

$a,r,s \in \mathbb{R}$, $a\neq 0$ if $c=1$, $a>0$ if $c\neq 1$.

In the sequel we will need some information about the collineation groups of the planes $\mathcal{P}_f$.

PROPOSITION: Let $f: \mathbb{R} \to \mathbb{R}$ be planar and continuous. Assume that $\mathcal{P}_f$ is nondesarguesian. Then the full collineation group of $\mathcal{A}_f$ (or eqivalently $\mathcal{P}_f$) is a semidirect product of $\mathbb{R}^2$ and a subgroup of the group of all mappings

$$\xi: \mathbb{R}^2 \to \mathbb{R}^2: (x,u) \to (ax, cx+du)$$

$a,c,d \in \mathbb{R}$, $a\neq 0$, $d>0$.

If ξ defines a collineation of $\mathcal{A}_f$ it maps $L(m,r)$ to $L(am+\lambda, cm+dr+\mu)$ where $\lambda=-f_a^{-1}(d(f(1)-f(0))+c)$ and $\mu=df(0)-f(-\lambda)$.

PROOF: $\mathbb{R}^2$ is a normal subgroup of Γ_f. If this were not true the dimension of Γ_f would be at least 3 but in this case $\mathbb{R}^2$ obviously is a normal subgroup of Γ_f.

Because $\mathbb{R}^2$ is a regular normal subgroup of Γ_f and every collineation of $\mathcal{A}_f$ is continuous [7: 3.5] the group Γ_f is a semidirect product of $\mathbb{R}^2$ and a subgroup of its group of continuous automorphisms, i.e. of $GL(2,\mathbb{R})$.

The projective plane $\mathcal{P}_f$ always admits a polarity which maps the point (x,u) to the line $L(-x,-u)$, the ideal point of the line $L(m,r)$ to the line $L(-m)$ and and the ideal point of $L(c)$ to the ideal line [2: Th.2; 6: p.260]. It follows that if $\xi \in GL(2,\mathbb{R})$ induces a collineation of $\mathcal{A}_f$ it is necessarily of the form

$$\xi: \mathbb{R}^2 \to \mathbb{R}^2: (x,u) \to (ax, cx+du)$$

$a, c, d \in \mathbb{R}$, $a, d \neq 0$. Furthermore there are $\alpha, \gamma, \delta, \lambda, \mu \in \mathbb{R}$ such that

$$\xi(L(m,r)) = L(\alpha m+\lambda, \gamma m+\delta r+\mu) \quad \text{for all } m, r \in \mathbb{R}.$$

This yields the following equation:

$$\frac{cx}{a} + df\left(\frac{x}{a}-m\right) + dr = f(x-\alpha m-\lambda) + \gamma m + \delta r + \mu \quad \text{for all } x, m, r \in \mathbb{R}.$$

By setting $m=\frac{x}{a}$ it is easily seen that $\delta=d$, $\alpha=a$ and $\gamma=c$.

By setting $x=m=r=0$ resp. $x=r=0$, $m=1$ we get

$$\mu = df(0) - f(-\lambda) \quad \text{and} \quad \lambda = -f_a^{-1}(df(1)-f(0))+c).$$

If we set $r=m=0$ we get $\frac{cx}{a} + df\left(\frac{x}{a}\right) = f(x-\lambda)$.

If d were <0 one side of this expression would tend to $+\infty$ and the other one to $-\infty$ if $|x|$ tended to ∞ because f fulfilles (ii), so we must have $d>0$.

EXAMPLE: The function $f: \mathbb{R} \to \mathbb{R}: x \to x^2 + \sin(x)$ is differentiable with strictly increasing derivative from $-\infty$ to $+\infty$. The resulting plane admits the collineation $(x,u) \to (x, 4\pi x+u)$. This shows that in general $c \neq 0$.

3. THE EXTENSION PROCESS

THEOREM: Let $f: \mathbb{R} \to \mathbb{R}$ be planar and continuous. Then the mapping

$$\hat{f}: \mathbb{R}^2 \to \mathbb{R}^2: (x,y) \to (f(x)-\varepsilon y^2, 2xy) = (f_1(x,y), f_2(x,y)),$$

where $\varepsilon=+1$ if f is strictly convex and -1 otherwise, is planar and the resulting plane $\mathcal{A}_{\hat{f}} = I(\mathbb{R}^2, \mathbb{R}^2; \hat{f})$ is a locally compact 4-dimensional topological affine plane.

Set $L(m,n,r,s) := \{(x, y, f_1(x-m, y-n)+r, f_2(x-m, y-n)+s) \mid (x,y) \in \mathbb{R}^2\}$ for $m, n, r, s \in \mathbb{R}$.

Let $\xi: \mathbb{R}^2 \to \mathbb{R}^2: (x,u) \to (ax, cx+du)$ be a collineation of $\mathcal{A}_f$. Set $\lambda = -f_a^{-1}(d(f(1)-f(0))+c)$ and $\mu = df(0) - f(-\lambda)$. Then

$$\hat{\xi}: \mathbb{R}^4 \to \mathbb{R}^4: (x,y,u,v) \to (ax, \sqrt{d}y, cx+du, -2\sqrt{d}\lambda y + a\sqrt{d}v)$$

is a collineation of $\mathcal{A}_{\hat{f}}$. It maps $L(m,n,r,s)$ to $L(am+\lambda, \sqrt{d}n, cm+dr+\mu, -2\sqrt{d}\lambda n + a\sqrt{d}s)$. $\sqrt{d}$ may be chosen either positive or negative.

$\beta: \mathbb{R}^4 \to \mathbb{R}^4: (x,y,u,v) \to (x,-y,u,-v)$ is a Baer involution of $\mathcal{A}_{\hat{f}}$ whose corresponding Baer subplane is isomorphic to $\mathcal{A}_f$.

The mappings $\sigma_t: \mathbb{R}^4 \to \mathbb{R}^4: (x,y,u,v) \to (x, y, u-ty, v+tx)$, $t \in \mathbb{R}$, are collineations of $\mathcal{A}_{\hat{f}}$.

Every collineation of $A_{\hat{f}}$ is continuous.

In the proof we will need the following simple

LEMMA: Let K be a field and $g_1, g_2\colon K \to K$ arbitrary. Define $f\colon K^2 \to K^2$ by $f(x,y) := (g_1(x), y+g_2(x))$. Then f is a bijection iff g_1 is one.

PROOF: straightforward.

PROOF of Theorem: We show first that $\hat{f}$ is planar. Up to isomorphism, we may assume that f is strictly convex, i.e. $\varepsilon=+1$.

Let $(r,s) \in \mathbb{R}^2 - \{(0,0)\}$. Then we have

$$\hat{f}_{(r,s)}(x,y) = \hat{f}(x+r, y+s) - \hat{f}(x,y) =$$

$$(f(x+r) - f(x) - (y+s)^2 + y^2, 2(x+r)(y+s) - 2xy) =$$

$$(x,y)\begin{pmatrix} & 2s \\ -2s & 2r \end{pmatrix} + (f(x+r)-f(x),0) + (-s^2, 2rs)$$

$\hat{f}_{(r,0)}$ is bijective because f_r is, so we may assume $s \neq 0$.

In this case the linear mapping $A_{(r,s)} = \begin{pmatrix} & 2s \\ -2s & 2r \end{pmatrix}$ is regular and so $\hat{f}_{(r,s)}$ is bijective iff $\hat{f}_{(r,s)}A^{-1}_{(r,s)}$ is. We have

$$\hat{f}_{(r,s)}A^{-1}_{(r,s)}(x,y) = (x,y) + (\tfrac{r}{2s^2}(f(x+r)-f(x)), \tfrac{-1}{2s}(f(x+r)-f(x)) + (\tfrac{r}{2}, \tfrac{s}{2})$$

By the preceding lemma this mapping is bijective iff

$$g\colon \mathbb{R} \to \mathbb{R}\colon\ x \to x + \frac{r}{2s^2}(f(x+r)-f(x)) + \frac{r}{2}$$

is. Because f is strictly convex the mapping $x \to f(x+r) - f(x)$ is strictly increasing if $r > 0$ and strictly decreasing otherwise, so $x \to \frac{r}{2s^2}(f(x+r)-f(x))$ is strictly increasing in either case. The sum of two strictly increasing bijections of $\mathbb{R}$ has the same property, so g is bijective and $\hat{f}$ is planar.

Pick an element $e \in \mathbb{R}^2 - \{(0,0)\}$. Up to isomorphism we may assume that $\hat{f}(0) = \hat{f}(e) = 0$. Define a multiplication $*\colon \mathbb{R}^2 \times \mathbb{R}^2 \to \mathbb{R}^2$ by the rule

$$z*w := \hat{f}(\hat{f}_e^{-1}(z) + \hat{f}_e^{-1}(w)) - \hat{f}(\hat{f}_e^{-1}(z)) - \hat{f}(\hat{f}_e^{-1}(w)).$$

Then $(\mathbb{R}^2, +, *)$ is a coordinatizing ternary field for $A_{\hat{f}}$ by [2: Th.6] . Because $\hat{f}$ is continuous $*$ is a continuous operation and so $(\mathbb{R}^2, +, *)$ is a topological ternary field by the result of [4] and $A_{\hat{f}}$ is a topological affine plane [7: 7.15].

Let $\xi\colon \mathbb{R}^2 \to \mathbb{R}^2\colon (x,u) \to (ax, cx+du)$ be a collineation of A_f and set λ and μ as in the theorem. To prove that $\hat{\xi}$ is a collineation of $A_{\hat{f}}$ we have to check the validity of the following equations:

$$\frac{cx}{a} + d(f(\frac{x}{a} - m) - (\frac{y}{\sqrt{d}} - n)^2 + r) = f(x - am - \lambda) - (y - \sqrt{d}n)^2 + cm + dr + \mu$$

$$-2\lambda y+a\sqrt{d}\,(2(\frac{x}{a}-m))(\frac{y}{\sqrt{d}}-n)+s=2(x-am-\lambda)(y-\sqrt{d}n)-2\sqrt{d}\lambda n+a\sqrt{d}s$$

for all $x, y, m, n, r, s \in \mathbb{R}$.

The first equation holds because ξ is a collineation of A_f (compare the proof of Proposition) and the second equation always holds.

β is the extension of id: $\mathbb{R}^2 \to \mathbb{R}^2$ if one chooses the negative square root of d and so is a collineation of $A_{\hat{f}}$. The fixed point set of β consists of the points $(x,0,u,0)$, $x,u \in \mathbb{R}$, and from $\hat{f}(x,0)=(f(x),0)$ it follows that this is the point set of an affine Baer subplane of $A_{\hat{f}}$ which is isomorphic to A_f.

An easy calculation shows that σ_t maps the line $L(m,n,r,s)$ to $L(m,n-\frac{t}{2},r-tn+\frac{t^2}{4},s-tm)$ and so is a shear of $A_{\hat{f}}$. Because $A_{\hat{f}}$ possesses a point with a 3-dimensional group of shears all collineations of $A_{\hat{f}}$ are continuous [5].

REFERENCES

[1] Betten,D., Komplexe Schiefparabelebenen. Abh.Math.Sem.Univ.Hamburg 48,76-88 (1979)

[2] Dembowski,P. and Ostrom,T.G., Planes of order n with collineation groups of order n^2. Math.Z. 103,239-258 (1968)

[3] Groh,H., Point homogeneous flat affine planes. J.Geom. 8, 145-161 (1976)

[4] Knarr,N. und Weigand,C., Ein Kriterium für topologische Ternärkörper. Arch. Math. 46, 368-370 (1986)

[5] Knarr,N., Unstetige Kollineationen 4-dimensionaler Ebenen. Arch.Math. 48, 548-549 (1987)

[6] Salzmann,H., Zur Klassifikation topologischer Ebenen III. Abh.Math.Sem.Univ. Hamburg 28, 250-261 (1965)

[7] Salzmann,H., Topological planes. Adv. in Math. 2, 1-60 (1967)

Annals of Discrete Mathematics 37 (1988) 231–248

PRODUCTS OF INVOLUTIONS IN ORTHOGONAL GROUPS

Frieder KNÜPPEL

Mathematisches Seminar der Universität, Olshausenstrasse 40, 23 Kiel, W. Germany

Let G be a group and S a set of involutions generating G. (a) What is the minimal number k such that every element of G is a product of at most k elements of S ? (b) Given some $\pi \in G$, what is the minimal number k such that π is a product of k elements of S ? We discuss both questions for particular groups G and subsets S. Here we call σ an involution if $\sigma^2=1$.

Throughout this article V denotes a finite-dimensional vector space over a commutative field K of characteristic distinct from 2. We call V an orthogonal vector space if a symmetric bilinear-form $f : V \times V \to K$ is given. V is called regular if f is non-degenerate. Let $O^+(V) := \{\pi \in O(V) : \det\pi = 1\}$ denote the special orthogonal group. $O^*(V) := \{\pi \in O(V) : v\pi = v \text{ for every } v \in \operatorname{rad}V\}$ is the weak orthogonal group. For $\pi \in GL(V)$ let $B(\pi) := V(\pi-1)$ denote the path of π and $F(\pi) := \operatorname{kernel}(\pi-1)$ the fix of π. A symmetry σ is an involution $\sigma \in O(V)$ such that $B(\sigma)$ is regular and $\dim B(\sigma) = 1$.

The results quoted in this article are due to F. Knüppel and Klaus Nielsen, unless particular reference is given.

The following theorems are well-known.

Theorem S [16]. Let V be a regular orthogonal vector space and $\pi \in O(V)$. Let $k := \dim B(\pi)$ provided $B(\pi)$ is not totally isotropic, else $k := \dim B(\pi) + 2$. Then $\pi = \sigma_1 \cdot \ldots \cdot \sigma_k$ for some symmetries σ_i, and k is the minimal number of symmetries needed.

Theorem WD[3,18]. Let V be a regular orthogonal vector space. Every $\pi \in O(V)$ is a product of 2 involutions.

Theorem GHR [8]. In $GL^{\pm}(V) := \{\pi \in GL(V) : \det\pi = 1 \text{ or } \det\pi = -1\}$ every element is a product of 4 involutions.

We want to discuss questions (a) and (b) in $O^{\pm}(V) := \{\pi \in O(V) : \det\pi = 1 \text{ or } \det\pi = -1\}$ and in $O^*(V)$ (V not necessarily regular), and also in $O^+(V)$ (V regular), where S is in each case the whole set of involutions of the group considered.

Furthermore, we study $O^{+}(V)$ (V regular) where S is the set of half-turns (i.e. $S = \{\sigma \in O(V) : \sigma$ involutory and $\dim B(\sigma) = 2\}$).

2. SOME ELEMENTARY CONCLUSIONS.

Let V be an orthogonal vector space. Let $R := \operatorname{rad} V := \{v \in V : f(v,w) = 0$ for every $w \in V\}$ denote the radical.

2.1 Lemma. Let U and W be subspaces of V such that $U \oplus R = V = W \oplus R$. Then $U\sigma = W$ and $V = B(\sigma) \oplus R$ for some involution $\sigma \in O^*(V)$.

Since this useful lemma does not appear in the literature we give a

Proof. Clearly, U and W are regular. Let $a_1,\dots a_k$ be an orthogonal basis of U. From $W \cap (\langle a_i\rangle + R) \neq 0$ we infer that $b_i = -a_i + r_i \in W$ for some $r_i \in R$. Let $c_i := a_i - b_i$. The c_i's are anisotropic vectors. Let σ_i denote the symmetry with $B(\sigma_i) = \langle c_i\rangle$. We have $f(c_i,c_j) = 0$ for $i \neq j$, hence any two σ_i's commute and $\sigma := \sigma_1 \cdot \ldots \cdot \sigma_k$ is an involution of $O^*(V)$. Now, $B(\sigma) = \langle c_1,\dots,c_k\rangle$ is a k-dimensional regular subspace, hence $V = B(\sigma) \oplus R$. The equations $c_i\sigma_i = -c_i$ and $(a_i + b_i)\sigma_i = a_i + b_i$ yield $a_i\sigma_i = b_i$. Finally, $a_j\sigma_i = a_j$ and $b_j\sigma_i = b_j$ for $i \neq j$, hence $a_i\sigma = b_i$ and $U\sigma = \langle b_1,\dots,b_k\rangle = W$.

The following proposition is proved in [14], and the previous lemma supplies a further proof.

2.2 Proposition. In $O^*(V)$ every element is a product of 3 involutions.

Proof. Let $\pi \in O^*(V)$. Select $U \leq V$ such that $V = U \oplus R$. U and $U\pi$ are regular subspaces and we have $V = U\pi \oplus R$, hence $U\pi\sigma = U$ for some involution $\sigma \in O^*(V)$; cf. 2.1. Theorem (WD) supplies involutions $\rho',\tau' \in O(U)$ such that $\pi\sigma|_U = \rho'\tau'$. Extending ρ',τ' one obtains involutions $\rho,\sigma \in O^*(V)$. Clearly, $\pi\sigma = \rho\tau$ holds. Thus $\pi = \rho\tau\sigma$.

2.3 Let $O^{\pm}(V) := \{\pi \in O(V) : \det\pi \in \{1,-1\}\}$. Every element of $O^{\pm}(V)$ is a product of at most 5 involutions.

Proof. Let $\pi \in O^{\pm}(V)$. Select $U \leq V$ such that $V = U \oplus R$. Take an involution $\sigma \in O^*(V)$ such that $U\pi = U\sigma$; cf. 2.1. We have $\pi\sigma|_U \in O(U)$, hence $\pi\sigma|_U = \rho_1'\rho_2'$ for some involutions $\rho_i' \in O(U)$ by theorem (WD). Clearly $\pi|_R \in GL^{\pm}(R)$, hence $\pi|_R = \omega_1'\omega_2'\omega_3'\omega_4'$ for some involutions $\omega_i \in GL^{\pm}(R)$ by theorem [GHR] . Let ρ_i and ω_i denote the extensions of the corresponding '-mappings such that $\rho_i \in O^*(V)$ and $\omega_i \in GL^{\pm}(V)$ with $\omega_i|_U = 1_U$. The inclusions $B(\rho_i) \subseteq U \subseteq F(\omega_i)$ and $B(\omega_i) \subseteq R \subseteq F(\rho_i)$ imply that every ρ_i commutes with every ω_j. Thus we have $\pi\sigma = \rho_1\rho_2\omega_1\omega_2\omega_3\omega_4 = (\rho_1\omega_1)(\rho_2\omega_2)\omega_3\omega_4$. Hence, $\pi\sigma$ is a product of 4 involutions of $O^{\pm}(V)$, and π is a product of 5 involutions of $O^{\pm}(V)$.

It can happen that $\pi \in O^{\pm}(V)$ is not a product of less than 4 involutions. However, the question is still open wether every $\pi \in O^{\pm}(V)$ is a product of 4 involutions.

3. ORTHOGONAL DECOMPOSITIONS.

Let V be an orthogonal vector-space. A decomposition $V = V_1 + \ldots + V_k$ is called orthogonal if $V_i \perp V_j$ for $i \neq j$. In this section we will assume that V is a regular orthogonal vector space. Hence, if $V = V_1 + \ldots + V_k$ is an orthogonal decomposition then $V = V_1 \oplus \ldots \oplus V_k$ (direct sum and $V_i \perp V_j$ for $i \neq j$).

Let $\pi \in O(V)$. An orthogonal decomposition $V = V_1 \oplus \ldots \oplus V_k$ into π-modules is called complete if each V_i does not admit a proper orthogonal decomposition into π-modules.

A classification of orthogonally indecomposable π-modules has been given by various authors [19, 18, 17, 3,9 ,10] . In the sequel we use [9] and [10] and compile some results.

Let $\pi \in O(V)$ and suppose that V does not admit a proper orthogonal decomposition into π-modules. Then π belongs to one of the following types.

Type (1) $V = U \oplus W$, where U and W are indecomposable π-modules possessing the same minimum polynomial $(x-1)^{\alpha}$ (type (1_-)) or $(x+1)^{\alpha}$ (type 1_+)) with an even α. In particular, U and W are π-cyclic modules having the same dimension α. Hence $\dim V = 2\alpha \equiv 0 \bmod 4$.

Type (2) V is a cyclic π-module whose minimum polynomial is $p(x)^{\beta}$, where $p(x)$ is irreducibel and self-reciprocal. If $p(x) = x+1$ or $p(x) = x-1$ then β is odd; else the degree of $p(x)$ is even.

Type (3) $V = U \oplus W$ where U and W are cyclic π-modules whose minimum polynomials are $p(x)^\gamma$ and $p^*(x)^\gamma$, respectively, such that $p(x)$ is irreducibel and prime to its reciprocal $p^*(x)$. In particular, U and W have the same dimension, $\dim V$ is even, and V is π-cyclic.

If V is a regular orthogonal vectorspace and $\pi \in O(V)$ then one may find distinct complete decompositions $V = V_1 \mathbin{\bot} \ldots \mathbin{\bot} V_k$. Using the Krull-Remak-Schmidt-theorem one can prove that the number of V_i's belonging to a certain type does not depend on the particular decomposition; cf. [12].

4. INVOLUTIONS IN O(V), WHERE V IS REGULAR.

Let V be a regular orthogonal vector space. We present some lemmas needed in the following sections. Simultaneously we obtain a simple proof of theorem (WD). For every $\pi \in O(V)$ one has $B(\pi)^\perp = F(\pi)$. If $\sigma \in GL(V)$ is an involution then $B(\sigma) = Neg(\sigma) :=$ kernel $(\sigma+1)$ and $V = F(\sigma) \oplus B(\sigma)$. If $\sigma \in O(V)$ is an involution then $B(\sigma)$ and $F(\sigma)$ are regular subspaces, and one has $V = F(\sigma) \mathbin{\bot} B(\sigma)$. If $\dim B(\sigma) = 1$ then σ is called a symmetry.

4.1 Let $\pi \in O(V)$. If $\rho,\sigma \in GL(V)$ are involutions such that $\pi = \rho\sigma$ and V is π-cyclic then $\rho,\sigma \in O(V)$.

Proof. We have $V = F(\rho) \oplus B(\rho)$, hence $V = \langle v\rangle_\pi$ for some $v \in F(\rho) \cup B(\rho)$. Using $\pi^i\rho = \rho\pi^{-i}$ we obtain $f(v\pi^i\rho, v\pi^j\rho) = f(v\rho\pi^{-i}, v\rho\pi^{-j}) = f(v\pi^{-i}, v\pi^{-j}) = f(v\pi^j, v\pi^i)$.

4.2 Let $\pi \in GL(V)$. If V is a π-cyclic module and if π and π^{-1} possess the same minimum polynomial then $\pi = \rho\sigma$ for some involutions $\rho,\sigma \in GL(V)$ such that $B(\pi) = B(\rho) \oplus B(\sigma)$.

Proof[1]. Let $v, v\pi, \dots v\pi^{n-1}$ be a basis of V. Define $\rho, \sigma \in GL(V)$ by the properties

$$v\pi^i\rho := v\pi^{n-i} \quad \text{for } i = 0,\dots,n-1, \text{ and}$$
$$v\pi^i\sigma := v\pi^{n+1-i} \quad \text{for } i = 0,\dots,n-1.$$

The mappings ρ and σ are involutions since π and π^{-1} have the same minimum polynomial. The pathcondition is easy to check.

4.2' Let V be a (not necessarily regular) orthogonal vector space and $\pi, \rho, \sigma \in O^*(V)$ such that ρ and σ are involutions with $\pi = \rho\sigma$. If $B(\rho) \subseteq B(\pi)$ then $B(\rho) \cap B(\sigma) = 0$.

Namely, from the assumptions we achieve $B(\rho) \cap B(\sigma) \subseteq F(\rho)$.

4.3 Let $\pi \in O(V)$. If V is π-cyclic then $\pi = \rho\sigma$ and $B(\pi) = B(\rho) \oplus B(\sigma)$ for some involutions $\rho, \sigma \in O(V)$.

Proof.[1] $\pi \in O(V)$ implies that π is conjugate to π^{-1} in GL(V). Hence π and π^{-1} have the same minimum polynomial. 4.2 and 4.1 yield the assertion.

4.4 Let $\pi \in O(V)$ be an orthogonally indecomposable π-module of type (1). Then $\pi = \rho\sigma$ for some involutions $\rho, \sigma \in O(V)$. One can choose ρ and σ so that U and also W (see § 3, type (1)) are invariant under both ρ and σ.

Proof.[2] We have $V = U \oplus W$, where U and W have both the same minimum polynomial $(x-1)^\alpha$ or $(x+1)^\alpha$. From 4.3 we obtain involutions $\rho', \sigma' \in O(U)$ such that $\pi|_U = \rho'\sigma'$ holds. The properties

$$f(u\rho', w) = f(u, w\rho'') \text{ and } f(u\sigma', w) = f(u, w\sigma'')$$

for all $u \in U$, $w \in W$ define unique isometries $\rho'', \sigma'' \in O(W)$. Clearly, ρ'' and σ'' are involutions satisfying $\pi|_W = \rho''\sigma''$. Let $\rho := \rho' \oplus \rho''$ and $\sigma := \sigma' \oplus \sigma''$. Clearly, ρ and σ are involutions and $\pi = \rho\sigma$ holds. Finally, $f(u\rho, w\rho) = f(u\rho', w\rho'') = f(u\rho'\rho', w) = f(u,w)$ for every $u \in U$, $w \in W$. Thus, ρ and also σ are isometries.
Now 4.3, 4.4 and the classification given in § 3 yield immediately

4.5 Theorem (WD). Let V be a regular orthogonal vector space. Every element of O(V) is a product of 2 involutions of O(V).

---

1) Compare [18,4]. 2) Compare [3].

Namely, given $\pi \in O(V)$ we take a complete orthogonal decomposition $V = V_1 \perp \dots \perp V_k$ into π-modules and apply 4.3 or 4.4 to each V_i.

4.6 Proposition. Let $\pi \in O(V)$. If a complete decomposition of V into π-modules does not contain a type (1_-) term then $\pi = \rho\sigma$ and $B(\pi) = B(\rho) \oplus B(\sigma)$ for some involutions $\rho, \sigma \in O(V)$.

Proof. We take a complete decomposition $V = V_1 \perp \dots \perp V_k$ into π-modules. If V_i is π-cyclic then $\pi_i = \rho_i\sigma_i$ and $B(\pi_i) = B(\rho_i) \oplus B(\sigma_i)$ for some involutions $\rho_i, \sigma_i \in O(V)$; cf. 4.3. There remains the case that V_i has type (1_+). 4.4 supplies involutions $\rho_i, \sigma_i \in O(V_i)$ such that $\pi_i = \rho_i\sigma_i$, and $B(\rho_i) \oplus B(\sigma_i) = B(\pi_i)$ is true because of $V_i = B(\pi_i)$ and 4.2'.

4.7 Let $\pi \in O(V)$ and suppose that V is an orthogonally indecomposable π-module of type (1_-). Suppose that $\rho, \sigma \in O(V)$ are involutions such that $\pi = \rho\sigma$. Then $B(\pi)$ is properly contained in $B(\rho) + B(\sigma)$.

Proof. Let $(x-1)^{2\alpha}$ be the minimum polynomial of π. $T := V(\pi-1)^{\alpha}$ is a totally isotropic π-invariant subspace with $\dim T = \frac{1}{2}\dim V$. From $\pi = \rho\sigma$ one concludes $\dim B(\sigma|_T) = \frac{1}{2}\dim B(\sigma)$: Hence $\dim B(\sigma)$ and also $\dim B(\rho)$ is even. If $B(\pi) = B(\rho) + B(\sigma)$ then $B(\pi) = B(\rho) \oplus B(\sigma)$ (cf. 4.2'), thus $\dim B(\pi) \equiv 0 \bmod 4$. But $\dim B(\pi) = \dim V - 2 \equiv 2 \bmod 4$; cf. § 3.

5. PRODUCTS OF 2 INVOLUTIONS IN O(V), WHERE V IS REGULAR.

We continue to assume that V is a regular orthogonal vector space. We want to prove the following theorems.

5.1 Theorem. Let $\pi, \rho, \sigma \in O(V)$ such that $\pi = \rho\sigma$ and ρ, σ are involutions. There is a complete orthogonal decomposition of V into π-modules which are invariant under ρ and σ.

5.2 Theorem. Let $\pi \in O(V)$. Then (i) and (ii) are equivalent.

(i) A complete orthogonal decomposition of V into π-modules does not contain a term of type (1_-).

(ii) There are involutions $\rho, \sigma \in O(V)$ such that $\pi = \rho\sigma$ and $B(\pi) = B(\rho) \oplus B(\sigma)$.

Proof of 5.2 from 5.1. "(i) $\Rightarrow$ (ii)" is just 4.6. (ii) $\Rightarrow$ (i). 5.1 supplies a complete orthogonal decomposition of V into π-modules V_i which are invariant under ρ and σ. The restrictions ρ_i, σ_i, π_i to V_i satisfy $\pi_i = \rho_i \sigma_i$ and $B(\rho_i) + B(\sigma_i) \subseteq B(\pi) \cap V_i = B(\pi_i)$. But if V_i were of type (1_-) this condition would be violated; cf. 4.7.

Now we outline a proof of 5.1.
We apply induction over $\dim V$.
Let $m(x) = r_1(x)^{s_1} \cdot \ldots \cdot r_k(x)^{s_k}$ be the decomposition of the minimum polynomial of π into a product of self-reciprocal factors $r_i(x)^{s_i}$. Let $V_i := \mathrm{kernel}(r_i(x))^{s_i}$. Then $V = V_1 \oplus \ldots \oplus V_k$ and the V_i's are invariant under ρ and σ (not necessarily orthogonally indecomposable π-modules). So we can assume from the beginning that $m(x) = r(x)^s$ where $r(x)$ is a self-reciprocal polynomial which is not a proper product of self-reciprocal polynomials. Two cases can occur.

(A) $m(x) = p(x)^s$ where $p(x)$ is an irreducible self-reciprocal polynomial, or

(B) $m(x) = p(x)^s p^*(x)^s$, where $p(x)$ is an irreducible polynomial whose reciprocal $p^*(x)$ is prime to $p(x)$.

Furthermore we can assume

(*) If M is a regular subspace of V which is invariant under π and ρ then $M = 0$ or $M = V$.

Namely, if M is a proper subspace of V satisfying the assumptions then $V = M \oplus M^{\perp}$ and we can apply the induction hypothesis to M and $M^{\perp}$.

Case (A). As $V = F(\rho) \oplus B(\rho)$ there is some $u \in F(\rho) \cup B(\rho)$ such that $up(\pi)^{s-1} \neq 0$. Since $\langle up(\pi)^{s-1} \rangle^{\perp} \neq V = \langle F(\rho) \cup B(\rho) \rangle$ we have

(a) $f(w, up(\pi)^{s-1}) \neq 0$ for some $w \in F(\rho) \cup B(\rho)$.

(b) Both $U := \langle u \rangle_\pi$ and $W := \langle w \rangle_\pi$ have minimum polynomial $p(x)^s$. U and W are invariant under π and ρ and do not admit a proper decomposition into π-modules. We can assume that $V \neq U, W$.

Due to (*) we have

(c) U and W are non-regular subspaces.

From (a),(b) and (c) one can conclude:
$U \cap W = 0$ and $U \oplus W$ is a regular subspace.
Thus, $U \oplus W = V$. Now one can show that V is an orthogonally indecomposable π-module, and case (A) is finished.

Case (B). Let $U := \mathrm{kernel}(p(\pi)^s$, $W := \mathrm{kernel}(p^*(\pi)^s)$.
Then one can prove the following statements.

(a) $V = U \oplus W$. The subspaces U and W are totally isotropic. $U\rho = W$ and $W\rho = U$.

(b) $f(up(\pi)^{s-1}, u\rho) \neq 0$ for some $u \in U$.

Now let $v := u + u\rho$. Clearly, $\langle v\rangle_\pi$ is invariant under π, ρ and σ. Furthermore, one can prove

(c) $\langle v\rangle_\pi$ is a regular subspace and does not admit a proper orthogonal decomposition into π-modules.

Now (*) yields $V = \langle v\rangle_\pi$ and the proof is finished.

More detailed proofs are given in [12].

6. INVOLUTIONS IN O*(V).

Let V be an orthogonal vector-space and $R := \mathrm{rad}\, V$. $O^*(V) := \{\pi \in O(V) : R \subseteq F(\pi)\}$ denotes the proper orthogonal group. We select a regular closure $\overline{V}$ of V. Note that $\dim \overline{V} = \dim V + \dim R$ and $R^\perp = V$, where $\perp$ is taken in $\overline{V}$. For $\pi \in O(V)$ let $\overline{\pi} \in O(\overline{V})$ denote an arbitrary extension of π. For $\pi \in O(V)$ let $\tilde{\pi} \in O(\tilde{V})$ denote the induced isometry, where $\tilde{V} := V/R$. $\tilde{V}$ is a regular orthogonal vector-space. The following statement is from [14] .

6.1 Theorem. The following statements are equivalent.
(i) $\dim(\mathrm{rad}\, V) \leq 1$ or $\mathrm{ind}\, V \leq 1$.
(ii) In O*(V) every element is a product of 2 involutions.

We prove (i) $\Rightarrow$ (ii) but omit some details. Let $\pi \in O^*(V)$. Take a splitting $V = V_1 + \ldots + V_k$ into pairwise orthogonal π-modules V_i which does not admit a refinement. Clearly,

$V_i \cap V_j \subseteq \mathrm{rad}\, V_i \cap \mathrm{rad}\, V_j \subseteq R := \mathrm{rad}\, V$, and $\mathrm{ind}\, V_i \leq 1$ for every i or $\dim \mathrm{rad}\, V_i \leq 1$ for every i. Suppose there are involutions $\rho_i, \sigma_i \in O^*(V_i)$ available such that $\pi_i = \rho_i\sigma_i$ for every i (where π_i denotes the restriction of π to V_i). Combining the ρ_i's and σ_i's one obtains involutions $\rho, \sigma \in O^*(V)$ such that $\pi = \rho\sigma$ holds. Thus we can assume

(1) V does not admit a proper orthogonal decomposition $V = U + W$, $U \perp W$, into π-modules. In particular, $R \subseteq B(\pi)$.

This implies

(2) $\tilde{V}$ is an orthogonally indecomposable $\tilde{\pi}$-module.

Some simple arguments yield

(3) $\dim F(\overline{\pi}) = \dim F(\pi) = \dim F(\tilde{\pi}) \leq 2$.

If $\operatorname{ind} \tilde{V} = \operatorname{ind} V \leq 1$ then $\tilde{\pi}$ is not of type $(1\_)$ (see § 3). Hence $\dim F(\tilde{\pi}) \leq 1$, and (3) implies $\dim R \leq 1$. The case $R = 0$ is done; cf. 4.5. There remains the case

(4) $\dim R = 1$.

In particular $x-1$ divides the minimum polynomial of π, and because of (1) the minimum polynomial of π is $(x-1)^k$ for some k. Together with $B(\overline{\pi}) \subseteq V$ follows $\overline{V}(\overline{\pi}-1)^{k+1} = 0$, i.e.

(5) the minimum polynomial of $\overline{\pi}$ is a power of $x-1$.

Taking (3) and (5) into account we have to face two possibilities for a complete decomposition of $\overline{V}$ into $\overline{\pi}$-modules.

(a) $\overline{V}$ is indecomposable of type $(1\_)$,

(b) $\overline{V}$ is the orthogonal sum of at most 2 $\overline{\pi}$-modules of type (2).

In case (a) a review of the proof of 4.4 supplies $\overline{\rho}, \overline{\sigma} \in O(\overline{V})$ such that $\overline{\pi} = \overline{\rho\sigma}$ and $R \subseteq F(\overline{\rho}) \cap F(\overline{\sigma})$ holds. In case (b) 4.6 supplies involutions $\overline{\rho}, \overline{\sigma} \in O(\overline{V})$ such that $B(\overline{\rho})$, $B(\overline{\sigma}) \subseteq B(\overline{\pi}) \subseteq V$. In both cases $V = R^{\perp}$ is invariant under $\overline{\rho}$ and $\overline{\sigma}$, and the restrictions $\rho, \sigma \in O^*(V)$ are involutions with the property $\pi = \rho\sigma$.

7. INVOLUTIONS IN THE SPECIAL ORTHOGONAL GROUP $O^+(V)$.

Let V be a regular orthogonal vector space. Let $\mathcal{H}_3$ denote the hyperbolic plane over GF(3). Observe that $O^+(\mathcal{H}_3) = \{1, -1\}$.

7.1 Theorem. In $O^+(V)$ every element is a product of 3 involutions, except when $\dim V = 2$ and $V \neq \mathcal{H}_3$. In $O^+(V)$ every element is a product of just 2 involutions if and only if $\dim V \not\equiv 2 \bmod 4$ or $V = \mathcal{H}_3$.

7.2 Theorem. An isometry $\pi \in O^+(V)$ is a product of 2 involutions in $O^+(V)$ if and only if $\dim V \not\equiv 2 \bmod 4$ or an orthogonal decomposition of V into orthogonally indecomposable π-modules contains an odd-dimensional term.

In order to prove the first assertion of 7.1 we need

7.3 Let $\pi = \rho\sigma$ where $\rho, \sigma \in O(V)$ are involutions with $\det \rho = -1 = \det \sigma$. If $v \in F(\rho) \cup B(\rho)$, $w \in F(\sigma) \cup B(\sigma)$ and $f(v,w) = 0$ for some pair v,w then π is a product of 3 involutions in $O^+(V)$.

Namely, one has $\pi = \rho\sigma = (\rho\tau)(\tau\omega)(\omega\sigma)$, where τ and ω denote the symmetries whose pathes are $\langle v\rangle$, $\langle w\rangle$, respectively.

Using 7.3 and 4.5 one proves easily that in $O^+(V)$ every element is a product of at most 3 involutions except when $\dim V = 2$. Now we outline a proof of the other assertions of 7.1.

Let us assume that $\dim V \equiv 2 \bmod 4$ and $V \neq \mathcal{H}_3$. We want to construct an isometry $\pi \in O^+(V)$ which is not a product of 2 involutions of $O^+(V)$.

Select a regular subspace U such that $\dim U = \frac{1}{2}\dim V$. One can find a regular subspace W such that $U \oplus W = V = U^\perp \oplus W$. Define involutions $\rho, \sigma \in O(V)$ by the properties $B(\rho) = U$ and $B(\sigma) = W$ and let $\pi := \rho\sigma$. A simple calculation shows that $F(\pi) = 0 = \mathrm{Neg}(\pi)$. Now suppose that $\pi = \omega\tau$ for some involutions $\omega, \tau \in O^+(V)$. The inclusion $F(\omega) \cap B(\tau) \subseteq \mathrm{Neg}(\pi) = 0$ yields $\dim F(\omega) \leq \dim F(\tau)$. A similar reasoning supplies $\dim F(\omega) \leq \dim B(\tau)$. Thus we obtained $\dim F(\omega) \leq \frac{1}{2}\dim V$. A similar argument yields $\dim B(\omega) \leq \frac{1}{2}\dim V$. Hence we proved that $\dim B(\omega) = \frac{1}{2}\dim V$, contradicting $\omega \in O^+(V)$ since $\frac{1}{2}\dim V$ is odd.

Now let us assume $\dim V \not\equiv 2 \bmod 4$. We claim that in $O^+(V)$ every element is a product of 2 involutions. Suppose that V is a counterexample having minimal dimension. Choose $\pi \in O^+(V)$ such that π is not a product of 2 involutions of $O^+(V)$. Let $V = V_1 \perp \ldots \perp V_k$ be a complete orthogonal decomposition into π-moduls V_i. There are involutions $\rho_i, \sigma_i \in O(V_i)$ such that $\pi_i = \rho_i\sigma_i$ holds. Combining the ρ_i's and σ_i's one gets involutions $\rho, \sigma \in O(V)$ such that $\pi = \rho\sigma$. By our choice of V we have $\rho, \sigma \notin O^+(V)$.

A study of the equation $\pi_i = \rho_i\sigma_i$ in the orthogonally indecomposable π_i-module V_i using 4.7, 5.2 and the classification of § 3 yields: If $\dim V_i$ is even then $\dim B(\rho_i) = \frac{1}{2}\dim V_i = \dim B(\sigma_i)$. Taking into account the choice of V one concludes that $\dim V_i \equiv 0 \bmod 4$ is impossible for any i. Furthermore, $\dim V_i$ cannot be an odd number; else we might replace ρ_i, σ_i by $-\rho_i$, $-\sigma_i$ and obtained $\rho, \sigma \in O^+(V)$. Thus we have

(*) $\dim V_i \equiv 2 \bmod 4$ for every i.

Hence k is even (remember $\dim V \not\equiv 2 \bmod 4$).
Since $\dim B(\rho_i) = \dim B(\sigma_i)$ is odd for every i we conclude $\rho, \sigma \in O^+(V)$, a contradiction.
The proof of theorem 7.1 is finished.

Let us turn to theorem 7.2.
Let $\pi \in O^+(V)$. If $\dim V \not\equiv 2 \bmod 4$ then π is a product of 2 involutions of $O^+(V)$ by theorem 7.1. If a complete orthogonal decomposition of V into π-modules contains a term of odd dimension then we construct $\rho_i, \sigma_i, \rho, \sigma$ as in the preceding proof and eventually apply the change-of-sign-argument explained in this proof in order to achieve $\rho, \sigma \in O^+(V)$.

Now we want to prove the converse assertion of 7.2.
Let us assume that the assertion is wrong. We pick a counterexample π having minimal decomposition length, i.e. the number k of terms occuring in a complete orthogonal decomposition $V = V_1 \oplus \ldots \oplus V_k$ (where $\pi \in O^+(V)$) into π-modules is minimal.
We have $\dim V \equiv 2 \bmod 4$, each V_i has even dimension, and $\pi = \rho\sigma$ for some involutions $\rho, \sigma \in O^+(V)$.
By 5.1 we may assume that every V_i is invariant under ρ and σ. Hence we can restrict the equation $\pi = \rho\sigma$ to the V_i's: $\pi_i = \rho_i\sigma_i$. As we mentioned already, one can prove $\dim B(\rho_i) = \frac{1}{2}\dim V_i = \dim B(\sigma_i)$.
If, say, $\dim V_1 \equiv 0 \bmod 4$ then the restriction of π to $V_1^{\perp}$ satisfies our assumptions but its decomposition length is shorter. Hence $\dim V_i \equiv 2 \bmod 4$ for every i. A look at the possible types (§ 3) yields $F(\pi) = 0 = \mathrm{Neg}(\pi)$. This implies that $\dim B(\rho) = \frac{1}{2}\dim V$ is odd, contradicting $\rho \in O^+(V)$.
Detailed proofs of 7.1 and 7.2 are given in [13] .

8. HALF-TURNS.

Let V be a regular orthogonal vector-space.
In [1] E. Artin proves that $O^+(V)$ is generated by the set of half-turns (i.e. involutions σ such that $\dim B(\sigma) = 2$), provided that $\dim V \geq 3$.
Ishibashi [11] and Ellers [6] proved the following

Theorem. Let $\dim V \geq 3$. Let $\pi \in O^+(V)$ and $b := \dim B(\pi)$. Then the minimal number k such that $\pi = \eta_1 \cdot \ldots \cdot \eta_k$ for some half-turns η_i is

$$k = \frac{b}{2} \text{ or } k = \frac{b}{2} + 1 \qquad \text{if } \dim(B(\pi)/\mathrm{rad}\, B(\pi)) \geq 2,$$

$$k = \frac{b}{2} + 1 \text{ or } k = \frac{b}{2} + 2 \quad \text{if } \dim(B(\pi)/\mathrm{rad}\, B(\pi)) \leq 1.$$

We want to find the precise number k.

8.1 Let $\pi \in O(V)$ with $\dim B(\pi) \geq 3$. Suppose that a complete orthogonal decomposition of V into π-modules does not contain a term of type $(1\_)$ (see § 3). Then one can find a half-turn η such that $\dim B(\pi\eta) = \dim B(\pi) - 2$, $B(\pi\eta) \subseteq B(\pi)$, and a complete orthogonal decomposition of V into $\pi\eta$-modules does not contain a term of type $(1\_)$.

Proof. 5.2 supplies involutions $\rho, \sigma \in O(V)$ such that $\pi = \rho\sigma$ and $B(\pi) = B(\rho) \oplus B(\sigma)$. We can assume $\dim B(\sigma) \geq 2$. Choose a 2-dimensional regular subspace $W \leq B(\sigma)$ and define a half-turn η by $B(\eta) = W$. Clearly, $\sigma\eta$ is an involution. We have $B(\sigma\eta) \oplus B(\eta) = B(\sigma)$ and $B(\eta) \perp B(\sigma\eta)$, hence

$$(*) \quad B(\pi\eta) = B(\rho(\sigma\eta)) = B(\rho) \oplus B(\sigma\eta), \text{ as}$$
$$B(\pi) = B(\rho) \oplus B(\sigma) = B(\rho) \oplus B(\sigma\eta) \oplus B(\eta).$$

(*) and 5.2 imply that a complete orthogonal decomposition of V into $\pi\eta$-modules does not involve a $\pi\eta$-module of type $(1\_)$.

8.2 Let $\pi \in O^+(V) \setminus 1$ such that a complete orthogonal decomposition of V into π-modules does not contain a $(1\_)$-term (see § 3). Then π is a product of $\frac{1}{2} \dim B(\pi)$ half-turns.
Exception: $\dim B(\pi) = 2$.

Proof. Suppose that $b := \dim B(\pi) \geq 4$. Applying $l := \frac{b-4}{2}$ - times 8.1 we obtain half-turns $\eta_1,\ldots,\eta_l$ such that $\dim B(\omega) = 4$ where $\omega := \eta_l \cdot \ldots \cdot \eta_1\pi$, and ω does not involve type $(1\_)$. Using 5.2 we obtain involutions $\rho,\sigma \in O(V)$ such that $\omega = \rho\sigma$ and $B(\omega) = B(\rho) \oplus B(\sigma)$ and $\dim B(\rho) = \frac{1}{2}\dim V = \dim B(\sigma)$. Hence ρ and σ are half-turns, and π is a product of $\frac{1}{2}\dim B(\pi)$ half-turns.

8.3 Let $\pi \in O(V)$ and let σ be a symmetry.

(a) If $B(\sigma) \subseteq B(\pi)$ then $B(\pi) = B(\pi\sigma) \oplus B(\sigma)$,
$F(\pi\sigma) = F(\pi) \oplus \langle y\rangle$ where $\langle y(\pi-1)\rangle = B(\sigma)$, and
$B(\pi\sigma) = B(\pi) \cap y^{\perp}$.

(b) If $B(\sigma) \subseteq B(\pi)$ but $B(\sigma) \not\subseteq B^2(\pi) := B(\pi)(\pi-1)$
then one has additionally $\operatorname{rad} B(\pi\sigma) = \operatorname{rad} B(\pi) \cap y^{\perp}$
and $\dim(\operatorname{rad} B(\pi\sigma)) = \dim(\operatorname{rad} B(\pi))-1$.

Since (a) is well-known and easy to prove we treat only (b). $\operatorname{rad} B(\pi\sigma) = B(\pi\sigma) \cap F(\pi\sigma) = B(\pi) \cap y^{\perp} \cap (F(\pi) + \langle y\rangle)$; cf. (a). Let $a \in F(\pi), \lambda \in K$ and $v = a + \lambda y \in \operatorname{rad} B(\pi\sigma)$. This implies that $\lambda y(\pi-1) = v(\pi-1) \in B^2(\pi)$, hence $\lambda = 0$ as $y(\pi-1) \notin B^2(\pi)$. We obtained $\operatorname{rad} B(\pi\sigma) = B(\pi) \cap F(\pi) \cap y^{\perp} = \operatorname{rad} B(\pi) \cap y^{\perp}$. We claim $\operatorname{rad} B(\pi) \not\subseteq y^{\perp}$. Else $y \in (\operatorname{rad} B(\pi))^{\perp} = B(\pi) + F(\pi)$; say $y = w(\pi-1) + b$ where $w \in V$, $b \in F(\pi)$. However, this yields the contradiction $y(\pi-1) = w(\pi-1)^2 \in B^2(\pi)$.

8.4 Let $\pi \in O^+(V)$ and $\dim B(\pi) \geq 6$. Suppose that $\dim(B(\pi)/\operatorname{rad} B(\pi)) \geq 2$. Then π is a product of $\frac{1}{2}\dim B(\pi)$ half-turns or one can find a half-turn η such that $\dim B(\pi\eta) = \dim B(\pi) - 2$, $B(\pi\eta) \subseteq B(\pi)$, and $\dim(B(\pi\eta)/\operatorname{rad} B(\pi\eta)) = \dim(B(\pi)/\operatorname{rad} B(\pi))$.

Proof. Let $V = V_1 \perp \ldots \perp V_k$ be a complete orthogonal decomposition of V into π-modules. We may assume that V_1 has type $(1\_)$; cf. 8.2. Hence,

(1) $\dim(\operatorname{rad} B(\pi)) = \dim(B(\pi) \cap F(\pi)) \geq 2$,
and $\dim V/B(\pi) \geq 2$, $\dim(B(\pi)/B^2(\pi)) \geq 2$.

$B(\pi)$ is not totally isotropic. Thus we can choose a symmetry σ such that $B(\sigma) \subset B(\pi)$ but $B(\sigma) \not\subseteq B^2(\pi)$. Now 8.3 yields $B(\omega) = B(\pi) \cap y^{\perp}$, where $\langle y(\pi-1)\rangle = B(\sigma)$ and $\omega := \pi\sigma$.

(2) $B(\omega) \cap F(\sigma)$ is not totally isotropic.

Proof. From 8.3 follows that $B(\omega) = U \oplus \operatorname{rad} B(\omega)$ and $B(\pi) = U \oplus \operatorname{rad} B(\pi)$ for some regular subspace U. Choose $u \in U$ and $r \in \operatorname{rad} B(\pi)$ such that $B(\sigma) = \langle u + r\rangle$. Then u is anisotropic and $U \cap u^{\perp}$ is regular. From $\dim U = \dim(B(\pi)/\operatorname{rad} B(\pi)) \geq 2$ follows $0 \neq U \cap u^{\perp} = U \cap \langle u + r\rangle^{\perp} = U \cap F(\sigma) \subseteq B(\omega) \cap F(\sigma)$.

Now we distinguish two cases. First we assume that

(3) $B(\omega) \cap F(\sigma) \not\subseteq B^2(\omega)$.

Using (2) we select a symmetry ρ such that $B(\rho) \subseteq B(\omega) \cap F(\sigma)$ but $B(\rho) \not\subseteq B^2(\omega)$.
Clearly, $\eta := \sigma\rho$ is a half-turn, and from 8.3 we deduce that η fullfills all of our assumptions.

Secondly we consider the case

($\neg$3) $B(\omega) \cap F(\sigma) \subseteq B^2(\omega)$.

We have $B(\omega) \cap F(\omega) \neq 0$, hence $B^2(\omega)$ is properly contained in $B(\omega)$, and ($\neg$3) implies that

$$B(\omega) \cap F(\sigma) = B^2(\omega),$$

in particular $\dim(B(\omega)/B^2(\omega)) = 1$. Therefore, a complete orthogonal decomposition of V into ω-modules does not involve a $(1\_)$-term. Thus 5.2 supplies involutions $x, \tau \in O(V)$ such that

(4) $\omega = x\tau$ and $B(\omega) = B(x) \oplus B(\tau)$.

One can assume $\dim B(\tau) \geq 3$ as $\dim B(\omega) \geq 5$. $B(\tau) \cap F(\sigma)$ is not totally isotropic, since $B(\tau)$ is regular and $\dim(B(\tau) \cap F(\sigma)) > \frac{1}{2}\dim B(\tau)$. We select a symmetry $\rho \in O(V)$ such that $B(\rho) \subseteq B(\tau) \cap F(\sigma)$. Clearly, $\tau\rho$ is an involution and $\eta := \rho\sigma$ is a half-turn. We achieve $B(\pi) = B(\omega) \oplus B(\sigma) = B(x) \oplus B(\tau) \oplus B(\sigma) = B(x) \oplus (B(\tau\rho) \oplus B(\rho)) \oplus B(\sigma)$ and $\pi\eta = \omega\sigma\eta = x(\tau\rho)$, $B(\pi\eta) = B(x) \oplus B(\tau\rho)$. Now, 5.2 states that a complete decomposition of V into $\pi\eta$-modules does not contain a $(1\_)$-type module. Thus, 8.2 supplies half-turns $\eta_1, \ldots, \eta_k$ such that $\pi\eta = \eta_1 \cdot \ldots \cdot \eta_k$ and $k = \frac{1}{2}\dim B(\pi\eta) = \frac{1}{2}(\dim B(\pi) - 2)$. Finally, $\pi = \eta_1 \cdot \ldots \cdot \eta_k\eta$.

8.5 Theorem. Let $\pi \in O^+(V)$. If $\dim(B(\pi)/\operatorname{rad} B(\pi)) \geq 3$ then π is a product of $\frac{1}{2}\dim B(\pi)$ half-turns.

Proof. Since $\dim B(\pi)$ is even we have $\dim B(\pi) \geq 4$, and due to 8.2 we can assume $\dim(\operatorname{rad} B(\pi)) \geq 2$, hence $\dim B(\pi) \geq 6$. If 8.4 does not supply the assertion immediately then we get a half-turn η_1 such that $\dim B(\pi\eta_1) = \dim B(\pi)-2$ and $\dim(B(\pi)/\operatorname{rad} B(\pi)) = \dim(B(\pi\eta_1)/\operatorname{rad} B(\pi\eta_1))$. We continue this process (if it does not end up with the assertion) and obtain half-turns $\eta_1,\dots,\eta_k$ such that $k = \frac{1}{2}\dim B(\pi)-2$, $\omega := \pi\eta_1 \cdot \ldots \cdot \eta_k$ satisfies $\dim B(\omega) = 4$, and $\dim(B(\omega)/\operatorname{rad} B(\omega)) \geq 3$. Hence $\dim(\operatorname{rad} B(\omega)) \leq 1$ and ω cannot involve an (1_-)-type module. Hence $\omega = \eta_{k+2}\eta_{k+1}$ for 2 half-turns; cf. 8.2. Finally, $\pi = \eta_{k+2} \cdot \ldots \cdot \eta_1$ and $k+2 = \frac{1}{2}\dim B(\pi)$.

8.6 Proposition. Let $\dim V \geq 3$. Let $\pi \in O^+(V)$ such that $\dim(B(\pi)/\operatorname{rad} B(\pi)) \geq 1$. Then there are half-turns $\eta_1,\dots,\eta_k$ such that $\pi = \eta_1 \cdot \ldots \cdot \eta_k$ and $k \leq \frac{1}{2}\dim B(\pi) + 1$.

Proof. First we consider the exceptional case that $\dim B(\pi) = 2$. As $B(\pi)$ is not totally isotropic we have $\pi = \nu\tau$ for some symmetries ν,τ. If $\dim V = 3$ then $-\nu,-\tau$ are half-turns such that $\pi = (-\nu)(-\tau)$. Else $F(\nu) \cap F(\tau)$ is not totally isotropic, hence $\pi = (\nu\sigma)(\sigma\tau)$ for some symmetry σ with $B(\sigma) \subseteq F(\nu) \cap F(\tau)$.

In the sequel we can assume

(1) $\dim B(\pi) \geq 4$. Furthermore, $\dim(\operatorname{rad} B(\pi)) \geq 1$; cf. 8.5.

Choose $b \in \operatorname{rad} B(\pi) \setminus 0$ and $a \in V$ such that

(2) a is anisotropic and $a \notin B(\pi) \cup b^\perp = b^\perp$.

Let σ be the symmetry with $B(\sigma) = \langle a\rangle$ and let $\omega := \pi\sigma$.

We have

(3) $B(\omega) = B(\pi) \oplus B(\sigma) = B(\pi) \oplus \langle a\rangle$.

This implies that $\operatorname{rad} B(\omega) = B(\omega) \cap B(\omega)^\perp = (B(\pi) + \langle a\rangle) \cap B(\pi)^\perp \cap a^\perp$. Let $v = w+\lambda a \in \operatorname{rad} B(\omega)$, where $w \in B(\pi)$ and $\lambda \in K$. From the statements $v \in B(\pi)^\perp \subseteq b^\perp$, $w \in B(\pi) \subseteq b^\perp$ and $a \notin b^\perp$ one obtains $\lambda = 0$ and $v \in B(\pi)$. Hence $\operatorname{rad} B(\omega) = \operatorname{rad} B(\pi) \cap a^\perp$. Furthermore, $b \notin a^\perp$ implies that $\operatorname{rad} B(\pi) \nsubseteq a^\perp$. Thus we proved $\dim B(\omega) = \dim B(\pi) + 1$ and $\dim(\operatorname{rad} B(\omega)) = \dim(\operatorname{rad} B(\pi)) - 1$. We conclude

(4) $\dim(B(\omega)/\operatorname{rad} B(\omega)) \geq 3$.

(5) $B(\omega) \cap F(\sigma)$ is not totally isotropic,

since the regular subspace $F(\sigma) \cap \langle a,b\rangle = a^\perp \cap \langle a,b\rangle$ is contained in $F(\sigma) \cap B(\omega)$.

Now we proceed as in the proof of 8.4.
If $B(\omega) \cap F(\sigma) \not\subseteq B^2(\omega)$ then we can find a symmetry ρ such that $\eta := \sigma\rho$ is a half-turn satisfying $\dim B(\pi\eta) = \dim B(\omega\rho) = \dim B(\omega)-1 = \dim B(\pi)$ and $\dim(B(\pi\eta)/\operatorname{rad} B(\pi\eta)) = \dim(B(\omega)/\operatorname{rad} B(\omega)) \geq 3$. 8.5 supplies half-turns $\eta_1,\ldots,\eta_t$ such that $\pi\eta = \eta_1 \cdot \ldots \cdot \eta_t$ and $t = \frac{1}{2}\dim B(\pi)$.

If $B(\omega) \cap B(\sigma) \subseteq B^2(\omega)$ then the argument in the proof of 8.4 (observe that $\dim B(\omega) \geq 5$) supplies a half-turn η such that $\pi\eta$ does not involve a $(1\_)$-type $\pi\eta$-module and $\dim B(\pi\eta) = \dim B(\pi) \geq 4$, hence $\pi\eta = \eta_1 \cdot \ldots \cdot \eta_t$ for some half-turns η_i with $t = \frac{1}{2}\dim B(\pi)$.

We collect our results.

8.7 Theorem. Let V be a regular orthogonal vector space such that $\dim V \geq 3$. Let $\pi \in O^+(V)$, and let k denote the minimal number such that $\pi = \eta_1 \cdot \ldots \cdot \eta_k$ for some half-turns η_i. Then

$k = \frac{1}{2}\dim B(\pi)$ in the case of $\dim(B(\pi)/\operatorname{rad} B(\pi)) \geq 3$,

$k = \frac{1}{2}\dim B(\pi)+1$ in the case of $\dim(B(\pi)/\operatorname{rad} B(\pi)) = 1$.

Proof. The existence of a suitable number of half-turns follows from 8.5 and 8.6. One needs at least $\frac{1}{2}\dim B(\pi)$ half-turns; namely $\pi = \eta_1 \cdot \ldots \cdot \eta_k$ implies that $B(\pi) \subseteq B(\eta_1) + \ldots + B(\eta_k)$. For $k = \frac{1}{2}\dim B(\pi)$ one has $B(\pi) = B(\eta_1) \oplus \ldots \oplus B(\eta_k)$. Such an identity cannot exist in the second case since $B(\pi)$ does not contain a 2-dimensional regular subspace. Hence in this case $k = \frac{1}{2}\dim B(\pi) + 1$ is the minimal number possible.

The remaining cases $(\dim(B(\pi)/\operatorname{rad} B(\pi)) = 0$ or $2)$ will be treated elsewhere.

It is not difficult to generalize our results to non-regular orthogonal vector spaces V where $O^+(V) := \{\pi \in O(V) : \operatorname{rad} V \subseteq F(\pi)$ and $\det \pi = 1\}$. The first step of this generalization is

8.8 Corollary. Let V be an orthogonal vector space which is not necessarily regular, such that $\dim(V/\operatorname{rad} V) \geq 3$. Let $\pi \in O^+(V)$ (in the above sense) such that $B(\pi) \cap \operatorname{rad} V = 0$. Then the assertions of 8.7 are valid.

Proof. Take a complement U such that $V = U \oplus \mathrm{rad}\, V$ and $B(\pi) \subseteq U$. Clearly, $\dim U \geq 3$ and $B(\pi) = B(\pi|_U)$. The previous theorem supplies half-turns $\eta_i' \in O^+(U)$ such that $\pi|_U = \eta_1' \cdot \ldots \cdot \eta_k'$ where k is the number specified in 8.7. Every η_i' admits a unique extension $\eta_i \in O^+(V)$. The η_i's are half-turns such that $\pi = \eta_1 \cdot \ldots \cdot \eta_k$.

REFERENCES

[1] Artin, E., Geometric Algebra. Wiley-Interscience, New York 1957.

[2] Djoković, D.Z., Product of two involutions. Arch. Math. 18(1967), 582-584.

[3] Djoković, D.Z., The product of two involutions in the unitary group of hermitian forms. Indiana Univ. Math. J. 21 (1971), 449-456.

[4] Ellers, E.W., Bireflectionality in classical groups. Can. J. Math. 29 (1977), 1157-1162.

[5] Ellers, E.W., Decomposition of orthogonal, symplectic, and unitary isometries into simple isometries. Abh. Math. Sem. Univ. Hamburg 46 (1977), 97-127.

[6] Ellers, E.W., Products of half-turns. Journal of Algebra 99 (1986), 275-294.

[7] Ellers, E.W., Frank, R., Nolte, W., Bireflectionality of the weak orthonogal and the weak symplectic groups. Arch. Vol. 39 (1982), 113-118.

[8] Gustavson, W.H., Halmos, P.R., Radjavi, H., Products of involutions. Lin. Alg. and Appl. 13 (1976), 157-162.

[9] Huppert, B., Isometrien von Vektorräumen I. Arch. Math. 35 (1980), 164-176.

[10] Huppert, B., Isometrien von Vektorräumen II. Math. Z. 175 (1980), 5-20.

[11] Ishibashi, H., On some system of generators of the orthogonal groups. Sci. Rep. Tokyo Kyoiku Daigaku, Sect. A 11, No. 287 (1972), 96-105.

[12] Knüppel, F. and Nielsen, K., On products of two involutions in the orthogonal group of a vector space. To appear in J. of Lin. Alg. and Appl.

[13] Knüppel, F. and Nielsen, K., Products of involutions in $O^+(V)$. To appear in J. of Lin. Alg. and Appl.

[14] Nielsen, K., On bireflectionality and trireflectionality of orthogonal groups. To appear in J. of Lin. Alg. and Appl.

[15] Nielsen, K., Involutionen in orthogonalen und symplektischen Gruppen. Dissertation, Universität Kiel 1986.

[16] Scherk, P., On the decomposition of orthogonalities into symmetries. Proc. Amer. Math. Soc. 1 (1950), 481-491.

[17] Wonenburger, M.J., A decomposition of orthogonal transformations. Can. Math., Bull. Vol. 7 No. 3 (1964), 379-383.

[18] Wonenburger, M.J., Transformations which are products of two involutions. J. Math. Mech. 16 (1966), 327-338.

[19] Zassenhaus, H., On a normal form of the orthogonal transformation. Can. Math. Bull. Vol. 1 (1958), Part I 31-39, Part II 101-111, Part III 183-191.

Annals of Discrete Mathematics 37 (1988) 249–250

EXAMPLES OF OVOIDAL MÖBIUS PLANES OF HERING CLASS II1

Hans-Joachim Kroll

Mathematisches Institut, Technische Universität München, Postfach 20 24 20, D-8000 München 2

Studying the ovoidal Möbius planes of Hering classes II-IV Yaqub [2] found representation theorems for a subclass of ovoidal Möbius planes of Hering class II1 and IV1 respectively. In both cases the ovoid in question is defined in the same way, and it depends only on the field and constants used whether the construction yields a Möbius plane of class II1 or IV1. But, while Yaqub gave an example for the second type she gave none for the first type. The purpose of this note is to present a class of such examples.

The ovoids yielding the above mentioned Möbius planes are defined according to the following theorem.

THEOREM A (Yaqub [2], theorem 4). Let F be a commutative field with char $F \neq 2$. Let H be a subgroup of index 2 in $F^* := F \setminus \{0\}$, $a_1, a_2 \in F^*$ with $a_1 \neq a_2$ and $\Phi : F \times F \longrightarrow F$ the mapping defined by $\Phi(x,y) = a_1x^2 + y^2$ if $x \in H$, $\Phi(x,y) = a_2x^2 + y^2$ if $x \notin H$. In the 3-dimensional projective space $PG(3,F)$ the set $\mathcal{O}(a_1,a_2) := \{F^*(0,0,1,0)\} \cup \{F^*(x,y,z,1) \mid x,y \in F,\ z = \Phi(x,y)\}$ is an ovoid with tangent plane $z = 0$ at $F^*(0,0,0,1)$ if and only if the following three conditions hold:

(i) $R := \{(a_i + y^2)(a_j + y^2)^{-1} \mid y \in F,\ \{i,j\} = \{1,2\}\} \subset H$.

(ii) If $v \in F^* \setminus H$ and $r \in R$ then $(v+1)^2 - 4vr \in F^{(2)} \cup \{0\}$ [1].

(iii) If $v \in H$, $r \in R$ and $u \in F^* \setminus H$ then $u^2 - (v+1)u + vr \neq 0$.

By [2], theorem 3 the Möbius plane M defined by the ovoid $\mathcal{O}(a_1,a_2)$ is of Hering class IV1 if and only if $\{\Phi(x,y) \mid x,y \in F,\ (x,y) \neq (0,0)\} \subset H$, otherwise M is of class II1.

Now we are able to present our example. Let K be any commutative field with char $K \neq 2$ and $F = K((t))$ the field of formal Laurent series over K. For $x = \sum_{i=n}^{\infty} x_i t^i \in F$ let $o(x)$ denote the order of x, i.e. $o(x) = \infty$ if $x = 0$ and $o(x) = \min\{i \in \mathbb{Z} \mid x_i \neq 0\}$ if $x \neq 0$. The function o satisfies, for all $x,y \in F$: $o(xy) = o(x) + o(y)$,

---

[1]) $F^{(2)} = \{x^2 \mid x \in F^*\}$

$o(x+y) \geq \min\{o(x),o(y)\}$, and we have $o(x+y) > \min\{o(x),o(y)\}$ if and only if $o(x) = o(y)$ and $x_{o(x)} = -y_{o(x)}$. The subgroup $H := \{x \in F \mid o(x) \in 2\mathbb{Z}\}$ of F^* has index 2. Put $a_1 = t$ and $a_2 = -t$. As in theorem A let
$R := \{(t+y^2)(-t+y^2)^{-1} \mid y \in F\} \cup \{(-t+y^2)(t+y^2)^{-1} \mid y \in F\}$.

We show that the conditions (i), (ii) and (iii) of theorem A are satisfied.

(i) First we note $0 \notin R$ because of $t,-t \notin F^{(2)}$. For $x,y \in F^*$ we have $o(xy^{-1}) = o(x) - o(y)$. Since $o(t+y^2) = o(-t+y^2)$ we obtain for every $r \in R$ that $o(r) = 0$, hence $R \subset H$.

(ii) Let $v \in F^* \setminus H$, $r \in R$ and $x := (v+1)^2 - 4vr$. Because of $v \in F^* \setminus H$ we have $o(v) \neq 0$, hence $o((v+1)^2) < o(v) = o(4vr)$, since $o(r) = 0$. Therefore, $o(x) = o((v+1)^2) = 2o(v+1) \in 2\mathbb{Z}$, and $x_{o(x)} \in K^{(2)}$, hence $x \in F^{(2)}$.

(iii) Let $v \in H$, $r \in R$, $u \in F^* \setminus H$ and $y := u^2 - (v+1)\cdot u + vr$. Let us suppose $y = 0$. This implies $o(u^2) = o((v+1)u)$ or $o((v+1)u) = o(vr)$ or $o(u^2) = o(vr)$, hence we have to discuss the following three cases:

Case 1: $o(u^2) = o((v+1)u)$, hence $2o(u) = o(v+1) + o(u)$ and therefore $o(u) = o(v+1)$. Because of $o(u) \in 2\mathbb{Z}+1$, $o(v), o(1) \in 2\mathbb{Z}$ we obtain $o(v+1) > \min\{o(v),o(1)\}$, hence $o(v) = o(1) = 0$, $v_0 = -1$ and $o(u) = o(v+1) \geq 1$, thus $o(u^2 + (v+1)u) \geq 2$ and $0 = o(v) = o(vr)$, hence $o(y) = 0$ contrary to the supposition $y = 0$.

Case 2: $o((v+1)u) = o(vr)$, hence $o(v+1) + o(u) = o(v)$ and therefore $o(v+1) \in 2\mathbb{Z}+1$. Thus $o(v) = 0$, $v_0 = -1$, $o(v+1) > 0$ and $o(u) < 0$, which implies $o((v+1)u) = o(vr) = 0$ and $0 > 2o(u) = o(u^2) = o(y)$ in contradiction to $y = 0$.

Case 3: $o(u^2) = o(vr)$, hence $2o(u) = o(v)$ and therefore $o(v) \neq 0$, thus $o(v+1) = \min\{o(v),0\} \leq 0$. The assumption $y = 0$ now furthermore implies $o(v+1) + o(u) = o((v+1)u) \geq o(u^2) = 2o(u)$, hence $o(u) \leq o(v+1) \leq 0$ and $2o(u) = o(v) < 0$, thus $0 > o(v) = o(v+1) > o(u)$ and $o(v) = 2o(u) < o(u)$, but this is a contradiction.

By theorem A the set $\mathcal{O} = \mathcal{O}(t,-t)$ is an ovoid. Since $\phi(1,0) = t \notin H$, the geometry of the plane sections of $\mathcal{O}$ is a Möbius plane of Hering class II1.

REFERENCES

[1] HERING, C., Eine Klassifikation der Möbius-Ebenen. Math. Z. 87 (1965) 252-262.

[2] YAQUB, J.C.D.S., Strongly ovoidal Möbius planes. Math. Z. 142 (1975) 281-292.

Annals of Discrete Mathematics 37 (1988) 251–256
Elsevier Science Publishers B.V. (North-Holland)

A CONSTRUCTION OF PAIRS AND TRIPLES OF k-INCOMPLETE ORTHOGONAL ARRAYS

C. Pellegrino and P. Lancellotti

In this note we construct pairs of orthogonal $IA(q+hk,hk)$ where q is a prime power, k is a proper divisor of $q-1$ and $h\leqq s+1$ for a suitable integer s (depending on q). Further we construct triples of mutually orthogonal $IA(3^u+2h,2h)$ with $h\leqq s+1$ for a suitable integer s.

1. A k-incomplete array of order n based on (X,Y) is a latin square Q of order n based on the set $S=X\cup Y$ with $X\cap Y=\emptyset$, $|X|=n-k$, $|Y|=k$ and with a missing subsquare Q' of order k based on the set Y. We shall say for short that Q is an $IA(n,k)$ based on (X,Y).

Two $IA(n,k)$ based on (X,Y) are said to be orthogonal if when superimposed all the pairs of the set $(S\times S)\setminus(Y\times Y)$ are obtained.

Yamamoto in [1], Hedayat and Seiden in [2], and more extensively in [3] suggest a method for the construction of pairs of mutually orthogonal latin squares with orthogonal subsquares: the method is based on the projection of transversals of latin squares and is known in the literature as "method of sum composition" and has been later considered by many other authors, including Horton who constructed in [4] pairs of orthogonal $IA(q+k,k)$ for a prime power q and a proper divisor k of $q-1$, while Seiden and Wu in [5] construct triples of mutually orthogonal $IA(q+k,k)$ with q, k as above, $k\geqq 4$. Expecially interesting seems to be the search for triples of mutually orthogonal $IA(n,2)$, as these can be used as a "building block" for other values of k: for instance Stinson in [6] has constructed orthogonal pairs of $IA(8,2)$ and $IA(10,2)$.

In this note, using the above method of projecting transversals in a latin square, we construct pairs of orthogonal $IA(q+hk,hk)$ for a prime power q, a proper divisor k of $q-1$ and with $h\leqq s+1$ for a suitable value of s (cf. Theorems 1 and 2). Furthermore we construct triples of mutually orthogonal $IA(3^u+2h,2h)$ with $h\leqq s+1$ for a suitable integer s (cf. Theorems 3 and 4).

2. Consider the field $GF(q)=\{0,1,\dots,q-1\}$; to each non-zero element x of the field define the latin square $L(x)=[a^x_{i,j}]$ $(i,y=0,1,\dots,q-1)$ in the following way:

$$a^x_{i,j}=\begin{cases} j-i & \text{if } x=1 \\ i+x(j-i) & \text{if } x\neq 1 \end{cases}$$

As x varies in $GF(q)\setminus\{\emptyset\}$ we obtain a complete set of mutually orthogonal latin squares. For all $a\in GF(q)$ we denote by $t(a)$ the set of all cells of $L(1)$ containing the element a, that is the set of cells (i,j) such that $j-i=a$. Obviously $t(a)$ is a common transversal of the squares $L(x)$ for $x\neq 1$. Let a,b,x,y be elements of $GF(q)$ such that $x,y\neq 0,1$ and $x\neq y$. Similarly to [3] we project onto the same row [column] the transversal $t(a)$ of $L(x)$ and $t(b)$ of $L(y)$ and we thus reconstruct pairs which correspond in $L(1)$ to cells containing one and the same element c [d]; we get the relations

$$(y-x)c = by-ax-b+a$$

$$(y-x)d = by-ax ,$$

or equivalently, solving the system for a and b,

$$a = d+y(c-d) \tag{1}$$

$$b = d+x(c-d) . \tag{2}$$

Let k be a proper divisor of $q-1$ and let ε be an element of order k in $GF(q)$. For each index $r=1,2,\ldots,k$ define

$$E(r) = 1+\varepsilon+\ldots+\varepsilon^{r-1} .$$

Let x be a non-zero element of $GF(q)$ with the properties:

i) $x\neq 1-\varepsilon$,

ii) $x\neq 1/E(r)$ $\quad(r=1,2,\ldots,k-1)$,

iii) $x\neq -\varepsilon^{r}/E(r)$ $\quad(r=1,2,\ldots,k-1)$.

We set $y=x+\varepsilon$ and consider the system of $2k$ equations

$$\left\{\begin{array}{l} a_{i+1} = b_i + y(a_i-b_i) \\ \\ b_{i+1} = b_i + x(a_i-b_i) \end{array}\right. \qquad (i=0,1,\ldots,k-1) \tag{3}$$

(where indices are to be taken mod k). For arbitrary distinct elements a_0 and b_0 in $GF(q)$ we obtain the following solution

$$\left\{\begin{array}{l} a_i = b_0 + [xE(i)+\varepsilon^{i}]\,(a_0-b_0) \\ \\ b_i = b_0 + xE(i)\,(a_0-b_0) . \end{array}\right. \qquad (i=0,1,\ldots,k-1) \tag{4}$$

It follows immediately from i), ii) and iii) that the elements a_i and b_i $(i=0,1,\ldots,k-1)$ are all distinct. As a consequence of (1) and (2) we have that after projecting the transversals $t(a_i)$ in $L(x)$ and $t(b_i)$ in $L(y)$ onto the same row and column we find the pairs resulting from the superimposition of the transversals $t(a_{i-1})$ and $t(b_{i-1})$. We can now prove the following:

Theorem 1. If q is a prime power and k is a proper divisor of $q-1$, then there exists a pair of orthogonal $IA(q+k,k)$.

Proof. By what we said before it is sufficient to show the existence of an element x in $GF(q)\setminus\{0\}$ satisfying i), ii) and iii). Let $t\notin\{0,1,\varepsilon,\dots,\varepsilon^{k-1}\}$ and set $x=(\varepsilon-1)/(t-1)$. A straightforward calculation shows that x satisfies the required properties □

We now prove that under certain conditions it is possible to construct recursively pairs of orthogonal $IA(q+hk,hk)$ for a proper divisor k of $q-1$ and with $h\leq s+1$ for a suitable integer s.

Let A and B be subsets of a finite group G with $|A|=m$, $|B|=n$ and $0\in B$. If $p,r\in B$, we set

$$A_p = A+p = \{x+p:x\in A\}$$

$$D_{p,r} = \{x-y:x\in A_p \text{ and } y\in A_r\}$$

$$D(A,B) = \bigcup_{p,r\in B} D_{p,r}.$$

If $D(A)=\{x-y:x,y\in A\}$ we have $D_{p,p}=D(A)$ for all p of B, whence the inequalities

$$|D_{p,p}|\leq m(m-1)+1$$

and

$$|D(A,B)|\leq[m(m-1)+1]\cdot[n(n-1)+1].$$

Theorem 2. Let q,k and s be respectively a prime power, a proper divisor of $q-1$ and a positive integer; if $q>[2k(2k-1)+1]\cdot[s(s-1)+1]$ then for each $h=1,2,\dots,s+1$ there exists a pair of orthogonal $IA(q+hk,hk)$.

Proof. Consider a pair of orthogonal $IA(q+k,k)$ (cf. Theorem 1) obtained from the latin squares $L(x)$ and $L(y)$ in the way indicated above by projecting the transversals $t(a_{0,0}), t(a_{1,0}), \dots, t(a_{k-1,0})$ of $L(x)$ and $t(b_{0,0}), t(b_{1,0}), \dots, t(b_{k-1,0})$ of $L(y)$. Define

$$A = \{a_{0,0}, a_{1,0}, \dots, a_{k-1,0}, b_{0,0}, b_{1,0}, \dots, b_{k-1,0}\}$$

and let p_1 be an element of $GF(q)$. If we set

$$a_{0,1} = a_{0,0}+p_1 \quad\text{and}\quad b_{0,1} = b_{0,0}+p_1$$

we have that

$$a_{i,1} = a_{i,0}+p_1 \quad\text{and}\quad b_{i,1} = b_{i,0}+p_1 \quad (i=0,1,\dots,k-1)$$

yield a solution of system (3).

If $p_1\notin D(A)$ then the elements $a_{i,1}$ and $b_{i,1}$ $(i=0,1,\dots,k-1)$ are pairwise distinct and are also distinct from the elements $a_{i,0}$ and $b_{i,0}$. Assuming $q>2k(2k-1)+1$ we can thus construct a pair of orthogonal IA(q+2k,2k) by projecting the transversals $t(a_{i,0})$ of L(x) and $t(b_{i,0})$ of L(y) $(i=0,1,\ldots,k-1)$ into the first k new columns and rows and projecting the transversals $t(a_{i,1})$ of L(x) and $t(b_{i,1})$ of L(y) into the remaining k columns and rows.

Set $B_1=\{p_1\}$ and assume that we have defined $B_j=\{p_1,p_2,\ldots,p_j\}$ for $j<s$ in such a way that the elements of GF(q)

$$a_{i,h}=a_{i,0}+p_h \quad ; \quad b_{i,h}=b_{i,0}+p_h \qquad (i=0,1,\ldots,k-1\ ;\ h=1,2,\ldots,j)$$

are all distinct. Then we have

$$|D(A,B_j)|<[2k(2k-1)+1]\cdot[j(j-1)+1]$$

and by hypothesis there exists an element p_{j+1} in $D(A,B_j)$. We set

$$a_{0,j+1}=a_{0,0}+p_{j+1} \quad ; \quad b_{0,j+1}=b_{0,0}+p_{j+1}$$

these yield a solution of (3) with the further property that the elements $a_{i,j+1}$ and $b_{i,j+1}$ $(i=0,1,\ldots,k-1)$ are all distinct and distinct from the elements $a_{i,h}$ and $b_{i,h}$ $(i=0,1,\ldots,k-1\ ;\ h=0,1,\ldots,j)$.
We are thus able to construct a pair of orthogonal IA(q+(j+2)k,(j+2)k) by projecting in the correct sequence the transversals $t(a_{i,k})$ of L(x) and $t(b_{i,k})$ of L(y) $(i=0,1,\ldots,k-1\ ;\ h=0,1,\ldots,j+1)$ hence the theorem is proved □

In special cases the above result can be considerably improved because it happens quite often that $|D(A)|$ is smaller than $2k(2k-1)+1$. For instance if $k=2$ then $|D(A)|\leq 9<13$, and with $q=11$ we can find a pair of orthogonal IA(15,4), the existence of which is not guaranteed by the previous theorem. Moreover we remark that here $2k$ is not a divisor of $q-1$.

Theorem 3. For each integer $u\geq 2$ there exists a triple of mutually orthogonal $IA(3^u+2,2)$.

Proof. Given the field $GF(3^u)$ with $u\geq 2$ let t by an element of $GF(3^u)\setminus GF(3)$. We consider the following three elements of $GF(3^u)$

$$x=1/(t-1) \quad ; \quad y=x+2 \quad ; \quad z=x+1$$

and the three latin squares L(x), L(y) and L(z). For a_0 in $GF(3^u)$ we define $b_0=a_0+2$ and $c_0=a_0+1$. From L(x) and L(y) making use of (3) we can thus construct a pair A_1, A_2 of orthogonal $IA(3^u+2,2)$ by projecting the transversals $t(a_0)$, $t(a_1)$ of L(x) respectively $t(b_0)$, $t(b_1)$ of L(y). Similarly we can construct a pair B_1,B_2 of orthogonal $IA(3^u+2,2)$ by projecting the transversals $t(b_0)$, $t(b'_1)$ of $L(y)$ respectively $t(c_0)$, $t(c'_1)$ of $L(z)$ and finally a pair C_1, C_2 of orthogonal $IA(3^u+2,2)$ by projecting the transversals $t(c_0)$, $t(c_1)$ of $L(z)$ respectively $t(a_0)$, $t(a'_1)$ of $L(x)$. It can be easily seen that

$$a_1 = a'_1 \quad ; \quad b_1 = b'_1 \quad ; \quad c_1 = c'_1$$

and so $A_2 = B_1$, $B_2 = C_1$ and $C_2 = A_1$. Hence A_1, B_1 and C_1 form a triple of mutually orthogonal $IA(3^u+2,2)$ □

Theorem 4. For $u \geq 2$ if $3^u > 9[s(s-1)+1]$ then for each $h=1,2,\ldots,s+1$ there exists a triple of mutually orthogonal $IA(3^u+2h,2h)$.

Proof. With $x,y,z,a_0,a_1,b_0,b_1,c_0,c_1$ as in theorem 3 set $A = \{a_0,a_1,b_0,b_1,c_0,c_1\}$ and observe that $|D(A)| \leq 9$. The rest of the proof is then as in theorem 2 □

REFERENCES

[1] Yamamoto, K., *Generation principles of Latin Squares"*, Bull. Inst. Internat. Statist., 38 (1961) pp.73-76.

[2] Hedayat, A. and Seiden, E., *On a method of sum composition of orthogonal Latin Squares*, Atti del Convegno di Geometria Combinatoria e sue Applicazioni, Perugia, (1971) pp.239-256.

[3] Hedayat, A. and Seiden, E., *On the theory and applications of sum composition of Latin Squares and orthogonal Latin Squares*, Pacif. Journ. of Math., 54 (1974) pp.86-113.

[4] Horton, J.D., *Sub-Latin Squares and Incomplete Orthogonal Arrays*, Journ. Combin. Theory, (A) 16 (1974) pp.23-33.

[5] Seiden, E. and Wu, C.J., *On construction of three mutually orthogonal Latin Squares by the Method of Sum Composition*, Essays in Probability and Statistics, Ikeda, S. and others (eds.) (1976) pp.57-64.

[6] Stinson, D.R., *The equivalence of certain incomplete transversal designs and frames*, (private communication).

Pellegrino, C. and Lancellotti P.
Dipartimento di Matematica
Università di Modena
Via Campi 213/B
41100 MODENA (Italy)

Annals of Discrete Mathematics 37 (1988) 257–264

RELATIVE INFINITY IN PROJECTIVE DE SITTER SPACETIME AND ITS RELATION TO PROPER TIME

J. A. LESTER

Department of Mathematics and Computer Science
California State University at Los Angeles
5151 State University Drive
Los Angeles, California 90032, U.S.A.

1. INTRODUCTION

The idea of a relative spacetime infinity (i.e., the idea that points of spacetime that appear finitely close to one observer need not appear so to another) was discussed in [**3**] and [**4**], where some of its consequences for Minkowski spacetime (in particular its effect on proper time) were examined. Here we describe a relative infinity for de Sitter spacetime, and obtain a similar modification for proper time.

A simple spatial example may help to clarify the notion of a relative infinity (see [**3**] for another based on the real projective plane). The real Möbius plane $\mathbb{M}_2$ can be constructed from the real Euclidean plane $\mathbb{E}_2$ by adjoining a single point at infinity. This point is assumed to be on all lines, which are then taken to be "circles through the point at infinity". The crux of the idea of a relative infinity is the *interpretation* the "inhabitants" of $\mathbb{M}_2$ give this added point.

If the point retains its special status, the geometry of the plane is little changed from that of $\mathbb{E}_2$. Its inhabitants have distinct notions of lines and circles (the lines are the ones through the special point at infinity) and agree with each other on which are which. If the added point has no special status whatsoever, the inhabitants of the plane again agree completely: all points of the plane are alike, and there are no lines distinct from circles. The "relativistic viewpoint" lies between the two extremes: it assumes that any inhabitant of the plane designates a point to be at infinity (and so *for himself* distinguishes lines from circles), but that different inhabitants need not agree on *which* point is infinite. The location of infinity and the derived distinction between lines and circles become *relative, observer-dependent* concepts. (A religious analogy: it's as if all denizens of the first plane believe in god, and moreover, in the *same* god, while those of the second all agree they are atheists. But each individual in the relativist plane has his *own* god, his neighbour another, and their different gods give them different views of the world.)

In §2 which follows, we introduce a similar relative infinity for de Sitter spacetime. Such a relative infinity will in fact result from two natural physical asumptions:

i) Each observer coordinatizes the events of spacetime *of which he is aware* as those of ordinary de Sitter spacetime.

ii) For events which they have in common, any two observers agree on which pairs of events can be connected by unreflected light-signals.

Then, in §3, we derive an appropriate invariant proper time parameter (to replace the usual proper time parameter, which is not invariant under changes of infinity). Finally, in §4, we discuss some of the physical implications of relative infinity and this new proper time parameter.

2. PROJECTIVE DE SITTER SPACETIME

(Ordinary) de Sitter spacetime can be cordinatized as follows (see [1] or [5]): let $\mathcal{V}$ denote $\mathbb{R}^5$ equipped with the metric

$$(\mathbf{r}_1,\mathbf{r}_2) := -v_1v_2 + w_1w_2 + x_2x_2 + y_1y_2 + z_1z_2$$

for all $\mathbf{r}_1 := (v_1,w_1,x_1,y_1,z_1)$ and $\mathbf{r}_2 := (v_2,w_2,x_2,y_2,z_2)$ in $\mathcal{V}$. Then we identify de Sitter spacetime with the hyperboloid $\mathcal{L} := \{\mathbf{r} \in \mathcal{V} | (\mathbf{r},\mathbf{r}) = 1\}$, and the group of de Sitter transformations with the transformations of $\mathcal{L}$ induced by the isometries (linear, metric-preserving bijections) of $\mathcal{V}$. In this coordinate system an event $\mathbf{r} \to (v,w,x,y,z)$ is assigned a time coordinate $t := \sinh^{-1}v$ relative to some clock, and spatial coordinates (w, x, y, z) on the sphere $w^2 + x^2 + y^2 + z^2 = 1 + v^2 = \cosh^2 t$. (The geometry of space is thus assumed to be that of a three-sphere whose radius is changing in time.) The events with coordinates $\mathbf{r}_1$ and $\mathbf{r}_2$ are connected by an unreflected light-signal iff $(\mathbf{r}_1,\mathbf{r}_2) = 1$ (see [5]). This condition is clearly preserved by the de Sitter group; conversely, it characterizes the de Sitter group: all bijections $\mathcal{L} \to \mathcal{L}$ which preserve the relation $(\mathbf{r}_1,\mathbf{r}_2) = 1$ must be de Sitter transformations [5].

We now embed the affine quadric $\mathcal{E}$ in a projective quadric. Define homogeneous coordinates $\mathbf{R} \propto (V,W,X,Y,Z,H) \propto (v,w,x,y,z,1)$, i.e.

$$H \neq 0,\ V := Hv,\ W := Hw,\ X := Hx,\ Y := Hy,\ Z := Hz.$$

These coordinates satisfy $[\mathbf{R},\mathbf{R}] = 0$, where $[\ ,\ ]$ is the metric on $\mathbb{R}^6$ given by

$$[\mathbf{R},\mathbf{R}] := -V^2 + W^2 + X^2 + Y^2 + Z^2 - H^2,$$

and also satisfy $[\mathbf{R},\mathbf{N}] = -H \neq 0$ where $\mathbf{N} := (0,0,0,0,0,1)$. The metric $[\ ,\ ]$ makes $\mathbb{R}^6$ into a real metric vector space $\bar{\mathcal{V}}$ of signature (4,2), called the *coordinate space* of $\mathcal{E}$. The de Sitter transformations are induced on $\mathcal{E}$ by those isometries of $\bar{\mathcal{V}}$ with eigenvector $\mathbf{N}$.

The relation $(\mathbf{r}_1,\mathbf{r}_2) = 1$ between events of $\mathcal{E}$ connected by a light-signal translates into $[\mathbf{R}_1,\mathbf{R}_2] = 0$ for the corresponding projective coordinates. *All* isometries of $\bar{\mathcal{V}}$ (even those without eigenvector $\mathbf{N}$) preserve this relation; in fact, it can be used to characterize the isometries of $\bar{\mathcal{V}}$ as follows.

Theorem: Let $\mathcal{D}$ be an open connected subset of $\mathcal{E}$, and let $\mathbf{r} \to \bar{\mathbf{r}}$ be a mapping from $\mathcal{D}$ to $\mathcal{E}$ such that for all $\mathbf{r}_1$ and $\mathbf{r}_2$ in $\mathcal{D}$,

$$(\mathbf{r}_1, \mathbf{r}_2) = 1 \quad \text{iff} \quad (\bar{\mathbf{r}}_1, \bar{\mathbf{r}}_2) = 1.$$

Then the mapping $\mathbf{r} \to \bar{\mathbf{r}}$ is induced on $\mathcal{E}$ by an isometry of $\bar{\mathcal{V}}$.

This theorem is an easy corollary to the main theorem of [2]: it holds because the coordinate space $\bar{\mathcal{V}}$ has the same signature as the coordinate space of conformal Minkowski space. (See [5] for a proof of a similar corollary.)

Those $\mathbf{R}$ in $\bar{\mathcal{V}}$ with $[\mathbf{R},\mathbf{R}] = 0$ and $[\mathbf{R},\mathbf{N}] = 0$ (i.e. with $H = 0$) do not coordinatize points of de Sitter spacetime; they coordinatize "points at infinity". We can extend de Sitter spacetime to include such points; in fact, from above theorem, this extension can be obtained directly from the assumption that all coordinate systems agree on which pairs of events are connected by unreflected light signals, provided we do not restrict the coordinate systems to share the same infinity.

Definition 2.1: *Projective de Sitter spacetime* is a set $\mathcal{S}$ of elements called *events*, together with a maximal collection of *coordinate systems* (f_α,S_α), where $S_\alpha \subseteq \mathcal{S}$ and $f_\alpha: S_\alpha \to \mathcal{E}$ is bijective, such that for any two coordinate systems (f_α,S_α) and (f_β,S_β), if $S_\alpha \cap S_\beta$ is not empty, then $f_\alpha(S_\alpha \cap S_\beta)$ is open and connected in $\mathcal{E}$ and $f_\beta \circ f_\alpha^{-1}$ preserves the relation $[\mathbf{R}_1,\mathbf{R}_2] = 0$ for all $\mathbf{R}_1$ and $\mathbf{R}_2$ in $f_\alpha(S_\alpha \cap S_\beta)$.

It follows from the theorem above that the coordinate transformations $f_\beta \circ f_\alpha^{-1}$ are induced by the isometries of $\bar{\mathbb{V}}$, so that $\mathcal{S}$ has the structure of a covering space of (extended) de Sitter spacetime.

We now interpret these coordinate systems. With every coordinate system (f_α, S_α) we associate an *observer*. To each event $\mathbf{e} \in S_\alpha$ with $f_\alpha(\mathbf{e}) := \mathbf{R} \propto (V,W,X,Y,Z,H)$ and $H \neq 0$, this observer assigns de Sitter coordinates $\mathbf{r} \to (v,w,x,y,z) = H^{-1}(V,W,X,Y,Z)$, and describes the geometry of spacetime via these coordinates as that of ordinary de Sitter spacetime as above. He is oblivious to points with $H = 0$; they are "at infinity". (They need *not* be at infinity to other observers.)

3. OBSERVER WORLDLINES AND PROPER TIME

We assume that any observer describes all positions in space (as he sees it) relative to himself; i.e., he assumes his own position to satisfy $x = y = z = 0$, $w > 0$ in his own coordinates.

Definition 3.1: The *worldline* of an observer is that set of points in $\mathcal{S}$ of the form $\mathbf{R} \propto (v,w,0,0,0,1)$, $w > 0$ in the observer's own coordinate system.

An observer's worldline may be parametrized by the time coordinate he assigns each point. (Note that $w = +(1+v^2)^{1/2}$ for points on his worldline.)

Definition 3.2: *Proper time* is the parameter $\tau := \sinh^{-1} v$ for points $\mathbf{R} \propto (v,w,0,0,0,1)$ along an observer's worldline. His worldline thus has parametric form $\mathbf{R} \propto \mathbf{F}(\tau) := (\sinh\tau, \cosh\tau,0,0,0,1)$.

It follows that

$$\mathbf{F}(\tau) = \mathbf{F}(0) + \sinh\tau\,\dot{\mathbf{F}}(0) + (\cosh\tau - 1)\ddot{\mathbf{F}}(0),$$

and that

$$\begin{aligned} [\mathbf{F},\mathbf{F}] &= [\dot{\mathbf{F}},\mathbf{F}] = [\dot{\mathbf{F}},\ddot{\mathbf{F}}] = 0, \\ [\mathbf{F},\ddot{\mathbf{F}}] &= [\ddot{\mathbf{F}},\ddot{\mathbf{F}}] = 1 = -[\dot{\mathbf{F}},\dot{\mathbf{F}}]. \end{aligned} \tag{3.1}$$

Note also that $\mathbf{F}(\tau) - \ddot{\mathbf{F}}(\tau) \propto \mathbf{N}$ for all τ. (3.2)

The points of the form $\mathbf{R} \propto (V,W,0,0,0,0)$ (i.e. with $H = 0$) can also be considered to lie on the observer's worldline. Since $V^2 - W^2 = 0$, there are two of these: $\mathbf{F}_1 \propto (1,1,0,0,0,0)$ and $\mathbf{F}_2 \propto (-1,1,0,0,0,0)$. For finite points, we have that $\mathbf{F}(\tau) \propto (\tanh\tau, 1, 0, 0, 0, \operatorname{sech}\tau)$, so since $\mathbf{F}(\tau) \to \mathbf{F}_1$ as $\tau \to +\infty$ and $\mathbf{F}(\tau) \to \mathbf{F}_2$ as $\tau \to -\infty$, the points $\mathbf{F}_1$ and $\mathbf{F}_2$ may be parametrized by proper times $\pm\infty$.

Now consider our observer's worldline as seen by another observer, i.e. in another coordinate system . In *this* system, his worldline has the form $\mathbf{R} \propto \mathbf{G}(\tau) := \mathcal{J}\mathbf{F}(\tau)$, where $\mathcal{J}$ is the isometry of $\mathbb{V}$ relating the two coordinate systems; thus

$$[\mathbf{G},\mathbf{G}] = [\dot{\mathbf{G}},\mathbf{G}] = [\dot{\mathbf{G}},\ddot{\mathbf{G}}] = 0,$$
$$[\mathbf{G},\ddot{\mathbf{G}}] = [\ddot{\mathbf{G}},\ddot{\mathbf{G}}] = 1 = -[\dot{\mathbf{G}},\dot{\mathbf{G}}],$$

since the same results hold for $\mathbf{F}$. The second observer may also describe the first's worldline as $\mathbf{R} \propto \mathbf{S}(\lambda)$ for some 6-vector function of the arbitrary parameter λ, determined up to a proportionality factor possibly depending on λ. He can then recover the first observer's proper time via the folowing theorem.

Theorem: Suppose that, in some coordinate system, an observer worldline is given by $\mathbf{R} \propto \mathbf{S}(\lambda)$, where $\mathbf{S}$ is a twice differentiable function of the parameter λ. Then proper time τ along the worldline is given by

$$[\mathbf{S}_0'',\mathbf{S}_0''] = \frac{3\tau''^2 + \tau'^4 - 2\tau'''\tau'}{\tau'^2}$$

where $\mathbf{S}_0 = \pm[\mathbf{S}',\mathbf{S}']^{-1/2}\,\mathbf{S}$.

Proof: We have that $\mathbf{G}(\tau) = k(\lambda)\mathbf{S}(\lambda)$ for some proportionality factor $k := k(\lambda)$. Also $\lambda = \lambda(\tau)$ where $\dot{\lambda} \neq 0$. Define $\mathbf{S}_0 = \dot{\lambda}\mathbf{G} = \dot{\lambda}k\mathbf{S}$, then differentiating $\mathbf{S}_0$ with respect to τ yields $\mathbf{S}_0'\dot{\lambda} = \ddot{\lambda}\mathbf{G} + \dot{\lambda}\dot{\mathbf{G}}$, from which

$$\dot{\lambda}^2[\mathbf{S}_0',\mathbf{S}_0'] = \ddot{\lambda}^2[\mathbf{G},\mathbf{G}] + 2\ddot{\lambda}\dot{\lambda}[\mathbf{G},\dot{\mathbf{G}}] + \dot{\lambda}^2[\dot{\mathbf{G}},\dot{\mathbf{G}}] = -\dot{\lambda}^2.$$

It follows that $[\mathbf{S}_0',\mathbf{S}_0'] = -1$.

Differentiating $\mathbf{S}_0$ with respct to λ yields $\mathbf{S}_0' = (\dot{\lambda}k\mathbf{S})' = (\dot{\lambda}k)'\mathbf{S} + \dot{\lambda}k\mathbf{S}'$, so

$$[\mathbf{S}_0',\mathbf{S}_0'] = (\dot{\lambda}k)'^2[\mathbf{S},\mathbf{S}] + 2(\dot{\lambda}k)'(\dot{\lambda}k)[\mathbf{S},\mathbf{S}'] + (\dot{\lambda}k)^2[\mathbf{S}',\mathbf{S}'].$$

Since $[\mathbf{S},\mathbf{S}] = 0$ implies that $2[\mathbf{S},\mathbf{S}'] = 0$, we have that $-1 = (\dot{\lambda}k)^2[\mathbf{S}',\mathbf{S}']$, from which $\dot{\lambda}k = \pm[\mathbf{S}',\mathbf{S}']^{-1/2}$ and $\mathbf{S}_0 = \pm[\mathbf{S}',\mathbf{S}']^{-1/2}\,\mathbf{S}$.

Differentiating $\mathbf{S}_0 = \dot{\lambda}\mathbf{G}$ twice with respect to τ gives

$$\ddot{\mathbf{S}}_0 = \dddot{\lambda}\mathbf{G} + 2\ddot{\lambda}\dot{\mathbf{G}} + \dot{\lambda}\ddot{\mathbf{G}},$$

thus

$$[\ddot{\mathbf{S}}_0,\ddot{\mathbf{S}}_0] = -4\ddot{\lambda}^2 + \dot{\lambda}^2 + 2\dddot{\lambda}\dot{\lambda}.$$

Differentiating $\mathbf{S}_0 = \mathbf{S}_0[\lambda(\tau)]$ twice with respect to τ also gives

$\ddot{\mathbf{S}}_0 = \ddot{\lambda}\mathbf{S}'_0 + \dot{\lambda}^2\mathbf{S}''_0$, so

$$[\ddot{\mathbf{S}}_0, \ddot{\mathbf{S}}_0] = -\ddot{\lambda}^2 + \dot{\lambda}^4[\mathbf{S}''_0, \mathbf{S}''_0].$$

Comparing the two expressions, we obtain

$$\dot{\lambda}^4[\mathbf{S}''_0, \mathbf{S}''_0] = -3\ddot{\lambda}^2 + \dot{\lambda}^2 + 2\dddot{\lambda}\dot{\lambda},$$

from which the required result can be obtained by replacing $\dot{\lambda}$ by $1/\tau'$. ∎

Now suppose we have an *arbitrary* timelike curve $\mathbf{R} \propto \mathbf{S}(\lambda)$. (The curve is timelike iff $[\mathbf{S}', \mathbf{S}'] < 0$, since in differential form, the metric is given by $ds^2 := (\mathbf{dr}, \mathbf{dr}) = H^{-2}[\mathbf{dR}, \mathbf{dR}] = H^{-2}k^2[\mathbf{S}', \mathbf{S}']d\lambda^2$ for some non-zero $k := k(\lambda)$.) We can extend the formula above to define proper time along this curve as well.

Definition 3.3: Let $\mathbf{R} \propto \mathbf{S}(\lambda)$ be an arbitrary timelike curve in $\mathcal{S}$, where $\mathbf{S}(\lambda)$ is a twice differentiable vector function of the parameter λ. Then proper time along the curve is given by

$$[\mathbf{S}''_0, \mathbf{S}''_0] = \frac{3\tau''^2 + \tau'^4 - 2\tau'''\tau'}{\tau'^2}, \qquad (3.3)$$

where $\mathbf{S}_0 = \pm [\mathbf{S}', \mathbf{S}']^{-1/2}\,\mathbf{S}$.

We note that if the above curve is given in the finite form $\mathbf{r}(\lambda) := (v(\lambda), w(\lambda), x(\lambda), y(\lambda), x(\lambda))$, then direct calculation shows that the formula above takes the form

$$\frac{3\tau''^2 + \tau'^4 - 2\tau'''\tau'}{\tau'^2} = \frac{3(\mathbf{r}'', \mathbf{r}'')(\mathbf{r}', \mathbf{r}') + 2(\mathbf{r}', \mathbf{r}''')(\mathbf{r}', \mathbf{r}') - 6(\mathbf{r}', \mathbf{r}'')^2}{(\mathbf{r}', \mathbf{r}')^2}.$$

(Compare this with the corresponding formula for ordinary de Sitter spacetime: $\tau'^2 = (\mathbf{r}', \mathbf{r}')$.)

Any 6-vector function $\mathbf{R} \propto \mathbf{F}(\tau)$ satisfying $[\mathbf{F}, \mathbf{F}] = 0$, $[\dot{\mathbf{F}}, \dot{\mathbf{F}}] = -1$ and $[\ddot{\mathbf{F}}, \ddot{\mathbf{F}}] = 1$ satisfies the remaining conditions of (3.1) (by differentiation), and thus represents an observer worldline parametrized by proper time. Conversely, for a solution τ of (3.3) for some timelike curve $\mathbf{R} \propto \mathbf{S}(\lambda)$, there exists a (unique up to sign) 6-vector function $\mathbf{R} \propto \mathbf{F}(\tau) \propto \mathbf{S}(\lambda)$ satisfying (3.1). The three conditions on $\mathbf{F}$ can be interpreted as follows.

i) The condition $[\mathbf{F}, \mathbf{F}] = 0$ ensures that $\mathbf{R} \propto \mathbf{F}(\tau)$ represents a point of $\mathcal{S}$. If $H \neq 0$, the point is finite in the corresponding coordinate system, where it has the form $\mathbf{R} \propto H(\mathbf{r}, 1)$ for some $\mathbf{r} \in \mathcal{E}$.

ii) The condition $[\dot{\mathbf{F}}, \dot{\mathbf{F}}] = -1$ gives a choice for H: $H = \pm\{-(\dot{\mathbf{r}}, \dot{\mathbf{r}})\}^{-1/2}$.

iii) The condition $[\ddot{\mathbf{F}}, \ddot{\mathbf{F}}] = 1$ (it is essentially (3.3)) then states that τ is proper time. It can be rewritten (with the help of i) and ii)) as

$$2\ddot{H} = H - H^3(\ddot{\mathbf{r}}, \ddot{\mathbf{r}}).$$

Any observer worldline has the form

$$\mathbf{R} \propto \mathbf{F}(\tau) - \mathbf{F}(0) + \sinh\tau\, \dot{\mathbf{F}}(0) + (\cosh\tau - 1)\ddot{\mathbf{F}}(0), \qquad (3.4)$$

(in *any* coordinate system) and satisfies $\mathbf{F}(\tau) - \ddot{\mathbf{F}}(\tau) = \mathbf{F}(0) - \ddot{\mathbf{F}}(0) =$ = constant. The observer's infinity is thus given by $[\mathbf{R},\bar{\mathbf{N}}] = 0$ for $\bar{\mathbf{N}} \propto$ $\propto \mathbf{F}(0) - \ddot{\mathbf{F}}(0)$. (cf. (3.2). Note that $\mathbf{F}(0) - \ddot{\mathbf{F}}(0) \propto \mathbf{N}$ iff $\ddot{\mathbf{r}} = \mathbf{r}$, i.e., the observer has the same infinity as the coordinate system iff his worldline satisfies $\ddot{\mathbf{r}} = \mathbf{r}$ in this system.) For an arbitrary timelike curve in $\mathcal{S}$, $\mathbf{F}(\tau) - \ddot{\mathbf{F}}(\tau)$ is not constant: it can thus be considered to give the *instantaneous infinity* at time τ along the curve, i.e., the infinity of the coordinate system whose observer worldline best approximates the curve at that point.

4. DISCUSSION

Our spacetime model shares some of the problems of the model in [3]; fortunately, it also shares some of the possible solutions to these problems.

A direct calculation shows that formula (3.3) is invariant under linear fractional transformations, of the form

$$\tau \rightarrow (\gamma\tau + \delta)^{-1}(\alpha\tau + \beta), \qquad \alpha\delta - \beta\gamma \neq 0. \qquad (4.1)$$

for constants α, β, γ, and δ determined up to proportionality. This means that (unlike in ordinary de Sitter spacetime) *three* choices of initial/boundry conditions are necessary to specify τ uniquely along a timelike curve. We could, for instance, choose an "infinity" (i.e. choose the points with $\tau \rightarrow \pm\infty$), an origin (the point with $\tau = 0$) and a unit of measurement (the point with $\tau = 1$). Without additional assumptions, this appears to be physically unreasonable: time measured by real-world clocks is determined up to an arbitrary origin and an arbitrary unit only.

Another problem arises with the initial conditions for observer worldlines: if these worldlines are to take over the role of the timelike geodesics of ordinary de Sitter spacetime (as representing "free" observers), they should be be determined uniquely by the initial position and velocity of the observer, or equivalently, by $\mathbf{r}(0)$ and $\dot{\mathbf{r}}(0)$ subject to $(\mathbf{r}(0),\mathbf{r}(0)) = 1$, $(\mathbf{r}(0),\dot{\mathbf{r}}(0)) = 0$ and $(\dot{\mathbf{r}}(0),\dot{\mathbf{r}}(0)) = -1$. But *our* observer worldlines (given by 3.4) are subject to (3.1) at $\tau = 0$. This is equivalent to specifying $\mathbf{r}(0)$, $\dot{\mathbf{r}}(0)$ and $\ddot{\mathbf{r}}(0)$ subject to $(\mathbf{r}(0),\mathbf{r}(0)) = 1$, $(\mathbf{r}(0),\dot{\mathbf{r}}(0)) = 0$ and $(\mathbf{r}(0),\ddot{\mathbf{r}}(0)) + (\dot{\mathbf{r}}(0),\dot{\mathbf{r}}(0)) = 0$. (The remaining conditions of (3.1) then define appropriate initial conditions for H.) For compatibility, then, some condition specifying $\ddot{r}(0)$ must be found.

One obvious solution to both problems is to assume that all allowable observers are *co-infinite*, (i.e., that they share the same infinity). From §3, this would imply that $\ddot{r}(\tau) = r(\tau)$, and since the linear fractional transformation (4.1) would then map the values $\pm\infty$ into themselves, it would take the form of an *affine* transformation:

$$\tau \to \delta^{-1}(\alpha\tau + \beta), \qquad \alpha\delta^{-1} \neq 0,$$

which specifies τ up to a change in origin and unit of measurement. This assumption gives us nothing new, however; it brings us back again to ordinary de Sitter spacetime.

It would be more useful to assume that all allowable observers are *locally* co-infinite, i.e., to restrict our coordinate transformations to those which map the infinities of neighbouring observers into "neighbouring infinities" (assuming that some suitable definition of "neighbouring" can be found). We would then have $\ddot{r}(\tau)$ *approximately* equal to $r(\tau)$, and proper times along neighbouring worldlines would be related *approximately* by affine transformations. (This is a reasonable situation: physical clock consistency has been verified only for nearby clocks, and may not even be meaningful for more distant ones.)

We plan to describe such a model in a future article.

REFERENCES

[1] Coxeter, H. S. M., *A Geometrical Background for de Sitter's World,* Amer. Math. Monthly **50** (1943), 217 - 228

[2] Lester, J. A., *A Physical Characterization of Conformal Transformations of Minkowski Spacetime.* Ann. Discrete Math. **18** (1983) 567 - 574

[3] Lester, J. A., *Conformal Minkowski Spacetime I: Relative Infinity and Proper Time.* Il Nuovo Cimento **72B** (1982) 261 - 272

[4] Lester, J. A., *Conformal Minkowski Spacetime II: A Cosmological Model.* Il Nuovo Cimento **73B** (1983) 139 - 149

[5] Lester, J. A., *Separation Preserving Transformations of de Sitter Spacetime.* Abh. Math. Sem. Univ. Hamburg **53** (1983) 217 - 224

Annals of Discrete Mathematics 37 (1988) 265–276

AFFINE HJELMSLEV RINGS AND PLANES

J.W. (Michael) Lorimer*
University of Toronto

INTRODUCTION

If L is a local ring (with unique maximal right ideal J, the Jacobson radical) then we can construct the *affine* and *projective Klingenberg planes* over L, denoted by $A(L)$ and $P(L)$ respectively ([4]). Such planes can have two points with either no joining line, exactly one joining line or many joining lines. If H is a *right chain ring* (any two right ideals are comparable) and every non-unit is a left and right zero-divisor, then H is an *affine Hjelmslev ring* (AH-ring), and $A(H)$ is the *AH-plane* over H, where two points now always have at least one joining line. If H is also a left chain ring, then H is a *H-ring* and $P(H)$ is the *PH-plane* over H.

There are examples of AH-rings which are not H-rings ([10]). In this note, we study AH-planes $A(H)$, of height n, or equivalently AH-rings with nilpotent radical (3.4), and obtain characterizations of such rings analogous to Törner's results [14] for H-rings (2.3). We next prove that an AH-ring with nilpotent radical is an H-ring if and only if the radical J is a principal left ideal (3.9); and then describe a geometric condition equivalent to the property that J is left principal (3.10). We end, using results of Bacon [4], by showing that an AH-plane $A(L)$ of height n is also of level n.

1. Notation, Definitions and Preliminary Results

All rings R have a unit element, and $J(R) = J$ is the *(Jacobson) radical.* P is a *prime ideal* of a ring R if $aRb \subseteq P$ implies that $a \in P$ or $b \in P$. An element $x \in R$ is a *left(right) zerodivisor* if there exists an element $y \neq 0$ in R so that $xy = 0 (yx = 0)$.

Definition 1.1.

(a) A ring R is *left(right) invariant* if all left(right) ideals are ideals or equivalently $aR \subseteq Ra$ $(Ra \subseteq aR)$ for all $a \in R$. R is *invariant,* if it is both left and right invariant.

(b) A ring R is a *right(left) chain ring* if the lattice of right(left) ideals is totally ordered. R is a *chain ring* if it is both a right and left chain ring.

*The author wishes to acknowledge the hospitality of the University of Tübingen as well as the financial support of both the Alexander von Humboldt Foundation and the national science and research council of Canada.

(c) A chain ring with no zerodivisors is called a *valuation ring* ([11], [15]).

Remark 1.2. The valuation rings of skew fields, as defined in [13], are, in our terminology, invariant valuation rings.

Lemma 1.3. Let R be a ring. Then, R is a right chain ring if and only if any two principal right ideals are comparable.

Proof. Suppose any two principal right ideals are comparable. Let I_1, and I_2 be two right ideals, and suppose there exists $x_1 \in I_1 \setminus I_2$ and $x_2 \in I_2 \setminus I_1$. Then $x_1R \subseteq x_2R$ or $x_2R \subseteq x_1R$. Hence, $x_1 \in I_2$ or $x_2 \in I_1$, a contradiction.

Lemma 1.4. [9] Let R be a ring with radical J. The following statements are equivalent.

(a) R has a unique maximal left(right) ideal.

(b) The set of non-units forms an ideal distinct from R.

(c) The set of non-units is contained in an ideal distinct from R.

(d) For each element r, of R, r or $1-r$ is a left(right) inverse.

(e) R/J is a skew field.

A ring with one of the properties above is a *local ring*. In such a ring, $xy = 1 \Leftrightarrow yx = 1$; and the ideal in (b) and (c) is J.

Lemma 1.5. If L is a local ring and J is nilpotent, then every non-unit is a left and right zerodivisor.

Proof. Every non-unit, by Lemma 1.4, is nilpotent and hence a two-sided zerodivisor.

Lemma 1.6. If R is a right chain ring, then R is local and every finitely generated right ideal is principal.

The next result is due essentially to Törner [14].

Lemma 1.7. If R is a right(left) chain ring and $J = Ra$ (aR) then $J = aR$ (Ra).

Proof. Since J is an ideal, $aR \subseteq Ra$. Suppose $aR \subsetneq Ra$ and take $ra \notin aR$. Hence, $aR \subseteq (ra)R$, as R is a right chain ring, and so $a = ras$. If $s \notin J$, then by 1.4 and 1.6 s has an inverse and so $as^{-1} = ra \in aR$, a contradiction. Thus, $s \in J = Ra$ and so $s = ta$. Then, $a = rata$ or $(rat - 1)a = 0$. Since $rat \in J$, 1.4(d) yields $a = 0$, a contradiction.

Lemma 1.8. Let R be a right chain ring with $J \neq (0)$.

(a) If $J^2 \subsetneq J$, then $J = aR$ for all $a \in J \setminus J^2$.

(b) If $\bigcap_{n=1}^{\infty} J^n = (0)$, then $J^2 \subsetneq J$ and $J = aR$ for all $a \in J \setminus J^2$

Proof.

(a) Take $a \in J \setminus J^2$. Then, $J^2 \underset{+}{\subset} aR \subseteq J$. We show that $aR = J$. Suppose $b \in J \setminus aR$. Then, $aR \subseteq bR$ and so $a = bu$ where u is a unit or else $a \in J^2$. Hence, $b = au^{-1} \in aR$, a contradiction.

(b) If $J^2 = J$, then $J^n = J$ for all $n \geq 1$ and so $J = (0)$, a contradiction. The result now follows from (a).

Lemma 1.9. If R is a right invariant ring, and (0) is a prime ideal, then R has no zerodivisors.

Proof. (0) is prime if $aRb = (0)$ implies $a = 0$ or $b = 0$. Since R is right invariant $Rz \subseteq zR$ for all z. Hence, $xy = 0$ implies $xRy \subseteq x(yR) = xyR = (0)$ and so $x = 0$ or $y = 0$.

Proposition 1.10. A right noetherian right chain ring is right invariant [7, Lemma 1].

Proposition 1.11. Let R be a right noetherian right chain ring with $\bigcap_1^\infty J^n = (0)$. Then, either $J^n = (0)$ for some n and R is an AH-ring with exactly one prime ideal and $J = aR$ or R is a right invariant ring with no zerodivisors, $J = aR$ is not nilpotent, and (0) and J are the only proper prime ideals.

Proof. By 1.8 and 1.10 R is right invariant and $J = aR$. From the proof of [7, "theorem"] the right ideals are exactly those in the descending chain

$$R \supset J = aR \supset a^2R \supset \ldots \supset a^nR \supset \ldots \supset \bigcap_1^\infty J^n = (0).$$

If J is nilpotent, then R is an AH-ring by 1.5., and J is the only prime ideal by the "theorem" mentioned above. If $J^n \neq (0)$ for all n, then R has countably many right ideals and so by part (4) of the same "theorem" our result follows.

Lemma 1.12. Let R be a ring and $I_i (i = 1, 2)$ two ideals. If $I_i = a_iR$ $(i = 1, 2)$, then $I_1 \cdot I_2 = (a_1 a_2) R$.

Proof. Clearly, $(a_1 a_2)R \subseteq I_1 \cdot I_2$. If $x \in I_1 \cdot I_2$, then $x = \sum_{i=1}^n (a_1 s_i)(a_2 t_i) = \sum_{i=1}^n a_1[s_i a_2 t_i]$. Now, $s_i a_2 t_i \in R \cdot I_2 = I_2$ and so $s_i a_2 t_i = a_2 u_i$. Then, $x = \sum_{i=1}^n a_1[a_2 u_i] \in (a_1 a_2)R$.

Theorem 1.13. Let R be a local ring, $J = aR \neq (0)$ and $\bigcap_1^n J^n = (0)$. Then, every non-zero right ideal has the form J^i, and for each $x \in J \setminus J^2$, $x^k R = J^k$. Moreover, either J is nilpotent and R is an AH-ring or R is a right invariant right chain ring with no zerodivisors, J is not nilpotent, and (0) and J are the only proper prime ideals.

Proof. As in 1.8(b), $\bigcap_{1}^{\infty} J^n = (0)$ means $J^2 \subsetneq J$. Take $x \in J \setminus J^2$. Then, $x \in J = aR$ implies $x = au$ where u is a unit, and so $xR = aR = J$. By induction 1.12 yields $x^k R = J^k$. Now take I a non-zero right ideal. Then, $I \subseteq J$, and since $\bigcap_{1}^{\infty} J^n = (0)$ there is a smallest $k \geq 1$ so that $I \subseteq J^k$ but $I \nsubseteq J^{k+1}$. Take $b \in I \setminus J^{k+1}$. Then, $b = x^k \cdot u$ and u is a unit or else $b \in J^{k+1}$. Thus, $x^k = bu^{-1} \in I$, and so $J^k = x^k R \subseteq I \subseteq J^k$ or $J^k = I = x^k R$. Hence R is a right noetherian right chain ring and the theorem now follows from 1.11.

2. AH-rings with nilpotent radicals

Definition 2.1.

(a) A ring R is a *right(left) E-ring* if there is an ideal I so that all right(left) ideals have the form I^i. The smallest n so that $I^n = (0)$ is the *rank of* R.

(b) An *E-ring* is a right and left E-ring ([8]).

Example 2.2. Let K be a field, K' a proper subfield and $\sigma : K \to K'$ an automorphism. $R = K \times K$ has the usual vector addition and multiplication is given by $(a, b)(c,d) = (ac, a^\sigma d + bc)$. Then, the right ideals are $((0, 0)) \subset (0, 1)R = \{(0, b) : b \in K\} = J \subset R$ and $J^2 = J$. But, if $b \in K \setminus K'$, then $R(0, b) \nsubseteq R(0, b^\sigma) \nsubseteq R(0, b)$. Moreover, J is not a principal left ideal. For if $b = a^\sigma \in K'$, then $R(0, b) \subseteq \{0\} \times K' \subsetneq J$; and if $b \notin K'$, then $x^\sigma \cdot b \notin K'$ for all $x \in K$ and so again $R(0, b) = \{(0, x^\sigma \cdot b) : x \in K\} \subsetneq J$. Hence, R is a right E-ring of rank two and J is a principal right ideal; but R is not a left E-ring (or even a left chain ring) and J is not a left principal ideal.

This example shows that the assumption that J is principal is essential in 1.13.

We next obtain an affine version of Törner's characterizations of H-rings with nilpotent radicals (see [14]).

Theorem 2.3. Let R be a ring with $J \neq (0)$. The following statements are equivalent.

(1) R is a right E-ring.

(2) R is a right artinian right chain ring.

(3) R is a right noetherian AH-ring with $\bigcap^{\infty} J^n = (0)$.

(4) R is a right chain ring with exactly one prime ideal and $J = aR$.

(5) R is a local ring and $J = aR$ is nilpotent.

(6) R is a right chain ring with nilpotent radical.

(7) R is an AH-ring with nilpotent radical.

Proof. [(1) => (2)] is clear. A right artinian ring is right noetherian [9, page 69] with a nilpotent radical [12, page 120] and so 1.5 yields [(2) => (3)]. Since $J \neq (0)$, 1.8 and 1.11 immediately give [(3) => (4)].

(4) => (5). Clearly, R is local. $J \neq (0)$, means J is the only prime ideal and so is the prime radical. Thus, J is a nil ideal [12; 4.21] and so $a^n = 0$ for some n. By 1.13, $J^n = a^n R = (0)$. [(5) => (6)] is immediate from 1.13 and 1.5 and [(6) => (7)] follows from 1.5 and 1.6.

(7) => (1). By 1.8, $J = aR$. Then, 1.13 yields (1).

It is not necessarily true that a quotient ring of an AH-ring is an AH-ring. However, for right E-rings this is the case, as the next result shows.

Theorem 2.4. Let R be a right E-ring of rank n (n > 1). Then, for all $a \in J \setminus J^2$, $J^k = a^k R$ and $J^{k+1} = a^k J = a^{k+1} R$ $(1 \leq k \leq n)$. Also, R/J^i is a right E-ring of rank i, $1 \leq i \leq n$.

Proof. Take $a \in J \setminus J^2$. By 1.13, $J^k = a^k R$. Thus, $a^k J = a^k(aR) = a^{k+1}R = J^{k+1}$. Finally, R/J^i is local with radical J/J^i. The right ideals of R/J^i are then K/J^i where K is a right ideal of R and $J^i \subseteq K$. Hence, $K = J^k$ $(1 \leq i \leq k)$ and so $K/J^i = J^k/J^i = (J/J^i)^k$. But $(J/J^i)^i = J^i/J^i = (0)$ and so R/J^i is a right E-ring of rank i.

3. Desarguesian AH-planes of Height n

Let L be a local ring with radical J. *The AK-plane over* L, $A(L) = \langle \mathbf{P}, \mathbf{L}, \|, \sim, \in \rangle$ is defined as follows ([5]):

(i) The *point set* is $\mathbf{P} = L \times L$.

(ii) The *line set*, $\mathbf{L}$, consists of the linear equations $xa + yb + c = 0$ where $(a,b) \notin J \times J$. Thus, there are *two types* of lines; $[m, n] = \{(x,y) \in L \times L : y = xm + n\}$ with $(m,n) \in L \times L$ and *slope* m; *and* $[u, v]' = \{(x,y) \in L \times L : x = yu + v\}$ with $(u,v) \in J \times L$ and *slope* u.

(iii) Two lines α,β are *parallel* $(\alpha \parallel \beta)$ if they are of the same type and have the same slope.

(iv) Two points $P = (p_1, p_2)$, $Q = (q_1, q_2)$ are *neighbours* $(P \sim Q)$ if $p_1 - q_1, p_2 - q_2 \in J$; and two lines α,β are *neighbours* if they are of the same type, say $\alpha = [m_1, n_1]$, $\beta = [m_2, n_2]$, and $m_1 - m_2, n_1 - n_2 \in J$.

Then, $A(L)$ has the following properties: $\sim$ and $\parallel$ are equivalence relations and

(AK1) If $P \nsim Q$, then there is a unique line $P \vee Q$ through P and Q, for all points P,Q.

(AK2) If $P \in \alpha,\beta$ and $\alpha \nsim \beta$, then P is the unique point incident with both α and β for all lines α,β. We write $P = \alpha \wedge \beta$.

(AK3) Given a point P and a line α there is a unique line β parallel to α and containing P.

(AK4) There exists an affine plane A^* and an incidence structure epimorphism $\pi : A(L) \to A^*$ with the properties:

(i) $P \sim Q \Leftrightarrow \pi(P) = \pi(Q)$ for all points P, Q.

(ii) $\alpha \sim \beta \Leftrightarrow \pi(\alpha) = \pi(\beta)$ for all lines α,β.

(iii) If $\alpha \wedge \beta = \phi$, then $\pi(\alpha) \parallel \pi(\beta)$ for all lines α,β

In our particular case, A^* is the affine plane over the skew field L/J and π is the map induced by the ring homomorphism $v : L \to L/J$. Any incidence structure $A = \langle \mathbf{P}, \mathbf{L}, \sim, \parallel, \in \rangle$ with the four properties above is an *AK-plane*. Now, for two points P, Q in an AK-plane, [P,Q] denotes the number of lines through P and Q ; and for two lines $\alpha,\beta,[\alpha,\beta]$ is the number of points lying on both α and β. The following chart illustrates the interaction between the geometric properties of $A(L)$ and the algebraic properties of L ([4] , [15]).

	Properties of *A(L)*		**Properties of** *L*
1.	For any two points P,Q : $[P,Q] \geq 1$	1.	L is a right chain ring.
2.	For any two points P,Q : $[P,Q] = 1 \Rightarrow P \not\sim Q$.	2.	Every non-unit of L is a right zerodivisor.
3.	For any two lines α,β and a point $P \in \alpha \cap \beta : [\alpha,\beta] = 1 \Rightarrow \alpha \not\sim \beta$.	3.	Every non-unit of L is a left zerodivisor.

An AK-plane is an *AH-plane* if it satisfies the 3 conditions above. Notice then that L is an AH-ring.

By a *homomorphism* of AK-planes we mean an incidence structure homomorphism which preserves the neighbour and parallel relations.

Definitions 3.1. ([6]) A homomorphism of K-planes, $\psi : A \to A'$ is an *eumorphism* if and only if $P \sim Q \Leftrightarrow \psi(P) \sim \psi(Q)$ for all points P,Q and $\alpha \sim \beta \Leftrightarrow \psi(\alpha) \sim \psi(\beta)$ for all lines α,β. ψ is *homogeneous* if and only if $P \in \alpha,\beta$ for a point P and lines α,β, always implies $\psi^{-1}\psi(P) \cap \alpha = \psi^{-1}\psi(P) \cap \beta$ and dually.

Definition 3.2. ([6]) A sequence $\{\psi_i\}_1^{n-1}$ of homogeneous eumorphisms between AH-planes,

$$A = A_n \xrightarrow{\psi_{n-1}} A_{n-1} \to \cdots \xrightarrow{\psi_2} A_2 \xrightarrow{\psi_1} A_1 = A^*$$

is a *homogeneous H-resolution of length* n for A.

Remark 3.3. If $\{\psi_i\}^{n-1}$, $\{\lambda_i\}_1^{n-1}$ are two homogeneous H-resolutions for an AH-plane A, then n = m and there is a sequence of isomorphisms $\{\theta_i\}_1^n$ with $\theta_i : A_i \to A_i'$ and $\theta_i \circ \psi_i = \lambda_i\theta_{i+1}$. ([6; 14.1.13]).

We can then say, without ambiguity, that an AH-plane *is of height* n if and only if it possesses a homogeneous H-resolution of length n.

Theorem 3.4. Let L be a local ring and $A(L)$ the AK-plane over L. The following statements are equivalent.

(1) $A(L)$ is an AH-plane of height n.

(2) L is a right E-ring of rank n.

(3) L is an AH-ring with $J^n = (0)$.

Proof. (2) and (3) are equivalent by 2.3.

(3) => (1). By (2) the only ideals of L are the chain $(0) \underset{+}{\subset} J^{n-1} \underset{+}{\subset} \cdots \underset{+}{\subset} J \underset{+}{\subset} L$. Then, each quotient ring L/J^i is an AH-ring by 2.4 and the obvious ring homomorphisms $v_i : L/J^{i+1} \to L/J^i$ induce eumorphisms $\psi_i : A_{i+1} \to A_i$ where $A_i = A(L/J^i)$. By [6; 16.3] $\{\psi_i\}_1^{n-1}$ is a homogeneous H-resolution.

(1) => (3). If $A(L)$ is an AH-plane, then from our chart, L is an AH-ring. If $A(L)$ is of height n, then L possesses only n proper ideals [6; 14.1.13 and section 15] and hence, $J^m = (0)$ for some m. But, then L is a right E-ring of rank m and so m = n (2.3).

Now, for a local ring L, the *PK-plane over L, P(L),* has *points* $<xyz> = \{\lambda(x,y,z) : \lambda \notin J\}$ *(where* $(x,y,z) \notin J \times J \times J$*), lines* $[uvw] = \{(u, v, w)\lambda : \lambda \notin J\}$ *(where* $(u, v, w) \notin J \times J \times J$*) and incidence* given by: $<xyz>$ lies on $[uvw]$ <=> $xu + yv + zw = 0$. Moreover, *two points,* $<xyz>$ and $<x'y'z'>$ are *neighbours* <=> there is a $\lambda \notin J$ so that $(x',y',z') - \lambda(x,y,z) \in J \times J \times J$; and dually for lines.

Then, $P(L)$ satisfies the following conditions: the neighbour relations ~ (on points and lines) are equivalence relations and

(PK1) If $P \vdash Q$, then there is a unique line $P \vee Q$ through P and Q for any two points P,Q.

(PK2) If $\alpha \vdash \beta$, then there is a unique point $\alpha \wedge \beta$, lying on both α and β, for any two lines α,β.

(PK3) There exists a projective plane P^* and an epimorphism $\pi : P(L) \to P^*$ with the properties:

(i) $P \sim Q$ <=> $\pi(P) = \pi(Q)$ for all points P,Q.

(ii) $\alpha \sim \beta$ <=> $\pi(\alpha) = \pi(\beta)$ for all lines α,β.

For the rest of this paper, L *is always (at least) a local ring, and A(L) and P(L) are the AK- and PK-planes respectively over* L*.* A K-plane is *Desarguesian* if it is isomorphic to a K-plane over some local ring L. $A(L)$ or $P(L)$ is *proper* if $J \neq (0)$.

If P is a PK-plane, and g is any line, let $A(P,g)$ be the corresponding *derived AK-plane* of P ([5]). Recall that the points of $A(P,g)$ are the points $P \vdash g$ (i.e. P is not a neighbour to any point of g), the lines, are the lines $h \vdash g$; and two lines α,β are parallel exactly when $\alpha \wedge g = \beta \wedge g$. Then, if $P = P(L)$, $A(L)$ is clearly isomorphic to $A(P,[001])$, and we identify them from now on without explicit mention.

Now, as evinced by example 2.2, if $A(L)$ is an AH-plane, it does not follow that $P(L)$ is a PH-plane. *The rest of this paper is concerned with finding algebraic conditions on* L *and* *geometric conditions on $A(L)$ so that $P(L)$ is a PH-plane.* In future papers we will consider this question again, first for compact H-planes, and then later for locally compact (non-compact) H-planes.

Definition 3.5. An AH-plane has a *projective parallelism* if and only if for any two lines α,β : $\alpha \sim \beta$ and $\alpha \nparallel \beta$ implies $[\alpha,\beta] \neq 0$.

In [3] Bacon called an AH-plane *projectively uniform* if it was uniform [height 2] with a projective parallelism. For Desarguesian AH-planes, $A(L)$, we next observe that the existence of a projective parallelism already implies that the plane is uniform [i.e. $J^2 = (0)$].

Proposition 3.6. Let $A(L)$ be a proper AH-plane. The following are equivalent.

(a) $A(L)$ has a projective parallelism.

(b) $J = La$ for all $a \neq 0$, $a \in J$.

(c) L is a H-ring with $J^2 = (0)$.

(d) $P(L)$ is a uniform PH-plane.

Proof. (c) is equivalent to (d) by [1,2]. We show
(a) => (b) => (c) => (a).

(a) => (b). Take $a \neq 0$, $a \in J$. Then, $La \subseteq J$. Now, if $b \in J$, then $[a, -b] \sim [O, O]$ and $[a, -b] \nparallel [O, O]$. Hence, there exists (x,y) lying on both $[a, -b]$ and $[O, O]$ and so $O = y = xa - b$ or $b = xa \in La$. Hence, $J = La$.

(b) => (c). We first show that L is a left chain ring. Take, Lx, Ly. Then, $y \in J \setminus \{0\}$ implies $Ly = J$ and so $Lx \subseteq Ly$ if $x \in J$; and $Ly \subseteq Lx = L$ if $x \notin J$. Now suppose $J^2 \neq (0)$. Take, $b \in J^2 \setminus \{0\}$. Then, $Lb \subseteq J^2 \subseteq J$ as J^2 is an ideal. But, by (b), $Lb = J$ and so $J^2 = J$. By 1.12, $J^2 = Lb^2 = Lb$. Hence, $b = xb^2$ or $(xb - 1)b = 0$. Then, $xb - 1$ is a unit and so $b = 0$, a contradiction. Hence, $J^2 = (0)$ and (c) follows from 1.5.

(c) => (a). By 2.4, and 1.8, $J = Lx$ for all $x \in J \setminus \{0\}$. Now, without loss of generality take two lines $[a, b] \sim [c, d]$ with $[a, b] \nparallel [c, d]$. Then, $a - c \neq 0$ and $a - c, d - b \in J$. Hence, $d - b \in L(a - c) = J$ and so $d - b = x(a - c)$ for some x in L. Clearly, if $y = xa + b$ then (x, y) lies on both $[a, b]$ and $[c, d]$.

In view of 2.4, to obtain a result analogous to 3.6 for AH-planes of height n, we require a geometric statement equivalent to the algebraic statement: $J = La$ for all $a \in J \setminus J^2$. We do this next.

Now, if A is a H-plane of height n with homogeneous H-resolution $\{\psi_i\}_{i=1}^{n-1}$, then let $\chi_i = \psi_i \circ \psi_{i+1} \circ \dots \circ \psi_{n-1} : A \to A_i$ $(1 \leq i \leq n-1)$. We then have,

Proposition 3.7. Let P be a PH-plane of height n. Then, each derived AH plane, $A(P, g)$ is an AH-plane of height n with the property: for any two lines α, β if $\alpha \sim \beta$ and $\chi_2(\alpha) \nparallel \chi_2(\beta)$ then there exists a point P lying on both α and β.

Proof. Let P have homogeneous, H-resolution $\{\psi_i\}_1^{n-1}$. If γ is any line, put $A = A(P, \gamma)$, and $A_i = A(P_i, \chi_i(\gamma))$. Then, $\{\psi_i\}_1^{n-1}$ easily induces a homogeneous H-resolution $A \to A_{n-1} \to \ldots \to A_2 \to A^*$. We write ψ_i also for the induced map $A_{i+1} \to A_i$. Now, take α, β two (derived) lines in A so that $\alpha \sim \beta$ and $\chi_2(\alpha) \nparallel \chi_2(\beta)$. If $n = 2$, then χ_2 is just the identity map and the result follows from [3, 1.3]. We proceed by induction. Suppose our claim is true for PH-planes of height k, $k < n$. Then, P_{n-1} is of height $n-1$ with resolution $\{\psi_i\}_1^{n-2}$ and $\hat{\chi}_2 = \psi_2 \circ \ldots \circ \psi_{n-2} : P_{n-1} \to P_2$. Then, $\psi_{n-1}(\alpha)$ and $\psi_{n-1}(\beta)$ are neighbours in P_{n-1} and $\chi_2(\alpha) = \hat{\chi}_2(\psi_{n-1}(\alpha))$ is not parallel to $\hat{\chi}_2(\psi_{n-1}(\beta)) = \chi_2(\beta)$. By the induction hypothesis there is a point $\psi_{n-1}(P)$ in $A(P_{n-1}, \psi_{n-1}(\gamma))$ lying on both $\psi_{n-1}(\alpha)$ and $\psi_{n-1}(\beta)$. Now, we show that there is a point in $A(P, \gamma)$ lying on both α and β.

Suppose not. Then, there is a point P_∞ in $P \setminus A(P, \gamma)$ so that P_∞ lies on α, β and $P_\infty \sim \gamma$. Now $\chi_2(\alpha) \nparallel \chi_2(\beta)$ in $A(P_2, \chi_2(\gamma))$ and so $\chi_2(\alpha)$ and $\chi_2(\beta)$ are distinct lines in P_2. Then, $\chi_2(P_\infty)$, $\chi_2(P)$ both lie on $\chi_2(\alpha)$ and $\chi_2(\beta)$ and so $\chi_2(P_\infty) \sim \chi_2(P)$. Hence, $P_\infty \sim P$, and so $\psi_{n-1}(P) \sim \psi_{n-1}(\gamma)$ or equivalently $P \sim \gamma$, a contradiction.

Proposition 3.8. Let L be a local ring with radical $J \neq J^2$ and $\chi_2 : A(L) \to A(L/J^2)$ the eumorphism induced by the natural homorphism from L to L/J^2. Then, the following statements are equivalent:

(i) $J = La$ for all $a \in J \setminus J^2$.

(ii). For any two lines α, β in $A(L)$:

if $\alpha \sim \beta$ and $\chi_2(\alpha) \nparallel \chi_2(\beta)$, then there is a point P lying on both α and β.

Proof. (i) => (ii). Take $\alpha \sim \beta$ and $\chi_2(\alpha) \nparallel \chi_2(\beta)$. Without loss of generality, $\alpha = [m, n]$, $\beta = [u, v]$ where $m - u, v - n \in J$ and $m + J^2 \neq u + J^2$. Hence $m - u \in J \setminus J^2$ and so $v - n \in L(m - u)$ or $v - n = x(m - u)$ for some $x \in L$. Then. with $y = xm + n$, (x, y) lies on both α and β.

(ii) => (i). Let $a \in J \setminus J^2$. Then, $La \subseteq J$. Take $b \in J$. Then, $[a, -b] \sim [0, 0]$ and $\chi_2([a, -b]) = [a + J^2, -b + J^2] \nparallel [J^2, J^2] = \chi_2([0, 0])$. Hence there is a point (x, y) lying on both $[a, -b]$ and $[0, 0]$ and so $0 = xa - b$ or $b = xa \in La$. Thus, $J = La$.

Theorem 3.9. Let R be a ring with radical J. R is an E-ring if an only if R is a local ring and $J = aR = Rb$ is nilpotent.

Proof. This is now immediate from 2.3.

Corollary. Let H be an AH-ring with $J^n = (0)$. The following are equivalent.

(a) H is a H-ring.

(b) $J = Ha$ for some $a \in J \setminus \{0\}$.

(c) $P(H)$ is a PH-plane.

Next, we present equivalent conditions for an AH-plane, $A(L)$, of height n to have a PH-projective extension $P(L)$.

Theorem 3.10. Let $A(L)$ be a proper AK-plane. The following statements are equivalent:

(i) L is an E-ring.

(ii) L is a local ring and $J = aL = Lb$ is nilpotent.

(iii) L is a right E-ring and $J = La$.

(iv) $A(L)$ is an AH-plane of height n and for any two lines α, β : if $\alpha \sim \beta$ and $\chi_2(\alpha) \nparallel \chi_2(\beta)$ then there exists a point P lying on both α and β.

(v) $P(L)$ is a PH-plane of height n.

Proof. The first three conditions are equivalent by 3.9 and 2.3 and (i) is equivalent to (v) by [1, 2]. By 3.7, (v) => (iv). Finally 3.4 and 3.8 show that (iv) implies (iii).

Next, we recall that an AH-plane of height n with homogeneous H-resolution $\{\psi_i\}_1^{n-1}$ satisfies the *axiom of reciprocal segments* when the following condition holds [(6)]: (RS). For a point P lying on lines α, β : $\chi_i(\alpha) = \chi_i(\beta)$ but $\chi_{i+1}(\alpha) \neq \chi_{i+1}(\beta)$ <=> $\alpha \cap \beta = \chi_{n-i}^{-1}\chi_{n-i}(P) \cap \alpha$ for all i, $0 \leq i \leq n$.

By [6; 16.4.2], an AK-plane, $A(L)$ with ideals $(0) = I_n \subsetneq I_{n-1} \subsetneq \dots \subsetneq I_1 = J$ satisfies; (RS) <=> for all i, $1 \leq i \leq n$: $x \in L$, $y \in I_i \setminus I_{i+1}$ => [$x \in I_{n-i}$ <=> $xy = 0$]. For a right E-ring of rank n, the ideals are $(0) = J^n \subsetneq J^{n-1} \subsetneq \dots \subsetneq J$, and these clearly satisfy the preceding condition.

Now, an AH-plane of height n, A, with homogeneous H-resolution $\{\psi_i\}^{n-1}$ is *of level* n if and only if each A_i ($A_{i+1} \xrightarrow{\psi_i} A_i$) satisfies (RS).

We end by stating the, now obvious

Theorem 3.10. Every Desarguesian AH-plane of height n is of level n.

References

[1] B. Artmann, Desarguessche Hjelmslev-Ebenen n-ter Stufe, *Mitt. Math. Sem. Giessen* **91** (1971) 1-19.

[2] B. Artmann, Geometric aspects of primary lattices, *Pacific J. Math.* **43** (1972) 15-25.

[3] P.Y. Bacon, On extensions of projectively uniform affine Hjelmslev planes, *Abh. Math. Sem. Univ. Hamburg* 41 (1974), 185-189.

[4] P.Y. Bacon, Desarguesian Klingenberg planes, *Trams. Amer. Math. Soc.* **241** (1978) 343-355.

[5] P.Y. Bacon, An Introduction to Klingenberg Planes, Vol. I, published by the author, Gainsville (1976).

[6] P.Y. Bacon, An Introduction to Klingenberg Planes, Vol. II, Vol. III, published by the author, Gainsville (1979).

[7] H.H. Brungs, Generalized discrete valuation rings, *Canad. J. Math.* **21** (1969) 1404-1408.

[8] B. Jonsson and G.S. Monk, Representations of primary arguesian lattices, *Pacific J. Math.* **30** (1969) 95-139.

[9] J. Lambek, Lectures on rings and modules, London Blaisdell Publ. Comp. (1966).

[10] J.W. Lorimer and N.D. Lane, Desarguesian affine Hjelmslev planes, *Journal für die reine und angewandte Mathematik,* Band 278/279, 336-352 (1976).

[11] K. Mathiak, Berwertungen nicht kommutativer Körper, *J. Algebra* **48** (1977) 217-235.

[12] Neal H. McCoy, The theory of ring, Chelsea Publ. Company Bronx N.Y. (1972).

[13] O.F. Shilling, The theory of valuations, New York Amer. Math. Soc., Math. Surveys IV (1950).

[14] G. Törner, Eine Klassifizierung von Hjelmslev-Ringen und Hjelmslev-Ebenen, *Mitt. Math. Sem. Giessen Heft* **107** (1974).

[15] F.D. Veldkamp, Projective Ring Planes: some special cases, Rendiconti del Seminaro Matematico di Brescia Vol. 7, Atti del Convegno Geometria combinatoria e di indidenza fondamenti e appliazioni pub. della Univ. Cattolina Milano, 609-615 (1984).

Annals of Discrete Mathematics 37 (1988) 277–280

IRREDUCIBLE REPRESENTATIONS OF HECKE ALGEBRAS OF RANK 2 GEOMETRIES

Stefan LÖWE

Institut für Geometrie
TU Braunschweig
Pockelsstr. 14
3300 Braunschweig, BRD

We give a constructive proof that the irreducible representations of the Hecke algebra $\mathbb{H}$ of a rank 2 geometry are at most two-dimensional. Furthermore, we give a primitive decomposition of unity of $\mathbb{H}$.

1. DEFINITIONS

Let $G = (P,L,\mathbb{F})$ be a finite regular geometry with parameters q_0 , q_1 , i.e. through any point there are $q_0 + 1$ lines, on any line there are $q_1 + 1$ points. The vector space $V_G := \{f \mid f: \mathbb{F} \longrightarrow \mathbb{C}\}$ is called *standard module* of G. The elements of $\mathbb{F}$ are called *flags*. We call two flags (p,l), (p',l') *o-adjacent* (*1-adjacent*) if they are different and p = p' (l = l'), and denote this relation by $\underset{0}{\sim}$ ($\underset{1}{\sim}$). Let $\sigma_i: V_G \to V_G$, $(\sigma_i f)(\phi) := \sum_{\zeta \underset{i}{\sim} \phi} f(\zeta)$; i = 0, 1; and $\mathbb{H}(G)$ $:= \langle 1,\sigma_1,\sigma_0\rangle \leqslant \mathrm{End}(V)$. $\mathbb{H}(G)$ is called *Hecke algebra* of G. We obtain at once the quadratic relations $\sigma_i^2 = (q_i - 1)\sigma_i + q_i 1$. Usually there are more relations for σ_0, σ_1, which reflect the structure of the geometry G. For example, if G is a projective plane then $\sigma_0\sigma_1\sigma_0 = \sigma_1\sigma_0\sigma_1$.

2. IRREDUCIBLE REPRESENTATIONS OF $\mathbb{H}(G)$

Let $\mathbb{K}$ be the unitary algebra generated by s_0, s_1, and defining relations $s_i^2 = (q_i-1)s_i + q_i 1$; i = 0, 1, and q_0, q_1 as above. For every geometry G with parameters q_0, q_1 there is a natural projection $\pi_G: \mathbb{K} \longrightarrow \mathbb{H}(G)$, induced by $s_i \mapsto \sigma_i$.

We first determine the irreducible representations of $\mathbb{K}$.

2.1. THEOREM (Ott)

Up to equivalence there are the following irreducible representations of $\mathbb{K}$:

4 one-dimensional:

	χ_1	χ_2	χ_3	χ_4
s_0	-1	-1	q_0	q_0
s_1	-1	q_1	-1	q_1

χ_1 *is usually called Steinberg character,* χ_2 *index character.*

A series R_λ, $\lambda \in \mathbb{C} \setminus \{0,1,-q_o,-q_1,q_oq_1\}$ *of two-dimensional representations*

$$R_\lambda(s_o) = \begin{pmatrix} 0 & \lambda \\ q_o\lambda^{-1} & q_o-1 \end{pmatrix} \qquad R_\lambda(s_1) = \begin{pmatrix} q_1-1 & q_1 \\ 1 & 0 \end{pmatrix}$$

where $R_\lambda \equiv R_\mu$ *iff* $\lambda\mu = q_oq_1$.

PROOF. We only have to show that a finite dimensional irreducible representation π of $\mathbb{K}$ which is not one-dimensional is equivalent to one of the R_λ. Let v be an eigenvector of $\pi(s_os_1)$ with eigenvalue λ. We claim that $W := \langle v,\pi(s_1)v\rangle$ is a two-dimensional subspace. Obviously W is invariant under $\pi(s_1)$. We also have $\pi(s_o)\pi(s_1)v = \lambda v$. Finally $\pi(s_o)v = {}^1/_\lambda\pi(s_o)\pi(s_os_1)v = {}^1/_\lambda((q_o-1)\pi(s_o) + q_o1)\cdot\pi(s_1)v = (q_o-1)v + {}^1/_\lambda\, q_o\pi(s_1)v \in W$. Thus W is a $\mathbb{K}$-module. We finish the proof by showing that $\pi \equiv R_\mu$, where $\mu = q_oq_1\lambda^{-1}$. For typographical convenience we write row vectors instead of column vectors we had to use.

The vector (0,1) is eigenvector of $R_\mu(s_os_1)$ with eigenvalue λ. We define a linear map $\psi: W \longrightarrow \mathbb{C}^2$ by $v \mapsto (0,1)$, $\pi(s_1)v \longmapsto (q_1,0)$ and linear continuation. To show that ψ is the wanted interwining map we only have to show that $R_\mu(s_i)\psi = \psi\pi(s_i)$; i = 0, 1.

$$R_\mu(s_o)\psi v = (\mu,q_o-1) = (q_o-1)\psi v + {}^{q_o}/_\lambda\psi\pi(s_1)v = \psi\pi(s_o)v.$$
$$R_\mu(s_o)\psi\pi(s_1)v = (0,q_oq_1\mu^{-1}) = (0,\lambda) = \lambda\psi v = \pi(s_o)\pi(s_1)v.$$
$$R_\mu(s_1)\psi v = (q_1,0) = \psi\pi(s_1)v.$$
$$R_\mu(s_1)\psi\pi(s_1)v = q_1(q_1-1,1) = (q_1-1)(q_1,0) + q_1(0,1) =$$
$$\psi((q_1-1)\pi(s_1) + q_11)v = \psi\pi(s_1)\pi(s_1)v.$$ +++

As a consequence we obtain the following

2.2. Corollary

Let π *be a representation of* $\mathbb{H}(G)$. *To know the irreducible constituents of* π *it is sufficient to know the eigenvectors of* $\pi(s_os_1)$. +++

It is possible now to find all the irreducible representations of a Hecke algebra $\mathbb{H}(G)$.

2.3. Lemma

Let $\{B_k(s_o,s_1) : k \in I\}$ *be a generating set of* $ke(\pi_G)$. *Then* R_λ *yields an irreducible representation of* $\mathbb{H}(G)$ *iff*

$$\forall k \in I: B_k(R_\lambda(s_o),R_\lambda(s_1)) = 0.$$ +++

2.4. Example

Let G be a generalized n-gon. Then $ke(\pi_G)$ is generated by $(s_os_1)^{n/2} - (s_1s_o)^{n/2}$ if n is even, and $(s_os_1)^{(n-1)/2}s_o - (s_1s_o)^{(n-1)/2}s_1$ if n is odd. In both cases we obtain $\lambda(\varepsilon) = \sqrt{q_oq_1}\,\varepsilon$, ε a n-th root of unity.

3. AN APPLICATION

We use the theorem to determine an orthogonal primitive decomposition of unity in $\mathbb{H}$.

Let R be the left regular representation of a Hecke algebra $\mathbb{H}(G)$. Let $a_1, \ldots, a_m$ be a complete set of eigenvectors of $T = R(\sigma_0\sigma_1)$ with eigenvalues $\lambda_1, \ldots, \lambda_m$. Let $b_i := R(\sigma_1)a_i$; $1 \leq i \leq m$. Then $W_i := \langle a_i, b_i \rangle$ are the irreducible constituents of R, and $\mathbb{H}$ is the direct sum of $\{W_i\}_{1 \leq i \leq m}$. To obtain a decomposition of 1 explicitly , we proceed in three steps.

First step. It is clear that a decomposition $1 = \sum_{i=1}^{m} f_i$, $f_i \in W_i$ exists.

Second step. Suppose that W_i and W_j are not equivalent, and $y \in W_j$. Then $T(a_i y) = \lambda_i a_i y \in W_j$, which implies $a_i y = 0$ since W_j cannot contain an eigenvector of eigenvalue λ_i. Thus $W_i W_j = 0$.

Third step. Suppose $V_i := W_{k_i}$; $1 \leq i \leq n$; $n \geq 2$; is a maximal set of pairwise equivalent constituents of R belonging to the eigenvalue λ of T. Let $0 \neq c_i \in V_i$ such that $T(c_i) = \lambda c_i$; $d_i := R(\sigma_1)c_i$; $1 \leq i \leq n$. Then $V_i = \langle c_i, d_i \rangle$. Since $T(c_i c_j) = \lambda c_i c_j \in V_j$, and $T(c_i d_j) = \lambda c_i d_j \in V_j$, there exist $\mu_{ij}, \nu_{ij} \in \mathbb{C}$ such that $c_i c_j = \mu_{ij} c_j$, and $c_i d_j = \nu_{ij} c_j$. Consequently, $d_i c_j = \mu_{ij} d_j$, and $d_i d_j = \nu_{ij} d_j$. Let $e_i := f_{k_i}$; $1 \leq i \leq n$. Then there exist $x_i, y_i \in \mathbb{C}$ such that $e_i = x_i c_i + y_i d_i$; $1 \leq i \leq n$. We have $e_i e_j = x_i(x_j \mu_{ij} + y_j \nu_{ij})c_j + y_i(x_j \mu_{ij} + y_j \nu_{ij})d_j = \delta_{ij} e_i$ (Kronecker's δ). Hence x_i, y_i solve the linear system

$$(*) \quad x_i \mu_{ij} + y_i \nu_{ij} = \delta_{ij}.$$

On the other hand, any solution x_i', y_i' of (*) yields a complete orthogonal set of primitive idempotents $x_i' c_i + y_i' d_i$, and thus gives rise to a decomposition of unity.

REFERENCES

[1] Liebler, R.,Tactical Configurations and their Generic Ring, Preprint Colorado State University.

[2] Ott,U., Private communication.

Annals of Discrete Mathematics 37 (1988) 281–292

A CHARACTERIZATION OF PAPPIAN AFFINE HJELMSLEV PLANES

Helmut Mäurer and Wolfgang Nolte

In [7], Pickert showed that an affine plane is isomorphic to an affine plane over a field if the dual Pappus theorem holds for two fixed pencils of parallel lines (see also [2]). Here we ask whether a similar result is true for an affine Hjelmslev plane Ω. The geometric structure Ω will be defined in such a generality that the algebraic models of affine Hjelmslev planes over commutative local rings (not only those over Hjelmslev rings) are included. We consider two distinguished not-neighbouring pencils V,W of parallel lines and assume the validity of the configuration conditions (DP), (SD). The main result is the following: If the epimorphic image $\alpha(\Omega)$ of Ω is not a Fano plane then the incidence structure of Ω is isomorphic to that of an affine Hjelmslev plane $\Omega(R)$ over a local ring R. In R the element 2 is invertible. All pencils of parallel lines in Ω correspond to pencils of parallel lines in $\Omega(R)$ with the possible exception of those neighbouring V or W.

1. AFFINE HJELMSLEV PLANES

An *incidence structure* $\Omega := (\Pi, \Lambda, \in)$ consists of a set $\Pi \neq \emptyset$ of *points* and a set $\Lambda \neq \emptyset$ of non-empty subsets of Π.

An equivalence relation $\|$ on the set Λ of *lines* is called a *parallel relation* if each equivalence class is a partition of Π. Let $\Omega_\nu := (\Pi_\nu, \Lambda_\nu, \|_\nu, \in)$, $(\nu = 1,2)$ be incidence structures with parallel relations. A mapping $\alpha : \Pi_1 \longrightarrow \Pi_2$ is called a *homomorphism* if $g, h \in \Lambda_1$; $g \|_1 h$ implies $\alpha(g), \alpha(h) \in \Lambda_2$ and $\alpha(g) \|_2 \alpha(h)$. A pair $(\Omega_1, \alpha : \Omega_1 \longrightarrow \Omega_2)$ - shortly denoted by (Ω_1, α) - is an affine *Hjelmslev plane* if α has the following properties:

> *α is a surjection, and $\alpha(\Omega_1) = \Omega_2$ is an affine plane; for $P,Q \in \Pi_1$ with $\alpha(P) \neq \alpha(Q)$ there is a unique line $P \vee Q$ containing P,Q; for $g, h \in \Lambda_1$ with $\alpha(g) \nparallel_2 \alpha(h)$ there is a unique point $g \wedge h$ contained in $g \cap h$.*

These properties of α canonically induce surjections $\Lambda_1 \longrightarrow \Lambda_2$ and $\Lambda_1/\|_1 \longrightarrow \Lambda_2/\|_2$, both denoted by α. The elements $x, y \in \Pi_1 \cup \Lambda_1$ are called *neighbouring* or *neighbour elements* if $\alpha(x) = \alpha(y)$ or if $\alpha(x)$ is incident with $\alpha(y)$; notation $x \sim y$. Similarly, for $B_1, B_2 \in {}^{\Lambda_1}/_{\|_1}$ we call B_1, B_2 *neighbouring*, and write $B_1 \sim B_2$ if $\alpha(B_1) = \alpha(B_2)$. Let $P \in \Pi_1$, $g \in \Lambda_1$. The pencil B_g is defined by $g \in B_g \in {}^{\Lambda_1}/_{\|_1}$, and the line $P \vee B_g$ by $P \in (P \vee B_g) \in B_g$.

EXAMPLE. The affine Hjelmslev plane $(\Omega(R), \alpha)$ *over a local ring.*
Let R be a commutative local ring, I its maximal ideal. Then $(\Omega(R), \alpha)$ is an affine Hjelmslev plane if we define:

$$\Omega(R) := (R^2, \Lambda, \|, \epsilon) ,$$
$$\Lambda := \{\varphi^{-1}(r) \mid r \in R, \varphi \in \mathrm{Hom}_R(R^2,R) \text{ with } \varphi(R^2) \not\subset I\} ,$$
$$\varphi_1^{-1}(r_1) \,\|\, \varphi_2^{-1}(r_2) \text{ if } R\varphi_1 = R\varphi_2 ,$$
$$\alpha : R^2 \longrightarrow ({}^R/_I)^2 \text{ where } (r_1,r_2) \longmapsto (r_1 + I, r_2 + I) .$$

2. THE CONFIGURATION THEOREMS (DP) AND (SD)

In the following $\Omega = (\Pi, \Lambda, \|, \epsilon)$ denotes an affine Hjelmslev plane with respect to a homomorphism α. We distinguish two non-neighbouring pencils $V, W \in {}^{\Lambda}/_{\|}$. V, W are not changed throughout this paper.

Notations: $\Lambda_0 := \{g \in \Lambda \mid \alpha(g) \not\subset \alpha(V) \cup \alpha(W)\}$,
$V_P' := \{v \in V \mid P \not\sim v\}$, $W_P' := \{w \in W \mid P \not\sim w\}$ for $P \in \Pi$,
$\|_0$ the restriction of $\|$ to the lines of $\Lambda_0 \cup V \cup W$,
$\Omega_0 := (\Pi, \Lambda_0 \cup V \cup W, \|_0, \epsilon)$.

The *dual Pappus theorem* for the pencils V, W is the following configuration condition:

(DP) *Let* P *be a point and let* $(w_2,v_3,w_1,v_2,w_3,v_1)$ *be a cyclically ordered set of distinct lines with* $v_i \in V$, $w_i \in W$ $(i = 1,2,3)$ *where at least three successive lines are non-neighbours of* P. *If two of the three sets* $M_i := \{P, v_j \wedge w_k, v_k \wedge w_j\}$ *with* $\{i,j,k\} = \{1,2,3\}$ *consist of collinear points then also the third one, or* $\alpha(M_1) \cup \alpha(M_2) \cup \alpha(M_3)$ *is collinear.*

There are several other names for the dual Pappus theorem in affine planes in the literature, such as the "Theorem of Pappus Brianchon" or the "Thomsen condition". The validity of (DP) in an affine Hjelmslev plane Ω implies the validity of (DP) in $\alpha(\Omega)$. By Pickert's result, $\alpha(\Omega)$ then is a pappian affine plane.

We remark that the assumptions of (DP) imply that $|\alpha(M_i)| > 1$ and that $M_i \cup M_j$ is not collinear for $i \neq j$.

The following configuration condition is a special case of the Desargues theorem.

(SD) *Let* $v_1,v_2,v_3 \in V$; $w_1,w_2,w_3 \in W$ *be six distinct lines with* $v_1 \nparallel v_2$ *and* $w_1 \nparallel w_2$, *further let* $g := (v_1 \wedge w_1) \vee (v_2 \wedge w_2)$, $h_3 := (v_1 \wedge w_2) \vee (v_2 \wedge w_1)$ *and* $h_2 := (v_1 \wedge w_3) \vee B_{h_3}$. *If* $v_3 \wedge w_3 \in g$ *then* $v_3 \wedge w_1 \in h_2$.

It is easy to deduce from (SD) that conversely under the same assumptions $v_3 \wedge w_1 \in h_2$ implies $v_3 \wedge g_3 \in g$.

DEFINITION. A quadruple (A,B,C,D) of four pairwise non-neighbouring points is called a (V,W)-*parallelogram* if $A \vee D,\ B \vee C \in V$ and $A \vee B,\ C \vee D \in W$. The Λ_0-lines $A \vee C,\ B \vee D$ are called the *diagonals* of the (V,W)-parallelogram.

(SD) implies:

LEMMA 1. *The* Λ_0*-lines* h_2,h_3 *are parallel if and only if there exist two* (V,W)*-parallelograms with a common point* A *and a common diagonal* g *through* A *such that* g,h_2 *and* g,h_3 *are the diagonals of the two parallelograms.*

By lemma 1 the $\parallel$-relation in Λ_0 is intrinsically determined by the incidence structure of Ω.

3. LINE REFLECTIONS

Throughout this paper we assume that $\alpha(\Omega)$ has order ≥ 3.
For each line $g \in \Lambda_0$ there exists a unique collineation σ_g of the incidence structure $(\Pi, V \cup W, \in)$ such that g is the set of fixed points of σ_g. σ_g has the order 2, it interchanges V and W and respects the neighbour relation.

LEMMA 2. *Assume* (DP). *Let* $g_1,g_2,g_3 \in \Lambda_0$ *be lines through a common point* P, *then the mapping* $\pi := (\sigma_{g_3}\sigma_{g_2}\sigma_{g_1})^2$ *fixes all lines of* $V'_P \cup W'_P$.

PROOF. If g_1,g_2,g_3 are different and if the three α-images of g_1,g_2,g_3 are not the same than we construct $v_i \in V'_P$, $w_i \in W'_P$ $(i = 1,2,3)$ in the following way: We start with $w_2 \in W'_P$ (resp. $v_2 \in V'_P$) and use the sequence

$$w_2 \xrightarrow[\sigma_{g_1}]{} v_3 \xrightarrow[\sigma_{g_2}]{} w_1 \xrightarrow[\sigma_{g_3}]{} v_2 \xrightarrow[\sigma_{g_1}]{} w_3 \xrightarrow[\sigma_{g_2}]{} v_1$$

in order to obtain elements $v_i \in V$ and $w_i \in W$ $(i = 1,2,3)$. Obviously the v_i and w_j are distinct lines. Using (DP) we get $v_1 \wedge w_2 \in g_3 = P \vee (w_1 \wedge v_2)$, and therefore $\sigma_{g_3}(v_1) = w_2$. This implies $\pi(w_2) = w_2$ (analogous $\pi(v_2) = v_2$). If g_1,g_2,g_3 are not different then $\pi = id$ is clear. In the case $\alpha(g_1) = \alpha(g_2) = \alpha(g_3)$ there exists a Λ_0-line g_4 through P with $\alpha(g_4) \neq \alpha(g_1)$. The first part of this proof yields

$$\sigma_{g_3}\sigma_{g_2}\sigma_{g_1}(w_2) = \sigma_{g_3}\sigma_{g_2}\sigma_{g_4}\sigma_{g_4}\sigma_{g_1}(w_2) = \sigma_{g_4}\sigma_{g_2}\sigma_{g_3}\sigma_{g_4}\sigma_{g_1}(w_2) =$$
$$\sigma_{g_4}\sigma_{g_2}\sigma_{g_1}\sigma_{g_4}\sigma_{g_3}(w_2) = \sigma_{g_4}\sigma_{g_4}\sigma_{g_1}\sigma_{g_2}\sigma_{g_3}(w_2) = \sigma_{g_1}\sigma_{g_2}\sigma_{g_3}(w_2) ,$$

hence $\pi(w_2) = w_2$ (and analogous $\pi(v_2) = v_2$) .

The following lemma is an immediate consequence of (SD).

LEMMA 3. *Assume* (SD).

(a) *Let* (A,B,C,D) *be a* (V,W)*-parallelogram with the diagonals* $g = A \vee C$, $h = B \vee D$. *Then we have* $g = \{\sigma_x(A) \mid x \in B_h\}$.

(b) *If* $h \in \Lambda_0$ *and* $A \nmid h$ *then* $A \vee \sigma_h(A) = \{\sigma_x(A) \mid x \in B_h\} \in \Lambda_0$.

LEMMA 4. *For two lines* $g_1, g_2 \in \Lambda_0$ *with a common point* P (DP) *implies: The image* $\sigma_{g_2}(g_1)$ *is a* Λ_0*-line, and* $\sigma_{g_1}\sigma_{g_2}\sigma_{g_1} = \sigma_{\sigma_{g_1}(g_2)}$.

PROOF. We choose points $X, A \in g_1 \setminus \{P\}$ with $\alpha(A) \neq \alpha(P), \alpha(X)$, and put $w_3 := A \vee W$, $w_2 := X \vee W$, $v_3 := X \vee V$, $v_2 := A \vee V$, $v_1 := \sigma_{g_2}(w_3)$, $w_1 := \sigma_{g_2}(v_3)$, $w := \sigma_{g_2}(v_2)$, $v := \sigma_{g_2}(w_2)$.
The lines $\alpha(v_2)$, $\alpha(w_3)$, $\alpha(v)$ don't contain the point $\alpha(P)$. Since $\bigcup_{i=1}^{3}\alpha(M_i)$ ist not collinear, (DP) applied on $(w_3,v_1,w_2,v_3,w_1,v_2)$ shows that the set M_3 is contained in a line g_3. P ist not a neighbour of one of the successive lines v_2,w_3,v_1 of the cyclically ordered set (w,v_2,w_1,v,w_2,v_1). The α-images of the points $w \wedge v_2$, $v_2 \wedge w_1$, $w_1 \wedge v$, $v \wedge w_2$, $w_2 \wedge v_1$ and $v_1 \wedge w$ are not contained in a line. Hence (DP) yields $v \wedge w_1 \in g :=$ $P \vee (w \wedge v_1) \in \Lambda_0$. From the result just proved the equation $\sigma_{g_2}(g_1) = g$ is easy to deduce. The second assertion of the lemma is obvious.

PROPOSITION 1. (DP) *and* (SD) *imply that for each* $g \in \Lambda_0$, *the line reflection* σ_g *is an automorphism of* Ω_0.

σ_g induces an involutory collineation σ_{g*} of the incidence structure $\alpha(\Omega)$ with $\alpha(g)$ as the set of fixed points of σ_{g*}. Without using (SD) we show:

(i) *If the order of* $\alpha(\Omega)$ *is at least* 4 *then* σ_g *is a collineation of* $(\Pi, V \cup W \cup \Lambda_0, \in)$.

PROOF OF (i). There exists a line $k_1 \in \alpha(\Lambda_0)$ such that $k_1 \nparallel \alpha(g)$ and $\sigma_{g*}(k_1) \neq k_1$. Let $k \in \Lambda$, $\alpha(k) = k_1$. Then we have $\sigma_{k*}(\alpha(g)) \nparallel \alpha(g)$ and $k \cap g \neq \emptyset$. Lemma 4 yields $k' := \sigma_g(k) \in \Lambda_0$ and $\sigma_g = \sigma_{k'}\sigma_g\sigma_k$.

In order to show $\sigma_g(h) \in \Lambda_o$ for any $h \in \Lambda_o$ we may assume $h \cap g = \emptyset$, hence $\alpha(h) \nparallel \alpha(k)$ and $\sigma_k(h) \in \Lambda_o$. From $\sigma_{k*}(\alpha(g)) \nparallel \alpha(g)$ we obtain $\alpha(\sigma_k(h)) \nparallel \alpha(g)$ and, by lemma 4, $l := \sigma_g(\sigma_k(h)) \in \Lambda_o$. Now $l \cap k' \neq \emptyset$ implies $\sigma_{k'}(l) = (\sigma_{k'}\sigma_g\sigma_k)(h) = \sigma_g(h) \in \Lambda_o$.

PROOF OF THE PROPOSITION. Of course, σ_g permutes the (V,W)-parallelograms. If the order of $\alpha(\Omega)$ is at least 4, then, by (i) and lemma 1, σ_g preserves the relation $\parallel_o$.
In the remaining case "order $\alpha(\Omega) = 3$" we consider a line $h \in \Lambda_o$ with $\alpha(h) \nparallel \alpha(g)$ and a point $A \in g$. By lemma 3, $\{\sigma_x(A) \mid x \in B_h\}$ is a line l, and B_h is the set of the diagonals $x \neq l$ of those (V,W)-parallelograms one diagonal of which is l. Note that $\sigma_g(l) \in \Lambda_o$ and $\sigma_g(B_h) \subseteq \Lambda_o$, so the internal description of B_h by means of (V,W)-parallelograms is transmitted by σ_g into an analogous one. We see that $\sigma_g(B_h)$ is a pencil of parallel lines.
Now let $k \in \Lambda_o$ be a line with $\alpha(k) \parallel \alpha(g)$. Consider a (V,W)-parallelogram one diagonal of which is k, the other one may be h. The assumption "order $\alpha(\Omega) = 3$" yields $\alpha(h) \nparallel \alpha(g)$ and, by the first part of the proof, we get $\sigma_g(B_h) = B_{\sigma_g(h)}$. By lemma 3 a) $k = \{\sigma_x(h \wedge k) \mid x \in B_h\}$. Lemma 4 yields $\sigma_g(k) = \{\sigma_g\sigma_x\sigma_g\sigma_g(h \wedge k) \mid x \in B_h\} = \{\sigma_y(\sigma_g(h \wedge k)) \mid y \in B_{\sigma_g(h)}\}$.
With lemma 3 b) we obtain $\sigma_g(k) \in \Lambda_o$.

In the following we always assume that (Ω, α) satisfies (DP), (SD) and that $\alpha(\Omega)$ is not a Fano plane unless the contrary is explicitely stated. If g,h are the diagonals of a (V,W)-parallelogram then σ_g fixes all lines $x \in B_h$. Assume $\sigma_g(f) = f$ and $f \notin B_h$. If f would contain a point P with $P \nsim g$ then $f = P \vee \sigma_g(P) \in B_h$ by (SD) - a contradiction. Hence $\alpha(f) = \alpha(g)$. Furthermore $\alpha(g) \neq \alpha(h)$ since $\alpha(\Omega)$ is not a Fano plane. Thus for each $F \in f$ we have $F = f \wedge (F \vee B_h)$; this implies $\sigma_g(F) = F$, $F \in g$ and finally $f = g$, a contradiction. Thus we have shown that the set of Λ_o-lines fixed by σ_g is $\{g\} \cup g^\perp$ where $g^\perp := B_h$ is a unique pencil of parallel lines. The *orthogonality relation* $\perp := \{(g,h) \in \Lambda_o \times \Lambda_o \mid h \in g^\perp\}$ is symmetrical on Λ_o because $g \perp h$ is equivalent to the property that $\sigma_g\sigma_h$ has order 2. Summarizing we obtain:

LEMMA 5.

(a) *The orthogonality relation* $\perp$ *is a symmetrical relation on* Λ_o.

(b) *Let* $g,h,l \in \Lambda_o$ *and* $g \perp h$. *Then* $1 \neq \sigma_g\sigma_h = \sigma_h\sigma_g$, *and* $l \parallel h$ *if and only if* $l \perp g$.

4. THE ROTATION GROUP D_P AND THE TRANSLATION GROUP T_B

Notations:
$\Sigma := \{\sigma_g \mid g \in \Lambda_o\}$,
$\Sigma_P := \{\sigma_g \mid P \in g \in \Lambda_o\}$ for a point P ,
$\Sigma_B := \{\sigma_g \mid g \in B\}$ for a pencil $B \in \Lambda_o/_{||}$
$D_P := \langle\Sigma_P\Sigma_P\rangle$, the *rotation group of* P, generated by the products $\sigma_a\sigma_b$, where $\sigma_a, \sigma_b \in \Sigma_P$,
$T_B := \langle\Sigma_B\Sigma_B\rangle$, the *translation group of* $B \in \Lambda_o/_{||}$
$T := \langle\bigcup_{B \in \Lambda_o/_{||}} T_B\rangle$, the *translation group.*

PROPOSITION 2. D_P *is equal to* $\Sigma_P\Sigma_P$ *for each* $P \in \Pi$. D_P *is an abelian group, it acts regularly and faithfully on each of the sets* V_P' *and* W_P' . *Furthermore* $\Sigma_P^3 := \Sigma_P\Sigma_P\Sigma_P \subseteq \Sigma_P$, *and* $\sigma_g\Sigma_P = \Sigma_P\sigma_g = D_P$ *for* $\sigma_g \in \Sigma_P$.

PROOF. Let π be the square of an element of Σ_P^3. By lemma 2, π fixes each line $l \in V_P' \cup W_P'$, and so also each point $x \wedge y \in M := (\bigcup_{x \in V_P'}) \cap (\bigcup_{y \in W_P'})$.
If $Q \sim P$ for $Q \in \Pi$ then there are two non-neighbouring Λ_o-lines through Q. Each of these two lines carries two non-neighbouring points of M. Hence π fixes Q, $Q \vee V$, $Q \vee W$, and π must be the identity. Since each element of Σ_P^3 has order ≤ 2 the group D_P is abelian.

If $\delta \in D_P$ fixes all lines of V_P' then for any $\sigma \in \Sigma_P$, $\delta^{-1} = \sigma\delta\sigma$ fixes all lines of W_P'. The same arguments as above now yield $\delta = 1$, i.e. D_P acts faithfully on V_P'. Analogously, we see that D_P acts faithfully on W_P'.

The proof of the remaining assertions is easy; it uses the fact that for any $x \in V_P'$ and $y \in W_P'$ there is a unique $\sigma \in \Sigma$ that interchanges x and y.

PROPOSITION 3. T_B *is equal to* $\Sigma_B\Sigma_B$ *for* $B \in \Lambda_o/_{||}$. *The group* T_B *is abelian and acts regularly and faithfully on each of the pencils* V *and* W. *The orbits of* T_B *are the lines of the pencil* $B^\perp := \{x \in \Lambda_o \mid x \perp b$ *for all* $b \in B\}$. *Furthermore* $\Sigma_B^3 \subseteq \Sigma_B$ *and* $\sigma_b\Sigma_B = \Sigma_B\sigma_b = T_B$ *for each* $b \in B$.

PROOF. By lemma 5 b) there is a unique pencil $B^\perp \in \Lambda_o/_{||}$ such that $\sigma_x\sigma_b = \sigma_b\sigma_x$ for all $b \in B$ and all $x \in B^\perp$. Using lemma 3 we obtain from this fact that the lines of $B^\perp$ are the orbits of T_B. Let $v_o \in V$ and $\rho \in \Sigma_B^3$, then $\rho(v_o) \in W$. For $F := v_o \wedge \rho(v_o)$ the mapping ρ fixes $F \vee B^\perp$, hence we get $\rho(F) = F$. Now let $b := F \vee B$ and $\pi := \sigma_b\rho$. Then π fixes each point on $b = \{\rho_x(F) \mid x \in B^\perp\}$. Considering $\pi(V) = V$ and $\pi(W) = W$ we obtain $\pi = 1$, $\rho = \sigma_b$ and $\Sigma_B^3 \subseteq \Sigma_B$. The last inclusion implies that the equation $T_B = \Sigma_B\Sigma_B$ holds and that T_B is an abelian group. T_B acts faithfully on V and also on W since only $\tau = 1$ fixes all lines of $V \cup B^\perp$ and of $W \cup B^\perp$, respectively.

For arbitrary elements $v \in V$ and $w \in W$ there is a unique $\sigma_c \in \Sigma_B$ with $\sigma_c(v) = w$. By using this fact it is easy to prove the remaining assertions of the lemma.

For $P, Q \in \Pi$ and $B \in \Lambda_o/_{\|}$ there exist $\tau_1 \in T_B$ and $\tau_2 \in T_{B^\perp}$ with $\tau_1(P) = (P \vee B) \wedge (Q \vee B^\perp) = \tau_2(Q)$. The product $\tau_2^{-1}\tau_1 \in T$ maps P onto Q. Hence the translation group T acts transitively on Π.

5. POINT REFLECTIONS

LEMMA 6. *For each point* P *the rotation group* D_P *contains a unique element* σ_P *of order* 2, *called the reflection at the point* P. *Furthermore* $\sigma_P(h) \parallel h$ *holds for all* $h \in \Lambda_o$.

PROOF. Let $\sigma_g \in \Sigma_P$. By proposition 2 each element $\delta \in D_P$ is a product $\delta = \sigma_h\sigma_g$ with $\sigma_h \in \Sigma_P$; its order is 2 if and only if $h = P \vee g^\perp$. Hence $\sigma_{P \vee g^\perp}\sigma_g =: \sigma_P$ is the only element of order 2 in D_P. For $x \in \Lambda_o$ and $g := P \vee B_x$ we obtain $\sigma_P(g) = g$, and by proposition 2 $\sigma_P(x) \parallel x$.

We introduce the notation $\Sigma_\Pi := \{\sigma_P \mid P \in \Pi\}$.

PROPOSITION 4. $\Sigma_\Pi^3 \subseteq \Sigma_\Pi$ *and* $T = \Sigma_\Pi^2$. T *is an abelian group acting regularly on* Π. *For each point* P *we have* $T = \sigma_P\Sigma_\Pi = \Sigma_\Pi\sigma_P$.

PROOF. Let $P_i \in \Pi$ $(i = 1,2,3)$ and $B \in \Lambda_o/_{\|}$. Then $\sigma_{P_i} \in \Sigma_B\Sigma_{B^\perp}$. Each element of Σ_B commutes with each element of $\Sigma_{B^\perp}$. Now lemma 7 yields $\sigma_{P_3}\sigma_{P_2}\sigma_{P_1} \in \Sigma_B\Sigma_{B^\perp} \subset \Sigma_\Pi$. Since $T_B = \Sigma_B\Sigma_B \subseteq (\Sigma_B\Sigma_{B^\perp})(\Sigma_{B^\perp}\Sigma_B) = \Sigma_\Pi\Sigma_\Pi$ the transitive translation group T is a subgroup of the commutative group $\Sigma_\Pi\Sigma_\Pi$. Hence T acts regularly on Π. The remaining assertion of the lemma is established by the inclusions $\sigma_P\Sigma_\Pi \subseteq \Sigma_\Pi\Sigma_\Pi = T$ and $\sigma_P T = \sigma_P\Sigma_\Pi\Sigma_\Pi \subseteq \Sigma_\Pi^3 \subseteq \Sigma_\Pi$.

PROPOSITION 5. *If* $P, Q \in g$ *and* $g \in V \cup W \cup \Lambda_o$ *then every line* $h \in B_g$ *is fixed by the translation* $\sigma_Q\sigma_P$. *The group* $\Gamma := \langle\Sigma\Sigma\rangle$ *(generated by the products of two line reflections) is the semidirect product of the normal subgroup* T *and any rotation group* D_P. Γ *is a group of automorphisms of* Ω_o.

PROOF. Using proposition 3 we see immediately that $\{\tau(g) \mid \tau \in T\} = B_g$. T is an abelian group and $\sigma_Q\sigma_P$ fixes g, hence $\sigma_Q\sigma_P$ fixes also each $h \in B_g$. Σ consists of automorphisms of Ω_o, and therefore Γ. Σ_Π is a conjugacy class in Γ, hence $T = \Sigma_\Pi\Sigma_\Pi$ is a normal subgroup of Γ. The transitivity of T on Π yields that $\Sigma^2 \subseteq D_P T$ and $\Gamma = D_P T$.

6. THE ACTION OF THE GROUP Γ ON V AND ON W

For each $\varphi \in \langle\Sigma\rangle$ there is a unique collineation $\tilde{\varphi}$ of $\alpha(\Omega)$ such that $\alpha \circ \varphi = \tilde{\varphi} \circ \alpha$. Clearly $\sim$ is a homomorphism from the group $\langle\Sigma\rangle$ into the collineation group of $\alpha(\Omega)$. Its kernel N consists of those elements which map every point onto a neighbouring one.
Each mapping $\varphi \in \Gamma$ and each subgroup Φ of Γ acts in a natural way on the pencils V and W. We denote this action by $\varphi_{|V}$, $\Phi_{|V}$ or $\varphi_{|W}$, $\Phi_{|W}$.

PROPOSITION 6.
(a) *For each* $g \in \Lambda_0$ *the pair* $(i_g, \sigma_{g|V})$ *with*

$$i_g : \left\{ \begin{array}{l} \Gamma \longrightarrow \Gamma \\ \gamma \longrightarrow \sigma_g \gamma \sigma_g \end{array} \right. \quad \textit{and} \quad \sigma_{g|V} : \left\{ \begin{array}{l} V \longrightarrow W \\ v \longrightarrow \sigma_g(v) \end{array} \right.$$

is an isomorphism of the transformation group (Γ,V) *onto the transformation group* (Γ,W).
(b) *The abelian group* $T_{|V}$ *acts regularly on* V. *Let* $P \in \Pi$. *Then* $V \setminus V'_P$ *is the orbit of the element* $P \vee V$ *under the subgroup* $T \cap N$. *The group* D_P *acts regularly on* V'_P.

PROOF. For $v \in V \setminus V'_P$ the points P and $Q := v \wedge (P \vee W)$ are neighbour elements. By proposition 4 there is a translation τ with $\tau(P) = Q$. The equation $\alpha(P) = \alpha(Q)$ implies $\tau \in T \cap N$. Hence $V \setminus V'_P$ is the orbit of the element $P \vee V$ under the group $T \cap N$. The last assertion has already been proved in proposition 2.

7. A CHARACTERIZATION OF THE GROUP OF AFFINE TRANSFORMATIONS IN A LOCAL RING

THEOREM 1. *Let* 0,1 *be distinct elements of a set* R *and let* U_*, T_*, D_* *be commutative groups of permutations of* R *such that*
(1) T_* *acts transitively on* R,
(2) U_* *is a subgroup of* T_* *with* $1 \notin U_*(0)$,
(3) T_* *is normalized by* D_*,
(4) $D_*(0) = \{0\}$ *and* $D_*(1) = R \setminus U_*(0)$.
Then there exist binary operations $+, \cdot : R \times R \longrightarrow R$ *such that the following holds:*
$(R,+,\cdot)$ *is a commutative local ring with the zero element* 0, *the unit element* 1 *and the maximal ideal* $I := U_*(0)$. T_* *is the group of all translations* $x \longmapsto x + r$, *where* $r \in R$, *and* D_* *is the group of all dilatations* $x \longmapsto ax$, *where* $a \in R \setminus I$.

PROOF. By (1), for each $r \in R$ there is a unique $\tau_r \in T_*$ with $\tau_r(0) = r$. This enables us to define an addition $+ : R \times R \longrightarrow R$ by $(x,y) \longmapsto x + y := (\tau_x \circ \tau_y)(0) = \tau_x(y)$. Of course, $r \longmapsto \tau_r$ is a group isomorphism $(R,+) \longrightarrow (T_*,\circ)$. The properties (3), (4) imply that D_* is an abelian group of automorphisms of $(R,+)$. Therefore D_* generates a commutative subring R_* of the endomorphism ring of the abelian group $(R,+)$. $D_*(1) = R \setminus U_*(0)$ is contained in the submodule $R_* \cdot 1$ of the R_*-module R. Hence $R = R_* \cdot 1$, and $\rho \longrightarrow \rho(1)$ is an additive bijection from R_* onto R. There exists a unique multiplication $\cdot : R \times R \longrightarrow R$ such that this bijection is a ring isomorphism from $(R_*,+,\circ)$ onto $(R,+,\cdot)$. Now the other statements of the theorem are obvious.

8. COORDINATIZATION OF THE AFFINE HJELMSLEV PLANE (Ω, α)

We choose pairwise non-neighbouring lines $v_0, v_1 \in V$; $w_0, w_1 \in W$, and then put $P_i := v_i \wedge w_i$ for $i \in \{0,1\}$, and $e := P_0 \vee P_1$. We intend to use theorem 1 for $R = V$, $0 = v_0$, $1 = v_1$, $T_* = T_{|V}$, $U_* := (T \cap N)_{|V}$ and $D_* := D_{P_0 |V}$. By proposition 6 b) we obtain $U_*(0) = (T \cap N)(v_0) = V \setminus V'_P$ and $U_*(1) = D_{P_0}(v_1) = V'_{P_0} = V \setminus U_*(0)$. With theorem 1 we get a local ring $(V,+,\cdot)$ with the zero element v_0, and the unit element v_1. V'_{P_0} is the set of invertible elements, and $I := V \setminus V'_{P_0}$ is the maximal ideal of this local ring $(V,+,\cdot)$.

The equation $\sigma_{P_0} \tau \sigma_{P_0} = \tau^{-1}$ for all $\tau \in T$ yields $\sigma_{P_0}(v) = -v$ for all $v \in V$. In the affine plane $\alpha(\Omega)$ the mapping $\widetilde{\sigma_{P_0}}$ does not fix $\alpha(v_1)$. Therefore $v_1 \not\sim -v_1$ and $v_1 + v_1 \not\sim v_0$. This shows that $v_1 + v_1$ is invertible and that the quotient field $V/_I$ has characteristic $\neq 2$. We define surjections $\omega_1, \omega_2 : \Pi \longrightarrow V$ by $\omega_1(P) := P \vee V$ and $\omega_2(P) := \sigma_e(P \vee W)$. Then $\omega(P) := (\omega_1(P), \omega_2(P))$ defines a bijection $\omega : \Pi \longrightarrow V \times V$. Two points P,Q are neighbouring if and only if $\omega_i(P) - \omega_i(Q) \in I$ for $i = 1,2$. What is the ω-image of a line of Ω_0 ? For $v \in V$ the image $\omega(v)$ is the set $\{v\} \times V$. For $w \in W$ we get $\omega(w) = V \times \{\sigma_e(w)\}$. For $P_0 \in g \in \Lambda_0$ the point P is on g if and only if $\omega_2(P) = (\sigma_e \sigma_g)(\omega_1(P))$. The product $\sigma_e \sigma_g$ belongs to D_{P_0}, and $(\sigma_e \sigma_g)(v_1)$ is invertible in the local ring $(V,+,\cdot)$. Hence $\omega(g) = \{(v, [(\sigma_e \sigma_g)(v_1)]\, v) \mid v \in V\}$ is the set of solutions $(\xi,\eta) \in V \times V$ of the equation $\eta = m\xi$ with $m := (\sigma_e \sigma_g)(v_1) \in V \setminus I$. Now let g be an arbitrary Λ_0-line. By proposition 4 there is a unique translation $\tau = \sigma_Q \sigma_{P_0} \in T$ with $\tau(P_0) = \sigma_Q(P_0) = \omega_0 \wedge g$. So we have $\sigma_Q(P_0 \vee W) = P_0 \vee W$. Lemma 6 yields $Q \in P_0 \vee W$. From proposition 5 we know that the translation $\tau = \sigma_Q \sigma_{P_0}$ fixes each line of W since it fixes

$P_0 \vee W$. Thus $\omega\tau\omega^{-1} : V \times V \longrightarrow V \times V$ is a transformation $(\xi,\eta) \longrightarrow (\xi - \xi_0,\eta)$ where $\xi_0 := (\omega_0 \wedge g) \vee V$. The line $\tau(g)$ is parallel to g, and it contains P_0. We just have proved that $(\omega\tau)(g)$ is described by an equation $\eta = m\xi$ where $m \in V \setminus I$. Therefore $\omega(g)$ is the set of solutions $(\xi,\eta) \in V \times V$ of the equation $\eta = m(\xi - \xi_0) = m\xi - m\xi_0$.

The incidence structure $(V \times V, \omega(\Lambda), \in)$ satisfies the configuration condition (DP). Furthermore we have found an algebraic description of the point sets $\omega(g)$ with $g \in V \cup W \cup \Lambda_0$. Combining these two facts, an easy calculation shows that for $g \in \Lambda \setminus (V \cup W \cup \Lambda_0)$ the set $\omega(g)$ is of the form $\{(\xi,\eta) \in V \times V \mid a_1\xi + a_2\eta = a_0\}$ where $a_0,a_1,a_2 \in V$ and exactly one of the coefficients a_1,a_2 is invertible.

Summarizing these results we obtain:

THEOREM 2. *Let* V,W *be two pencils of parallel lines in an affine Hjelmslev plane* (Ω,α) *with* $\alpha(V) \neq \alpha(W)$. *Suppose that the configuration conditions* (DP) *and* (SD) *hold and that* $\alpha(\Omega)$ *is not a Fano plane. Then there is a commutative local ring* R *with the properties:*

(a) $1 + 1 \in R \setminus I$ *where* I *is the maximal ideal in* R.

(b) *Up to a bijection the point set of* Ω *can be identified with* $R \times R$ *such that the following holds:*

(b1) *The lines of* V,W *respectively are the sets* $\{r\} \times R$, $R \times \{r\}$ *where* $r \in R$.

(b2) *If the line* g *is not a neighbour of a line* $x \in V \cup W$, $g = \{(\xi,\eta) \in R \times R \mid \eta = m(\xi - \xi_0)\}$ *where* $\xi_0 \in R$ *and* $m \in R \setminus I$.

(b3) *The lines neighbouring a line of* V,W *respectively are the sets* $\{(\xi,\eta) \in R \times R \mid \xi = m\eta + r\}$, $\{(\xi,\eta) \in R \times R \mid \eta = m\xi + r\}$ *where* $m \in I$, $r \in R$.

(b4) *Two lines of the type described in (b2) are parallel if and only if the corresponding coefficients* m *are the same.*

(b5) *Two points have the same* α*-image if and only if their difference belongs to* $I \times I$.

Final remarks.

1. The result formulated in theorem 2 can be extended to the $\|$ - relation of the lines $x \in \Lambda \setminus (\Lambda_0 \cup V \cup W)$. In order to do this we can take a configuration condition that is valid in $\Omega(R)$ and that determines the $\|$ - relation of the lines in question intrinsically by the incidence structure of Ω (similar to (SD) and the Λ_0-lines in lemma 1).
2. Let $\alpha(\Omega)$ be a Fano plane of order $\neq 2$. Then theorem 1 and the methods used for theorem 2 also lead to a coordinatization if the group T and D_P have the properties needed in chapter 8.

REFERENCES

[1] BRUEN, R.P.: Bol Quasi-Fields and Pappus' Theorem. Math. Z. 105 (1968), 351-364.

[2] KARZEL, H. / SÖRENSEN, K.: Die lokalen Sätze von Pappus und Pascal. Mitt. Math. Gesellsch. Hamburg 10, 1 (1971), 28-55.

[3] KLINGENBERG, W.: Projektive und affine Ebenen mit Nachbarelementen. Math. Z. 60 (1954), 384-406.

[4] LÜCK, H.H.: Projektive Hjelmslevräume. Journal f. r. u. a. Math. 243 (1970), 121-158.

[5] LÜNEBURG, H.: Affine Hjelmslev-Ebenen mit transitiver Translationsgruppe. Math. Z. 79 (1962), 260-288.

[6] LÜNEBURG, H.: Lectures on Projective Planes. Lecture Notes, University of Illinois 1969, Chapt. VIII.

[7] PICKERT, G.: Der Satz von Pappos mit Festelementen. Arch. Math. 10 (1959), 56-61.

[8] SEIER, W.: Der kleine Satz von Desargues in affinen Hjelmslev-Ebenen. Geom. Ded. 3 (1974), 215-219.

[9] VELDKAMP, F.D.: Projective planes over rings of stable rank 2. Geometriae Dedicata 11 (1981), 285-308.

Prof. Dr. Helmut Mäurer
Fachbereich Mathematik
Technische Hochschule Darmstadt
Schloßgartenstr. 7
D - 6100 Darmstadt

Prof. Dr. Wolfgang Nolte
Fachbereich Mathematik
Technische Hochschule Darmstadt
Schloßgartenstr. 7
D - 6100 Darmstadt

Annals of Discrete Mathematics 37 (1988) 293–296

EMBEDDING LOCALLY PROJECTIVE PLANAR SPACES INTO PROJECTIVE SPACES

Klaus Metsch

Johannes-Gutenberg Universität
Fachbereich Mathemetik
Saarstr. 21
D-6500 Mainz, Federal Republic of Germany*

We shall show that a 3-dimensional locally projective planar space of finite order n can be embedded into a 3-dimensional projective space of order n, if it has at least n^3 points.

1. INTRODUCTION

A *linear space* is a pair $S=(p,L)$ of a set p of *points* and a set L of subsets of p, called *lines*, such that any two distinct points are contained in a unique line, and every line has at least two points. A *subspace* of S is a set E of points with the property that every line is either contained in E, or has at most one point in common with E. The linear space S is called a *planar space*, if there is a family of at least two subspaces, called *planes*, such that any three noncollinear points are contained in a unique line, and each plane contains some three noncollinear points. For every point p of a planar space S, we denote by S/p the incidence structure, whose points are the lines of S through p, and whose lines are the planes of S containing p. S is said to be a *3-dimensional locally projective* planar space, if S/p is a projective plane for every point p of S. The (common) order of these projective planes is then also called the *order* of S.

Examples of such spaces can be obtained by removing a certain part of the structure of a 3-dimensional projective space. J. Kahn [2] proved that a 3-dimensional locally projective planar space arises in this way if and only if it satisfy the *Bundle Theorem*. In general this needs not to be true. As noticed by W.M. Kantor [3], the Witt space W_{22} [4] is a counterexample of order 4. But it seems likely that the number of points of such a counterexample of finite order n is small compared with number n^3+n^2+n+1 of points of a 3-dimensional projective space of the same order, if it exists. It is therefore of interest to find bounds for the number of points implying embeddability. To do a first step in this direction, we want to prove the following theorem.

THEOREM. Let S be a finite 3-dimensional locally projective planar space and denote by n the order of S. If S has at least n^3 points, then it is embeddable into a 3-dimensional projective space of the same order n.

REMARK. This theorem was proved by A. Beutelspacher [1] under the additional condition that any two distinct planes intersect in a line.

* This paper was written while the author was visiting the university of Florence.

2. PROOF OF THE THEOREM

From now on, **S** always denotes a finite 3-dimensional locally projective planar space, and n denotes the order of **S**. In particular, any two distinct planes E and F intersect in a line $E \cap F$. If H and L are distinct coplanar lines, then $\langle H,L \rangle$ denotes the unique plane containing H and L. Similary, if p is a point outside a line L, then $\langle p,L \rangle$ denotes the unique plane containing p and L. A *parallel class* of **S** (or of a plane E of **S**) is a set M of lines of **S** (or of lines contained in E) such that every point of **S** (or E, respectively) lies on a unique line of M. The *degree* of a line is the number of points on it.

DEFINITION 1. a) A *bundle* is a quadrupel (L_1, L_2, L_3, L_4) of four mutually disjoint lines of **S** such that L_i and L_j are coplanar for $i \neq j$ and $\{i,j\} \neq \{3,4\}$ and such that no three of the lines L_i are coplanar.
b) **S** is said to satisfy the *Bundle Theorem*, if L_3 and L_4 are coplanar for every bundle (L_1, L_2, L_3, L_4).

We shall show that **S** satisfies the Bundle Theorem, if there are at least n^3 points.

LEMMA 2. Suppose N is a line of degree n. If L_3 and L_4 are two lines disjoint and coplanar to N, then L_3 and L_4 are coplanar.
PROOF. W.l.o.g. we may assume that $L_3 \cap L_4 = \emptyset$. Set $E_j = \langle N, L_j \rangle$, $j=3,4$. Clearly, L_3 and L_4 are coplanar, if $E_3 = E_4$. We may therefore assume that $E_3 \cap E_4 = N$. Denote by p a point of L_4, and set $E = \langle p, L_3 \rangle$. Since p is a point of E_4 and E, E_4 and E intersect in a line L containing p. Because of $p \notin N$, $N \neq L$.
Assume by way of contradiction that L and N intersect in a point q. Then $\langle N, L_3 \rangle = \langle q, L_3 \rangle = \langle L, L_3 \rangle$ so that $\langle L_3, N \rangle = \langle L, N \rangle = \langle p, N \rangle = \langle L_4, N \rangle$, a contradiction.
Consequently, N and L are disjoint. Because N has degree n, there is a unique line in $\langle p, N \rangle$, which is disjoint to N and contains p. Hence, $L = L_4$ so that L_3 and L_4 are coplanar.■

LEMMA 3. Suppose L_1, L_2 and N are three distinct, mutually coplanar lines. Suppose furthermore that L is a line disjoint and coplanar to L_1 and L_2, disjoint to N, and not contained in $\langle L_1, L_2 \rangle$. If N has degree n, then N and L are coplanar.
PROOF. Fix a point p of L. Clearly, $p \notin \langle L_1, L_2 \rangle$ and consequently $L = \langle p, L_1 \rangle \cap \langle p, L_2 \rangle$. Set $E = \langle p, N \rangle$.
If $p \in \langle N, L_j \rangle$, then $\langle N, L_j \rangle = \langle p, L_j \rangle = \langle L, L_j \rangle$ so that N and L are coplanar. We may therefore assume that p is not contained in $\langle N, L_j \rangle$, $j=1,2$. In particular, E and $\langle p, L_j \rangle$ are distinct planes and intersect therefore in a line H_j, which contains p. We have $H_j \cap N = \emptyset$ (if H_j and N would intersect in a point q, then we would get the contradiction $p \in \langle p, L_j \rangle = \langle H_j, L_j \rangle = \langle q, L_j \rangle = \langle N, L_j \rangle$). Since N has degree n, this shows that H_j is the unique line of E, which is disjoint to N and passes through p. It follows $H_1 = H_2$. Hence, $H_1 = \langle p, L_1 \rangle \cap \langle p, L_2 \rangle = L$, and L is therefore coplanar to N.■

LEMMA 4. Suppose L_1 and L_2 are disjoint and coplanar lines of degree at most n-1. If **S** has at least n^3 points, then it exists a line N of degree n, which is disjoint and coplanar to L_1 and L_2.
PROOF. Set $E_1 = \langle L_1, L_2 \rangle$, and let $E_2, \ldots, E_{n+1}$ denote the other planes containing L_1. For $j=2,\ldots,n+1$,

$$M_j := \{E_j \cap F \mid F \text{ is a plane containing } L_2 \text{ and with } E_j \cap F \neq \emptyset\}$$

is a parallel class of E_j, which contains L_1 and has at most n+1 elements. Let G_2 be a line other than L_1 of M_2, and set

$$M_1 := \{E_1 \cap F \mid F \text{ is a plane containing } G_2 \text{ and with } E_1 \cap F \neq \emptyset\}.$$

Then M_1 is a parallel class of E_1, which contains L_1 and L_2 and has at most n+1 elements. Hence, $M := M_1 \cup \ldots \cup M_{n+1}$ is a parallel class of **S** with at most n^2+n+1 lines. Since no line of M can have degree n+1, and because **S** has at least n^3 points, M contains a line N of degree n. By the construction of M, N is disjoint and coplanar to L_1 and L_2.■

PROPOSITION 5. If **S** has at least n^3 points, then it satisfies the Bundle Theorem.

PROOF. Let (L_1, L_2, L_3, L_4) be a bundle. We have to show that L_3 and L_4 are coplanar. If L_1 or L_2 has degree n, this follows from Lemma 2. Since L_1 and L_2 can have degree at most n, we may therefore assume that they have degree at most n-1. By Lemma 4, it exists a line N of degree n, which is disjoint and coplanar to L_1 and L_2. If N is disjoint to L_3 and L_4, then Lemma 3 shows that N is coplanar to L_3 and L_4, so that L_3 and L_4 are coplanar by Lemma 2. Thus, it is no loss of generality to assume that N and L_3 have a point p in common. Because (L_1, L_2, L_3, L_4) is a bundle, p is not contained in the plane $\langle L_1, L_2 \rangle$. Consequently, it exists just one line through p, which is coplanar to L_1 and L_2. Hence, N = L_3, and now Lemma 3 shows that L_3 and L_4 are coplanar.■

In view of Proposition 5, our theorem follows at once from the result of Kahn [2] mentioned in the introduction.

REMARK. If **S** has at least n^3 points, then it is easy to see that **S** is either a 3-dimensional affine plane of order n, or it has exactly n^3+n^2+n+1 planes. Thus, in the preceeding proof we only need a quite special case of the theorem of Kahn.

REFERENCES

[1] Beutelspacher, A.: Embedding finite planar spaces in projective spaces. Finite Geometries (C.A. Baker and L.M. Batten ed.), papers presented at a conference held in Winnipeg at Saint John's College, July 9-18, 1984. M. Dekker 1985.

[2] Kahn, J.: Locally projective-planar lattices which satisfy the Bundle Theorem. Math. Z. 175 (1980), 219-247.

[3] Kantor, W.M.: Dimension and embedding theorems for geometric lattices. J. Comb. Theory (A) 17 (1974), 173-195.

[4] Witt, E.: Die 5-fach transitiven Gruppen von Mathieu. Abh. Math. Sem. Univ. Hamburg 13 (1938), 256-264.

Annals of Discrete Mathematics 37 (1988) 297–300

ON TOPOLOGICAL INCIDENCE GROUPOIDS

Rita MEYER, Jürgen MISFELD, Elena ZIZIOLI*

Universität Hannover, Institut für Mathematik
Welfengarten 1
3000 Hannover 1, Germany

Università Cattolica del Sacro Cuore
Dipartimento di Matematica
via Trieste 17
25121 Brescia, Italy

In this note the notion of a topological incidence groupoid is introduced. We prove an algebraic representation theorem and characterize the classical incidence loops over the quaternions and the octonions.

1. INTRODUCTION

The rotation group O_3^+ of the real space fixing 0 is an example of a topological incidence group, that means O_3^+ is a group which carries a geometric structure (projective space of dimension 3) and a topological structure such that suitable compatibility conditions hold. This group is isomorphic to the factor group H^*/R^* of the real quaternions over the reals. Starting with the real octonions O one observes that, due to the lack of the associative law, the factor structure leads to a topological incidence groupoid. In this note the notion of a topological incidence groupoid is introduced and these structures are represented by topological algebras.

2. PROJECTIVE INCIDENCE GROUPOIDS

E. Ellers and H. Karzel [1] defined the notion of an incidence group in order to give a foundation of the theory of motion groups of absolute planes. It is shown that these groups can be described by near-fields. Later on this theory was extended to the more general sliced spaces; the algebraic structure was extended to groupoids (see H.Wähling [4]) .

Definition. A triple $(G,\cdot,\pi)$ is called a (left-) *incidence groupoid* if the following holds:

(i) $(G,\cdot)$ is a groupoid, i.e. "$\cdot$" is a binary operation on G,

(ii) (G,π) is a projective space,

(iii) for each $a \in G$ the left translation

$$a_\ell : \begin{cases} G \to G \\ x \to ax \end{cases}$$

* Research supported in part by the Italian Ministry of the Education (40% M.P.I.)

is a collineation of (G, π).
If the right translations a_r are also collineations then $(G,\cdot,\pi)$ is called *two-sided*. $(G,\cdot,\pi)$ is called *right regular* if every a_ℓ is injective.

Examples were constructed by H. Wähling [4] from (left-) near-algebras. Let (F,K) be a (left-) *near-algebra* ;i.e. F is a near-ring,K is a field, (F,K) is a (left-) vector space of rank at least 2, K is normal in F $(K^*(ab) = (k^*a)b = a(K^*b)$ for every $a,b\in F$), then Wähling [4] has shown:

(2.1) If (F,K) is a near-algebra without zero divisors,then the factor structure F^*/K^* is a right regular desarguesian projective incidence groupoid,where the groupoid operation is defined by $(K^*a)\cdot(K^*b) = K^*(ab)$ and the projective structure is given in the usual way by the vector space (F,K).

(2.2) If $(G,\cdot,\pi)$ is a right regular desarguesian projective incidence groupoid,then there exists a near-algebra (F,K) without zero divisors such that F^*/K^* and G are isomorphic (with respect to the binary operation and the projective structure).

Remarks.
1. (2.1) and (2.2) are generalizations of the results in [1] ,where is shown that every desarguesian projective incidence group can be represented by a normal near-field.
2. Taking a two-sided loop instead of a groupoid, Wähling has shown that there is a corresponding division algebra (A,K). If,in addition, the left cancellation law holds, (A,K) is a normal alternative field.

3. TOPOLOGICAL PROJECTIVE SPACES

In the mentioned classical example of an incidence groupoid the underlying projective space carries topologies such that the space is a topological projective space. This notion is introduced in [3] .
Let (G, π) be a desarguesian projective space of dimension n and let $\mathcal{T}_k$ be the set of k-dimensional subspaces , $0 \le k \le n-1$. We assume that every $\mathcal{T}_k$ is a topological space with topology τ_k (the trivial topologies, i.e. the discrete and indiscrete topology,are always excluded). (G, π) is called a *topological projective space* if joining of point and subspace:

$$h_k : \left| \begin{array}{lcl} (\mathcal{T}_o \times \mathcal{T}_k)^* & \longrightarrow & \mathcal{T}_{k+1} \\ (T_o, T_k) & \longrightarrow & \overline{\{T_o, T_k\}} \end{array} \right. , \quad 0 \le k < n-1 ,$$

(where $(\mathcal{T}_o \times \mathcal{T}_k)^* := \mathcal{T}_o \times \mathcal{T}_k \setminus \{(T_o,T_k) \mid T_o \in T_k\}$),

and intersection of hyperplane and subspace:

$$s_k : \left| \begin{array}{lcl} (\mathcal{T}_{n-1} \times \mathcal{T}_k)^* & \longrightarrow & \mathcal{T}_{k-1} \\ (T_{n-1}, T_k) & \longrightarrow & T_{n-1} \cap T_k \end{array} \right. , \quad 0 < k \le n-1 ,$$

(where $(\mathcal{T}_{n-1} \times \mathcal{T}_k)^* := \mathcal{T}_{n-1} \times \mathcal{T}_k \setminus \{(T_{n-1},T_k) \mid T_k \subset T_{n-1}\}$),

are continuous with respect to the topologies defined in an obvious way.

In [2] ,theorem 2, is shown that the topologies τ_k are defined in a unique way by the point set topology τ_o. Therefore we can write $((G,\pi),\tau_o)$ or shorter $((G,\pi),\tau)$.

Here and in the following is always assumed that the dimension of the projective space and the vector space is finite.
The representation theorem in [3] states:

(3.1) Let (V,K) be a topological vector space.Then V^*/K^* is a desarguesian topological projective space with respect to the projective structure of (V,K) and the quotient topology.

(3.2) Let (G,π,τ) be a desarguesian topological projective space.Then there exists(up to isomorphism) an unique topological vector space (V,K) such that G and V^*/K^* are isomorphic (with respect to the geometric and topological structure).

Further it is shown that every locally compact connected topological projective space is isomorphic to a projective space over the field of real numbers,complex numbers or the quaternions.

4. TOPOLOGICAL INCIDENCE GROUPOIDS

Definition. A quadruple $(G,\cdot,\pi,\tau)$ is called a *topological incidence groupoid* if G is a groupoid,a projective space and a topological space with respect to "$\cdot$", π and τ ,such that holds:
(i) $(G,\cdot,\pi)$ is an incidence groupoid,
(ii) $(G,\cdot,\tau)$ is a topological groupoid , i.e. the multiplication is continuous,
(iii) (G,π,τ) is a topological projective space.

Examples can be constructed from topological near-algebras. A near-algebra (F,K) is called a *topological near-algebra* if (F,K) is a topological vector space, such that the topology in F is the product topology of the field topology in K (canonical vector space topology),and if the mapping

$$\begin{cases} (F^*/K^*)\times(F^*/K^*) \longrightarrow F^*/K^* \\ (K^*a\ ,\ K^*b) \longrightarrow K^*ab \end{cases}$$

is continuous with respect to the quotient topology induced by the canonical surjection $\varphi : F \rightarrow F^*/K^*$.
We remark that (F,K) is not assumed to be a topological algebra,i.e. the multiplication in the algebra F may not be continuous.
Obviously we can prove

(4.1) If (F,K) is a topological near-algebra without zero divisors then F^*/K^* is a right regular topological desarguesian projective incidence groupoid.
To every right regular topological desarguesian projective incidence groupoid exists a topological near-algebra wich represents this groupoid.

The proof is the same as that for a theorem given by Karzel [2] in the special case of groups.
The multiplication in F need not be continuous,only the multiplication in the factor group F^*/K^* is continuous. But for the spacial case of two-sided loop it can be shown that there is a corresponding *topological division algebra*

(see remark 2 in 2.),i.e. that the multiplication in F is continuous and that the equations $ax = b$ and $xa = b$ have continuous solutions.
Here we need the following Lemma:

(4.2) Let (V,K) be a topological vector space over a commutative field with the canonical vector space topology.
Then every bilinear mapping $b : V \times V \to V$ is continuous.

Proof. b is continuous if the coordinates of $b(x,y) \varepsilon V$ are continuous functions of $(x,y) \varepsilon V \times V$. Therefore it must be shown that every bilinear form $f : V \times V \to K$ is continuous. Such a bilinear form f is a polynomial in the coordinates of $(x,y) \varepsilon V$. Now K is a topological field. So for $x,y \varepsilon V$ with $x = \sum_{i=1}^{n} x_i e_i$, $y = \sum_{i=1}^{n} y_i e_i$ we have $f(x,y) = \sum_{i,k=1}^{n} x_i y_k a_{i,k}$ with $a_{i,k} = f(e_i, e_k) \varepsilon K$, from which follows that f is continuous.

From (4.1) , (4.2) and remark 2 in 2. follows the

Theorem. Let $(G, \cdot, \pi, \tau)$ be a two-sided topological desarguesian projective incidence loop. Then there exists (up to isomorphism) exactly one topological division algebra (A,K) such that G and A^*/K^* are isomorphic (with respect to the loop structure,to the projective structure and to the topological structure).

If the topologies are locally compact and connected the underlying field is either isomorphic to the reals,the complex numbers or the quaternions (see [3]). As we assume the dimension of the projective space to be at least two,such topological incidence loops are isomorphic to those over the quaternions H or the octonions O.Being more exact we have the

Corollary. Every locally compact connected two-sided topological desarguesian projective incidence loop with left cancellation law is either isomorphic to H^*/R^* (in the group case) or to O^*/R^* (in the loop case) .

ACKNOWLEDGEMENTS

This note has been carried out while the third named author was visiting the Mathematics Institute of the University of Hannover; she wishes to thank the whole staff for their hospitality.

REFERENCES

[1] Ellers,E. and Karzel H., Kennzeichnung elliptischer Gruppenräume, Abh. Math. Sem. Univ. Hamburg 26 (1963) , 55-77.

[2] Karzel,H. , Beziehungen zwischen topologischen Inzidenzgruppen und topologischen Fastkörpern, in: Celebrazioni archimedee del secolo XX , Simposio di topologia , Gubbio (1964), 75-84 .

[3] Misfeld,J. , Topologische projektive Räume , Abh.Math. Sem. Univ. Hamburg 32 (1968), 232-262.

[4] Wähling,H. , Projektive Inzidenzgruppoide und Fastalgebren, J.of Geometry 9 (1977) , 109-126.

Annals of Discrete Mathematics 37 (1988) 301–310

ISOMORPHISMS OF FINITE HYPERGROUPOIDS

Renato MIGLIORATO (*)

SUMMARY

We characterize a particular class of finite commutative hypergroupoids, called *minimal hypergroupoids*, such that for every finite commutative hypergroupoid H, there exists one and only one minimal hypergroupoid isomorphic to H. We denote it with MIN(H). We utilize the concept of minimal hypergroupoid in the study of two problems on the isomorphisms of finite hypergroupoids.

INTRODUCTION

The content of this paper owes its origin to two problems which are of particular interest in the electronic elaboration on finite hypergroupoids.

It should be noted that a hypergroupoid is a non-empty set H structured by a hyperoperation $\circ$ (i.e. an application $\circ : H \times H \to \mathcal{P}^*(H)$ where $\mathcal{P}^*(H) = \mathcal{P}(H) - \phi$). A hypergroupoid H is called a semi-hypergroup if $\forall\, x,y,z \in H$, $(x \circ y) \circ z = x \circ (y \circ z)$; a semi-hypergroup H is called hypergroup if, we have

$$\forall\, x \in H,\ x \circ H = H \circ x = H \quad \text{(see [1], [2])}.$$

Some methods have been given for the construction of a hypergroup starting from a known hypergroup of minor cardinality [4], but, generally speaking, the construction of examples of hypergroups is very difficult, above all as regards the verifying of the associative propertiy. This is one reason because electronic elaboration can be useful.

However, electronic elaboration can also produce serious problems because of the large number of hypergroups that can be constructed on one set, even when it is of small cardinality [3], [5].

The first problem regards the proof of possible isomorphisms between two given hypergroupoids H and H'.

Since the isomorphisms $f : \langle H, \circ \rangle \to \langle H, * \rangle$ of a finite hypergroupoid of cardinality n are exactly the permutations on the n elements, the direct checking of the isomorphism would require n! transformations. This work reduces the problem to that of the construction of the minimal hypergroupoids MIN(H) of a given hypergroupoid H.

The second problem regards the generation of a class C of hypergroupoids which are pairwise nonisomorphic. For this purpose it is enough to generate all the elements of C which are minimal.

(*) Address: Dipartimento di Matematica, Università di Messina, Contrada Papardo, 98010 S. Agata, Messina (Italy).

1. MINIMAL HYPERGROUPOIDS

If $H = \{0, 1, \ldots n-1\}$ let us define the application $V\colon \mathcal{P}^*(H) \longrightarrow N$ (from the set of the subsets of H into the set of natural numbers) such that

$$V(\emptyset) = 0,$$

$$\forall A \in \mathcal{P}^*(H), \quad V(A) = \sum_{x\in A} 2^x$$

It follows immediately that

$$V(H) = \sum_{i=1}^{n-1} 2^i = 2^n - 1,$$

and that the application $V\colon \mathcal{P}^*(H) \longrightarrow \{0, 1, \ldots, 2^n-1\}$ is bijective.

If H is a finite commutative hypergroupoid of cardinality n, we can indicate with $0, 1, \ldots, n-1$ the elements of H; therefore in the rest we will suppose that the support of H is the set $\{0, 1, \ldots, n-1\}$.

By commutativity it is sufficient, to consider only the hyperproducts $x \circ y$ with $x \leq y$. We give therefore the following

DEFINITION 1.1. H^2_+ *denotes the set of the pairs* $(x,y) \in H^2$ *such that* $x \leq y$ *. In* H^2_+, $<$ *denotes the relation such that*

(1.1) $\quad y < y' \Longrightarrow (x,y) < (x',y'),$

(1.2) $\quad (x,y) < (x',y) \Longleftrightarrow x < x'$

The relation thus defined is clearly a relation of strict total order in H^2_+.

DEFINITION 1.2. $\forall y \in H$ *the evaluation of* y *in* H, *denoted by* $V_H(y)$, *is here defined as the number*

$$V_H(y) = \sum_{x=0}^{y} 2^{n(y-x)} V(x \circ y).$$

It follows immediately that $V_H(y)$ identifies univocally all the hyperproducts $x \circ y$ with $x \leq y$.

DEFINITION 1.3. *If* H *and* H' *are two hypergroupoids on the same support* $\{0, \ldots, n-1\}$, H *is said to be* $H < H'$ *if and only if* $\exists (\bar{x},\bar{y}) \in H^2_+$ *such that, if* $\circ$ *denotes the hyperproduct in* H *and* $*$ *the hyperproduct in* H', *one has*

(1.3) $\quad \forall (x,y) \in H^2, \ (x,y) < (\bar{x},\bar{y}) \Longrightarrow x \circ y = x * y,$

(1.4) $\quad V(\bar{x} \circ \bar{y}) < V(\bar{x} * \bar{y}).$

From definitions 1.2 and 1.3 it immediately follows that $H < H'$ *if and only if* $\exists \bar{y} \in H$ *such that*

(1.5) $\quad \forall y \in H : y < \bar{y}, \ V_H(y) = V_{H'}(y),$

(1.6) $\quad V_H(\bar{y}) = V_{H'}(\bar{y}).$

From definition 1.2 and from the bijectivity of the application V, it furthermore follows that if $H \neq H'$, then $H < H'$ or $H' < H$.

The relation $\leq$ is therefore a relation of total order in the set of the hypergroupoids with the same support $\{0, \ldots, n-1\}$. In particular it is a relation of total order in the class of all the hypergroupoids isomorphic to a given hypergroupoid H.

Then, the following, definition can be given:

DEFINITION 1.4. *A hypergroupoid* H *is called minimal if and only if for every* H' *isomorphic to* H, *it is* $H \leq H'$. MIN(K) *denotes the minimal hypergroupoid isomorphic to a hypergroupoid* K.

From the fact that MIN(H) is unique the following proposition derives immediately,

PROPOSITION 1.1. *If* H *and* K *are two finite commutative hypergroupoids, they are isomorphic if and only if*

$$\mathrm{MIN}(H) = \mathrm{MIN}(K)$$

At this point we are faced with the following problems:

I. Given a hypergroupoid H, to recognize whether H is minimal without comparing it with all the n! hypergroupoids isomorphic to H.

II. To find an algorithm that, given H, allows the construction of MIN(H) without having to construct all the n! hypergroupoids isomorphic to H.

The rest of this paper aims at the study of these two problems.

2. CLASS $C_{(x,y)}$

DEFINITION 2.1. $\forall (\bar{x},\bar{y}) \in H_+^2$, $\mathcal{N}_H(\bar{x},\bar{y})$ *denotes the subset of* H *such that*

$$(2.1) \qquad z \in \mathcal{N}_H(\bar{x},\bar{y}) \Longleftrightarrow \begin{cases} z \in \bar{x}\circ\bar{y} \cup \{\bar{x},\bar{y}\} \\ \forall (x,y) \in H_+^2, \ (x,y) < (\bar{x},\bar{y}) \Longrightarrow z \notin x\circ y \cup \{x,y\} . \end{cases}$$

The elements of $\mathcal{N}_H(\bar{x},\bar{y})$ *are called new elements of* $\bar{x}\circ\bar{y}$.

From definition 2.1 it follows immediately that

$$(2.2) \qquad z \in \mathcal{N}_H(\bar{x},\bar{y}) \Longrightarrow z > \bar{y} .$$

In fact if $z \leq y$, then $x \leq z$, $z \in x\circ y \cup \{x,y\}$.

THEOREM 2.1. *If* $(\bar{x},\bar{y}) \in H_+^2$ *and* $z \in \bar{x}\circ\bar{y} \cup \{\bar{x},\bar{y}\}$, *then* $\exists (x,y) \leq (\bar{x},\bar{y})$ *such that*

$$(2.3) \qquad z \in \mathcal{N}_H(x,y) \cup \{x,y\} .$$

PROOF. If $\forall (x,y) \in H_+^2$, $(x,y) \leq (\bar{x},\bar{y}) \Longrightarrow z \notin x\circ y \cup \{x,y\}$, then clearly $z \geq \bar{y}$ and therefore by the definition 2.1, we have $z \in \mathcal{N}_H(\bar{x},\bar{y}) \cup \{\bar{x},\bar{y}\}$. On the other hand, if (x,y) is the first pair of H_+^2 such that $z \in x\circ y \cup \{x,y\}$, then $z \in \mathcal{N}_H(x,y) \cup \cup \{x,y\}$.

DEFINITION 2.3. *If* $(\bar{\bar{x}},\bar{\bar{y}})\in H_+^2$, *we say that* $H\in C_{(\bar{\bar{x}},\bar{\bar{y}})}$, *if and only if* $\forall(\bar{x},\bar{y})\in H_+^2, (\bar{x},\bar{y})\leq(\bar{\bar{x}},\bar{\bar{y}})$,

$$\forall z\in N_H(\bar{x},\bar{y}),\ \forall z' < z,\ \exists(x,y)\leq(\bar{x},\bar{y}) : z'\in N_H(x,y)\cup\{x,y\}. \tag{2.4}$$

THEOREM 2.2. *If* $(\bar{\bar{x}},\bar{\bar{y}})\in H_+^2$, *then* $H\in C_{(\bar{\bar{x}},\bar{\bar{y}})}$ *if and only if* $r = 0$ *or*

$$N_H(\bar{x},\bar{y}) = \bigcup_{i=1,\dots,r}\{m+i\}, \tag{2.5}$$

where is

$$\begin{cases} m = 0, \text{ if } \bar{y} = 0 \\ m = \max\left(\bigcup_{(x,y)<(\bar{x},\bar{y})}(x\circ y\cup\{y\})\right), \text{ if } \bar{y}\neq 0, \end{cases} \tag{2.6}$$

$$r = |N_H(\bar{x},\bar{y})|. \tag{2.7}$$

PROOF.

I. *The condition is sufficient* . If $r \neq 0$, then $N_H(x,y)\neq\emptyset$ and therefore (2.4) is obviously true. Let $r\neq 0$ and suppose that (2.5) is valid. If $\bar{y} = 0$, (2.4) follows immediately from (2.5) and from $N_H(\bar{x},\bar{y})\cup\{\bar{x},\bar{y}\} = \{0, 1, \dots, r\}$. Let us suppose that $\bar{y}\neq 0$. Let $z\in N_H(\bar{x},\bar{y})$, $z'<z$. Suppose that

$$\forall(x,y)\in H_+^2 : (x,y) < (\bar{x},\bar{y}),\quad z'\notin N_H(x,y)\cup\{x,y\}. \tag{2.8}$$

Immediately we have $z' > \bar{y}$ because otherwise $z'\in\{x,z'\}$ and (2.8) could not subsist.
If $z' = m$, by (2.6), since $\bar{y}\neq 0$, $\exists(x,y) < (\bar{x},\bar{y})$ such that $z'\in x\circ y\cup\{y\}$ and therefore (2.8) is false.
From (2.8) and (2.5) it follows that

$$\bar{y} < z' < m. \tag{2.9}$$

From (2.6), since $\bar{y}\neq 0$, it follows also that $\exists(x,y) < (\bar{x},\bar{y})$ such that $m = \max(x\circ y\cup\{y\})$.
But $y<\bar{y}<m$ and therefore $m\neq y$, thus $m = \max(x\circ y)$, and indeed, by (2.9) and theorem 2.1,

$$\exists(x_1,y_1) : m = \max N_H(x_1,y_1).$$

From (2.5) we have

$$N_H(x_1,y_1) = \bigcup_{i=1,\dots,r}\{m_1 + i\},\quad m_1 + r_1 = m,\quad r_1\neq 0,$$

where m_1 and r_1 are defined by (2.6) and (2.7), where (x_1,y_1) is in the place of $(\bar{x},\bar{y})$.
If z' is one of the numbers $m_1 + 1, \dots, m_1 + r_1 = m$, then $z'\in x_1\circ y_1$ and consequently, from theorem 2.1, (2.8) is absurd.

Let $z' < m_1$. Then

$$\bar{y} < z' < m_1 < m.$$

Repeating the preceding reasoning we find that $\exists(x_2, y_2)$ such that $m_1 = \max N_H(x_2, y_2)$ and that

$$N_H(x_2,y_2) = \bigcup_{i=1,\ldots,r} \{m_2 + i\}, \quad m_2 + r_2 = m_1, \quad r_2 \neq 0.$$

If z' is one of the numbers $m_2 + 1, \ldots, m_2 + r_2 = m_1$, then $z' \in x_2 \circ y_2$ and (2.8) is absurd from theorem 2.1.
If $z' < m$, proceding in the same way, we find that

$$\bar{y} < \ldots \; m_2 < \ldots < m_2 < m_1 < m.$$

Since it is not possible to proceed indefinitely, we must conclude that (2.8) is absurd. From (2.5) therefore. It follows (2.4).

II. *The condition is necessary.* Let us suppose (2.4). From (2.2) we have that $z \in N_H(\bar{x},\bar{y}) \Rightarrow z > \bar{y}$ and, so, from definition of $N_H(\bar{x},\bar{y})$ and (2.5),

$$z > m.$$

It follows from this that

$$(2.10) \qquad z \in N_H(\bar{x},\bar{y}) \Longrightarrow z = m + \rho, \; \rho \geq 1.$$

If $\bar{z} = \max N_H(\bar{x},\bar{y})$, and $\bar{z} = m + \bar{\rho}$, from (2.4) we have

$$\forall \rho' : 1 \leq \rho' < \bar{\rho}, \; \exists(x,y) \leq (\bar{x},\bar{y}) : m + \rho' \in N_H(x,y) \cup \{x,y\},$$

and therefore for any number $\alpha \in \{m + 1, \ldots, m + \bar{\rho}\}$, $\exists(x,y) \leq (\bar{x},\bar{y})$ such that $\alpha \in N_H(x,y) \cup \{x,y\}$, but $\alpha > m$ and than $\alpha \in N_H(\bar{x},\bar{y})$; indeed, from (2.10), $N_H(\bar{x},\bar{y}) = \{m + 1, \ldots, m + \bar{\rho}\}$.
In the rest we will call *segment* every subset $A \subseteq H$ such that

$$A = \{x \in H : \min A \leq x \leq \max A\}.$$

Now theorem 2.2 can be enunciated as follows.

THEOREM 2.2.1. $H \in C_{(\bar{\bar{x}},\bar{\bar{y}})}$ *if and only if* $\forall(\bar{x},\bar{y}) \in H^2_+$, $(\bar{x},\bar{y}) \leq (\bar{\bar{x}},\bar{\bar{y}})$ *implies that the set* $N_H(\bar{x},\bar{y})$, *when it is not empty, is a segment with minimum* $m+1$, *where* m *is defined by* (2.6).

THEOREM 2.3. *If* $H \in C_{(\bar{x},\bar{y})}$ *and* $(\bar{\bar{x}},\bar{\bar{y}})$ *is the pair successive to* $(\bar{x},\bar{y})$ *in* H^2_+, *then there exists an isomorphism* $f : H \longrightarrow H'$ *such that*

a) $H' \in C_{(\bar{\bar{x}},\bar{\bar{y}})}$,

b) $\forall(x,y) \in H^2_+$, $(x,y) \leq (\bar{x},\bar{y}) \Longrightarrow x * y = x \circ y$.

where $*$ denotes the hyperoperation in H'.

PROOF. If $N_H(\bar{\bar{x}},\bar{\bar{y}}) = \emptyset$, the statement is obviously true because it is enough to assume f as the identity.
Let $N_H(\bar{\bar{x}},\bar{\bar{y}}) = \{a_1, \ldots, a_r\}$, and let m be the number defined by (2.6) when

$(\bar{x},\bar{y})$ is substituted by $(\bar{\bar{x}},\bar{\bar{y}})$. Then it is sufficient to consider any permutation $f : H \longrightarrow H'$ that leaves every $z \leq m$ unvaried and such that

$$\forall i = 1, \ldots, r, \quad f(a_i) = m + i.$$

It follows immediately therefore from theorem 2.2 that $H' \in C_{(\bar{\bar{x}},\bar{\bar{y}})}$. Besides, since f leaves every $z \leq m$ unvaried, the hyperproducts $x \circ y$ with $(x,y) < (\bar{\bar{x}},\bar{\bar{y}})$, do not vary.

3. CLASS $C^*_{(\bar{x},\bar{y})}$

DEFINITION 3.1. *If* $(\bar{x},\bar{y}) \in H^2_+$, $A_H(\bar{x},\bar{y})$ *indicates the set such that*

$$(3.1) \qquad z \in A_H(\bar{x},\bar{y}) \Longleftrightarrow \begin{cases} z > \bar{y} \\ \exists (x,y) < (\bar{x},\bar{y}) \ : \ z \in \mathcal{N}_H(x,y) \end{cases}$$

DEFINITION 3.2. $\begin{bmatrix} \bar{x},\bar{y} \\ H \end{bmatrix}$ *denotes the relation in* $A_H(\bar{x},\bar{y})$ *such that*

$$(3.2) \qquad z \begin{bmatrix} \bar{x},\bar{y} \\ H \end{bmatrix} z' \Longleftrightarrow (\forall (x,y) < (\bar{x},\bar{y}), \ z \in x \circ y \Longleftrightarrow z' \in x \circ y).$$

$S^H_z(\bar{x},\bar{y})$ *denotes the class of equivalence mod* $\begin{bmatrix} \bar{x},\bar{y} \\ H \end{bmatrix}$ *to with* z *belongs.*

DEFINITION 3.3. *If* $H \in C_{(\bar{\bar{x}},\bar{\bar{y}})}$, *we can say that* $H \in C^*_{(\bar{\bar{x}},\bar{\bar{y}})}$ *if and only if* $\forall (\bar{x},\bar{y}) \in H^2_+$: $(\bar{x},\bar{y}) \leq (\bar{\bar{x}},\bar{\bar{y}})$, *if* $A_1, \ldots, A_r$ *denote the classes of equivalence mod* $\begin{bmatrix} \bar{x},\bar{y} \\ H \end{bmatrix}$, *then* $\forall i = 1, \ldots, r$, $A_i \cap \bar{x} \circ \bar{y}$ *is segment and* $\min(A_i \cap \bar{x} \circ \bar{y}) = \min A_i$.

THEOREM 3.1. *If* $H \in C^*_{(\bar{x},\bar{y})}$, *and* $(\bar{\bar{x}},\bar{\bar{y}})$ *is the pair successive to* $(\bar{x},\bar{y})$ *in* H^2_+, *then there exists an isomorphism* $f : H \to H'$ *such that*

i) $H' \in C^*_{(\bar{x},\bar{y})}$,

ii) $\forall (x,y) \leq (\bar{x},\bar{y})$, $x * y = x \circ y$,

where $*$ denotes the hyperoperation in H'.

We prove the following two lemmas before the proof of theorem 3.1.

LEMMA I. *If* $(x_0,y_0) < (x_1,y_1) < (x_2,y_2)$ *and* $z_0 \in \mathcal{N}_H(x_0,y_0)$, *then*

$$S^H_{z_0}(x_2,y_2) \subseteq S^H_{z_0}(x_1,y_1).$$

PROOF. From definitions 3.1 and 3.2 it follows immediately that

$$z \in S_{z_0}(x_2,y_2) \Longrightarrow \left\{ \begin{array}{l} z \in \mathcal{N}_H(x_0,y_0) \\ z > y_2 \\ (x,y) < (x_1,y_1) \Rightarrow (z \in x \circ y \Leftrightarrow z_0 \in x \circ y) \end{array} \right\} \Longrightarrow$$

$$\Longrightarrow \left\{\begin{array}{l} z\in \mathcal{N}_H(x_0,y_0) \\ z>y_1 \\ (x,y)<(x_1,y_1) \Longrightarrow (z\in x\circ y \Longleftrightarrow z\in x\circ y) \end{array}\right\} \Longrightarrow z\in S^H_{z_0}(x_1,y_1).$$

LEMMA II. *In the hypothesis of theorem 3.1.,* $z_0\in A^H(\bar{\bar{x}},\bar{\bar{y}})$, $S^H_{z_0}(\bar{\bar{x}},\bar{\bar{y}})$ *is a segment.*

PROOF. Let $(x_0,y_0)<(x_1,y_1)<(x,y)\leq(\bar{\bar{x}},\bar{\bar{y}})$, with (x_2,y_2) successive to (x_1,y_1) and let $z_0\in\mathcal{N}_H(x_0,y_0)$. We prove that if $S^H_{z_0}(x_1,y_1)$ is a segment, then $S^H_{z_0}(x_2,y_2)$ is also a segment. We can ignore the trivial case in which $S^H_{z_0}(x_2,y_2)=\emptyset$.

In these hypothesis, by lemma I, we have

$$S^H_{z_0}(x_2,y_2)\subseteq S^H_{z_0}(x_1,y_1)$$

and therefore $S^H_{z_0}(x_2,y_2)$ can be obtained from $S^H_{z_0}(x_1,y_1)$ eliminating if necessary some of its elements.

If $z\in S^H_{z_0}(x_1,y_1)-S^H_{z_0}(x_2,y_2)$, we have that

$$(3.3)\qquad z\in S^H_{z_0}(x_1,y_1)\Longleftrightarrow\left\{\begin{array}{l} z>y, \\ z\in\mathcal{N}_H(x_0,y_0) \\ (x,y)<(x_1,y_1)\Longrightarrow(z\in x\circ y\Longleftrightarrow z_0\in x\circ y), \end{array}\right.$$

and besides, since $z\in S^H_{z_0}(x_1,y_1)$,

$$(3.4)\qquad z\notin S^H_{z_0}(x_2,y_2)\Longleftrightarrow\left\{\begin{array}{l} z\leq y \\ \underline{\text{or}} \\ \exists(x,y)<(x_2,y_2)\ :\ z\in x\circ y,\quad z_0\notin x\circ y \\ \underline{\text{or}} \\ \exists(x,y)\in H^2_+:\ (x,y)<(x_2,y_2),\ z\notin x\circ y,\ z_0\in x\circ y \end{array}\right.$$

Taking into consideration (3.3) and (3.4) simultaneously, it follows that $z\in S^H_{z_0}(x_1,y_1)-S^H_{z_0}(x_2,y_2)$ if and only if at least one of the following conditions is satisfied.

a) $z=y_2$

b) $z\in x_1\circ y_1$, $z_0\notin x_1\circ y_1$,

c) $z\notin x_1\circ y_1$, $z_0\in x_1\circ y_1$.

Applying to $S^H_{z_0}(x_1,y_1)$ the condition (a) (as explained previously to obtain $S^H_{z_0}(x_2,y_2)$), this implies at the most the elimination of the element y_2; this, if y_2 is contained in $S^H_{z_0}(x_1,y_1)$, then it is the minimum of this set, because by definiction, $z \in S^H_{z_0}(x_1,y_1) \Longrightarrow z > y_1$ and from the fact that (x_2,y_2) is the pair successive to (x_1,y_1), it follows that $y_1 \leq y_2 \leq y_1 + 1$. Therefore if, as supposed, $S^H_{z_0}(x_1,y_1)$ is a segment, then (a), transforms it again into a segment.

In the case (b), since $z \in S^H_{z_0}(x_1,y_1)$, we have

$$z \in x_1 \circ y_1 \Longleftrightarrow z \in S^H_{z_0}(x_1,y_1) \cap x_1 \circ y_1,$$

and therefore

$$S^H_{z_0}(x_1,y_1) \cap x_1 \circ y_1 \subseteq S^H_{z_0}(x_1,y_1) - S^H_{z_0}(x_2,y_2)$$

According to the hypothesis that $(x_1,y_1) \leq (\bar{\bar{x}},\bar{\bar{y}})$ and supposing that $H \in C^*_{(\bar{\bar{x}},\bar{\bar{y}})}$, the set $S^H_{z_0}(x_1,y_1) \cap x_1 \circ y_1$ is a segment such that its minimum coincides with the minimum of $S^H_{z_0}(x_1,y_1)$; therefore in the passage from $S^H_{z_0}(x_1,y_1)$ to $S^H_{z_0}(x_2,y_2)$, the condiction (b) implies the elimination from $S^H_{z_0}(x_1,y_1)$, at the most, of the above mentioned segment. What remains is still a segment.

In the case (c), since $z_0 \in S^H_{z_0}(x_2,y_2)$,

$$z_0 \in x_1 \circ y_1 \Rightarrow z_0 \in S^H_{z_0}(x_2,y_2) \cap x_1 \circ y_1,$$

and therefore, since $z_0 \in S_{z_0}(x_2,y_2)$,

$$S^H_{z_0}(x_2,y_2) \subseteq x_1 \circ y_1 .$$

From this and from lemma I it follows that

$$S^H_{z_0}(x_2,y_2) \subseteq S^H_{z_0}(x_1,y_1) \cap x_1 \circ y_1.$$

In any case, $\forall z \in S^H_{z_0}(x_1,y_1) \cap x_1 \circ y_1$, (3.3) is satisfied and indeed,

$$\forall (x,y) \in H^2_+,\ (x,y) < (x_1,y_1) \Longrightarrow (z \in x \circ y \Longleftrightarrow z_0 \in x \circ y).$$

Hence if $z = y_2$, then $z \in S^H_{z_0}(x_2,y_2)$.

Therefore in the case (c) the set $S^H_{z_0}(x_2,y_2)$ either coincides with $S^H_{z_0}(x_1,y_1) \cap x_1 \circ y_1$ or is obtained from this by eliminating every element less or equal to y_2. Since $S^H_{z_0}(x_1,y_1) \cap x_1 \circ y_1$ or is a segment whose minimum is larger than y_1, it follows immediately that $S^H_{z_0}(x_2,y_2)$ is also a segment.

With an analogous procedure it can be proved that if (x_1,y_1) is the successive pair of (x_0,y_0) and if $\mathcal{N}_H(x_0,y_0)$ is a segment whose minimum is larger or equal to y_0, then $S^H_{z_0}(x_1,y_1)$ is also a segment. But $\mathcal{N}_H(x_0,y_0)$ has always the minimum larger or equal to y; besides, from hypothesis that $H \in C_{(\bar{x},\bar{y})}$ and from theorem 2.2, $\mathcal{N}_H(x_0,y_0)$ is a segment.

The thesis follows by induction.

PROOF OF THEOREM 3.1. By theorem 2.3, there exists an isomorphism $f : H \longrightarrow \bar{H}$ such that $\bar{H} \in C_{(\bar{x},\bar{y})}$ and for which the property (ii) is true. We can therefore suppose that $H \in C_{(\bar{x},\bar{y})}$.

Let $A_1, \ldots, A_r$, represent the classes of equivalence mod $\begin{bmatrix} \bar{\bar{x}},\bar{\bar{y}} \\ H \end{bmatrix}$ of $A^H(\bar{\bar{x}},\bar{\bar{y}})$. $\forall k = 1, \ldots, r$, if $z_k \in A_k$, then

$$A_k = S^H_{z_k}(\bar{\bar{x}},\bar{\bar{y}}).$$

By lemma I,

$$(3.5) \qquad A_k = \bigcup_{i=1,\dots,r_k} \{m_k + i - 1\},$$

where $m_k = \min A_k$, $r_k = |A_k|$.

If $A_k \cap \bar{\bar{x}} \circ \bar{\bar{y}} \neq \emptyset$, put $A_k \cap \bar{\bar{x}} \circ \bar{\bar{y}} = \{a_1^k, \dots, a_{s_k}^k\}$, $f: H \longrightarrow H'$ be a permutation such that

i) $\forall k : A_k \cap \bar{\bar{x}} \circ \bar{\bar{y}} = \emptyset$, *the elements of* A_k *are permuted in such a way that* $\forall i = 1, \dots, s_k$, $f(a_i^k) = m_k + i - 1$.

ii) $\forall x$ *which does not belong to any* A_k, $f(x) = x$.

H', thus obtained, satisfies the condition (i) and (ii) of the proposition. The condition (i) follows immediately from the construction of f and from (3.5). As regards to the condition (ii), this is guaranted, as well as by (ii), by the fact $\forall (x,y) \in H_+^2 : (x,y) < (\bar{\bar{x}}, \bar{\bar{y}})$, $x \circ y \cap A_k \neq \emptyset \Rightarrow A_k \subseteq x \circ y$, hence a permutation on the elements of A_k does not alter the hyperproducts $x \circ y$ if $(x,y) < (\bar{\bar{x}}, \bar{\bar{y}})$.

THEOREM 3.2. *If* H *is a finite commutative hypergroupoid, and* $(x,y) \in H_+^2$, *then there exists an isomorphism* $f : H \longrightarrow H'$ *such that* $H' \in C^*_{(x,y)}$.

PROOF. Let $(x,y) = (0,0)$. If $0 \circ 0 = \{0\}$, the proposition is trivial because it is sufficient to assume as f the identity.

If $N_H(0,0) = \{a_1, \dots, a_r\}$, it is enough to assume as f an application such that $f(0) = 0$ and $\forall i = 1, \dots, r$, $f(a_i) = i$.

From this and from theorems 2.3 and 3.1, the proposition follows by induction.

4. QUASI-MINIMAL HYPERGROUPOID

Let H be a hypergroupoid on the set $\{0, 1, \dots, n-1\}$.

DEFINITION 4.1. H *is said to be* 0-*minimal (zero-minimal) if and only if* $H \in C^*_{(0,0)}$ *and* $\forall H' \in C^*_{(0,0)}$ *such that* H' *is isomorphic to* H, $V_H(0) \leq V_{H'}(0)$. *If* $0 < \bar{y} \in H$, H *is said to be* $\bar{y}$-*minimal if and only if*

(i) $H \in C^*_{(\bar{y},\bar{y})}$,

(ii) H *is* $(\bar{y} - 1)$-*minimal*,

(iii) $\forall H'$ *satisfactory to* (i) *and* (ii), $V_H(\bar{y}) \leq V_{H'}(\bar{y})$.

H *is said to be quasi-minimal if and only if it is* $(n-1)$-*minimal*.

THEOREM 4.1. *Every minimal hypergroupoid is quasi-minimal.*

PROOF. Let H be minimal and not quasi-minimal. Let $\bar{y}$ be the minimum element of H such that H is not y-minimal. i.e. that $\forall y \in H : y < \bar{y}$, H is y-minimal, but not $\bar{y}$-minimal.

If $H \in C^*_{(\bar{x},\bar{y})}$, then there exists an isomorphism $f: H \longrightarrow H'$ which leaves unchanged all hyperproducts $x \circ y$ with $x \leq y < \bar{y}$, but such that $V_{H'}(\bar{y}) < V_H(\bar{y})$. But that is absurd since H is supposed to be minimal.

If $H \notin C^*_{(\bar{x},\bar{y})}$, let $(\bar{x},\bar{y})$ be the lowest pair such that $H \notin C^*_{(\bar{x},\bar{y})}$.

There exists, therefore, an isomorphism $f: H \longrightarrow H'$ which transform H in $H' \in C^*_{(\bar{x},\bar{y})}$ and such that

(a) $\forall (x,y) \in H^2_+ : (x,y) < (\bar{x},\bar{y}),\ x * y = x \circ y$ ($*$ = hyperoperation in H');

(b) $V(\bar{x} * \bar{y}) < V(\bar{x} \circ \bar{y})$.

This is absurd since H is supposed to be minimal.

As a consequence of the theorem just now proved, $\forall H$, MIN(H) is to be sought only among the hypergroupoids which are quasi-minimal and isomorphic to H.

If we call *minimal isomorphism of* H every isomorphism which transforms H into a quasi-minimal H', from theor. 4.1 it follows that

COROLLARY. *A quasi-minimal hypergroupoid* H *is minimal if and only if every minimal isomorphism transforms* H *in a hypergroupoid* H' *such that* $H \leq H'$.

REFERENCES

[1] M. KOSKAS, *Groupoides, demi-hypergroupes et hypergroupes*, J. Math. Pures et Appl., 49 (1970), pp. 155-192.

[2] P. CORSINI, *Ipergruppi semiregolari e regolari*, Rend. Sem. Mat. Univ. Torino, (1982), pp. 35-46.

[3] R. MIGLIORATO, *Una classe di semi-ipergruppi e di ipergruppi*, Atti Sem. Mat. Fis. Univ. Modena, XXXI, (1982), pp. 123-141.

[4] R. MIGLIORATO, *Ipergruppi di cardinalità 3 e isomorfismi d'ipergruppoidi t.r.*, Atti Convegno su Ipergruppi, semi-ipergruppi e altre strutture multivoche, Udine (1985), pp. 131-136.

[5] M. DE SALVO, *Ipergruppi finiti di lunghezza costante*, Rend. Ist. Lombardo, Accad. Sci. Lett. Sez. A, 120, (1986), pp. 41-56.

Annals of Discrete Mathematics 37 (1988) 311–314

SEMINVERSIVE PLANES

Domenico Olanda

Dipartimento di Matematica ed applicazioni "R.Caccioppoli" ,
Via Mezzocannone ,8 - 80134 Napoli (ITALY)*

Seminversive planes are defined and investigated. In the finite case such a plane is either an inversive plane or a punctured inversive plane.

1. INTRODUCTION

An H-semiaffine plane [1], where H is a finite set of non-negative integers, is a finite linear space (P,L) such that for any non-incident point-line pair (p, ℓ) the number $\pi(p,\ell)$ of lines on p not meeting ℓ belongs to H. Suppose (P,L) is an H-semiaffine plane and n+1 is the maximum number of lines on a point. Then the integer n is called the *order* of the plane. H-semiaffine planes have been investigated by several authors; here it turns out useful to recall the following theorem due to Oehler [3] concerning $\{1,2\}$-semiaffine planes.

THEOREM 1.1. Suppose (P,L) is a $\{1,2\}$-semiaffine plane of order $n\geq 5$. Then one the following holds :

(i) (P,L) is an affine plane of order n ;

(ii) (P,L) is a punctured affine plane of order n (i.e. an affine plane with one of its points deleted) ;

(iii) (P,L) is an affine plane of order n from which one line and all the points on it have been deleted.

This paper provides an application of the above theorem. We start with the following

DEFINITION. An H-inversive plane , H a set of positive integers, is a pair (Ω,C) where Ω is a set of *points* and C a family of subsets of Ω, called *circles* such that

1.1. any three distinct points lie on a unique circle;

1.2. given a circle B , a point $x\in B$ and a point $y\notin B$, the number of circles through x and y meeting B just at the point x belongs to H.

1.3. there exist at least two circles and every circle contains at least three points.

Obviously, when H = $\{1\}$ inversive plane are recovered. In this paper we shall investigate the $\{1,2\}$-inversive planes, we call *seminversive planes* under the assumption Ω is finite.

* Work supported by National Research Project on "Strutture Geometriche, Combinatoria loro applicazioni" of Italian M.P.I. and by G.N.S.A.G.A. of C.N.R.

Suppose (Ω, C) is a finite seminversive plane and set $n+1 = \{|B|, B \in C\}$ Then the integer n is defined to be the *order* of the plane. Clearly, the removing of one point from a finite inversive plane of order n results in a seminversive plane of order n. We shall show that the converse is also true; namely we shall prove the next

THEOREM 1.2. Suppose (Ω, C) is a finite seminversive plane of order $n > 5$. Then, (Ω, C) is either an inversive plane or a punctured inversive plane of order n.

2. SOME PROPERTIES OF THE FINITE SEMINVERSIVE PLANES.

(Ω, C) always denotes a finite seminversive plane of order $n > 5$. For any point $x \in \Omega$, let $\alpha_x = (\Omega_x, C_x)$ be defined as follows: $\Omega_x = \Omega - \{x\}$, $C_x = \{B - \{x\}, B \in C, x \in B\}$. Taking into account 1.1 and 1.2, α_x is a $\{1,2\}$-semiaffine plane for any point x. Let B_0 be a circle with n+1 points and x_0 one of them. Obviously, the $\{1,2\}$-semiaffine plane α_{x_0} has order n. Let y be a point other than x_0 since the circles through x_0 and y correspond to the lines on y in the plane α_{x_0} whose order is n through y there passes at least one circle of length n+1. Consequently, also the plane α_y has order n. Hence, the $\{1,2\}$-semiaffine planes α_x all have the same order n when x ranges over Ω; furthermore, all those planes are of the same type (i), (ii) or (iii) (cfr. theorem 1.1), as $|\Omega| = |\alpha_x| + 1$. Obviously, the next result holds.

PROPOSITION 2.1. If for any $x \in \Omega$, α_x is an affine plane of order n, then (Ω, C) is a finite inversive plane of order n.

Therefore, assume (Ω, C) is not an inversive plane that is, for any point x, the plane α_x is either a punctured affine plane or an affine plane with one of its lines deleted. Under these assumptions, since α_x is a $\{1,2\}$-semiaffine plane of order n, all the circles through x, with x deleted, have either n-1 or n points. Consequently,

PROPOSITION 2.2. On a circle of (Ω, C) either n or n+1 points lie.

In what follows a circle will be referred to as a *short* or a *long* circle according to its having size n or n+1.

Let x and y be any two distinct points. The circles on x and y are precisely the lines on y in the plane α_x; thus they number n+1. By (1.1) these circles with the points x and y deleted, partition $\Omega - \{x,y\}$. Denote λ and Λ the number of short and long circles on x and y respectively; then

$$(2.1)\qquad \begin{cases} \lambda(n-2) + \Lambda(n-1) = |\Omega| - 2 \\ \lambda + \Lambda = n+1. \end{cases}$$

In case $|\Omega| = n^2$, i.e. for any x, α_x is a punctured affine plane,

$$(2.2)\qquad \lambda = 1 \quad , \quad \Lambda = n.$$

If $|\Omega| = n^2-n+1$, i.e. for any point x, α_x is an affine plane with one of its lines deleted, then

$$(2.3)\qquad \lambda = n \quad , \quad \Lambda = 1.$$

Assume $|\Omega| = n^2-n+1$ and count in two ways the pairs $(\{x,y\}, L)$, $x \neq y$, $x,y \in L$, L a long circle; then,

(2.4) $$\binom{n^2-n+1}{2} = |\mathscr{L}| \binom{n+1}{2}$$

where $\mathscr{L}$ is the set of all long circles. By (2.4), n+1 divides $(n-1)(n^2-n+1)$, hence $n\leq 5$. Since $n>5$ is supposed, the following is proved.

PROPOSITION 2.3. If (Ω, C) is a finite seminversive plane of order n , then

1. $|\ \Omega\ | = n^2$,
2. through any two distinct points a unique short circle passes.

PROPOSITION 2.4. Through any point n+1 short circles pass.
Proof. Fix any point y and count in two ways the pairs $(\{x,y\}, S)$, $x\neq y$,$x,y\in S$, S a short circle. Thus $n^2 -1 = |\beta_y|(n-1)$ where β_y is the set of all short circles on y. The statement follows.

Let Δ denote any symbol and $(\Omega_0\ C_0)$ be defined by $\Omega_0 = \Omega \cup \{\Delta\}$, C_0 consists of all long circles of C and all the short ones to which Δ has been added.

PROPOSITION 2.5. (Ω_0, C_0) is an inversive plane of order n.
Proof. Through any three points in Ω_0 there passes a unique circle as a consequence of 1.1 and proposition 2.3 , 2. Let B be a circle of C_0, $x\in B$, $y\notin B$. We must show that there exists a unique circle, say D, through x and y which meets B exactly at the point x. Assume B is long and for any $x_j\in B$, $x_j\neq x$, denote L_j the unique circles through y, x, x_j . Such circles number n , any two of them are distinct and all meet B at two points.Since n+1 circles pass through x and y (even if $y=\Delta$, cfr. proposition 2.4) , the statement is true.If B is short and $x\neq \Delta$ the same argument applies. If $x=\Delta$, then, for any $x_j \neq \Delta$, the unique short circle L_j through y and x_j meets B at two points. Since there are n+1 short circles on y , n of which are secant, the statement follows.

Since the plane (Ω , C) is obtained deleting one point from the inversive plane (Ω_0, C_0) , theorem 1.2 is proved.

REFERENCES

[1] Beutelspacher, A. and Meinhardt,J., On finite h-semiaffine planes, in: Europ.J.Comb. 5,(1984) pp. 113-122.
[2] Dembowski,P., Finite geometries, Springer Verlag (1968).
[3] Oehler, M., Endliche biaffine inzidenzebenen , in: Geom.Ded.4,(1975).

Annals of Discrete Mathematics 37 (1988) 315–356

GEOMETRIC AND ALGEBRAIC METHODS IN THE CLASSIFICATION OF GEOMETRIES BELONGING TO LIE DIAGRAMS

Antonio PASINI

University of Siena. Department of Mathematics.
Siena. ITALY.

In this paper we give a survey of what is presently known on Tits geometries with Lie diagrams.

1. INTRODUCTION

The paper is divided into two parts. In Part I no finiteness or thickness assumption is made. This part does not contain any new result other than those given in [48] and [2]. But we have tried to exploit elementary methods as far as we could. In particular, we use coverings only for E_6, E_7 and E_8. We avoid them for all other diagrams.

Part II is devoted to the finite thick case and mainly to the diagrams C_n and F_4. As to A_n, D_n and E_6, all geometries belonging to any of these diagrams are buildings (Part I of this paper), so we have nothing else to say about them in Part II.

Some problems are proposed at the end.

The reader is referred to [48] and [4] for all definitions and basic results concerning Tits-Buekenhout geometries, diagrams, chamber systems and coverings. All geometries considered here are residually connected (hence, strongly connected by [7]). Given a geometry Γ, we denote the incidence relation and the type function of Γ by the symbols $*$ and τ respectively, as in [48]. Given a type 0 and an element x of Γ, the symbol $\sigma_0(x)$ denotes the 0-shadow of x (see [4]).

Here is the list of irreducible Lie diagrams:

A_n (n nodes, $n \geq 1$)
C_n (n nodes, $n \geq 2$)
D_n (n nodes, $n \geq 4$)
E_6
E_7
E_8
F_4
$G_2(6)$
$G_2(8)$

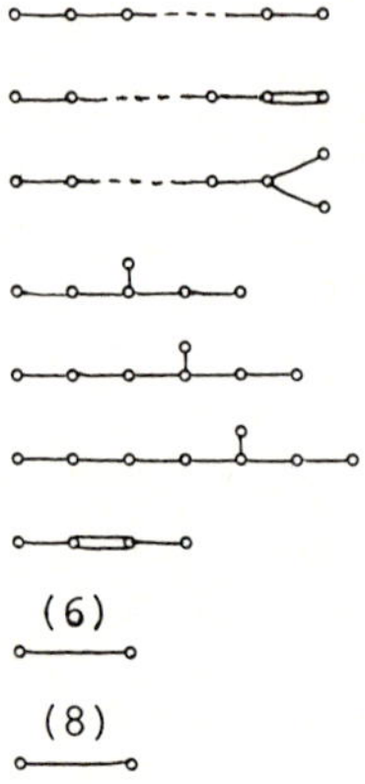

Lie diagrams are joins of irreducible Lie diagrams. They correspond to finite thick buildings (see [47] and [10]). We recall that a geometry with a disconnected diagram is the direct sum of irreducible subgeometries corresponding to the connected components of its diagram (see [4] and [48]). So only irreducible diagrams are worthy to be studied, in a geometric approach. We are not interested in the rank 2 case here. So, we shall not consider the diagrams $G_2(6)$ and $G_2(8)$.

2. PART I. THE GENERAL CASE

Exploiting elementary methods we shall deal with linear spaces and some other geometries which are not of Coxeter type. So, we have to recall some conventions and definitions on those geometries.

We use the symbols

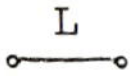

and
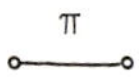

as in [3] and [4] to denote the class of linear spaces and the class of partial planes, respectively. Given a partial plane P, a set X of points of P is a *subspace* if every line which is not contained in X meets X in at most one point. Given a set X of points of P, the subspace $\bar{X}$ *spanned* by X is the smallest subspace of P containing X. A linear space L is a *linear plane* if, given any three non collinear points of L, they span the set of all

points of L. We denote the class of linear planes by the symbol
$$\underset{\circ\!\!-\!\!-\!\!-\!\!-\!\!\circ}{L}$$

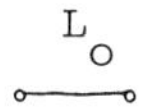

Affine planes of order greater than 2 and projective planes (even degenerate ones) are obvious examples of linear planes, whereas affine planes of order 2 (i.e. complete graphs on 4 vertices) are not such. Still, the system of points and lines of a matroid of dimension > 2 is a linear space but not a linear plane.

Given a geometry Γ belonging to the following diagram

$$\underset{0}{\circ}\overset{L}{-\!\!-\!\!-}\underset{1}{\circ}\overset{L}{-\!\!-\!\!-}\underset{2}{\circ}\overset{L}{-\!\!-\!\!-}\underset{3}{\circ}-\!\cdots\!-\underset{n-4}{\circ}\overset{L}{-\!\!-\!\!-}\underset{n-3}{\circ}\overset{\pi}{-\!\!-\!\!-}\underset{n-2}{\circ}\overset{\pi}{-\!\!-\!\!-}\underset{n-1}{\circ}$$

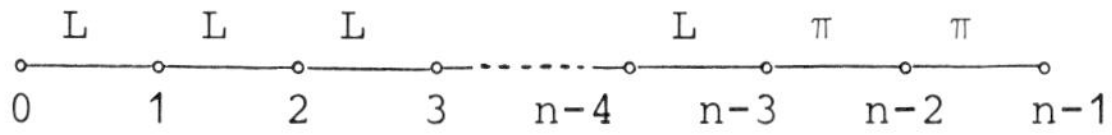

$(n \geq 3;\ 0, 1, \ldots n-1$ denote types)

elements of type 0 are called *points*, those of type 1 *lines* and those of type 2 *planes*. Two points are *collinear* if there is a line incident with both of them. If two points a and b are collinear then we write a $\underline{\ |\ }$ b. If $n > 3$, then elements of type n-1 are called *maximal subspaces* if their residues belong to the diagram

$$\underset{0}{\circ}\overset{L}{-\!\!-\!\!-}\underset{1}{\circ}\overset{L}{-\!\!-\!\!-}\underset{2}{\circ}-\!\cdots\!-\underset{n-3}{\circ}\overset{L}{-\!\!-\!\!-}\underset{n-2}{\circ}$$

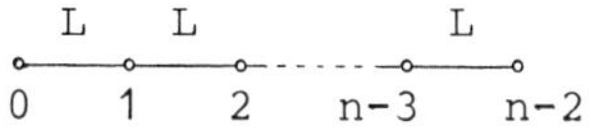

otherwise they are called *hyperlines*. This terminology is a mixture of that used in [48] and the one that is usual for polar spaces.

In particular, in the case of C_n those elements called hyperlines in [48] are called maximal subspaces here. In the case of a geometry Γ belonging to D_n the hyperlines of Γ in the meaning of [48] are the maximal subspaces of the 0-linearization Γ^{*0} of Γ. In the case of F_4, E_6, E_7 and E_8 our terminology is consistent with that of [48]. If Γ is a geometry belonging to E_6, E_7 or E_8, the hyperlines and the maximal subspaces of Γ in the usual meaning (that of [48]) are precisely the hyperlines and the maximal subspaces of the 0--linearization Γ^{*0} of Γ, in our meaning.

We have mentioned 0-linearizations. The reader is referred to [25] for the definition of the 0-*linearization* Γ^{*0} of a geometry Γ. We warn that the Intersection Property of [4] is assumed in [25], but it has not any essential role in that definition (see [26]).

Let Γ be still a geometry belonging to the above diagram. Let us assume that the *Intersection Property* (IP) holds in Γ (the reader can see [4] for the satement of that property). Then it is easily seen that the following properties hold in Γ:

(LL) *Given any two points, there is at most one line through them.*

(LH) *Given a line* x *and a hyperline* u, *if* $|\sigma_0(x) \cap \sigma_0(u)| \geq 2$, *then we have* x ∷ u.

(HH) *Given hyperlines* u,v *and points* a,b *such that* a ∷ u ∷ b ∷ v ∷ ∷ a, *if* $a \neq b$ *and* $u \neq v$, *then* $a \perp b$.

(we note that Theorem 7 of [3] is needed in order to prove (HH); see also Proposition 1 of the next paragraph).

2.1. The diagram A_n.

The following result is well known

THEOREM 1. (Tits [48], Proposition 6). *Geometries belonging to* A_n *are precisely* n-*dimensional generalized projective geometries (hence, buildings of type* A_n*).*

The proof of this statement is quite elementary. It essentially consists of two steps. We briefly recall them.

First step. Let Γ belong to A_n. Given any two points of Γ there is just one line through them. This is proved by induction on n. So, Γ forms a linear space. Moreover, 0-shadows are subspaces. This is proved again by induction on n. At this stage it is easily seen that 0-shadows uniquely determine those elements of which they are shadows and that incidences between elements of Γ can be viewed as inclusions of shadows as follows: given two elements x and y, we have $\sigma_0(x) \subseteq \sigma_0(y)$ iff x ∷ y and $\tau(x) \leq \tau(y)$. Moreover, intersections of shadows are still shadows. So Γ can be viewed as a system of subspaces of a linear space, closed under arbitrary intersections.

Second step. Now we can easily see that the system of subspaces constructed above is actually the system of all proper nonempty subspaces of an n-dimensional generalized projective geometry *G*. Showing this amounts to prove that the so-called 'triangle axiom' holds and that all subspaces of *G* actually occur as 0-shadows in Γ. All details of this proof are left to the reader. □

It is worthy of mention that we can prove the following more general result by almost the same argument:

PROPOSITION 1. (Buekenhout [3],Theorem 7). *Geometries belonging to the following diagram*

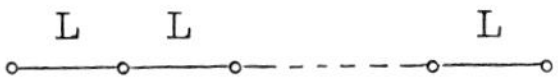

are precisely n-*dimensional matroids.*

Actually Theorem 1 is a trivial corollary of this Proposition. We warn that the Intersection Property is assumed in [3]. But the proof of Theorem 7 of [3] does not make any essential use of that property.

2.2. The diagram C_n.

Let us start from a more general diagram related to *bouquets of matroids* (in the meaning of Deza and Laurent [9])

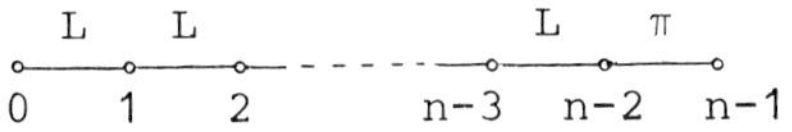

Let Γ be a geometry belonging to the diagram above. Let us consider the following property

$(LL)_{res}$ *The property* (LL) *holds in* Γ *and in the residue* Γ_F *of* F, *for every flag* F *of* Γ *of type* $\{0,1,\ldots i\}$ $(1 \leq i \leq n-4)$.

We have the following

PROPOSITION 2. *The geometry* Γ *is a bouquet of matroids (hence, the Intersection Property holds in it) if and only if the property* $(LL)_{res}$ *holds in* Γ.

The "only if" part is trivial. The "if" part can be proved by induction on n exploiting Proposition 1 and mimicking the first step of the proof of Theorem 1. The main point of the proof is to show that, given any two elements x,y of Γ, we have x :: y and $\tau(x) \leq \tau(y)$ if $\sigma_0(x) \subseteq \sigma_0(y)$. Assume that $\sigma_0(x) \subseteq \sigma_0(y)$. Take a point a in $\sigma_0(x)$. Given a line z incident with a and x and a point b in z, we have b :: y because b :: x and $\sigma_0(x) \subseteq \sigma_0(y)$. Let z' be the line in Γ_y incident with both b and a (see Proposition 1). We have z = = z' by (LL). Then z :: y. So $\sigma_1(a,x) \subseteq \sigma_1(a,y)$. So, we can apply the induction hypothesis on Γ_a (by $(LL)_{res}$) and we have the conclusion. From now on everything is easy. All details are left to the reader.

□

Proposition 2 already appears for C_n in [25] (Lemma 2) and [27] (Proposition 4). It appears also in [9] in a slightly different form (with (IP) instead of $(LL)_{res}$). The property $(LL)_{res}$ does not appear in [48]. Instead of it, the following property is considered there:

(O) *Let* x,y *be elements of type* $i < n-1$ *.If* $\sigma_0(x) = \sigma_0(y)$*, then* $x = y$.

(note that also this property is a consequence of the Intersection Property). Anyway, we have the following

LEMMA 1. *Let* Γ *be a geometry belonging to the following diagram*

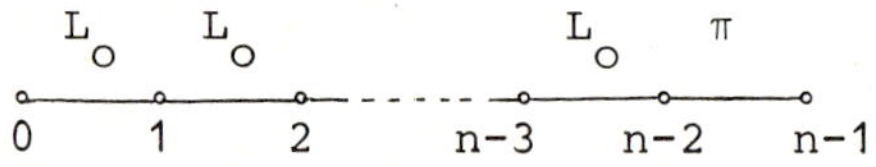

The property $(LL)_{res}$ *holds in* Γ *if and anly if both the properties* (LL) *and* (O) *hold in it.*

The property $(LL)_{res}$ implies ((LL) and) (O) because it implies (IP) by Proposition 2. Conversely, the properties (LL) and (O) imply $(LL)_{res}$. This can be proved mimicking the argument used in [8] (Section 6) to show that, if (LL) and (O) hold in a geometry $\bar{\Gamma}$ of type C_n, then they hold in the residue $\bar{\Gamma}_a$ for every point a of $\bar{\Gamma}$. Then an obvious inductive argument gives the conclusion. We have only to be careful in lifting (LL) from Γ to residues of points (the property (O) can be lifted to residues of points just as in [48]). Let us see this in detail. Let u,v be planes and x,y distinct lines through a point a and such that $x * u * y * v * x$. Let b,c be points on x and y respectively, different from a. Let d be any point in u. The lower part of Γ_u is a linear plane. So we can find a sequence $d_0, d_1, \ldots d_m = d$ of points in u such that $d_0, d_1 = a, b$ or c and d_i belongs to the line in u through d_j and d_k for suitable indices $j,k < i$ (this for $i = 2,3,\ldots m$). We can prove that $d_i * v$ for $i = 2,3,\ldots m$ by induction on i. Then $v * d = d_m$. Indeed, let j,k be as above. Let z be the line in u through d_j and d_k. We have $d_j * v$ and $d_k * v$ by the inductive hypothesis. So, let z' be the line in v through d_j and d_k. We have $z = z'$ by (LL). Then $d_i * v$ because $d_i * z = z'$. We are done. Then $d * v$. So we have $\sigma_0(u) = \sigma_0(v)$. Then $u = v$ by (O). Hence, (LL) holds in Γ_a. □

Let us come to C_n now. The following lemma has been proved in [28] (Lemmas 1 and 2 of [28]).

LEMMA 2. *Let* Γ *be a geometry belonging to* C_n

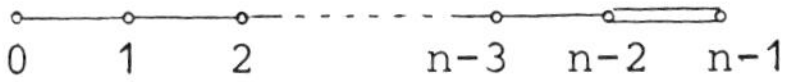

Then the following hold:

(i) Given elements x,y *of type* i *and* j *respectively, if* $j \geq i+1$ *then there are an element* u *of type* j *and an element* v *of type* j-i-1 *such that* $x * u * v * y$.

(ii) Given two distinct maximal subspaces, there is at most one element of type n-2 *incident with both of them.*

Now we are close to get the characterization of polar spaces given in Proposition 9 of [48]. But we shall not follow the methods of [48] and we avoid coverings. The following elementary lemma allows us to do that. We note that this lemma already appears in [48] (Section 6) for the rank 3 case.

LEMMA 3. *Polar spaces of rank* n *are precisely bouquets of matroids belonging to* C_n.

Of course, every polar space is a bouquet of matroids. Conversely, let Γ be a bouquet of matroids belonging to C_n. We must show that Γ is a polar space. We have to check axioms (P.1)-(P.4) of Chp.7 of [47]. Axioms (P.1) and (P.2) are trivial by Theorem 1. The existential part of (P.3) readily follows from (i) of Lemma 3. Let us prove the uniqueness part of (P.3). Let a be a point, u a maximal subspace and v,v' and w,w' maximal subspaces and elements of type n-2 respectively, such that $a * v * w * u$ and $a * v' * w' * u$. Let us assume that $(v,w) \neq (v',w')$. Then $v \neq v'$ by Lemma 2,(ii). If $w = w'$, then $a * w$ by the Intersection Property. So $a * u$ and we are done. If $w \neq w'$, let y be the element of type n-3 incident with both w and w' (y is uniquely determined in the residue of u). Let z be the element of Γ_v of type n-2 incident with both a and y and let z' be the analogous of z in $\Gamma_{v'}$. If $z \neq z'$, then a ⁚⁚ y by the Intersection Property. So a ⁚⁚ u and we are done. If z = z', then z,w,w' and v,v',u form a proper triangle in the upper part of the residue of y. This is impossible because that part is a generalized quadrangle. Axiom (P.3) is proved. We have still to prove Axiom (P.4). We need some preliminary steps.

Step 1. Let a,b,c be pairwise distinct points and x,y,z pairwise distinct lines such that a ⁚⁚ z ⁚⁚ b ⁚⁚ x ⁚⁚ c ⁚⁚ y ⁚⁚ a. Then there is a plane incident with all of x,y,z and a,b,c (provided that n > 2, of course). Indeed, let u be a maximal subspace incident with x. If a ⁚⁚ u, then both y and z are incident with u, by the Intersection Property. So we find a plane as above in Γ_u. Let us assume that a $\notin \sigma_0(u)$. In Γ_b we find a maximal subspace v and an element w of type n-2 such that z ⁚⁚ v ⁚⁚ w ⁚⁚ u. Similarly we find a maximal subspace v' and an element w' of type n-2 in Γ_c such y ⁚⁚ v' ⁚⁚ w' ⁚⁚ u. We have v = v' and w = w' by Axiom (P.3) on the pair (a,u). Then x ⁚⁚ w in Γ_u. Then x ⁚⁚ v. In Γ_v we find that plane which we are looking for.

Step 2. Let u be a maximal subspace and let a be a point not incident with u. Let v and w be the maximal subspace and the element of type n-2, respectively, such that a ⁚⁚ v ⁚⁚ w ⁚⁚ u (Axiom (P.3)). Then $\sigma_0(w) = \sigma_0(u) \cap a^{\perp}$, where $a^{\perp} = \{b \mid b \perp a \text{ and } b \text{ is a point}\}$. The inclusion $\sigma_0(w) \subseteq \sigma_0(u) \cap a^{\perp}$ is trivial. Let us prove the converse inclusion. Let $b \in \sigma_0(u) \cap a^{\perp}$. Let $c \in \sigma_0(w)$. Assume that $b \neq$ $\neq c$. Let x,y,z be the lines through b and c, through c and a and through a and b respectively. If two of them coincide, then they are all equal by the Intersection Property, so we have the contradiction a ⁚⁚ u. Then they are pairwise distinct. Then there is a plane p incident with all of x,y,z and a,b,c (Step 1). If n = 3, then p = v and w = x by Axiom (P.3). So b ⁚⁚ w. If n > 3, let v',w' be a maximal subspace and an element of type n-2 in Γ_x such that p ⁚⁚ v' ⁚⁚ w' ⁚⁚ u (recall that x ⁚⁚ u by the Intersection Property). We have a ⁚⁚ v'. Then v = v' and w = w' by Axiom (P.3). We have x ⁚⁚ w. Then b ⁚⁚ w. We are done.

Now we can prove Axiom (P.4). By induction on n. If $n = 2$ we have nothing to prove. Assume that $n > 2$. Assume that any two maximal subspaces meet in some point, by way of contradiction. Let u be any maximal subspace and let a be a point in u. By the inductive hypothesis we find in Γ_a some maximal subspace v such that $\sigma_0(u) \cap \sigma_0(v) = \{a\}$. Let us assume that there is some point b such that $b \not\perp a$. Then $b \notin \sigma_0(u) \cup \sigma_0(v)$. Let u', v' and x, y be maximal subspaces and elements of type n-2 respectively such that $b :: u' :: x :: u$ and $b :: v' :: y :: v$ (Axiom (P.3)). We have $\sigma_0(x) = \sigma_0(u) \cap b^{\perp}$ and $\sigma_0(y) = \sigma_0(v) \cap b^{\perp}$ (Step 2). Then $a \notin \sigma_0(x) \cup \sigma_0(y)$. So $\sigma_0(x) \cap \sigma_0(y) = \emptyset$ because $x :: u$, $y :: u$ and $\sigma_0(u) \cap \sigma_0(v) = \{a\}$. But we have $\sigma_0(v') \cap \sigma_0(u) \neq \emptyset$ according to our hypotheses. Let $c \in \sigma_0(v') \cap \sigma_0(u)$. Then $c \perp b$. Then $c \in \sigma_0(x)$ (Step 2). Let z be the line through b and c. The line z is incident with both u' and v', by the Intersection Property. Then it meets $\sigma_0(y)$ in a point d, different from both b and c (recall that $\sigma_0(x) \cap \sigma_0(y) = \emptyset$ and $c :: x$). Of course, we have both $a \perp c$ and $a \perp d$. Then (Step 1) there is a plane w incident with z and with the lines z_c through a and c and z_d through a and d. So $b :: w$ because $b :: z$: we have $b \perp a$ in Γ_w. But this contradicts the assumption that $a \not\perp b$. Then $a \perp b$ for every point b. But the point a was arbitrarily chosen. Then any two points are collinear. Then all maximal subspaces are incident with the same set of points, by Step 2. Then there is just one maximal subspace, by the Intersection Property. We have the final contradiction. □

Now we can state the following

THEOREM 2. (Tits [48], Proposition 9). *Let* Γ *be a geometry belonging to* C_n. *The geometry* Γ *is a polar space if and only if both the properties* (LL) *and* (O) *hold in it.*

Trivial, by Proposition 2 and Lemmas 1 and 3. □

2.3. Non-building geometries of type C_n and geometries 2-covered by polar spaces.

Several examples are known of geometries of type C_n which are not buildings. Some of them are got by free constructions (see [48]).

Geometries constructed in that way are infinite. A lot of finite non thick examples can be found in [34] and [39]. All of them have some thin lines (i.e. lines with exactly 2 points). Geometries of type C_n with some thin lines can be provided also by the construction of [40]. They are not buildings provided that some of the geometries from which we start is not such. Here is another non--building non thick example. It is infinite, but it has thick lines.

Example 1. Let $\overline{\Gamma}$ be a thick building of type D_n. Here n = 3 is allowed. In that case D_3 is the same as A_3

Let γ be a non special involutory automorphism of $\overline{\Gamma}$ that does not fix any flag of $\overline{\Gamma}$ (when n = 3 the automorphism γ is an involutory polarity without any absolute element). It is clear that γ exists only if the ground field of $\overline{\Gamma}$ is infinite when $n > 3$, and only if the ground division ring of $\overline{\Gamma}$ is an infinite field when n = 3. Let $\overline{\Gamma}^{*0}$ be the 0-linearization of $\overline{\Gamma}$. The quotient $\Gamma = \overline{\Gamma}^{*0}/\gamma$ is a geometry of type C_n that is not a polar space. Anyway, it is covered by a polar space, namely by the polar space $\overline{\Gamma}^{*0}$.

Example 2. The construction described above works even if we start from the Coxeter complex of type D_n instead of thick buildings of that type. In that case we find thin geometries of type C_n which are proper quotients of Coxeter complexes of type C_n. For instance in the case of n = 3 we get the quotient of the octohedron by the antipodality relation. This small C_3 geometry has the following feature, which will turn out to be fairly interesting in the forthcoming: all of its points are incident with all of its planes.

Example 3. A nice thick but infinite example of a non-building C_3 geometry Γ is given in [37], starting from a hyperbolic quadric Q in PG(5,R) (where R is an ordered field) and a plane P exterior to Q. The set of planes of Γ is one of the two families of planes of Q. We take the points of Q as lines of Γ and the points of P as points of Γ. Incidences are defined as follows. A point of Q (i.e. line of Γ) and a point of P (point of Γ) are incident if they are orthogonal in the bilinear form defining Q. Lines and planes of Γ are incident if they are incident as points and planes of Q. Every point of Γ is incident with all planes of Γ .So,we get a C_3 geometry which is not a building: indeed all of its points are incident with all of its planes. Anyway, it can be shown that Γ is covered by a building (see [37]).

Example 4 (the A_7-geometry). The construction described in Example 3 can be rearranged to get geometries of type C_3 over any field F, replacing the plane P with a suitable set of points X 'exterior' to Q (see [37]),provided that such a set exists.We warn that X cannot be a plane if F is not ordered.Lunardon [17] has proved that such sets do not exist when F is a finite field of order $\neq 2$ (see also [8] for the case of even characteristic; other partial results in the same direction of [17] appeared in [37] and [49]. But,when F = GF(2),such an 'exterior' set actually exists (see [37]). In that case we get a finite thick C_3-geometry Γ with parameters 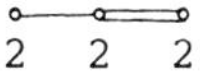

.

The geometry Γ constructed in this way has still the property that each of its points is incident with all of its planes. It is simply connected. So it is not covered by a building.The full automorphism group of Γ is the alternating group A_7. So Γ is called *A_7-geometry*. It is the only finite thick example of non-building C_n-geometry presently known.It appeared firstly in [19]. It can be characterized in several ways. A group theoretic construction can be found in [1]. From that construction we get the following elementary description. The alternating group has two orbits, each of size 15, on the set of 30 planes that can be drawn on a given set S of 7 objects. Take one of those two orbits as set of planes, S as set of points and all 3-subsets of S as lines. Define incidences in the natural way. We get the A_7-geometry.

We have met the following property very often in the examples of non-building C_3-geometries given above: all points are incident with all planes. More remarkably, that property holds in the only known example of finite thick non-building geometry of type C_3. So it deserves a name. We say that a geometry Γ belonging to C_n

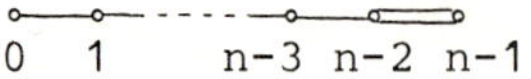

is *flat* if all elements of Γ of type less than n-2 are incident with all maximal subspaces of Γ.

Not all examples of non-building C_n geometries given above are so far from polar spaces. Indeed some of them are quotients of polar spaces. The following proposition gives us a practical criterion to check whether a C_n geometry is a quotient of a polar space or not.

PROPOSITION 3. (Tits [48]). *Let Γ be a geometry of type C_n. The universal 2-covering of Γ is a building if and only if all C_3-residues of Γ are covered by buildings.*

This proposition is a specialization of Theorem 1 of [48]. The proof is long and hard and it is not possible to give even a sketch of it here. We prefer to mention two results which can be proved with the aid of Proposition 3.

PROPOSITION 4. (Pasini and Rees [34]). *Let Γ be a geometry of type C_n and let us assume that all lines of Γ are thick and every residue of Γ of type C_3 is either a building or flat. Then either Γ is 2-covered by a polar space or it is flat.*

PROPOSITION 5. (Rees [38]). *Let Γ be a geometry of type C_n admitting parameters as below*

(x = ∞ is allowed)

Then Γ is 2-covered by a polar space.

Proposition 4 is proved in the following way. If all C_3 residues of Γ are flat, then Γ is flat. Otherwise, if there is some C_3 residue Γ' which is a polar space, then, given any flat residue Γ", we can construct a covering from Γ' to Γ" inside Γ. So we can apply Proposition 3 and we have the conclusion.

Proposition 5 is proved showing firstly that every C_3 geometry with parameters 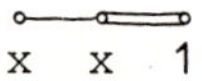

is a quotient of a polar space as in Examples 1 and 2 (Rees [38]). Then the conclusion follows from

Proposition 3.

2.4. The diagram D_n.

We give an elementary proof of the following result

THEOREM 3. (Timmesfeld [46], Lemma 3.3). *All geometries belonging to* D_n *are buildings.*

Let Γ be a geometry belonging to the diagram

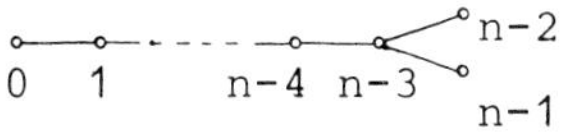

The 0-linearization Γ^{*0} of Γ belongs to C_n. Let us prove that $(LL)_{res}$ holds in Γ^{*0}. Having proved that (LL) holds will be enough (by an easy inductive argument: indeed we have $(\Gamma^{*0})_a = (\Gamma_a)^{*1}$ for every point a of Γ).So, let a,b be distinct points and x,y lines of Γ^{*0} incident with both a and b. In Γ^{*0} we can take a maximal subspace u incident with y. By Lemma 2 there are a maximal subspace v and an element w of type n-2 in the residue $(\Gamma^{*0})_a$ of a in Γ^{*0} such that $x * v * w * u$. But elements of type n-2 of Γ^{*0} are flags of type {n-2,n-1} in Γ.Then $v * u$ in Γ.The residue Γ_v of v in Γ is a projective geometry by Theorem 1. Then $x * u$ in Γ_v because both x and u are incident with both a and b and $a \neq b$. Then we have $x = y$ in the projective geometry Γ_u. So (LL) holds in Γ^{*0}. Then Γ^{*0} is a polar space by Proposition 2 and Lemma 3, and Γ is the oriflamme complex of the polar space Γ^{*0} (see [47],Chp.7). Hence Γ is a building. □

2.5. The diagram F_4.

We can still avoid coverings. Let us start from the following

LEMMA 4. *Let* Γ *be a geometry belonging to the following diagram*

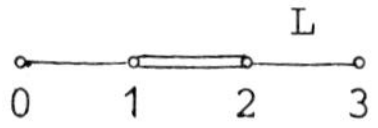

The Intersection Property (IP) *holds in* Γ *if and only if all the properties* (LL),(LH) *and* (HH) *hold in* Γ .

The "only if" part is trivial.As for the "if" part, first of all we observe that (IP) holds in a geometry Γ with string basic diagram if and only if the statement of (IP) holds with respect to the initial node of the diagram. The proof of this fact is implicit in Theorem 6 of [3]. So we have to prove only that the statement of (IP) holds with respect to the type 0. This can be done by elementary arguments. They are left to the reader. □

So we have the following

THEOREM 4. (Tits [48], Proposition 9). *A geometry Γ of type F_4 is a metasymplectic space (hence, a building) if and only if all the properties* (LL), (LH) *and* (HH) *hold in it.*
The "only if" part is well known. Let us prove the "if" part. Assume that (LL), (LH) and (HH) hold. Then (IP) holds by Lemma 4. Then Γ is a system of subspaces in the meaning of Theorem 6 of [3]. It is easily seen that all axioms given in Chp.10 of [47] hold. Then Γ is a metasymplectic space by n.10.3 of [47]. □

Few examples are presently known of non-building F_4 geometries. Some of them are constructed taking proper quotients of the Coxeter complex of type F_4. They are thin. Other examples can be constructed starting from a non-building geometry Γ of type C_4

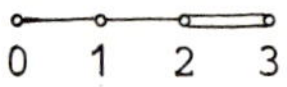

and taking the 1-linearization Γ^{*1} of Γ. It is worth observing that 2-coverings (or quotients) of Γ^{*1} are precisely 1-linearizations of 2-coverings (quotients, respectively) of Γ. So, geometries of type F_4 constructed in this way are 2-covered by buildings precisely when the C_4 geometry from which we start is such. Of course if we start from polar spaces then we get metasymplectic spaces. For instance, let $\overline{\Gamma}$ be a building of type D_4

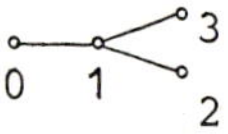

Let $\Gamma = \overline{\Gamma}^{*0}$ be the polar space of $\overline{\Gamma}$ (i.e., the 0-linearization of $\overline{\Gamma}$). Then $\Gamma^{*1} = \overline{\Gamma}^{*1}$. That is, Γ^{*1} is the usual metasimplectic space of $\overline{\Gamma}$. Assume that $\overline{\Gamma}$ admits a non special involutory automorphism γ as in Example 3 of §2.3 of this paper. Let us now set $\Gamma =$ $= \overline{\Gamma}^{*0}/\gamma$. Then $\Gamma^{*1} = \overline{\Gamma}^{*1}/\gamma$. So $\overline{\Gamma}^{*1}$ is the universal 2-covering of

Γ^{*1}.

Anyway, only non thick examples can be constructed in this way. It is worth remarking that no finite thick non-building F_4 geometry is presently known.

As in the case of C_n, it is interesting to know if a given F_4 geometry is a quotient of a metasymplectic space or not. The following specialization of Theorem 1 of [48] gives us a practical criterion to test that.

PROPOSITION 6. (Tits [48]). *Let Γ be a geometry of type F_4. The universal 2-covering of Γ is a building if and only if all C_3 residues of Γ are covered by buildings.*

2.6. The diagrams E_6, E_7 and E_8.

We mark types as below

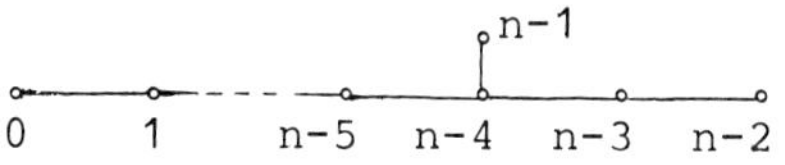

(n = 6, 7 or 8)

If Γ is a geometry belonging to the diagram above, then elements of Γ are said to be points, lines, planes, hyperlines or maximal subspaces according to whether they are such in the 0-linearization Γ^{*0} of Γ. We say that the properties (LL), (LH) or (HH) hold in Γ if they hold in Γ^{*0}.

THEOREM 5. (Brouwer and Cohen [2]). *All geometries of type E_6 are buildings. A geometry of type E_7 is a building if and only if the property* (LH) *holds in it. A geometry of type E_8 is a building if and only if both the properties* (LH) *and* (HH) *hold in it.*

The proof given in [2] essentially exploits the fact that the universal 2-covering of a geometry Γ of type E_6, E_7 or E_8 is a building (Tits [48], Theorem 1). So, let Γ be a geometry of type E_6, E_7 or E_8 and let $f: \overline{\Gamma} \to \Gamma$ be 'the' universal 2-covering of Γ. The geometry $\overline{\Gamma}$ is a building. If Γ is of type E_7 we assume (LH) on Γ. If Γ is of type E_8 we assume both (LH) and (HH). Let us see the proof of [2] in detail.

First of all we must prove that f induces isomorphisms on the residues of elements of $\overline{\Gamma}$ .This follows from Theorems 1 and 3 and from the fact that buildings are 2-connected (Tits [48],Theorem 1) in the case of E_6. The same argument works in the case of E_7, provided that we assume to have already proved that all geometries of type E_6 are buildings. In the case of E_8 we must in advance prove that (LH) holds in the residue of every point of Γ.This can be done exploiting (LH) in Γ.Indeed,let a be a point of Γ,let x,y be distinct lines through a and p and u a plane and a hyperline respectively, both incident with both of x and y. In Γ_p we see that every point of x is collinear with all points of y. So the same is true in Γ_u by (LH) in Γ.Then there is a plane q in Γ_u incident with both of x and y, because Γ_u is a building by Theorem 3. The residue $\Gamma_{a,x}$ of the flag {a,x} is a building of type E_6 (assume to have already proved our statement on E_6).Then,by a well known property of such buildings,there is a hyperline v incident with both p and q. But Γ_v is a building,by Theorem 3. Then we have p = q in Γ_v. So p * u. Then (LH) holds in Γ_a.

By the way, the last part of this argument actually proves a little bit more than what we have explicitly stated: namely,it proves that (LL) holds in all E_7 geometries.

Let us come back to E_8. As (LH) holds in Γ by assumption, f induces isomorphisms on residues of elements of Γ,because all buildings are 2-connected and all residues of elements of Γ are buildings,either by Theorem 1,or by Theorem 3, or by that part of the statement of the present theorem that is related to E_7, because (LH) holds in residues of points of Γ.

So,f is an isomorphism if it is injective on the set of points of $\overline{\Gamma}$ .Now everything is as in §6 of [48].In the case of E_6 the injectivity readily follows from the fact that,given any two points in a building of type E_6, there is always some hyperline incident with both of them.The 2-covering f induces isomorphisms on residues of hyperlines and we have the conclusion. Let Γ be of type E_7.Assume that there are two points a,b in $\overline{\Gamma}$ such that f(a) = f(b) but a $\neq$ b. There is not any hyperline in $\overline{\Gamma}$ incident with both a and b, because f induces isomorphisms on residues of hyperlines of

$\overline{\Gamma}$. Anyway, taken any hyperline u incident with b, there is a point c such that a $\perp$ c :: u, by a well known property of E_7 buildings. Let x be the line in $\overline{\Gamma}$ through a and c. We have f(u) :: f(c) in Γ by (LH). Then x :: u because f induces an isomorphism on $\overline{\Gamma}_c$. Then both a and b are incident with u. We have a contradiction.

A similar argument works in the case of E_8. Now we should exploit the following property of E_8 buildings: given a point a and a hyperline u, there are a point b and a hyperline v such that a :: v :: :: b :: u. All details are left to the reader. Anyway, the reader can find them in §6 of [48].

So far as to the "if" part in the cases of E_7 and E_8. The "only if" part follows from the fact that the Intersection Property holds in all buildings (Tits [47], Chp.12). □

An example of a geometry Γ of type E_7 that is a proper quotient of a building $\overline{\Gamma}$ is given in [2]. The building $\overline{\Gamma}$ is defined over the field of complex numbers. Hence, Γ is infinite.

The following proposition has already been used in the proof of Theorem 5. But we shall need it again later. So, we mention it explicitly.

PROPOSITION 7. (Tits [48]). *All geometries of type* E_7 *or* E_8 *are 2-covered by buildings.*

TABLE 1 (Summary of Part I).

A_n	building
D_n	building
E_6	building
E_7	building iff (LH)
E_8	building iff (LH) and (HH)
C_n	building iff (LL) and (O)
F_4	building iff (LL), (LH) and (HH)

2.7. Comments and remarks.

2.7.1. Attempts have lately been done to describe geometries of spherical type with some thin lines in terms of 'products' of small geometries whose types are involved in the diagram of the geometry that has to be described (see [43] and [41]). Those attempts develope ideas which already appear in [6] and [40] for C_n geometries.

2.7.2. An approach different from those followed here (namely,the elementary one,for A_n,D_n,C_n and F_4,and that one which exploits coverings,for E_6,E_7 and E_8) is used by Ott in [23],and developed by Hillebrandt in [11]. It still uses coverings, but it focuses onto apartments.

2.7.3. Many people will object that we have been wrong in chosing not to put enough emphasis on coverings and in particular on Tits' celebrated Theorem 1 of [48]. We might answer that we just wanted to point out that geometries with Lie diagrams have so particular properties (because of the very features of their diagrams) that even the most significant results in diagram geometry can sometimes be left in the background when we deal with such geometries.

3. PART II. THE FINITE THICK CASE

We recall that a geometry is *thick* if each of its flags of corank 1 is contained in at least three chambers. Polar spaces got from thick buildings of type D_n are not thick in this sense. Anyway, they are 'almost thick': all of their lines are thick (contain at least 3 points).And they are not so much different from thick polar spaces. So,C_n geometries with thick lines will be included in this section.

3.1. The work by Brouwer and Cohen. The diagrams E_7 and E_8.

THEOREM 6. (Brouwer and Cohen [2]). *All finite thick geometries of type E_7 or E_8 are buildings.*

The proof depends on a nice result on regular graphs. Brouwer and Cohen consider finite regular graphs G with certain additional regularity properties (the reader is referred to [2] for all details). They prove that every fixed-point-free automorphism of G maps some vertex onto a vertex adjacent with it. The methods used by Brouwer and Cohen are similar to those employed by Feit and Higmann in [10], and remind us the usual way to prove that every polarity in a finite projective plane admits some absolute elements. Actually, this latter statement is a corollary of that proved by Brouwer and Cohen. Indeed, if we take points and lines of a finite projective plane and define the adjacency by means of incidences in the projective plane, we get a graph satisfying all properties assumed in [2]. A polarity of the projective plane is a non identical automorphism of this graph. Then it maps some vertex onto a vertex adjacent with it. That is, it has some absolute elements.

In the case of a finite thick geometry Γ of type E_7 or E_8, the graph G is the collinearity graph of the universal 2-covering $\overline{\Gamma}$ of Γ. We already know that $\overline{\Gamma}$ is a (finite, thick) building (Proposition 7). Then G has all properties assumed in [2]. Then every fixed-point-free deck transformation for a 2-covering f: $\overline{\Gamma} \to \Gamma$ (see [48]) identifies a pair of collinear points. But f cannot identify collinear points. Then every deck transformation for f fixes some points. But f is a 6-covering in the case of E_7, by Theorems 1,3 and 5 and because buildings are 2-connected ([48], Theorem 1). It is a 7-covering in the case of E_8, by Theorems 1,3 and by the statement on E_7 of the present theorem (assume to have already proved it at this stage) and because buildings are 2-connected. Then a deck transformation is the identity if it fixes some point. Then there is just one deck transformation, namely the identity. So f is an isomorphism. □

By methods similar to those sketched above for E_7 and E_8, Brouwer and Cohen get also the following

PROPOSITION 8. (Brouwer and Cohen [2]). *Let Γ be a finite geometry of type C_n with thick lines or a finite thick geometry of type F_4. The geometry Γ is a building iff it is 2-covered by a building.*

In the case of C_n it is possible to give a proof easier than that sketched above for E_7 and E_8, exploiting only the fact that every polarity in a finite projective geometry admits some absolute points. Brouwer and Cohen used this argument in an earlier version of [2]. By the way, it will be clear that the same argument can work even if we replace the finiteness assumption with other assumptions (for instance, that the geometry is 2-covered by a building defined over an algebraic extension of a finite field), provided that they are sufficent to get the same conclusion on polarities.

The proof is by induction on n. If $n = 2$ there is nothing to prove. Assume $n > 2$. Let us assume that we have a 2-covering $f:\overline{\Gamma} \to \Gamma$ from a building $\overline{\Gamma}$ to Γ. Assume that f is not an isomorphism. Then f must identify two maximal subspaces u,v of $\overline{\Gamma}$, because f induces isomorphisms on residues of maximal subspaces (by Theorem 1 and because all buildings are 2-connected). We have $\sigma_0(u) \cap \sigma_0(v) = \emptyset$ by the inductive hypothesis and because buildings are 2-connected. For every element x of $\overline{\Gamma}_u$, let $\phi(x)$ be that element of $\overline{\Gamma}_v$ such that $f(x) = f(\phi(x))$ (the element $\phi(x)$ is uniquely determined because f induces isomorphisms from $\overline{\Gamma}_u$ and $\overline{\Gamma}_v$ to $\Gamma_{f(u)}$). Let $\alpha(x)$ be that element of $\overline{\Gamma}_u$ such that $\sigma_0(\alpha(x)) = \sigma_0(u) \cap (\phi(x))^{\perp}$. The element $\alpha(x)$ is uniquely determined because $\overline{\Gamma}$ is a building. It is easily seen that the mapping α: $x \to \alpha(x)$ is a polarity in the projective geometry $\overline{\Gamma}_u$. It admits some absolute point a. We have $a \perp \phi(a)$ by the definition of α. Then f identifies two collinear points, namely a and $\phi(a)$. This is impossible. Then f must be an isomorphism. □

Proposition 8 potentially has a lot of consequences. Here are two of them.

COROLLARY 1. (Pasini [28]). *Let Γ be a finite geometry of type C_n $(n \geq 4)$ with thick lines. Let us assume that every C_3 residue of Γ is either a building or flat. Then Γ is a polar space.*

This result follows from Propositions 4 and 8 and from the fact that finite flat geometries of type C_n with thick lines cannot exist if $n > 3$.

COROLLARY 2. *Let Γ be a finite geometry of type C_n with thick lines and let us assume that every element of Γ of type* $n-2$ *is incident with exactly two maximal subspaces. Then Γ is a polar space.*
Trivial, by Propositions 5 and 8.

3.2. The work by Aschbacher. Flag-transitive locally classical geometries of type C_n and F_4.

A finite geometry Γ of type C_3

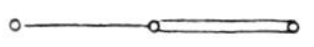

points lines planes

is *locally classical* (or of *classical type*) if residues of planes of Γ are desarguesian and residues of points of Γ are generalized quadrangles of one of the following Chevalley groups, or dual of such generalized quadrangles. We include also $P\Omega_4^+(q)\cdot 2$ in the following list, even if it is not a Chevalley group (actually, it is not included in the list of [1]). We do not give dual examples.

Group	parameters
$P\Omega_5(q) \simeq PSp_4(q)$	 q q
$P\Omega_6^-(q) \simeq PSU_4(q^2)$	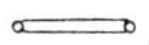 q q^2
$PSU_5(q^2)$	 q^2 q^3
$P\Omega_4^+(q)\cdot 2$	 q 1

In those cases of isomorphism quoted above, the two isomorphic groups give rise to the same generalized quadrangle (up to duality). It is worth mentioning also another (exceptional) isomorphism $PSp_4(3) \simeq PSU_4(2^2)$. But in this case we get non isomorphic generalized quadrangles, with parameters (3,3) and (4,2) respectively (the group $PSp_4(3) \simeq PSU_4(2^2)$ admits two non isomorphic BN-pairs).
A finite geometry of type C_n or F_4 is *locally classical* (or of *classical type*) if all of its C_3 residues are such. Of course, all locally classical finite geometries of type C_n have thick lines

and those of type F_4 are thick. In detail, the following table gives the parameters that can occur together with the Chevalley groups that give rise to buildings with those parameters, if such buildings exist. We list also parameters for which no building exists. In that case we write NOTHING in the group column. We shall see later that in those cases no geometry exist at all. We have inserted also the group $P\Omega^+_{2n}(q)\cdot 2$ in this list, even if it is not a Chevalley group. We omit dual cases for F_4. By the way, the following list gives all finite buildings of type C_n $(n \geq 3)$ with thick lines and all finite thick buildings of type F_4 (by Tits [47]).

TABLE 2

parameters (q is a prime power).	group.
 q q ... q q q	$PSp_{2n}(q)$ or $P\Omega_{2n+1}(q)$
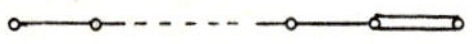 q q ... q q q^2	$P\Omega^-_{2n+2}(q)$
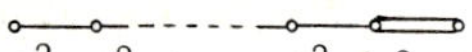 q^2 q^2 ... q^2 q^2 q	$PSU_{2n}(q^2)$
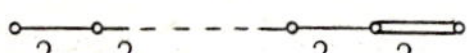 q^2 q^2 ... q^2 q^2 q^3	$PSU_{2n+1}(q^2)$
 q^3 q^3 ... q^3 q^3 q^2	NOTHING
 q q ... q q 1	$P\Omega^+_{2n}(q)\cdot 2$
 q q q q	$F_4(q)$
 q q q^2 q^2	${}^2E_6(q^2)$
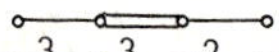 q^3 q^3 q^2 q^2	NOTHING

We warn that locally classical geometries are usually defined in a way more general than here. But the previous definition is sufficient here, in view of Theorems 1, 2, 3 and 6.

We have the following

THEOREM 7. (Aschbacher [1]). *Let Γ be a finite locally classical geometry of type* C_n *or* F_4 *and let us assume that* Aut(Γ) *is flag--transitive. Then Γ is either a building or the* A_7*-geometry.*
The reader is referred to Example 4 of Section 2.3 for the definition of the A_7-geometry. Aut(Γ) denotes the group of all special (i.e. type preserving) automorphisms of Γ. "Flag-transitive" is the usual way to say 'transitive on the set of chmabers'.
The main step of the proof is to prove the statement on C_3. Aschbacher does it considering a counterexample to that statement and getting a contradiction exploiting the classification of flag--transitive subgroups of Chevalley groups by Seitz [44]. He exploits that classification several times and his proof is not completely elementary. I give a sketch of a more elementary proof here. The reader can find all details in [33].
Given a point a and a line or a plane x, let θ(a,x) be the set of all lines (planes,respectively) in Γ_a that are not coplanar (collinear) with x in Γ_a. We have the following

(*) The stabilizer G_a of a in Aut(Γ) acts transitively on θ(a,x).

We can prove this using Seitz's Theorem [44] (and this is the only step where we really need that theorem).Here are the only two exceptional cases to examine (by [44]):

(1) The residue Γ_a of a is of type $PSp_4(2)$ and G_a acts as A_6 on it.

(2) The residue Γ_a of a is of type $PSp_4(3)$ and the action of G_a on it is either that of the stabilizer L in $PSU_4(2^2)$ of a line of the polar space of $PSU_4(2^2)$, or that of L•2.

In all other cases (*) follows from the fact that all classical generalized quadrangles are Moufang. In Case (1) the statement (*) follows from the action of A_6 on residues of points in the A_7-geometry. In Case (2) the actions on Γ_a and Γ_u of the stabilizer in Aut(Γ) of a point-plane flag {a,u} do not fit together. Then, Case (2) cannot occur.Then (*) holds. Now, by elementary geometric arguments, we can prove the following:

(**) Let Γ be a finite geometry of type C_3 with thick lines. Let us assume that Aut(Γ) is flag-transitive and that (*) holds. Then either Γ is a building or it is flat with unifor parameter 2

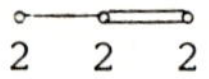

Now we can get the conclusion in two ways. Either we recognize the A_7-geometry inside our geometry exploiting the flag-transitivity of Aut(Γ), or we use the following result by Rees [36]: the A_7-geometry is the only flat geometry with uniform parameter 2.
Aschbacher uses again Seitz's Theorem in the case of C_n $(n \geq 4)$ and F_4. Anyway, the statement on C_n $(n \geq 4)$ easily follows from that on C_3 and Corollary 1 of §3.1 of this paper. The statement on F_4 can be proved exploiting that on C_3 and the following statement proved in [27] (§5): let Γ be a finite thick geometry of type F_4 with uniform parameter

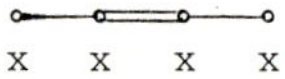

Then there is a point-hyperline flag $\{a,u\}$ such that both Γ_a and Γ_u are buildings. Then all C_3 residues of Γ are buildings by the flag-transitivity of Aut(Γ). Then Γ is covered by a building by Proposition 6. So, it is a building, by Proposition 8.
The case related to $P\Omega^+_{2n}(q)\cdot 2$ is not considered in [1], and it cannot be settled by Seitz's Theorem. Anyway, it has already been settled in Corollary 2 of this paper. □

3.3. The diagram C_3. Representation theory.

Propositions 3, 6 and 8, Corollary 1, the proof of Corollary 2 and that of Theorem 7 show that having got information about C_3 residues is often enough to describe geometries of type C_n or F_4. So, we will focus ourselves onto C_3 geometries. We have already observed that the A_7-geometry is flat. So, Theorem 7 and the fact that the A_7-geometry is the only example of non-building finite thick geometry presently known suggest the following conjecture:

CONJECTURE 1. *Every finite geometry of type* C_3 *with thick lines is either a building or flat.*

If this conjecture were true, then all finite geometries of type C_n $(n \geq 4)$ with thick lines would be buildings, by Corollary 1.
A concrete argument in support of Conjecture 1 is given by Ott in [22]. He proves that Conjecture 1 is true in the case of finite geometries of type C_3 admitting uniform parameter

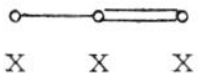

(x any positive integer)

Unfortunately we are not yet able to prove Conjecture 1 in the general case. Representation theory has been regarded as the main instrument to use in this inquiry. Anyway, it is the only one that we have nowadays, beside all kinds of geometric tools that we are able to invent, of course. That theory has been developed by Hoefsmit [12], Ott [20] and [21] and Liebler [15] (and Curtis, Iwahori, Steinberg and others, before them). Here we recall only its basic ideas. The reader is referred to [20], [21], [22], [15] and [12] for all details.
Given a finite geometry Γ over a set of types I, admitting parameters x_i $(i \in I)$ and belonging to a Coxeter diagram of matrix

$$(m_{i,j})_{i,j \in I}$$

let us take the set $\mathcal{C}$ of all chambers of Γ as a basis for a vector space V over the field of complex numbers. For every type $i \in I$, let us consider the endomorphism ϕ_i of V defined as below:

$$\phi_i(C) = \sum_{\substack{D \underset{i}{\sim} C \\ D \in \mathcal{C}}} D \qquad (C \in \mathcal{C},\ \underset{i}{\sim} \text{ means 'i-adjacent'})$$

The endomorphisms ϕ_i $(i \in I)$ together with the identity matrix $\underline{1}$ generate a subalgebra $H(\Gamma)$ of Hom(V,V), called *Hecke algebra* of Γ. It is worth observing that the following relations hold in $H(\Gamma)$:
(1) We have $\phi_i^2 = x_i\underline{1} + (x_i - 1)\phi_i$ (for every $i \in I$).
(2) We have $(\phi_i\phi_j)^{m_{ij}} = \underline{1}$ (for every choice of $i, j \in I$, $i \neq j$).
The Hecke algebra of Γ is semisimple (see [20]). In the case of C_3

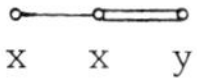

(x,y are parameters)

it is possible to compute all multiplicities of irreducible representations of $H(\Gamma)$ explicitly, starting from the parameters x,y and another number α, called *Ott-Liebler number* of Γ, which we shall describe in a moment. Ten distinct possible irreducible representations of $H(\Gamma)$ exist. They correspond to double partitions of the set $\{0,1,2\}$ of types of Γ (see [15] and [12]).Here is the complete list of their multiplicities.It has been found by Hoefsmit [12] at first,although not exactly in the form given here, but in an equivalent form.

TABLE 3

Double partition.	Representation (shortened name)	Multiplicity
$(\{0,1,2\};\emptyset)$	3/0	1
$(\{0,1\},\{2\};\emptyset)$	2,1/0	$\frac{(1+x^2y)(1+x)(x^3+\alpha)}{x(x+y)(1+\alpha)}$
$(\{0\},\{1\},\{2\};\emptyset)$	$1^3/0$	$\frac{(1+xy)(1+x^2y)(x^6+\alpha)}{(x^2+y)(x+y)(1+\alpha)}$
$(\{0,1\};\{2\})$	2/1	$\frac{(1+x+x^2)(1+xy)(x^2y-\alpha)}{x(x+y)(1+\alpha)}$
$(\{0\},\{1\};\{2\})$	$1^2/1$	$\frac{(1+x^2y)(1+x+x^2)(x^4y-\alpha)}{x(x^2+y)(1+\alpha)}$
$(\{0\};\{1\},\{2\})$	$1/1^2$	$\frac{(1+xy)(1+x+x^2)(x^4y^2+\alpha)}{x(x+y)(1+\alpha)}$
$(\{0\};\{1,2\})$	1/2	$\frac{(1+x^2y)(1+x+x^2)(x^2y^2+\alpha)}{x(x^2+y)(1+\alpha)}$
$(\emptyset;\{0\},\{1\},\{2\})$	$0/1^3$	$\frac{x^6y^3-\alpha}{1+\alpha}$
$(\emptyset;\{0,1\},\{2\})$	0/2,1	$\frac{(1+x)(1+x^2y)(x^3y^3-\alpha)}{x(x+y)(1+\alpha)}$
$(\emptyset;\{0,1,2\})$	0/3	$\frac{(1+xy)(1+x^2y)(y^3-\alpha)}{(x+y)(x^2+y)(1+\alpha)}$

Of course,some representation could disappear in some concrete examples, that is,it could have multiplicity 0. This happens if and only if the set of relations (1) and (2) above does not give a presentation of $H(\Gamma)$ in those cases.Some of the representations listed above are better known under special names.So, 3/0 is the *index representation*, $1^3/0$ is the *Steinberg representation* and 2/1 is the *reflection representation*.

Multiplicities have to be nonnegative integers.So,we get relations between x,y and α which are sometimes sufficent to get strong conclusions.Unfortunately they do not give us so much in general (see Lemma 6 below).

Let us give the definition of the Ott-Liebler number α,now.Let Γ be any C_3 geometry.Neither the finiteness nor the thickness nor parameters are assumed for the moment.Let (a,u) be an incident point-plane pair in Γ and let $\alpha(a,u)$ be the number of planes v distinct from u and such that the line z incident with both u and v does not pass through a (that line is uniquely determined by (ii) of Lemma 2).In [29] it is proved that the number $\alpha(a,u)$ does not depend on the choice of the incident point-plane pair (a,u).We set $\alpha = \alpha(a,u)$.α is the Ott-Liebler number of Γ.We observe that α has other interesting properties. For instance,given a non incident point-plane pair (b,v),the number of planes w incident with b and collinear with v is equal to $\alpha+1$ (see [28]; of course,we have $\alpha = \alpha+1$ if α is infinite).Here is another property of α.Given an incident point-line pair (a,z),let b be any point of z distinct from a and let $\ell(a,b)$ be the number of lines incident with both a and b and different from z. Then we have (see [30]):

$$\alpha = \sum_{\substack{b * z \\ b \neq a}} \ell(a,b)$$

More remarkably:

LEMMA 5. (Pasini [29]). *Let Γ be a geometry of type C_3 .We have $\alpha = 0$ if and only if Γ is a building.*
Let us assume that Γ is finite and admits parameters

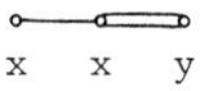

Then we have $\alpha \leq x^2y$ .We have $\alpha = x^2y$ if and only if Γ is flat.The number $\alpha+1$ divides both $(1+x+x^2)(x^2y+1)$ and $(x^2y+1)(xy+1)(y+1)$ and there are exactly $(1+x+x^2)(x^2y+1)/(\alpha+1)$, $(x^2y+1)(xy+1)(y+1)/(\alpha+1)$ and $(1+x+x^2)(x^2y+1)(xy+1)/(\alpha+1)$ points,planes and lines,respectively.

The reader can see [29] for the proof: it is quite elementary (in particular,no use of representation theory is made).The restriction $\alpha \leq x^2y$ can be improved by representation theory.

LEMMA 6. *Let* Γ *be a finite* C_3 *geometry admitting parameters*

$\circ$——$\circ$═══$\circ$
x x y

and let d *be the greatest common divisor of* x^2 *and* y, *and* m *the minimum of* y^3 *and* x^2y . *Then* xd *divides* α *and* $\alpha \leq m$.

The first relation is got by the formulas for the multiplicities of the representations $1^2/1$ or $1/2$ of Table 3. The second relation follows from the formula for 0/3 of Table 3. □

Lemmas 5 and 6 enable us to settle the case of finite C_3 geometries with known parameters. It turns out that Conjecture 1 cannot be proved by representation theory even in this case. We recall that a finite C_3 geometry has *known parameters* if it admits parameters

$\circ$——$\circ$═══$\circ$
x x y

and one of the following relations holds between them

(i) $x = y$

(ii) $y = x^2$ or (dually) $x = y^2$

(iii) $y^2 = x^3$ or (dually) $x^2 = y^3$

(iv) $y = x+2$ or (dually) $x = v+2$

(v) $x = 1$ or (dually) $y = 1$

A motivation for this definition is given by the fact that, in each of the known examples of finite generalized quadrangles admitting parameters, the parameters satisfy one of the relations listed above (Case (iv) occurs only in non classical examples). We warn that x and y are prime powers in all known examples satisfying (i), (ii) or (iii) and $(x+y)/2$ is a prime power in all known examples satisfying (iv). But we do not make such assumptions here.

THEOREM 8. (Ott [22], Rees and Scharlau [42], Pasini [29]). *Let* Γ *be a finite* C_3 *geometry with thick lines and known parameters. Then one of the following occurs:*

(i) The geometry Γ *is a building.*

(ii) The geometry Γ *is flat with parameters as below:*

either $\circ$——$\circ$═══$\circ$ (x x x) *or* $\circ$——$\circ$═══$\circ$ (t^2 t^2 t^3)

(iii) The geometry Γ *is neither a building nor flat, it has parameters as below*

$$\circ \!\!-\!\!\!-\!\!\circ \!\!-\!\!\!-\!\!\!\Rightarrow$$
$$y^2 \quad y^2 \quad y$$

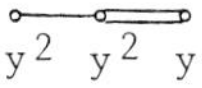

and we have $\alpha = y^3$.

The proof is nothing but a computation exploiting the information given by Lemmas 5 and 6 on α, x and y. See [29] for all details. The case of $x = y$ has been settled by Ott [22] at first. The cases of $y = x^2$, $y^2 = x^3$ and $x^2 = y^3$ have been examined by Rees and Scharlau [42]. The case of $y = 1$ is already covered by Corollary 1. All other cases are settled in [22]. We warn that no contradiction arises in Case (iii) of Theorem 8 with the formulas of Table 3: each of them gives us a nice integer, whatever y is. We observe also that Theorem 8, together with the classification of finite thick buildings of rank ≥ 3 given in [47], implies that no geometry exists with parameters x, x, y such that either $y = x+2$ and $x > 2$, or $x = y+2$ and $y > 2$ or $x^2 = y^3$ and $1 \neq x, y$. Then, 'a fortiori', no geometry of type C_n ($n \geq 3$) or F_4 exists with parameters as below

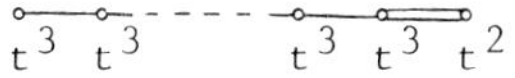

$$\circ\!\!-\!\!\circ\!-\!-\!-\!\circ\!\!-\!\!\!\Rightarrow$$
$$t^3 \quad t^3 \qquad t^3 \quad t^3 \quad t^2 \qquad (t > 1)$$

$$\circ\!\!-\!\!\circ\!-\!-\!-\!\circ\!\!-\!\!\!\Rightarrow$$
$$t-1 \quad t-1 \qquad t-1 \quad t-1 \quad t+1 \qquad (t > 3)$$

$$\circ\!\!-\!\!\circ\!-\!-\!-\!\circ\!\!-\!\!\!\Rightarrow$$
$$t+1 \quad t+1 \qquad t+1 \quad t+1 \quad t-1 \qquad (t > 3)$$

$$\circ\!\!-\!\!\circ\!\!-\!\!\circ\!\!-\!\!\!\Rightarrow$$
$$t^2 \quad t^2 \quad t^3 \quad t^3 \qquad \text{(or dually)} \quad (t > 1)$$

$$\circ\!\!-\!\!\!\Rightarrow\!\!-\!\!\circ$$
$$t-1 \quad t-1 \quad t+1 \quad t+1 \qquad \text{(or dually)} \quad (t > 3)$$

Examining Table 3 we see that, if Γ is a finite C_3 geometry admitting parameters $\circ\!\!-\!\!\!\Rightarrow\!\!-\!\!\circ$, then $\alpha = x^2y$ (flat case) and $\alpha = y^3$ are
$ x \quad x \quad y$
the only possibilities to lose some representation of $H(\Gamma)$. Actually, $\alpha = y^3$ is Case (iii) of Theorem 8. We have already observed that no representation is lost if and only if the set of relations (1) and (2) given before presents $H(\Gamma)$. Then we have:

COROLLARY 3. *Let* Γ *be a finite* C_3 *geometry with thick lines and known parameters*

$$\circ\!\!-\!\!\!\Rightarrow\!\!-\!\!\circ$$
$$x \quad x \quad y$$

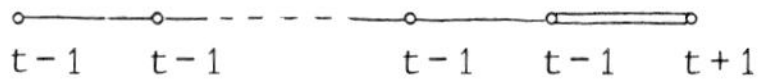

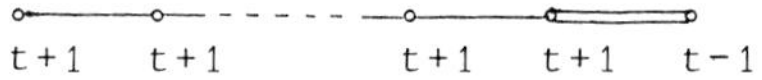

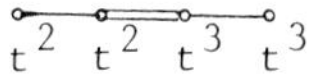

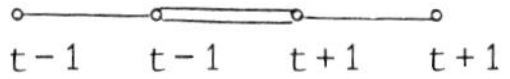

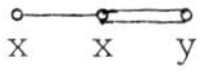

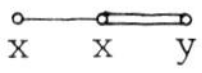

The geometry Γ *is a building if and only if its Hecke algebra* $H(\Gamma)$ *is presented by the following set of relations:*

$$\phi_i^2 = x\underline{1} + (x-1)\phi_i \quad (i = 0,1), \qquad \phi_2^2 = y\underline{1} + (y-1)\phi_2,$$

$$(\phi_0\phi_1)^3 = (\phi_1\phi_2)^4 = (\phi_2\phi_0)^2 = \underline{1}$$

We have already observed (Lemma 6) that α cannot exceed the minimum of x^2y and y^3. That is, if m_{xy} is the minimum of x and y, then $m_{xy}^2 y$ is the maximal value for α. Then Theorem 8 can be restated so:

COROLLARY 4. *Let Γ be a finite C_3 geometry with thick lines and known parameters*

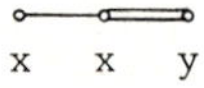

Then either $\alpha = 0$ or $\alpha = m_{xy}^2 y$ (maximal value).

The following lemma gives two geometric characterizations of the class of finite C_3 geometries that are either buildings or flat (that is,where either $\alpha = 0$ or $\alpha = x^2y$). The first of them will be necessary in the proof of Proposition 9. The other one may be fairly interesting in itself. It is worth observing that no geometric property is presently known that is equivalent to the condition $\alpha = y^3$ or to the condition that either $\alpha = m_{xy}^2 y$ or $\alpha = 0$.

We need some preliminary definitions.A point a of a C_3 geometry Γ is *homogeneous* if,given any point b collinear with a and different from a, the number $\ell(a,b)+1$ of lines through a and b does not depend on the choice of the point b as above. Let us consider also the following weaker versions of (LL) and (LH):

$(LL)_0$ *Two lines have the same set of points if they have more than one point in common.*

$(LH)_0$ *If a line* z *meets a plane* u *in more than one point, then all points of* z *belong to* u .

We observe that $(LH)_0$ is a consequence of $(LL)_0$.We shall not use $(LL)_0$ now (we shall refer to it later).

LEMMA 7. (Pasini [30]). *Let Γ be a finite C_3 geometry with thick lines. The geometry Γ is either a building or flat if and only if one of the following holds:*

(i) The geometry Γ possesses some homogeneous point.

(ii) The property $(LH)_0$ holds in Γ.

The reader is referred to [30] for the proof.

Trusting Conjecture 1, we say that a finite C_3 geometry with thick lines is *anomalous* if it is neither a building nor flat.

PROPOSITION 9. (Pasini [32]). *Let* Γ *be an anomalous finite* C_3 *geometry admitting parameters* $\underset{x}{\circ}\!-\!\underset{x}{\circ}\!=\!\underset{y}{\circ}$ *(where* $x > 1$*) and let* $\mathrm{Aut}(\Gamma)$ *be flag-transitive. Then the following hold:*

(i) *The number* x *is even,* $1+x+x^2$ *is prime and* $x+1 \equiv 0 \pmod{3}$. *Let* d *be the greatest common divisor of* x^2 *and* y. *Then we have* $x^2-x \geq y > x > d^2$, $y \geq (x-1)d^2+d$ *and* dx *divides* α. *Moreover,* $\alpha+1$, $(x+y)(\alpha+1)$ *and* $(x^2+y)(\alpha+1)$ *divide* $\frac{xy}{d} - \frac{\alpha}{dx}$, $(1+xy)(xy-\alpha/x)$ *and* $(1+x^2y)(x^3y-\alpha/x)$ *respectively.*

(ii) *Given a plane* u, *let* $\bar{G}_u$ *be the action over* Γ_u *of the stabilizer* G_u *of* u *in* $\mathrm{Aut}(\Gamma)$. *Then* $\bar{G}_u$ *is a Frobenius group of order* $(1+x+x^2)(1+x)$ *with Frobenius kernel cyclic of order* $1+x+x^2$ *and cyclic Frobenius complements of order* $1+x$. $\bar{G}_u$ *acts regularly over the set of flags of* Γ_u *and the Frobenius complements are stabilizers of antiflags of* Γ_u.

(iii) *Either* $\mathrm{Aut}(\Gamma)$ *acts imprimitively over the set of points of* Γ *or* y *is odd.*

The reader is referred to [32] for all details of the proof. Here we mention only its main ideas. Let G_u and $\bar{G}_u$ be defined as above. By the classification of flag-transitive projective planes by Kantor [14], we get that either $\bar{G}_u \geq PSL(3,x)$ or (ii) holds, x is even, $1+x+x^2$ is prime and $x+1 \equiv 0 \pmod{3}$. In the earlier case, every point of Γ is homogeneous. Then that case cannot occur, by Lemma 7. In the latter case, most of the remaining part of (i) is got via Table 3, exploiting the fact that $1+x+x^2$ is prime. And (i) collects all that Table 3 provides. As for (iii), it follows from the well known theorem by O'Nan and Scott on primitive groups, from an analysis of elements of order $1+x+x^2$ and of involutions and Sylow 2-subgroups of $\mathrm{Aut}(\Gamma)$, from theorems on strong embeddings of subgroups of finite simple groups and from the classification of primitive groups of odd degree (Theorem C of [14]).

It is worth observing that the arithmetical conditions listed in (i) above seem to be actually incompatible with the well known Bruck-Ryser condition on finite projective planes and with the condition that x+y divides xy(xy+1) (see [35]). I have tested them by a computer and it has turned out that they never hold together if $x \leq 1000$ (I stopped computations at x = 1000). We recall also that it is well known that the statement of (ii) holds on a finite projective plane P of order $x < 1600$ where $1+x+x^2$ is prime, only if x is a prime power and either P is non desarguesian or x = 2 or x = 8.

The following proposition completes Proposition 9.

PROPOSITION 10. (Pasini [32]). *Let Γ be a finite* C_3 *geometry with flag-transitive automorphism group and parameters* 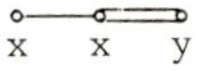

where $x > 1$. *Then one of the following holds:*

(i) *The geometry Γ is a building.*

(ii) *Γ is the* A_7*-geometry.*

(iii) *The geometry Γ is flat,* x *is a prime power properly dividing* y *and either* $y^2 = x^3$ *or* $y^4 = x^5$ *or* y *is not a prime power. We have* $x^2 - x \geq y$ *in any case. Given a plane* u, *the residue* Γ_u *of* u *is desarguesian and the action on* Γ_u *of the stabilizer of* u *in* Aut(Γ) *contains* PSL(3,x).

(iv) *The geometry Γ is anomalous as in Proposition 9.*

The reader is referred to [32] for the proof and to [18] for all details in the case of x = y. We give only the main ideas here. Kantor's classification of flag-transitive projective planes I14I still plays the central role in the proof. Only the flat case has to be examined,of course, and it is proved that the Frobenius case of Theorem A of [14] cannot occur, both if x = y and if $x < y$ (the classification of primitive groups of odd degree is used again here).Then only the PSL(3,x) case survives. Now,if x = y, a Klein quadric and an exterior set can be recognized from Γ. That is,Γ is constructed as in Example 4 of §2.3 of this paper. So,[17] can be applied. Then x = 2 and Γ is the A_7-geometry by a result of Rees [36]. In the case of $x < y$, it is easily seen that x divides y (this fact has been discovered by Ott [24] at first,by representation theoretic methods; but it can be proved by elementary methods).From this the remaining part of (iii) easily follows by Table 3.

The following corollaries are trivial consequences of Propositions 9 and 10.

COROLLARY 5. *Let Γ be a finite C_3 geometry with flag-transitive automorphism group and parameters* $\underset{x\ \ x\ \ y}{\circ\!\!-\!\!\!-\!\!\!-\!\!\circ\!\!=\!\!\!=\!\!\!=\!\!\circ}$ *where $x > 1$ and $x \geq y$. Then Γ is either a building or the A_7-geometry.*

COROLLARY 6. *Let Γ be a finite C_3 geometry with thick lines, known parameters and flag-transitive automorphism group. Then one of the following holds:*

(i)   *The geometry Γ is a building.*

(ii) *Γ is the A_7-geometry.*

(iii) *The geometry Γ is flat and we have $x = q^2$ and $y = q^3$ where q is a prime power. The residue Γ_u of any plane u is desarguesian and the action on Γ_u of the stabilizer in $\mathrm{Aut}(\Gamma)$ of u contains $\mathrm{PSL}(3,q^2)$.*

3.4. The diagrams C_n ($n \geq 4$) and F_4. Known parameters.

We recall that a geometry Γ of type C_n or F_4 admits *known parameters* if all of its C_3 residues have known parameters.

PROPOSITION 11. (Pasini [31]). *Let Γ be a finite geometry of type C_n ($n \geq 4$) with thick lines and known parameters. Then either Γ is a building or it has parameters as below*

$$\underset{y^2\ \ \ \ y^2}{\circ\!-\!\!-\!\circ}\cdots\cdots\underset{y^2\ \ \ y^2\ \ \ y}{\circ\!-\!\!-\!\circ\!=\!\!=\!\circ}$$

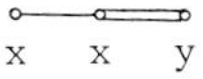

and the relation $\alpha = y^3$ holds in at least one of the C_3 residues of Γ.

The proof is trivial, by Theorem 8 and Corollary 1.

PROPOSITION 12. (Pasini [27],[31]). *Let Γ be a finite thick geometry of type F_4 with known parameters. Then one of the following holds:*

(i) *The geometry Γ is a building.*

(ii) *The geometry Γ has parameters as below*

$$\text{points}\underset{y^2\ \ y^2\ \ y\ \ \ y}{\circ\!-\!\!-\!\circ\!=\!\!=\!\circ\!-\!\!-\!\circ}\text{hyperlines} \quad \text{(or dually)}$$

it is not a building, all residues of points are buildings

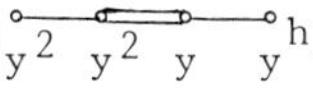

and the relation $\alpha = y^3$ *holds in* Γ_u *for some hyperline* u.

(iii) The geometry Γ *has uniform parameter*

points o———o═══o———o hyperlines
x x x x

it is not a building and all the following hold:

(iii.a) Every C_3 *residue is either a building or flat and some* C_3 *residue is flat.*

(iii.b) Given a point-hyperline flag {a,u}, *at least one of* Γ_a *and* Γ_u *is a building.*

(iii.c) If v *and* w *are distinct collinear points or distinct coplanar hyperlines, then at least one of* Γ_v *and* Γ_w *is a building.*

The reader is referred to [27] for the proof. It uses Propositions 6 and 8, Theorem 8 and some elementary lemmas stated in [27].

Remark. Because of a mistake made in [15], Liebler excluded Case (iii) of Theorem 8 in [15], so he excluded also the exceptional case of Proposition 11 and Case (ii) of Proposition 12. Actually, there is not any way to exclude them by merely rearranging arguments from [15], avoiding that mistake. Liebler proved in [15] also a relation between parameters of an F_4 geometry, using the assumption that all C_3 residues were either buildings or flat (he believed that this was a theorem, as we have remarked above). That relation would force x = 2, if it were true. Unfortunately, the proof by Liebler seems to be wrong.

3.5. The diagrams C_n ($n \geq 4$) and F_4. Flag-transitivity.

THEOREM 9. (Pasini [31]). *Let* Γ *be a finite geometry of type* C_n *(*$n \geq 4$*) with thick lines or a finite thick geometry of type* F_4*. The geometry* Γ *is a building iff* Aut(Γ) *is flag-transitive.*

The "only if" part is well known. Anyway, it follows from the classification of finite thick buildings of rank ≥ 3 by Tits [47]. Let us prove the "if" part. Let us see the case of C_n for the first. Let u be a maximal subspace of Γ and let G_u be the stabilizer of u in Aut(Γ). By Seitz [44], either the action of G_u on Γ_u contains PSL(n,x), where x is the first parameter of Γ, or n = 4, Γ has

parameters as below

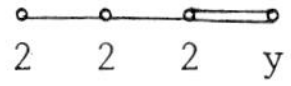

$$\circ\!\!-\!\!\!-\!\!\!-\!\!\circ\!\!-\!\!\!-\!\!\!-\!\!\circ\!\!=\!\!=\!\!=\!\!\circ \atop 2\quad 2\quad 2\quad y$$ (where y = 1,2 or 4)

and G_u acts as A_7 on Γ_u. Anyway, in the latter case the stabilizer in Aut(Γ) of a flag F consisting of point and a maximal subspace acts on the projective plane Γ_F as PSL(3,2). Then (ii) of Proposition 9 fails to hold in C_3 residues of Γ. Then every C_3 residue of Γ is either a building or flat, by Proposition 9. Then Γ is a building by Corollary 1.

The case of F_4 is even easier. Let x,x,y,y be the parameters of Γ

$$\text{points}\;\;\circ\!\!-\!\!\!-\!\!\circ\!\!=\!\!=\!\!\circ\!\!-\!\!\!-\!\!\circ\;\;\text{hyperlines} \atop \phantom{\text{points}\;\;}x\quad x\quad y\quad y$$

Let us assume that all C_3 residues of Γ are either buildings or flat. They cannot be all flat. Indeed in that case we have x = y by Lemma 6 and we get a contradiction with Proposition 12. Then some C_3 residues of Γ are buildings. Then Γ has known parameters. So, we can apply Proposition 12 and, by it and Corollary 6, we get that Γ is a building.

Let us assume that some C_3 residue of Γ is anomalous. For instance, let Γ_u be anomalous for some hyperline u. Then we have x < y by Proposition 9. Then Γ_a is a building for every point a, by Corollary 5. Then Γ has known parameters and we contradict Corollary 6. We are done. □

4. PROBLEMS

1. Is there any finite thick flat C_3 geometry different from the A_7-geometry? No effort has been made up to now to construct examples with parameters

$$\circ\!\!-\!\!\!-\!\!\circ\!\!=\!\!=\!\!\circ \atop t^2\quad t^2\quad t^3$$

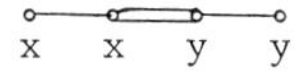

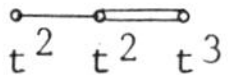

and we have not even much information about how such a geometry should be done. We know much more about the case of uniform parameter

$$\circ\!\!-\!\!\!-\!\!\circ\!\!=\!\!=\!\!\circ \atop x\quad x\quad x$$

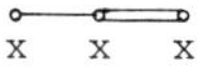

If a flat geometry existed different from the A_7-geometry and with uniform parameter x ≠ 1, then x > 2 (by [36]), Aut(Γ) could not be flag-transitive (by [18]; see Proposition 10 of this paper) and the

geometry could not be constructed as in Examples 3 and 4 of §2.3 (by [17] and [8]). However, this does not yet answer our question completely, even in this case. What about the case of parameters as below (see Proposition 10) ?

t^4 t^4 t^5

What about the general case ?

2. Is there any flat geometry of type C_n ($n \geq 4$) with thick lines ? Can we construct such a geometry as a quotient of a polar space ? We warn that such a geometry should be infinite (Corollary 1) and its 0-shadow space should coincide with the projective geometry Γ_u, where u is any maximal subspace of the geometry. In particular, Property $(LL)_0$ (see §3.3) should hold in it. Actually, this very feature leads to a contradiction in the finite thick case (but non thick examples can easily be constructed; see [34]). Anyway, if we consider also the case of $n = 3$, then we find at least one (infinite, but thick) example of a flat C_3 geometry which is done in that way (i.e.: $(LL)_0$ holds in it): namely, Example 3 of §2.3 (we recall that it is a quotient of a polar space).

3. Is it possible to construct a finite thick flat C_3-geometry where $(LL)_0$ holds ? We sometimes meet geometries of that kind as last possibilities in arguments by contradiction, just a moment before to get a final contradiction. But that is always got using also other facts. We observe that a lot of finite non thick examples of flat C_3 geometries exist where $(LL)_0$ holds (see [39]).

4. Try to construct a non flat proper quotient of a polar space of rank $n \geq 4$, with thick lines and such that all of its C_3 residues are either buildings or flat. Such a geometry should be infinite, by Corollary 1. Moreover, the polar space, which we start from, could not be defined over an algebraic extension of a finite field, because polar spaces defined over such fields do not admit proper quotients (see the proof of Proposition 8).

5. Which relations hold between the properties of the field over which a classical polar space Γ is defined and the properties of

the class of all proper quotients of Γ? (Of course, to be empty is a possible property of that class. To contain flat geometries would be another interesting property. By the way, the only infinite thick flat C_3 geometries presently known are got from polar spaces over ordered fields: see Example 3 of §2.3).

6. Let Γ be a finite thick non-building C_3 geometry with parameters as below (Theorem 8,(iii))

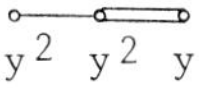

(provided that such a geometry exists).Corollary 5 states that Aut(Γ) cannot be flag-transitive.Lemma 7 implies that Γ is fairly 'irregular'. Liebler [16] has considered the case of y = 2. He takes the graph *G* defined over the set of planes of Γ by the collinearity relation and shows that certain irregularities must occur in *G*. So, Γ cannot be regular in standard ways. Can we find any trick to construct such an irregular example ? Can we do that when y = 2, for instance ?

7. Corollaries 3 and 4 and Theorem 8 suggest to consider also the following conjecture (weaker than Conjecture 1):

CONJECTURE 2. *Let Γ be a finite* C_3 *geometry with parameters*

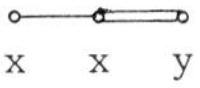

where $x > 1$. *Then either* $\alpha = 0$ *or* $\alpha = x^2y$ *or* $\alpha = y^3$ *(that is,either* $\alpha = 0$ *or* $\alpha = m^2_{xy}y$ *; that is,the geometry Γ is a building iff its Hecke algebra* H(Γ) *is presented by the set of relations given in Corollary 2).*

Conjecture 2 would look rather sensible if we had to rely mainly on algebraic methods (i.e.,on representation theory). We might have better reasons to trust Conjecture 2 instead of Conjecture 1 if we were able to prove the following generalization of Corollary 1: let Γ be a finite geometry of type C_n ($n \geq 4$) with thick lines and let us assume that, for every C_3 residue of Γ , α takes either the maximal value or the null one. Then Γ is a building. We observe that,if this claim were true, than the exceptional case of Proposition 11 would be ruled out immediately. Try to prove that claim.

8. The following problem is deeply related to Conjecture 2. Note that the equalities $\alpha = 0$ and $\alpha = x^2y$ have a clear geometric meaning, as well as the condition that one of the previous equalities holds (see Lemma 7 of this paper and the results of [29] and [30]). Can we find a geometric property that is equivalent to the condition that $\alpha = y^3$ or to the condition that $\alpha = m_{xy}^2 y$, or to the condition that either $\alpha = 0$ or $\alpha = m_{xy}^2 y$? Succeeding in this would give us one more reason to trust Conjecture 2.

9. Prove that the exceptional case of Proposition 11 is impossible. As for this, I have recently got a proof of the following statement (provided that nothing is wrong with my proof): if such an exceptional geometry exists in the rank 4 case, then its collinearity graph is complete. This is not yet a non-exsistence proof, of course, but it smells like that.

10. Can we find any number that can play in the C_n case ($n \geq 4$) the same role as α plays in the C_3 case ? What about F_4 ?

11. Improve Proposition 12.

12. Improve Propositions 9 and 10. The reader can find hints for this job in [32]. It would be too long to explain them here.

13. Several results listed in this paper give us the impression that proper quotients of buildings of Lie type are not so frequent. Anyway, they do not exist at all in the finite thick case. Probably the same is true if we consider also non-building geometries. For instance: is there any finite C_3 geometry with thick lines and admitting proper quotients ? By the way, the answer is "no" if we consider only geometries whith known parameters. Observe that, on the contrary, a lot of finite non thick C_3 geometries admit proper quotients (see [39] and [29]).

REFERENCES

1. M.ASCHBACHER, Finite geometries of type C_3 with flag transitive automorphism groups, Geom.Ded.,16 (1984), 195-200.

2. A.BROUWER and A.COHEN, Some remarks on Tits's geometries, Indag.Math.,45 (1983), 393-402.

3. F.BUEKENHOUT, Diagrams for geometries and groups, J.Comb.Th. A, 27 (1979), 121-151.

4. F.BUEKENHOUT, The basic diagram of a geometry, in "Geometries and Groups", L.N.893, Springer 1981, 1-29.

5. F.BUEKENHOUT, A.DELANDTSHEER and J.DOYEN, Finite linear spaces with flag-transitive and locally primitive groups,(to appear).

6. F.BUEKENHOUT and A.SPRAGUE, Polar spaces having some line of cardinality two, J.Comb.Th. A, 33 (1982), 223-228.

7. F.BUEKENHOUT and W.SCHWARTZ, A simplified version of strong connectivity in geometries, J.Comb.Th. A, 37 (1984), 73-75.

8. F.DE CLERK and J.THAS, Exterior sets with respect to the hyperbolic quadric in PG(2n-1,q), in "Finite Geometries", C.Baker and L.Batten ed., Dekker 1985, 83-90.

9. M.DEZA and M.LAURENT, Bouquets of matroids, d-injection geometries and diagrams, to appear in J.Geometry.

10. W.FEIT and G.HIGMAN, The nonexistence of certain generalized polygons, J.Algebra, 1 (1964), 114-131.

11. G.HILLEBRANDT, Über dünne geometrien vom sphärischen typ, Dissertation, University of Brunswick, January 1986.

12. P.HOEFSMIT, Representations of Hecke algebras of finite groups with BN-pairs of classical type. Ph.D.Thesis, University of British Columbia, 1974.

13. W.KANTOR, Generalized polygons, SCABs and GABs, in "Buildings and the Geometry of Diagrams", L.N.1181, Springer 1986, 79-158.

14. W.KANTOR, Primitive groups of odd degree and an application to finite projective planes, J.Algebra, 106 (1987),14-45.

15. R.LIEBLER, A representation theoretic approach to finite geometries of spherical type, preprint, April 1985.

16. R.LIEBLER, personal communication (March 1986).

17. G.LUNARDON, Exterior sets with respect to the hyperbolic quadric $Q^+(5,q)$, to appear.

18. G.LUNARDON and A.PASINI, A result on C_3 geometries, to appear.

19. A.NEUMAIER, Some sporadic geometries related to PG(3,2), Arch.Math.,42 (1984),89-96.

20. U.OTT, Some remarks on representation theory in finite geometry , in "Geometries and Groups", L.N.893, Springer 1981, 68-110.

21. U.OTT, Bericht über Hecke algebren und Coxeter algebren endlicher geometrien, in "Finite Geometries and Designs",Cameron Hirschfeld and Hughes ed.,Cambridge University Press 1981, 260-271.

22. U.OTT, On finite geometries of type B_3, J.Comb.Th. A, 39 (1985),209-221.

23. U.OTT, On geometries of type D_n, to appear.

24. U.OTT, personal communication (June 1986).

25. A.PASINI, Canonical linearization of pure geometries, J.Comb. Th. A, 35 (1983), 10-32.

26. A.PASINI, Some remarks on covers and apartments, in "Finite Geometries", C.Baker and L.Batten ed., Dekker 1985, 223-250.

27. A.PASINI, On certain geometries of type C_n and F_4, Discr.Math. 58 (1986),45-61.

28. A.PASINI, On Tits geometries or type C_n, Eur.J.Comb.,8 (1987) 45-54.

29. A.PASINI, On geometries of type C_3 that are either buildings or flat, to appear in Bull.Soc.Math. de Belgique.

30. A.PASINI, On finite geometries of type C_3 with thick lines, to appear in Note di Matematica.

31. A.PASINI, Geometries of type C_n and F_4 with flag-transitive automorphism groups, to appear in the Proceedings of the Workshop "Geometries and Groups,Finite and Algebraic",Leeuwenhorst, March 1986.

32. A.PASINI, Flag-transitive C_3 geometries, to appear.

33. A.PASINI, On a theorem by Aschbacher, to appear in J.Geometry

34. A.PASINI and S.REES, A theorem on Tits geometries of type C_n to appear in J.Geometry.

35. S.PAYNE and J.THAS, "Finite Generalized Quadrangles",Pitman 1984.

36. S.REES, On diagram geometry, Ph.D.Thesis, Oxford 1983.

37. S.REES, C_3 geometries arising from the Klein quadric, Geom. Ded.,18 (1985), 67-85.

38. S.REES, A classification of a class of C_3 geometries, to appear in J.Comb.Th.

39. S.REES, Finite C_3 geometris in which all lines are thin, Math.Zeit.,189 (1985),263-271.

40. S.REES, Products of Tits geometries, J.Geometry,25 (1985),77--87.

41. S.REES, Weak buildings of spherical type, to appear.

42. S.REES and R.SCARLAU, personal communication (March 1985).

43. R.SCHARLAU, A structure theorem for weak buildings of spherical type, to appear.

44. G.SEITZ, Flag-transitive subgroups of Chevalley groups, Ann. Math.,97 (1973),27-56.

45. J.THAS, lecture given at the Workshop "Geometries and Groups, Finite and Algebraic",Leeuwenhorst,March 1986.

46. F.TIMMESFELD, Tits's geometries and parabolic systems in finite groups, Math.Zeit.,184 (1983),377-396.

47. J.TITS, "Buildings of Spherical Type and Finite BN-pairs", L.N.386, Springer 1974.

48. J.TITS, A local approach to buildings, in "The Geometric Vein",Springer 1981, 519-547.

49. P.ZIESCHANG, The nonexistence of certain geometries of type C_3, to appear.

<u>ADDED IN PROOF</u>. A mistake has lately been found in [17]. So, the problem to prove that the A_7-geometry is the only finite non-building example that can be got starting from sets of points exterior to Klein quadrics (see this paper,§2.3,Example 4 and §4,Problem 1), is still open. Anyway, F.De Clerck and J.Thas have got interesting partial results on this problem, which might lead to a proof of the previous claim in a short time.

Annals of Discrete Mathematics 37 (1988) 357–366

THE THAS-FISHER GENERALIZED QUADRANGLES

Stanley E. PAYNE

Department of Mathematics
University of Colorado at Denver
1100 14th Street, Campus Box 170
Denver, CO 80202

Using a flock of a quadratic cone discovered by J. C. Fisher, J. A. Thas has recently shown the existence of a new generalized quadrangle of order (q^2,q), q an odd prime power, $q \geq 5$. Here we give a very explicit description of Fisher's flock and show that up to isomorphism there arises just one new generalized quadrangle for each q.

1. INTRODUCTION

During the past few years, work of W. M. Kantor (cf. [2,3,4]) and S. E. Payne (cf. [5,6,7]) has led to a recipe for the construction of generalized quadrangles (GQ) of order (q^2,q) beginning with a family M of 2×2 upper triangular matrices satisfying certain conditions. Recently, J. A. Thas [9] has shown that such a family M is equivalent to a flock of a quadratic cone in PG(3,q). He studied the known flocks and the known GQ of the corresponding type and discovered that previously there was not known a GQ corresponding to the flock discovered by J. C. Fisher and first described in [1].

The descriptions of Fisher's flock given previously seem to us to lack the necessary explicitness needed to facilitate a study of the corresponding GQ. Moreover, W. M. Kantor [4] has brought to our attention other reasons that a computationally explicit study of these flocks and the corresponding family M would be of interest. Hence we feel justified in presenting here one more treatment of Fisher's flocks.

The geometrical connection between a flock and its corresponding GQ is not at all well understood. For example, it is not even known whether or not an automorphism of the flock corresponds to an automorphism of the GQ! It is partly for this reason that we have had as one main goal in this essay to show that just one new GQ is obtained for each q. To begin, let q be any prime power, let $F = GF(q)$, and let K be the cone

$$K = \{\bar{x} = (x_0,x_1,x_2,x_3) \in PG(3,q): x_0x_1 = x_2^2\} = \{(0,0,0,1)\}$$

$$\cup \{(1,c^2,c,d) \in PG(3,q): c,d \in F\} \cup \{(0,1,0,d): d \in F\}. \qquad (1)$$

So K is the cone over the conic $\{(0,1,0,0)\} \cup \{(1,c^2,c,0): c \in F\}$, with vertex $V = (0,0,0,1)$. A plane π in PG(3,q) contains V precisely when it has coordinates $[a,b,c,0]^T$. Consider a set $\mathcal{F}$ of q planes π_i, $i = 1,\ldots,q$, none of which contains V. $\mathcal{F}$ determines a flock of K (i.e., a partition of $K-\{V\}$ into disjoint ovals) provided that $\pi_i \cap \pi_j \cap K = \phi$ whenever $i \neq j$. A flock $\mathcal{F}$ is linear provided all its planes contain a same line.

1.1. (J. A. Thas [9]). $M = \{\pi_1,\ldots,\pi_q\}$ with $\pi_i = [a_i,b_i,c_i,1]^T$ yields a flock of K if and only if $(a_i-a_j)x^2 + (c_i-c_j)x + b_i - b_j$ is irreducible over F whenever $i \neq j$. This condition is easily interpreted for odd or even q as follows:

(i) For odd q, $(c_i-c_j)^2 - 4(a_i-a_j)(b_i-b_j)$ is a nonsquare in F whenever $i \neq j$.

(ii) For even q, $(a_i-a_j)(b_i-b_j)(c_i-c_j)^{-2}$ has trace 1 whenever $i \neq j$.

Associated with $\pi_i = [a_i,b_i,c_i,1]^T$ is a 2×2 upper triangular matrix $A_i = \begin{vmatrix} a_i & c_i \\ 0 & b_i \end{vmatrix}$. The condition in 1.1 is then interpreted as follows, where "A is anisotropic" means $\alpha A \alpha^T = 0$ if and only if $\alpha = (0,0)$.

1.2. $\mathcal{F} = \{\phi_1,\ldots,\phi_q\}$ yields a flock of K if and only if A_i-A_j is anisotropic whenever $i \neq j$.

Let $K = \{(\alpha,c,\beta): \alpha,\beta \in F^2, c \in F\}$. Then K is turned into a group with the binary operation defined by

$$(\alpha,c,\beta)\cdot(\alpha^1,c^1,\beta^1) = (\alpha+\alpha^1,c+c^1+\beta\cdot\alpha^1,\beta+\beta^1), \qquad (2)$$

where $\beta\cdot\alpha^1$ is the usual dot product. $C = \{(0,c,0) \in K: c \in F\}$ is the center of K. $A(\infty) = \{(0,0,\beta) \in K: \beta \in F^2\}$ is a subgroup of K having order q^2.

Now suppose $A_i = \begin{vmatrix} a_i & c_i \\ 0 & b_i \end{vmatrix}$, $1 \leq i \leq q$, with A_i-A_j anisotropic whenever $i \neq j$. Put $K_i = A_i + A_i^T$. Put $A(i) = \{(\alpha,\alpha A_i\alpha^T,\alpha K_i) \in K: \alpha \in F^2\}$, $1 \leq i \leq q$. Put $J = \{A(i): i \in \{1,\ldots,q,\infty\}\}$. For $A \in J$, put $A^* = AC$. Then J is a family of $1+t = 1+q$ subgroups of K, each of order $s = q^2$. W. M. Kantor [3] (for odd q) and S. E. Payne [7] (for even q) used computations worked out in [5] to prove the following:

1.3. The family J is a 4-gonal family for K giving rise to a GQ of order (q^2,q); i.e., J satisfies the following axioms of Kantor [2]:

K1. $A_iA_j \cap A_k = 1$ if i, j, and k are distinct.

K2. $A_i^* \cap A_j = 1$ if $i \neq j$.

For the actual construction of the GQ and several related definitions and results, we suggest the monograph [8].

In Section 2 we derive an explicit description of Fisher's flocks for a cone $\hat{K}$ projectively equivalent to but different from K. Then in Section 3 this flock is transferred to a flock of K. In Section 4 we initiate study of the associated GQ first discovered by J. A. Thas [9]. The reader should note that we have borrowed quite freely from the work of Fisher, Thas and Kantor in deriving the present description of Fisher's flocks.

2. FLOCKS BY FISHER

Let q be an odd prime power, let $-\alpha$ be a nonsquare of $F = GF(q)$, and let $E = F(i)$, where $i^2 = -\alpha$. Let ζ be a primitive root of E and put $z = \zeta^{q-1} = a+bi$, so z is an arbitrary generator of the cyclic group of E^* having order q+1. For $x,y \in F$, the conjugate of $x+iy$ is $\overline{x+iy} = x-iy = (x+iy)^q$. And $(x+iy)(x-iy) = x^2+\alpha y^2$, so $\gamma = x+iy$ satisfies

$$\gamma\bar{\gamma} = \gamma^{q+1} = 1 \text{ iff } x^2 + y^2 = 1 \text{ iff } \gamma = z^j \text{ for some } j. \tag{3}$$

In particular,

$$a^2 + \alpha b^2 = 1 = z^i \cdot \bar{z}^i = (z\bar{z})^i. \tag{4}$$

$$z^{(q+1)/2} = -1 = z^{-(q+1)/2}; \quad z^{(q-1)/2} = -z^{-1}. \tag{5}$$

$$z^{j-k} - z^{j+k+1} = (z^j - z^{j+1})\cdot\lambda_k, \text{ where}$$

$$\lambda_k = (z^{-k} + \ldots + 1 + \ldots + z^k) \in F. \tag{6}$$

Equation 6 is especially important for what follows!

For each integer k, put

$$a_k = \frac{(z^{k+1} - z^{-(k+1)}) - (z^k - z^{-k})}{z - z^{-1}} \in F. \tag{7}$$

Similarly, put

$$b_k = \frac{i((z^{k+1} + z^{-(k+1)}) - (z^k + z^{-k}))}{z - z^{-1}} \in F. \tag{8}$$

We now collect several observations concerning the a_k's and b_k's, none of whose proofs is especially complicated.

2.1.

(i) $a_j = (z^{j+1}+z^{-j})/(z+1) = (z^j+z^{-j}) - (z^{j-1}+z^{-(j-1)}) + \ldots + (-1)^j$.

(ii) $a_j \neq a_k$ for $0 \leq j < k \leq (q-1)/2$.

(iii) $-a_j = a_{(q-1)/(2-j)}$, so $a_j = 0$ for some j (specifically for $j = (q-1)/4$) if and only if $q \equiv 1 \pmod 4$.

(iv) $a_j = a_{q-j}$, so $\{a_{2j}: 0 \leq j \leq (q-1)/2\} = \{a_j: 0 \leq 1 \leq (q-1)/2\}$.

(v) $a_0 = 1$; $a_{(q-1)/2} = -1$; $a_{-j} = a_{j-1}$.

(vi) $b_k = i(z^{k+1}-z^{-k})/(z+1) = i(z-1)\lambda_k/(z+1) = (a-1)\lambda_k/b = -\alpha bb_0/(a+1)$.

(vii) $aa_j+bb_j = a_{j+1}\backslash b_j = (a_{j+1}-a_{j-1})/2b$ and $aa_j-bb_j = a_{j-1}\backslash a_j = (a_{j+1}+a_{j-1})/2a$.

(viii) $b_k = -b_{q-k}$; $b_k = b_{(q-1)/(2-k)}$.

Put

$$A_{\hat{K}} = \begin{vmatrix} 1 & 0 & 0 & 0 \\ 0 & \alpha & 0 & 0 \\ 0 & 0 & -1 & 0 \\ 0 & 0 & 0 & 0 \end{vmatrix},$$

and let $\hat{K}$ be the cone defined by $\hat{K} = \{\bar{x}: \bar{x}A_{\hat{K}}\bar{x}^T = 0\} = \{\bar{x}: x_0^2+\alpha x_1^2 = x_2^2\} = \{V\} \cup \{(\gamma,1,t): \gamma \in E, \gamma\bar{\gamma} = 1\}$. Hence,

$$\hat{K} = \{V\} \cup \{P_{t,j} = (z^j,1,t): 0 \leq j \leq q,\ t \in F\}. \tag{9}$$

The planes $\hat{\pi}(t) = [0,0,t,-1]^T$, $t \in F$, all meet in the line $N = \{(x,y,0,0): x,y \in F; (x,y) \neq (0,0)\}$. Fix j modulo q+1. Then for $0 \leq k \leq (q-1)/2$, in the plane $\hat{\pi}(t)$ we have (using Equation 6):

$$P_{t,j-k} \cdot P_{t,j+k+1} \cap N = (z^{j-k}-z^{j+k+1},0,0) \equiv (z^{j}-z^{j+1},0,0). \tag{10}$$

It is clear that $\{\hat{\pi}(t): t \in F\}$ determines a linear flock of $\hat{K}$. For $0 \leq j \leq (q-1)/2$, let $\hat{\pi}'(j)$ be the plane spanned by $0 = (0,0,1,0)$, $P_{1,2j}$, and $P_{1,2j+1}$.

2.2. The planes $\hat{\pi}'(j), 0 \leq j \leq (q-1)/2$, all meet each of the planes $\hat{\pi}(a_k), 0 \leq k \leq (q-1)/2$, each time in a secant line of $\hat{K}$.

Proof: An easy computation shows that

$$P_{a_k,2j-k} = (1-a_k)0 + \frac{z^{k+1} - z^{-(k+1)}}{z - z^{-1}} \cdot P_{1,2j} + \frac{z^{-k} - z^{k}}{z - z^{-1}} \cdot P_{1,2j+1} \cdot \tag{11}$$

and

$$P_{a_k,2j+k+1} = P_{a_k,2j-k} + \lambda_k(P_{1,2j+1} - P_{1,2j}). \tag{12}$$

As j varies with k fixed, $P_{a_k,2j-k}$ and $P_{a_k,2j+k+1}$ yield all q+1 points of $\hat{K} \cap \hat{\pi}(a_k)$. Hence,

$$\cup\{\hat{\pi}(a_k) \cap \hat{K}: 0 \leq k \leq (q-1)/2\} = \cup\{\hat{\pi}'(j) \cap \hat{K}: 0 \leq j \leq (q-1)/2\}. \tag{13}$$

Equation 13 provides us with Fisher's flock for $\hat{K}$.

2.3. $\{\hat{\pi}'(j): 0 \leq j \leq (q-1)/2\} \cup \{\hat{\pi}(t): t \in F - \{a_0, a_1, \ldots, a_{(q-1)/2}\}\}$ yields a flock of $\hat{K}$ (which is nonlinear for $q \geq 5$; see [9]).

A point of $\hat{K} \cap \hat{\pi}(j)$ other than $P_{1,2j}$ or $P_{1,2j+1}$ must have the form $(0,0,1,0) + y(z^{2j+1},0,1) + (t-y)(z^{2j},0,1) = (tz^{2j}+y(z^{2j+1}-z^{2j}),1,t) = (z^k,1,t)$, with $z^k = tz^{2j} + y(z^{2j+1}-z^{2j})$. For a given $t \in F$, this may or may not hold for some $y \in F$ and some integer k modulo q+1. (Of course we already know that it holds precisely for t of the form a_i for some i.) But $z^k = tz^{2j} + y(z^{2j+1}-z^{2j})$ if and only if $z^{k-2j} = t + y(z-1)$, which has a solution if and only if $t + y(z-1) \in \langle z \rangle$ if and only if $1 = (t+y(z-1))(t+y^{-1}(z^{-1}-1))$ if and only if $y^2N(z-1) - ytN(z-1) + t^2 = 1$ has two solutions in F if and only if $t^2(N(z-1))^2 - 4N(z-1)(t^2-1)$ is a (necessarily nonzero) square.

(Here $N(\gamma)$ means the 'norm' of γ; i.e., $N(z-1) = (z-1)(z^{-1}-1)$.) After simplifying, this is if and only if $t^2 - 2/(1+a) = t^2 - 4/N(1+z)$ is a nonsquare in F (since

$a^2 - 1 = -\alpha b^2$ is a nonsquare). This gives us a revised description of Fisher's flock.

<u>2.4. Fisher's flock of $\hat{K}$ is determined by</u> $\{\hat{\pi}'(j): 0 \le j \le (q-1)/2\} \cup \{\hat{\pi}(t): t^2 - 4/N(1+z)$ <u>is a (necessarily nonzero) square in F</u>$\}$.

Let

$$[T] = \begin{vmatrix} a & b & 0 & 0 \\ -\alpha b & a & 0 & 0 \\ 0 & 0 & 1 & 0 \\ 0 & 0 & 0 & 1 \end{vmatrix},$$

and define a collineation T of PG(3,q) by

$$T: \bar{x} \to \bar{x}[T], \quad \text{for each point } \bar{x} \in PG(3,q). \tag{14}$$

Note that $T(\bar{x}) = (ax_0 - \alpha bx_1, ax_1+bx_0, x_2, x_3)$. Since $(x_0+x_1 i)(a+b_i) = (ax_0-\alpha bx_1) + (ax_1+bx_0)i$, it follows that T leaves invariant each point of the line VO, each plane $\hat{\pi}(t)$ through N, and maps $P_{t,j}$ to $P_{t,j+1}$. It follows that $T^2: \hat{\pi}'(j) \to \hat{\pi}'(j+1)$, $0 \le j \le (q-1)/2$ and

$$T^2: \hat{\pi}(t) \to \hat{\pi}(t), \quad t \in F - \{a_j: 0 \le j \le (q-1)/2\}. \tag{15}$$

It seems quite remarkable to us that we cannot seem to find a way to transfer the action of T^2 to the corresponding GQ! (The missing collineation has been found and will appear in a later paper.)

3. THE TRANSFER FROM $\hat{K}$ TO K

Put

$$P = \begin{vmatrix} 1 & 0 & 1 & 0 \\ -\alpha & 0 & \alpha & 0 \\ 0 & 2 & 0 & 0 \\ 0 & 0 & 0 & 1 \end{vmatrix}.$$

Then

$$PA_{\hat{K}}P^T = -4\alpha A_K. \tag{16}$$

So K is related to $\hat{K}$ by: $\bar{y} \in K$ if and only if $\bar{x} \in \hat{K}$, where $\bar{x} = \bar{y}P$. So we transfer from $\hat{K}$ to K by mapping a point $\bar{x}$ to $\bar{x}P^{-1}$ and a plane $\bar{y}^T$ to $P\bar{y}^T$. In particular, for $t^2 - 2/(1+a)$ a (nonzero) square in F, the plane $\hat{\pi}(t) = [0,0,t,-1]^T$ is mapped to

$$\pi(t) = P[0,0,t,-1]^T = [-t,-t\alpha,0,1]^T. \tag{17}$$

For $t^2 - 2/(1+a)$ a nonsquare, say $t = a_j$, $0 \le j \le (q-1)/2$, $\hat{\pi}'(j) = \langle P_{1,2j}, P_{1,2j+1}, 0\rangle$ is mapped (after some computation) to

$$\pi'(j) = \left\langle \left(\frac{N(1+z^{2j})}{4}, \frac{N(1-z^{2j})}{4\alpha}, \frac{z^{2j}-z^{-2j}}{4i}, 1\right), \right.$$

$$\left. \left(\frac{N(1+z^{2j+1})}{4}, \frac{N(1-z^{2j+1})}{4\alpha}, \frac{z^{2j+1}-z^{-(2j+1)}}{4i}, 1\right), \left(\frac{1}{2}, \frac{1}{2\alpha}, 0, 0\right)\right\rangle. \tag{18}$$

After more computation we see that

$$\pi'(j) = [-a_{2j}, a_{2j}, 2b_{2j}, 1]^T. \tag{19}$$

This completes a proof of the following:

<u>3.1. The set M of 2 x 2 upper triangular matrices corresponding to Fisher's flock for the standard cone K consists of all the matrices of the following two types</u>:

(i) $\begin{vmatrix} -t & 0 \\ 0 & -\alpha t \end{vmatrix}$, $t^2 - 4/N(1+z)$ a square in F.

(ii) $\begin{vmatrix} -a_{2j} & 2b_{2j} \\ 0 & \alpha a_{2j} \end{vmatrix}$, $0 \le j \le (q-1)/2$.

In 3.1 it would be quite convenient to have a formula for b_k in terms of a_k alone. It is easy to see that $b_k = \pm i\sqrt{a_k^2(z+1)^2-4z/(1+z)}$. So given a value t $(=a_k)$ such that $t^2 - 2/(1+a)$ is a nonsquare in F, the corresponding b_k is $\pm i\sqrt{t^2(z+1)^2-4z/(1+z)}$, but we do not see how to determine the sign of b_k. At least $\pm t$ is paired with $\pm i\sqrt{t^2(z+1)^2-4z/(1+z)}$.

The cone $\hat{K}$ is invariant under the projective collineation T, so we may consider what happens to Fisher's flock under T. Transferring to K, we see that if $\pi = [a,b,c,1]^T$ is a plane in Fisher's flock for K, then π is replaced by $P[T]^{-1}P^{-1}[a,b,c,1]^T$. So compute:

$$P[T]^{-1}P^{-1} = \begin{vmatrix} (a+1)/2 & (1-a)/2\alpha & -b/2 & 0 \\ \alpha(1-a)/2 & (a+1)/2 & \alpha b/2 & 0 \\ \alpha b & -b & a & 0 \\ 0 & 0 & 0 & 1 \end{vmatrix} . \tag{20}$$

Multiplying this matrix times $[-a_{2j},\alpha a_{2j},2b_{2j},1]^T$, we obtain

$$[-aa_{2j},-bb_{2j},a\alpha_{2j}+b\alpha b_{2j}2(-\alpha ba_{2j}+ab_{2j}),1]^T = [-a_{2j+1},\alpha a_{2j+1},2b_{2j+1},1]^T. \tag{21}$$

(Here we use $-\alpha b^2 = a^2-1, a_{2j-1} = 2aa_{2j}-a_{2j+1}, b_{2j+1} = (a_{2j+2}-a_{2j})/2b.$) But $a_{2j} = a_{q-2j} = a_{2k+1}$ if and only if $k = (q-1)/2 - j$, and for that k we have $b_{2j} = -b_{q-2j} = -b_{2k+1}$. This proves the following.

3.2. <u>The matrix $\begin{vmatrix} a & c \\ 0 & b \end{vmatrix}$ arises from Fisher's flock $\mathcal{F}$ for K (as in 3.1) if and only if the matrix $\begin{vmatrix} a & -c \\ 0 & b \end{vmatrix}$ arises from the image of $\mathcal{F}$ under the action of</u> $P[T]^{-1}P^{-1}$.

Now we ask: How is the set M of 3.1 affected by replacing z with some other generator z^r of $\langle z\rangle$; i.e., $\gcd(r,q+1) = 1$, so $r = 2j+1$ for some j?

$$(1+z^{2j+1})/(1+z) = z^j a_j. \tag{22}$$

From Equation 22, we have

$$N(1+z^{2j+1}) = N(1+z)(a_j)^2. \tag{23}$$

So $t^2 - 4/N(1+z)$ is a square in F if and only if $(t/a_j)^2 - 4/N(1+z^{2j+1})$ is a square. This implies that $t \to t/a_j$ maps the set $\{t \in F: t^2 - 4/N(1+z)$ is a nonsquare in F$\}$ to the set $\{t \in F: t^2 - 4/N(1+z^{2j+1})$ is a nonsquare in F$\}$.

We continue to use the notation a_k, b_k as a given in Section 2, and let $\hat{a}_k, \hat{b}_k$ be the corresponding elements of F obtained when z is replaced with z^{2j+1}, $\gcd(2j+1,q+1) = 1$. So if $t = a_{2i}$, then $t/a_j = \hat{a}_{2k}$ for some k determined modulo $(q+1)/2$. After some computation we see that

$$a_{2i} = a_j\hat{a}_{2k} \quad \text{iff} \quad (z^{2k(2j+1)+2i+j+1}-1)(z^{-2k(2j+1)+2i-j}-1) = 0. \tag{24}$$

Put $\mu \equiv (2j+1)^{-1} \pmod{q+1}$. Then the left factor is zero if and only if

$$k \equiv [-i-(j+1)/2]\mu \pmod{(q+1)/2}, \quad \text{if } j \text{ is odd.} \tag{25}$$

And the right factor is zero if and only if

$$k \equiv [i-j/2]\mu \pmod{(q+1)/2}, \quad \text{if } j \text{ is even.} \tag{26}$$

A similar computation shows that

$$b_{2i} = \pm a_j\cdot\hat{b}_{2k} \quad \text{iff} \quad (z^{2k(2j+1)+2i+j+1}\pm 1)(z^{-2k(2j+1)+2i-j}\mp 1) = 0. \tag{27}$$

It now follows that

$$b_{2i}/a_j = \begin{cases} -\hat{b}_{2k}, & j \text{ odd} \\ \hat{b}_{2k}, & j \text{ even.} \end{cases} \tag{28}$$

This essentially completes a proof of the following.

<u>3.3. Let M be the set of matrices arising from Fisher's flock for K starting with the generator z (as in 3.1). Let $\hat{M}$ be the corresponding set of matrices obtained when z is replaced with</u> z^{2j+1}, $\gcd(2j+1,q+1) = 1$. <u>Then</u>

$$\begin{vmatrix} a & c \\ 0 & b \end{vmatrix} \in M \quad \text{iff} \quad \begin{vmatrix} a & (-1)^j c \\ 0 & b \end{vmatrix} a_j^{-1} \in \hat{M}.$$

4. THE GQ OF THAS

Most of the details needed to prove the theorem of this section have already been given.

<u>4.1. For each odd prime power q only one GQ arises from Fisher's flock.</u>

<u>Proof</u>: The idea is that neither replacing the original flock with its image under T nor replacing z with z^r, $\gcd(r,q+1) = 1$, yields different GQ. In view of 3.2 and 3.3, the following results complete the proof.

Put $Q = \begin{vmatrix} 1 & 0 \\ 0 & -1 \end{vmatrix}$ and define $\Theta: K \to K$ by $\Theta: (\alpha, c, \beta) \to (\alpha Q, c, \beta Q)$. Then Θ is an automorphism of K for which $\Theta: (\alpha, \alpha A \alpha^T, \alpha(A+A^T)) \to (\alpha Q, \alpha Q(QAQ)(\alpha Q)^T, \alpha Q(QAQ+(QAQ)^T))$. Since $Q \begin{vmatrix} x & y \\ 0 & z \end{vmatrix} Q = \begin{vmatrix} x & -y \\ 0 & z \end{vmatrix}$, $A = \begin{vmatrix} x & y \\ 0 & z \end{vmatrix}$ is replaced with $QAQ = \begin{vmatrix} x & -y \\ 0 & z \end{vmatrix}$.

For $t \in F$, $t \neq 0$, define $\phi: K \to K$ by $\phi(\alpha, c, \beta) \to (\alpha, tc, t\beta)$. It is easy to check that ϕ is an automorphism of K that replaces A with tA.

REFERENCES

[1] Fisher, J. C. and Thas, J. A., Flocks in PG(3,q), Math. Z. 169 (1979) 1-11.

[2] Kantor, W. M., Generalized quadrangles associated with $G_2(q)$, Jour. Combin. Theory (A) 29 (1980) 212-219.

[3] Kantor, W. M., Some generalized quadrangles with parameters (q^2,q), Math. Z. 192 (1986) 45-50.

[4] Kantor, W. M., Generalized quadrangles and translation planes, Algebras, Groups and Geometries 3 (1985) 313-322.

[5] Payne, S. E., Generalized quadrangles as group coset geometries, Congressus Numerantium 29 (1980) 717-734.

[6] Payne, S. E., A garden of generalized quadrangles, Algebras, Groups and Geometries 3 (1985) 323-354.

[7] Payne, S. E., A new infinite family of generalized quadrangles, Congressus Numerantium 49 (1985) 115-128.

[8] Payne, S. E. and Thas, J. A., Finite Generalized Quadrangles, Research Notes in Mathematics No. 110 (Pitman Pub. Inc., 1984).

[9] Thas, J. A., Generalized quadrangles and flocks of cones, preprint.

Annals of Discrete Mathematics 37 (1988) 367–374

ON GROUP SPACES DEFINED BY SEMIDIRECT PRODUCTS OF GROUPS

J.Pfalzgraf

Fachbereich Mathematik
Universität des Saarlandes
D-6600 Saarbrücken
Fed.Rep.Germany

0. INTRODUCTION

The "classical" examples of noncommutative geometric spaces such as ray spaces, circle spaces, disc spaces, nearfiel spaces, spaces defined by the more general F-groups, all turn out to be group spaces with respect to special semidirect products of groups.
We introduce the concept of a semidirect product space (SPS) induced by the "affine action" of a semidirect product of groups (cf. 1.2).
It is possible to make explicit constructions of spaces with prescribed properties. This possibility is indicated by examples using free groups to show that the equivalence of the properties "semiaffine" and "tactical", in [A2] Satz 2.1, is typical for finite spaces (it does not hold for infinite spaces, in general, cf. 3.6).
As another application of SPS we present a geometric version of Witt's cancellation theorem for quadratic forms over fields (characteristic not 2) involving two geometric invariants which can be defined for a quadratic form (cf. 4.5).
Finally we mention that with respect to arbitrary group extensions $1 \to N \to E \to G \to 1$ further natural questions of geometric type arise.

1. NOTATION

Noncommutative geometric spaces $(X,\sqcup,\|)$ as introduced by J.André (cf. [A1] - [A4]) can be described by their parallel structure (cf. [P1]): each map $\langle,\rangle: X^2 \to R$, defining a <u>parallelism</u> on the point set X (R is the set of <u>directions</u> or <u>ideal points</u>), leads to a so-called <u>LP1'-space</u> $(X,\langle,\rangle,R)$, i.e. the axioms (L1),(L2),(P1') hold, and vice versa (see [P1] for more details).
<u>Lines</u> are defined by $x \square y := x \sqcup y \cup \{\langle x,y\rangle\}$; $x \sqcup y := \{x\} \cup \{z \mid \langle x,z\rangle = \langle x,y\rangle\}$ is the set of <u>proper points</u> of $x \square y$; $\langle x,y\rangle$ is the <u>ideal point</u> (<u>direction</u>) of the line. $x \square y$ is parallel to $u \square v$, iff $\langle x,y\rangle = \langle u,v\rangle$. Further axioms of a space X are expressed by corresponding equations in that "$\langle,\rangle$-calculus", cf. [P1] 2.5.

Now let $G \times X \to X$ be the action of a group G on a point set X. Then the <u>group space</u> (cf. [A1],[A2]), denoted by $\underline{V}(G,X)$, is defined by the parallel map (cf. [P2]):

(1.1) $\langle,\rangle: X^2 \to R$, with $R := G \backslash X^2$, $\langle x,y\rangle := G(x,y)$ the orbit of (x,y).

G acts componentwise on each pair of X^2. For lines we obtain $x \sqcup y = \{x\} \cup G_x.y$.

$\underline{V}(\ ,\ )$ is a covariant functor from "Group Operations" to "Noncommutative Geometric Spaces". For more information about group spaces see [A1] - [A4] and [P2]. Our interest here is concerned with group spaces associated with semidirect products of groups.

1.2 Definition

Let G,N be groups (multiplicatively written), G acts on N:

$\tau: G \to \mathrm{Aut}(N)$, denoted by ${}^{g}x := \tau_g(x)$, $x \in N$, $g \in G$.

We obtain the semidirect product $E := N \rtimes G$ with respect to that action.

This gives rise to the "affine action" of E on N, defined as $E \times N \to N$ with $[a,g].x := a.{}^{g}x$, $[a,g] \in E$, $x \in N$, leading to the group space $\underline{V}(E,N)$. We call this a semidirect product space (SPS for short).

We note here that all the 1-point stabilizers E_x , $x \in N$, can be calculated because we know the stabilizer group of $1 \in N$ which is $E_1 = (1,G)$, hence $E_x = [x,1]\, E_1 [x,1]^{-1} = \{ [x.{}^{g}x^{-1},g] \mid g \in G\}$.

It turns out that all the following "classical" examples of noncommutative geometric spaces are special SPS.

2. EXAMPLES

2.1 Ray Space $S\mathbb{R}^n$

$X := \mathbb{R}^n$, $R := S^{n-1} \cup \{0\}$, $\langle x,y\rangle := (y-x)/\|y-x\|$, if $x \neq y$ and $\langle x,y\rangle = 0$ otherwise. Then for $x \neq y$, $x \sqcup y$ is the ray $x + \mathbb{R}_{\geqq 0}(y-x)$.

We have the following presentation of the ray space $S\mathbb{R}^n$ as SPS:

$S\mathbb{R}^n = \underline{V}(\mathbb{R}^n \rtimes \mathbb{R}_{>0}, \mathbb{R}^n)$, with respect to the natural action of $\mathbb{R}_{>0}$ on $\mathbb{R}^n$.

2.2 Affine Space $AG(n,\mathbb{R})$

$AG(n,\mathbb{R})$ is obtained from $S\mathbb{R}^n$ by identification of antipodal directions, thus $R := \mathbb{R}P^{n-1} \cup \{0\}$ and $\langle,\rangle$ is defined on $X := \mathbb{R}^n$ by $\langle x,y\rangle := \{\pm (y-x)/\|y-x\|\}$, $x \neq y$ and $\langle x,y\rangle := 0$, $x = y$. Then $AG(n,\mathbb{R}) = \underline{V}(\mathbb{R}^n \rtimes \mathbb{R}^*, \mathbb{R}^n)$, $\mathbb{R}^* := \mathbb{R} \setminus \{0\}$.

2.3 Circle Space

$X := \mathbb{R}^2$, $R := \mathbb{R}_{\geq 0}$, $\langle x,y\rangle := \|x-y\|$; $x \sqcup y$ is the circle with radius $\|x-y\|$ and its midpoint x as basepoint of the line. X can be described by $\underline{V}(\mathbb{R}^2 \rtimes O(2,\mathbb{R}), \mathbb{R}^2)$ or by $\underline{V}(\mathbb{R}^2 \rtimes SO(2,\mathbb{R}), \mathbb{R}^2)$, alternatively. Thus we see that the covariant functor $\underline{V}(\ ,\ )$ is not faithful.

There is an obvious generalization of the above definition of circle space to $\mathbb{R}^n$ leading to spheres as lines.

2.4 Disc Space

$X := \mathbb{R}^2$, $R := \mathbb{R} \cup \{\infty\}$, $\langle x,y\rangle := y_1 - x_1$ for $x = (x_1,x_2) \neq (y_1,y_2) = y$ and

$\langle x,y\rangle := \infty$ for $x = y$. Then $x \sqcup y = \{x_1\} \times \mathbb{R}$, if $x_1 = y_1$ and $x \sqcup y = \{x\} \cup (\{y_1\} \times \mathbb{R})$, if $x_1 \neq y_1$. It holds $(X,\langle,\rangle,R) = \underline{V}(\mathbb{R}^2 \rtimes H, \mathbb{R}^2)$ with $H = \{\begin{pmatrix} 1 & 0 \\ a & \alpha \end{pmatrix}, a \in \mathbb{R}, \alpha \in \mathbb{R}^*\} < GL(2,\mathbb{R})$.
Of course, this definition can be generalized (cf. [P1],[P2]).

2.5 Remark
A lot of further examples can be constructed in this way (for example using $\mathbb{R}^n \rtimes G$, G a subgroup of GL(n,$\mathbb{R}$), etc.)
Evidently one can define the above spaces over more general coefficient domains. In this context we mention the following observation.
Consider the disc space 2.4 over a ring A with unit 1, instead of a field. Then, using the analogous definitions (A^* denotes the group of invertible elements (units) of A) we obtain the following result

(2.6) $(X,\langle,\rangle,R) = \underline{V}(A^2 \rtimes H, A^2)$ if,and only if, A is a field.

Proof: If A is a field, then the equality holds (cf. [P2]).
Conversely assume that A is not a field. Then there exists $c \in A$, $c \neq 0$, which is not a unit. Now consider the points $x = (0,0)$, $y = (0,c)$, $u = (1,1)$, $v = (1,0)$ in $X = A^2$ and show that the parallel classes $(A^2 \rtimes H).(x,y)$ and $P_{\langle x,y\rangle} = \{(x',y') \mid \langle x',y'\rangle = \langle x,y\rangle\}$ are distinct (hence the spaces are distinct, having different parallelisms): obviously $(u,v) \in P_{\langle x,y\rangle}$ since $\langle u,v\rangle = 0 = \langle x,y\rangle$, but $(u,v) \notin (A^2 \rtimes H).(x,y)$, otherwise there were $[z,h] \in A^2 \rtimes H$ such that $[z,h].(x,y) = (u,v)$, this means $u = z+h(x)$, $v = z+h(y)$, for suitable $h = \begin{pmatrix} 1 & 0 \\ a & \alpha \end{pmatrix}$, $a \in A$, $\alpha \in A^*$, and we obtain for the second coordinates:
$u_2 = z_2 + ax_1 + \alpha x_2$ and $v_2 = z_2 + ay_1 + \alpha y_2$, substituting the original values yields the equation $0 = 1 + \alpha c$, hence $c \in A^*$, a contradiction.

2.7 F-Groups and Nearfield Spaces
As mentioned in [A4] §3, the nearfield spaces (these are regular desarguesian nearaffine spaces and vice versa) are SPS. It turns out that even the more general F-groups (cf. [A3]) lead to SPS.
For convenience we recall the definition of a F-group and adopt the notation of [M] (this work is a well written continuation of [A3] dealing with several further topics of linear algebra over nearfields).

(1) A pair (F,V) is called F-group (cf. [A3],[M]) if the following four conditions hold:

(F1) (V,+) is a group and (F,.) a monoid.

(F2) F operates on V, i.e. $(\alpha\beta).x = \alpha(\beta.x)$ and $1.x = x$ and there exist $0,-1 \in F$ such that $0.x = 0$, $(-1).x = -x$ and furthermore $\alpha(x+y) = \alpha x + \alpha y$.

(F3) $F^* = F \setminus \{0\}$ is a (multiplicative) group.

(F4) Cancellation: $\alpha x = \beta x$ implies $x = 0$ or $\alpha = \beta$.

These conditions imply the following:
$(V,+)$ is abelian; $\alpha.0_V = 0_V$; $\alpha(-x) = -\alpha x$; $(\alpha(-1))x = ((-1)\alpha)x = -\alpha x$;
F^* acts like a group of automorphisms on $(V,+)$.

(1) The geometric spaces $A(F,V) = (V,\sqcup,\|)$ in [A3] p.295 (for more details see J.André, Affine Geometrien über Fastkörpern. Mitt.Math.Sem.Gießen 144(1975)), are defined by $\underline{L} := \{Fx + y \mid x \in V \setminus \{0\}, y \in V\}$ as set of lines and $x \sqcup y = F(y-x) + x = x + F(y-x)$ and parallelism $x \sqcup y \| x' \sqcup y'$, iff $F(y-x) = F(y'-x')$.
In terms of the $\langle,\rangle$-model we can describe $A(F,V)$ as follows:
$\langle,\rangle : V^2 \to R$, $R := \{F^*.x \mid x \in V \setminus \{0\}\} \cup \{0\}$; $\langle x,y\rangle := F^*(y-x)$, if $x \neq y$ and $\langle x,y\rangle := 0$, $x = y$.
Note: obviously $F^*v = F^*w$, iff [there exists $\alpha \neq 0$: $v = \alpha w$], iff $Fv = Fw$.
Since F^* acts like automorphisms on $(V,+)$ we can define the semidirect product $V \rtimes F^*$ and the SPS $\underline{V}(V \rtimes F^*, V)$. Now we state

(3) $\qquad A(F,V) = \underline{V}(V \rtimes F^*, V)$

Proof: It is readily seen that $\langle x,y\rangle = \langle u,v\rangle$ is equivalent to $(V \rtimes F^*)(x,y) = (V \rtimes F^*)(u,v)$, since the latter equation means [there exists $[w,\alpha] \in V \rtimes F^*$: $(x,y) = [w,\alpha](u,v)$], iff [$x = w + \alpha u$, $y = w + \alpha v$], iff [$y-x = \alpha v - \alpha u = \alpha(v-u)$], iff $F^*(y-x) = F^*(v-u)$, iff $\langle x,y\rangle = \langle u,v\rangle$.
Thus the parallel structures of both spaces are equal, hence (3) holds.

2.8 Remark
The group spaces $\underline{S}(F,U)$, F a nearfield, $U^* < F^*$, $U = U^* \cup \{0\}$, as mentioned in [A4] Prop.5.3 (they were originally studied by K.P. Rößler in his Diplomarbeit, Saarbrücken 1985) turn out to be SPS, too. Namely with respect to the semidirect product $F \rtimes U^*$.

With affine actions of semidirect products one can make explicit constructions of spaces with prescribed properties (exploiting the knowledge of the 1-point stabilizers). This possibility will be indicated in the next section.

3. A CONSTRUCTION WITH FREE GROUPS

Let $F_n = F(a_1,\ldots,a_n)$ be the free group of rank n ($n \geq 2$) with basis $\{a_1,\ldots,a_n\}$ and φ the automorphism defined by $\varphi(a_i) = a_{i+1}$, $1 \leq i < n$, $\varphi(a_n) = a_1$, then $G := \{id,\varphi,\ldots,\varphi^{n-1}\} < Aut(F_n)$ is cyclic and we obtain the SPS

(3.1) $\qquad \underline{V}(F_n \rtimes G, F_n)$, we recall from [P2] 3.5

(3.2) $\qquad$ for $1 \leq t \leq n-1$: φ^t has no fixed points $\neq 1$

(3.3) $\qquad$ for $x \in F_n \setminus \{1\}$: $1 \sqcup x = \{1\} \cup (F_n \rtimes G)_1 x = \{1\} \cup (1,G).x = \{1,x,\varphi x,\ldots,\varphi^{n-1}x\}$ and this implies the general case

(3.4) $\qquad |x \sqcup y| = n+1$, $x \neq y \in F_n$.

We make the following

3.5 Remark

(1) $\underline{V}(F_n \rtimes G, F_n)$ are examples of (n+1)-tactical desarguesian skewaffine spaces with Pappos-condition (cf. [P2]).

(2) By the way we note that these examples serve to show the independence of the parallelogram-condition (Pgm). Indeed, the commutator criterion for (Pgm) does not hold (cf. [P2] 3.6).

Satz 2.1 of [A2] shows that a finite imprimitive skewaffine space is k-tactical (i.e. $|x \sqcup y| = k$ for all lines $x \sqcup y$, $x \neq y$) if, and only if, it is a semi-affine space (i.e. the flat conditions (F1),(F2) for subspaces hold, cf. [A2] Def. 2.2 , conditions (S1),(S2)).
In the desarguesian case such a space is a group space with respect to an imprimitive Frobenius group and vice versa ([A2] Satz 4.1,4.2).
Subsequently we show that [A2] Satz 2.1 is a "finite space result", it does not remain true, in general, if X is an infinite point set. To see this, we take the tactical spaces $\underline{V}(F_n \rtimes G, F_n)$ and show that (F1) does not hold, hence they are not semiaffine.

3.6 Example

Flat condition (F1) says: if $U < X$ is a subspace of a space X and $L \not\subseteq U$ a line, then $|U \cap L| \leq 1$.
For $\underline{V}(F_n \rtimes G, F_n)$ we shall now construct a subspace $1 \in U < F_n$ and $x \notin U$ such that $|(x \sqcup 1) \cap U| > 1$:
(1) If $U < F_n$ is a subgroup on which G acts (by restriction), then we obtain the subgroup $U \rtimes G < F_n \rtimes G$ and the corresponding subspace $(U \rtimes G).1$, cf. [P2] 1.7. Obviously $(U \rtimes G).1 = U$.

(2) For $1 \neq x \in F_n$ the 1-point stabilizer is $(F_n \rtimes G)_x = \{[x.{}^g x^{-1}, g] \mid g \in G\}$, hence $x \sqcup 1 = \{x\} \cup (F_n \rtimes G)_x.1 = \{x\} \cup \{1, x\varphi(x^{-1}), x\varphi^2(x^{-1}), \dots, x\varphi^{n-1}(x^{-1})\}$.
With this notation:
(3) Let U be the subgroup of F_n generated by the elements $s := a_1 a_2^{-1}$, $\varphi(s) = a_2 a_3^{-1}$ $, \dots, \varphi^{n-1}(s) = a_n a_1^{-1}$. Then G acts on U and $x := a_1 \notin U$.
On the other hand: $1 \in U$ and $s \in (x \sqcup 1) \cap U$, hence $|(x \sqcup 1) \cap U| \geq 2$.
Therefore (F1) does not hold for U and $\underline{V}(F_n \rtimes G, F_n)$ is not semiaffine.
Another application of semidirect product spaces is presented in the next section.

4. TWO GEOMETRIC INVARIANTS FOR QUADRATIC FORMS

First we fix the general notations.
Let A be a commutative ring with 1, M an A-module and $q : M \to A$ a quadratic form (cf. [B]). For a quadratic module (M,q) we can define the following geometric space $(X,\langle,\rangle_q,R)$ as a geometric invariant:

(4.1) $X := M$, $R := A \cup \{\infty\}$, $\infty \notin A$, $\langle x,y\rangle_q := q(x-y)$ for $x \neq y$ and $\langle x,y\rangle_q := \infty$, $x = y$.

Let O(M,q) or O(q), for short, denote the orthogonal group consisting of all isometries $\sigma : M \to M$ (i.e. σ is a linear isomorphism with $q(\sigma x) = q(x)$).

4.2 Remark

(1) Obviously O(q) is the group of all dilatations of the geometric space $(X,\langle,\rangle_q)$ which are linear isomorphisms.

(2) If we had defined $\langle x,y\rangle_q := q(x-y)$ for all $x,y \in M$, then it is readily seen that q is anisotropic if, and only if, (L0) holds for $(X,\langle,\rangle_q)$.

With respect to the natural action of O(M,q) on M we obtain the SPS

(4.3) $\underline{V}(M \rtimes O(q),M)$

as another geometric invariant of the quadratic module (M,q).

Subsequently we show for quadratic forms over fields of characteristic not 2 that Witt's cancellation theorem can be expressed by the equality relation of both geometric invariants for all quadratic forms over that field. In [P2] 3.9, 3.10 the following is mentioned without proof.

4.4 Notations

(1) Let A = K be a field of characteristic $\neq 2$. X = M a finite dimensional vector space over K. We shall restrict our considerations to nondegenerate quadratic forms $q : X \to K$ (note that radicals of quadratic forms always can be cancelled).

(2) We say "The cancellation theorem holds over K" if the following is true for quadratic forms q,p_1,p_2 over K: $q \perp p_1 \simeq q \perp p_2$ implies $p_1 \simeq p_2$.

4.5 Theorem

With the above notations the following is equivalent

(1) The cancellation theorem holds over K

(2) For all quadratic forms (X,q) over K the geometric space $(X,\langle,\rangle_q)$ is equal to the semidirect product space $\underline{V}(X \rtimes O(q),X)$.

Proof: (1) $\Rightarrow$ (2): Let (X,q) be a quadratic form. We shall show that the parallel classes of both geometric spaces are identical. This is trivial for diagonal elements (x,x). Therefore let $x \neq y$, then obviously $(X \rtimes O(q)).(x,y) \subseteq P_{\langle x,y\rangle_q} = \{(u,v) \mid \langle u,v\rangle_q = \langle x,y\rangle_q$, i.e. $q(u-v) = q(x-y)\}$. Now we prove the reverse inclusion: let $(u,v) \in P_{\langle x,y\rangle_q}$, we set $w := u-v$ (hence $w \neq 0$) and $z := x-y$, then $q(w) = q(z)$. Now we show the existence of a $\sigma \in O(q)$ such that $\sigma(z) = w$.

i) $\underline{q(w)=q(z)\neq 0}$: for $a := q(w) = q(z)$ the two nondegenerate subspaces $(K.z,(a))$ and $(K.w,(a))$ of X lead to orthogonal decompositions of (X,q): $(K.z \perp X_1,(a) \perp q_1)$ and $(K.w \perp X_2,(a) \perp q_2)$ with q_1,q_2 nondegenerate. From the isometry $(a) \perp q_1 \simeq (a) \perp q_2$ we obtain by the cancellation theorem an isometry $\varphi : q_1 \simeq q_2$. Let $\sigma_o : K.z \to K.w$ be defined by $\sigma_o(z) := w$, then $\sigma := \sigma_o \perp \varphi : K.z \perp X_1 \simeq K.w \perp X_2$ is an element of $O(X,q)$ such that $\sigma(z) = w$.

ii) $\underline{q(w)=q(z)=0}$: it follows that $\dim(X) > 1$ and there are two hyperbolic planes $H_1 = (K.z \oplus K.z',h_1)$, $H_2 = (K.w \oplus K.w',h_2)$ belonging to z,w respectively (cf. [L] chI 3.2,3.4), with $h_1(z) = h_1(z') = 0$, $B_{h_1}(z,z') = 1$, B_{h_1} being the corresponding bilinear form of h_1 (similar with h_2). In the case $\dim(X) = 2$, i.e. $X = H_1 = H_2$ there exists obviously the isometry $\sigma(z) := w$, $\sigma(z') := w'$. If $\dim(X) > 2$, then $(X,q) = H_1 \perp (X_1,q_1)$ and we can apply lemma 4.2 of [B] chIII §4 to $H_2 \subset (X,q)$, implying the existence of an isometry $\sigma \in O(q)$ such that $\sigma(z) = w$ (and $\sigma(z') = w'$). Altogether we have proved the existence of a $\sigma \in O(q)$ with $\sigma(z) = w$ and we continue the proof: taking such an isometry σ the following holds for $[v-\sigma y,\sigma] \in X \rtimes O(q)$: $[v-\sigma y,\sigma].(x,y) = (v+\sigma(x-y),v) = (u,v)$.

(2) $\Rightarrow$ (1): Let q,p_1,p_2 be (nondegenerate) quadratic forms, $q = (a_1) \perp \ldots \perp (a_n)$, $a_j \neq 0$, such that $q \perp p_1 \simeq q \perp p_2$. By induction we may restrict ourselves to the case $n = 1$ (cf. also the remark in [L] chI,thm 4.2, step 3 and 4.7). Therefore we shall prove for $0 \neq a \in K$: $(a) \perp p_1 \simeq (a) \perp p_2$ implies $p_1 \simeq p_2$. Now let $\varphi : (K.x,(a)) \perp (X_1,p_1) \simeq (K.x,(a)) \perp (X_2,p_2)$ be an isometry. We set $X := K.x \perp X_2$, $q := (a) \perp p_2$, $y := \varphi(x)$, then $q(y) = a = q(x)$, hence $\langle 0,x\rangle_q = q(x) = q(y) = \langle 0,y\rangle_q$. By assumption the parallel classes $(X \rtimes O(q)).(0,x)$ and $P_{\langle 0,x\rangle_q}$ are equal, therefore $[t,\sigma] \in X \rtimes O(q)$ exists such that $(0,y) = [t,\sigma].(0,x)$ which implies $t = 0$ and $\sigma(x) = y = \varphi(x)$. For the resulting isometry $\varphi^{-1}\sigma : K.x \perp X_2 \simeq K.x \perp X_1$ we obtain $\varphi^{-1}\sigma(x) = x$ and finally $(X_2,p_2) \simeq (X_1,p_1)$. Cf. also [B] chIII §4, Cor. (4.3).

4.6 Remark

The foregoing theorem suggests an alternative formulation of cancellation in terms of the two geometric invariants which might be convenient for more general situations.

5. REMARKS ON GROUP EXTENSIONS

We close with some final remarks on group extensions

(5.1) $1 \to N \to E \to G \to 1$

The classification of group extensions involves the first three cohomology functors $H^i(G,-)$, $i = 1,2,3$ (cf. e.g. [Br] ch IV).

Let us restrict to the case $N = (A,+)$ a G-module and

(5.2) $0 \to A \to E \to G \to 1$

an extension which gives rise to the given G-action (cf. [Br] loc.cit.), then the second cohomology group $H^2(G,A)$ describes the equivalence classes of extensions (5.2). Such an extension is split if, and only if, E is the semidirect product $A \rtimes G$.

The classification of all conjugacy classes of splitting homomorphisms $s:G \to E$ involves $H^1(G,A) = Der(G,A)/Ider(G,A)$.

From a geometric point of view it would be natural to generalize SPSs in passing to arbitrary group extensions. In the case of (5.2) we have the group spaces $\underline{V}(G,A)$, $\underline{V}(E,A)$, E acting by conjugation (this is compatible with the original action of G on A).

J.André suggested to express the splitting of an extension in geometric terms. This finallv leads to characterize geometrically when a group space is a SPS. Since in the characterization of split extensions derivations and inner derivations play an essential role, it might be very interesting to study these objects from the geometric point of view. Moreover, these observations together with results on simplicial structures of geometric spaces give a motivation to try to introduce (co-)homological methods for our geometric spaces.

REFERENCES

[A1] André,J.,Zur Geometrie der Frobeniusgruppen. Math.Z.154(1977),159-168

[A2] André,J.,Eine geometrische Kennzeichnung imprimitiver Frobeniusgruppen. Abhandlungen Math.Sem.Univ.Hamburg 51(1981),120-135

[A3] André,J.,Lineare Algebra über Fastkörpern.Math.Z.136(1974),295-313

[A4] André,J.,Noncommutative geometry, nearrings and nearfields. Proceedings of a Conference on Nearrings and Nearfields, Tübingen 1985.

[B] Baeza,R.,Quadratic Forms Over Semilocal Rings. Springer Lecture Notes in Mathematics 655, Springer Verlag 1978

[Br] Brown,K.S.,Cohomology of Groups. Graduate Text in Math. 87, Springer Verlag 1982

[L] Lam,T.Y.,The Algebraic Theory of Quadratic Forms. W.A.Benjamin, 1973

[M] Müller,K.,Lineare Algebra über Fastkörpern.Diplomarbeit,Universität des Saarlandes, Saarbrücken 1984

[P1] Pfalzgraf,J.,On a model for noncommutative geometric spaces. J.Geometry 25(1985),147-163

[P2] Pfalzgraf,J.,On geometries associated with group operations. Geometriae Dedicata 21(1986),193-203

Annals of Discrete Mathematics 37 (1988) 375–376

ON PERMUTATION PROPERTIES FOR FINITELY GENERATED SEMIGROUPS

Giuseppe Pirillo
I.A.G.A. - I.A.M.I.
Consiglio Nazionale delle Ricerche
Viale Morgagni 67/A
Firenze (Italia)

Summary. We present a finitely generated semigroup which does not have the property P defined by Restivo and Reutenauer, but satisfies a weaker permutation property P^*.

In this paper, we improve a result of |1|, to which we refer for definitions of properties P_n, P_n^*, P and P^* concerning semigroups.
More precisely, we prove that P is stronger than P^*, even upon restriction to finitely generated semigroups. In fact, we have the following proposition.

Proposition 1. There exists a finitely generated semigroup which has the property P_5^* and does not have the property P.

Proof. Let a and b be two letters and A={a, b} the alphabet of the free semigroup A^+ (respectively, free monoid A^*).
Let the subset I of A^+ be defined as follows

$$I = \{w \in A^+ \ / \ w = u_1 v u_2 \ ; \ u_1, u_2 \in A^*; \ v = b^x a^i b^y a; \ x \geq y \geq 1; \ i \geq 1\}$$

The subset I of A^+ is an ideal of A^+ and the Rees quotient

$$S = A^+ / I$$

has the required properties.
In fact, we can show:

1) S has the property P_5^*;

2) for each $n \geq 2$, S does not have the property P_n.

1. Let u_1, u_2, u_3, u_4 and u_5 be arbitrary words of A^+.

i) If there exists a letter $c \in \{a, b\}$ such that for a suitable pair of indices i and j the words u_1 and u_j are powers of c, then

$$u_i . u_j = u_j . u_i .$$

Without loss of generality, we can suppose i = 1 and j = 2. Thus the words

$$u_1 . u_2 . u_3 . u_4 . u_5$$

and

$$u_2 \cdot u_1 \cdot u_3 \cdot u_4 \cdot u_5$$

coincide. Hence, in this case the result follows.

ii) If a letter with the property considered in i) does not exist, then four words among u_1, u_2, u_3, u_4 and u_5 contain the letter a and three of them contain a and b.

But in this case, from the definition of I, there are two different permutations f and g of $\{1, 2, 3, 4, 5\}$ such that

$$u_{f(1)} \cdot u_{f(2)} \cdot u_{f(3)} \cdot u_{f(4)} \cdot u_{f(5)}$$

and

$$u_{g(1)} \cdot u_{g(2)} \cdot u_{g(3)} \cdot u_{g(4)} \cdot u_{g(5)}$$

belong to I, i. e. they represent the same element of the Rees quotient $S = A^+/I$.

2. Consider the following n-tuple

$$ab, ab^2, \ldots, ab^n$$

of words of A^+.

It is easy to see that

$$(ab)\ (ab^2)\ldots.(ab^n) \notin I$$

and, for every $f \neq id$,

$$(ab^{f(1)})(ab^{f(2)})\ldots.(ab^{f(n)}) \in I.$$

So, for each $n \geq 2$, S does not have the property P_n.

REFERENCE

1. G. PIRILLO, On permutation properties for semigroups, Proceedings of the International Conference on Group Theory, Bressanone - Brixen, May 1986 (to appear in Springer Lecture Notes).

Annals of Discrete Mathematics 37 (1988) 377–384

ON k-SETS OF TYPE $(0,m,n)$ IN $S_{r,q}$ WITH THREE EXTERIOR HYPERPLANES

Rita PROCESI CIAMPI - Rosaria ROTA

Dipartimento di Matematica, Università degli Studi di Roma "La Sapienza", Piazzale Aldo Moro 5, 00185 Roma, Italia °

It is well known that the study of k-sets of type (m,n) in an affine space is equivalent to the study of k-sets of type $(0,m,n)$ in a projective space such that their exterior lines are only those belonging to a fixed hyperplane. This consideration suggested us the problem of studying the k-sets of type $(0,m,n)$ such that the exterior lines are exactly those belonging to two [1] or three hyperplanes. In this paper we present the results concerning the case of three exterior hyperplanes either belonging or not belonging to the same pencil. In the first case necessary conditions are stated for the existence of such k-sets. In the second case we obtain a non existence theorem for $r \geq 5$.

1. INTRODUCTION

Let K be a k-set of type $(0,m,n)$ in $S_{r,q}$ with three exterior hyperplanes S^1_{r-1}, S^2_{r-1}, S^3_{r-1} not belonging to the same pencil and let $S^h_{r-2} = S^i_{r-1} \cap S^j_{r-1}$, $i,j,h \in \{1,2,3\}$; such subspaces intersect in a $\hat{S}_{r-3}$. For an S'_{r-2} containing $\hat{S}_{r-3}$ and not in S^i_{r-1}, $K \cap S'_{r-2}$ is a k'-set of type $(0,m,n)$ in S'_{r-2} with exactly one exterior hyperplane. Furthermore for S''_{r-1} containing S^h_{r-2} and different from S^i_{r-1} and S^j_{r-1}, $K \cap S''_{r-1}$ is a k''-set of type $(0,m,n)$ in S''_{r-1} with two exterior hyperplanes. Therefore [1] :

1.1. There does not exist in $S_{r,q}$, $r \geq 5$, a k-set of type $(0,m,n)$ with three exterior hyperplanes not belonging to the same pencil.

We first observe that $K = S_{r,q} \setminus \{S^1_{r-1}, S^2_{r-1}, S^3_{r-1}\}$ is a k-set of type $(0,q-2,q-1,q)$. If we now consider the case in which the three exterior hyperplanes belong to the same pencil, $K = S_{r,q} \setminus \{S^1_{r-1}, S^2_{r-1}, S^3_{r-1}\}$ is of type $(0,q-2,q)$. Furthermore in this case, for any S'_{r-1} containing the $\hat{S}_{r-2}$ intersection of the three exterior hyperplanes and different from them, $K \cap S'_{r-1}$ is a k'-set of type $(0,m,n)$ with exactly one exterior hyperplane, thus, for $r>3$, [3] only three cases are possible for (m,n): $(0,1)$, $(q-1,q)$ and $((q-\sqrt{q})/2, (q+\sqrt{q})/2)$ q odd square, therefore:

° Research supported by G.N.S.A.G.A. - C.N.R.

1.2. If a k-set of type $(0,m,n)$ with three exterior hyperplanes belonging to a pencil exists in $S_{r,q}$, $r>3$, then q must be an odd square and $m=(q-\sqrt{q})/2$, $n=(q+\sqrt{q})/2$.

To study the cases $r=3$, $r=4$, we will need the results concerning the case $r=2$.

2. k-SETS OF TYPE $(0,m,n)$ IN π_q

Let K be a k-set of type $(0,m,n)$ in π_q with exactly three exterior lines not belonging to a pencil. As in [2] we can obtain for k the equation:

(2.1) $$k^2-k[1+(q+1)(n+m-1)]+mn(q^2+q-2) = 0$$

We must observe that $K = \pi_q \setminus \{r_1,r_2,r_3\}$ is a k-set of type $(0,q-2,q-1)$, $k=q^2-2q+1$ and this is the only possible case for $n-m=1$. In fact for $m=q-2$, $n=q-1$, $k = q^2-2q+1$ is the only acceptable solution of (2.1). Otherwise, for $m=n-1$ in (2.1) we obtain $\Delta = 4(n-1)[n(q+3)-q(q+1)]+1$, wich is greater or equal to zero if and only if $n\geq q(q+1)/(q+3)$, and thus $n=q-1$. Therefore:

2.1. If K is a k-set of type $(0,m,n)$ in π_q with three exterior lines not belonging to a pencil and $n-m=1$, then $K = \pi_q \setminus \{r_1,r_2,r_3\}$.

We shall now recall that [1] $u_m,u_n,u'_m,u'_n,w_m,w_n,v_m,v_n$ represent the numbers of m-secants and n-secants respectively for: $Q\in\pi_q \setminus \{K,r_1,r_2,r_3\}$, for $R\in r_i$ and $R\neq P_h = r_i\cap r_j$, $R\neq P_j=r_i\cap r_h$, for P_i, $i=1,2,3$, and for $T\in K$. Since the values of such numbers do not change in this case, we have:

2.2. For the existence of a k-set of type $(0,m,n)$ in π_q, k must satisfy (2.1) and $m(q+1)\leq k\leq min(n(q-1),n(q+1)-q)$; furthermore $m=s(n-m)$, $n=(s+1)(n-m)$ and $n-m$ divides q and k.

2.3. If K is a k-set of type $(0,m,n)$ in π_q with three exterior lines not belonging to a pencil and K is prime then $K = \pi_q \setminus \{r_1,r_2,r_3\}$.

We must now observe that:

$u_n = 0$ implies $k_1 = m(q+1)$, $k_2 = n(q+1)-q$, $n = q(q+1)/(q+3)$; such value of n is an integer only for $q = 3$, from which $m = 1$, $n = 2$ and $K = \pi_3 \setminus \{r_1,r_2,r_3\}$.

$w_m = 0$ implies $k_1 = n(q-1)$, $k_2 = m(q+1)-q+2n$, $m = 2n-q$.

$v_m = 0$ implies $q = 3$, $n = 2$, $m = 1$ which implies $v_m = 1$ (impossible).

From these and the analogues for the other values [1] we obtain:

2.4. Through any $P\neq P_h$, h=1,2,3 , at least one m-secant and one n-secant must pass. Through P_h at least one n-secant must pass, but not necessarily one m-secant. In fact $w_m=0$ implies $m=2n-q$, $k_1=n(q-1)$, $k_2=m(q+2)$.

The equality $m=2n-q$ can also be expressed as $n=q-a$, $m=q-2a$, $a|q$, and we have:

2.5. The values $m=q-2a$, $n=q-a$ lead to integers solutions of *(2.1)* for any a dividing q.

Furthermore from 2.4. the following is also true:

2.6. If K is a k-set of type $(0,m,q-1)$ then $m=q-2$ and $K=\pi_q\setminus\{r_1,r_2,r_3\}$.

Let now $n=q-2$ and then $1\leq u_{q-2}\leq 3$. The first case $u_{q-2}=1$ is easily excluded. For $u_{q-2}=2$ then $k_1=2q+mq-m-4$ and $k_2=q^2-4q+2m+2$, thus:

$$m^2(q-1)-m(2q^2-7q+7)+q^3-6q^2+10q-4 = 0 ,$$

of which the only possible integer solution is $m=q-4$. The values $m=q-4$, $n=q-2$ and $u_{q-2}=2$ give us $k=(q-4)(q+2)$ which implies $u_{q-2}=(q-4)/2$. Thus:

2.7. If a k-set of type $(0,m,q-2)$ with $u_{q-2}=2$ exists in π_q , then $q=8$, $m=4$ and $k=(q-4)(q+2)$.

For $u_{q-2}=3$, then $k_1=(m+3)(q-2)$ and $k_2=q^2-5q+3m+4$. With the same tecnique as before, together with the condition $w_m\geq 0$, we obtain:

2.8. If a k-set of type $(0,m,q-2)$ with $u_{q-2}=3$ exists in π_q then only two cases are possible: 1) $m=q-4$, $k=(q-1)(q-2)$ or 2) $q=10$, $m=q-4$, $k=(q-4)(q+2)$.

The following proposition gives the only possible example for case 1) in 2.8. .

2.9. If K is a k-set of type $(0,q-4,q-2)$ and $k=(q-1)(q-2)$ then $K=\pi_q\setminus\{\Gamma,r_1,r_2,r_3\}$ where Γ is an oval passing through the three intersecting points of r_1,r_2,r_3 and vice versa.

Proof. Let $\Gamma^\circ=\{Q_1,\dots,Q_{q-1}\}$, $Q_i\notin K\cup\{r_1,r_2,r_3\}$. We prove that $\Gamma=\Gamma^\circ\cup\{P_1,P_2,P_3\}$ is an oval in π_q. Let us consider a line r different from r_1,r_2,r_3 . If $P_i\in r$ then r is a $(q-2)$-secant since $w_{q-4}=0$, thus r must contain exactly one $Q_j\in\Gamma^\circ$. If $P_i\notin r$ and r is a $(q-2)$-secant, then r is exterior to Γ ; otherwise $r\cap\Gamma=\{Q_i,Q_j\}$. Thus Γ is an oval; the vice versa is obvious.

Let us now consider some special cases. If $n-m=2$, thus $q>7$ and q even, the discriminant of *(2.1)* is $\Delta=4m^2(q+3)-4m(q^2-q-6)+(q+2)^2$ and $\Delta\geq 0$ if and only if

$m \le A$, $m \ge B$ where: $A=(q^2-q-6-\sqrt{q^4-3q^3-18q^2-4q+24})/2(q+3)$ and $B=(q^2-q-6+\sqrt{q^4-3q^3-18q^2-4q+24})/2(q+3)$. It is easy to verify that $A<2$, $B>q-6$; thus:

2.10. If K is a k-set of type $(0,m,n)$ and $n-m=2$, then $m=q-4$, $n=q-2$.

Similarly for $n-m=3$, $q>8$:

2.11. If K is a k-set of type $(0,m,n)$ and $n-m=3$, then $m=q-6$, $n=q-3$.

In general for $n-m=a$ we obtain $\Delta \ge 0$ if and only if $m \le A$, $m \ge B$ where:

$A=(q^2+q-aq-3a-\sqrt{q^4+q^3-2q^2-a^2q^3-4a^2q^2-a^2q+6a^2})/2(q+3)$,

$B=(q^2+q-aq-3a+\sqrt{q^4+q^3-2q^2-a^2q^3-4a^2q^2-a^2q+6a^2})/2(q+3)$.

For $a=4$ and $q>23$ is $A<4$; thus:

2.12. If K is a k-set of type $(0,m,n)$ and $n-m=4$, $q>23$, then $m \ge B$.

The cases $q=16$, $q=20$ give respectively the pairs (4,8) or (8,12) and (12,16). At this point let us consider the case in which the exterior lines belong to a pencil. We must observe that the (2.1) does not change in this case as the formulas for $u_m, u_n, u'_m, u'_n, v_m, v_n$, while it results: $w_m=(n(q-2)-k)/(n-m)$ and $w_n = =(k-m(q-2))/(n-m)$. Proposition 2.4. thus becomes:

2.13. Through any $P \neq P_h$, $h=1,2,3$, at least one m-secant and one n-secant must pass. Through P_h at least one n-secant must pass, but not necessarily one m-secant. In fact $w_m=0$ implies $k_1=n(q-2)$, $k_2=3n+m(q+1)-q$, $m=(3nq-6n-q^2+2q)/2q$.

Because of the different formula for w_m we also obtain a different limitation for k; namely $m(q+1) \le k \le min(n(q-2),n(q+1)-q)$, thus proposition 2.5. becomes:

2.14. The values $m=q-2a$, $n=q-a$ give $k=m(q-2)$ as only possible solution of (2.1), for any a dividing q .

Following the previous results we obtain:

2.15. If K is a k-set of type $(0,m,q)$ with three exterior lines belonging to a pencil, then $m=q-2$ and $K=\pi_q \setminus \{r_1,r_2,r_3\}$.

2.16. There does not exist any k-set of type $(0,m,q-1)$ with three exterior lines belonging to a pencil.

Proof. Let $P \in \pi_q \setminus \{K,r_1,r_2,r_3\}$ and $P \in r$ which is a $(q-1)$-secant; then $r_1 \cap r_2 \cap r_3 = =S \in r$. This implies $u_{q-1}=1$, from which we have $k_1=(m+1)q-1$, $k_2=q^2-2q+m$ and thus $m^2q-m(2q^2-4q+3)+(q^3-3q^2+2q)=0$ with non acceptable solutions. Similarly:

2.17. There are no k-sets of type $(0,m,q-2)$ with three exterior lines belonging to a pencil.

From these results and those obtained in the case of the exterior lines not belonging to a pencil we have:

2.18. If a non trivial k-set of type $(0,m,n)$ with three exterior lines belonging to a pencil exists, then $n-m>2$. For $n-m=3$ must be $m=q-6$, $n=q-3$.

2.19. If q is a prime, in π_q there are no non trivial k-sets of type $(0,m,n)$ with three exterior lines belonging to a pencil.

3. k-SETS OF TYPE $(0,m,n)$ IN $S_{3,q}$ AND $S_{4,q}$

Let K be a k-set of type $(0,m,n)$ in $S_{3,q}$ with three exterior planes π^1, π^2, π^3 not belonging to a pencil. If we consider the points of K distributed over the planes of the pencil through an n-secant t intersecting the line $r_h=\pi^i\cap\pi^j$, we obtain the following relation: $|K|=\tilde{k}+Nk_1+(q-N)k_2-nq$, where k_1,k_2 are the solutions of (2.1), $\tilde{k}=|\tilde{K}|$, where $\tilde{K}=K\cap\tilde{\pi}$, and $\tilde{\pi}=t\cup r_h$, moreover N is the number of those planes through t intersecting K in a k_1-set. Similarly for an m-secant in $\tilde{\pi}$, we obtain: $|K|=\tilde{k}+Mk_1+(q-M)k_2-mq$. Thus $N-M=(n-m)q/\sqrt{\Delta}$ where Δ is the discriminant of (2.1). Since $n-m$ divides n,m then $N-M=q/\sqrt{\Delta'}$, $\Delta'=\Delta/(n-m)^2$ and, as $q=p^h$, $\Delta'=p^{2z}$, $1\le z\le h$. The case $\Delta'=1$ is easily excluded. We will also need the relation $\tilde{\Delta}=\Delta'-4s(s+1)$, where $\tilde{\Delta}$ is the discriminant of (9.1) in [1]. Summarizing the results of §2. and those obtained in [1] we get:

3.1. The necessary conditions for the existence in $S_{3,q}$ of a k-set of type $(0,m,n)$ with three exterior planes not belonging to a pencil are: $n-m$ divides q,m,n,k, $5\le m<n\le q-2$ and $n-m>2$. Furthermore if $n-m=3$ then $m=q-6$, $n=q-3$. If q is a prime there are no k-sets of type $(0,m,n)$ with three exterior planes not belonging to a pencil in $S_{3,q}$.

Moreover the number $k=|K|$ must satisfy:

$$(3.1)\qquad k^2-k[1+(q^2+q+1)(n+m-1)]+mn(q^4+q^3-q^2-2q+1) = 0 .$$

We now consider the following four cases: (α) q odd square, $m=(q-\sqrt{q})/2$, $n=(q+\sqrt{q})/2$; (β) $q=2^h$; (γ) $q=p^2$, $p\neq 2$; (δ) $q=p^h$, $p>2$, $h>2$.

(α) Solving (2.1), (3.1) and (3.1) of [1], we get:

$\tilde{k}=(q^2\pm q\sqrt{3q-2})/2$, $k=(q^3\pm\sqrt{q}\sqrt{2q^3+q^2-3q+1})/2$, $k'=(q\pm q\sqrt{2q-1})/2$.

Thus the following must be true: $2q-1=a^2$, $3q-2=b^2$, $2q^3+q^2-3q+1=c^2$, which proves to be impossible, then:

3.2. In $S_{3,q}$, q odd square, there are no k-sets of type $(0,(q-\sqrt{q})/2,(q+\sqrt{q})/2)$ with three exterior planes not belonging to a pencil.

(β) Since $\Delta'=\tilde{\Delta}+4s(s+1)=2^{2z}$, and $\tilde{\Delta}$ is odd (see [1]), it results:

3.3. In $S_{3,q}$, $q=2^h$, there are no k-sets of type $(0,m,n)$ with three exterior planes not belonging to a pencil.

(γ) In this case $\Delta'=(p^2+1)p^2+(2s+1)^2(p^2+1)-2p(p^2+1)(2s+1)+2(2s+1)^2-2+p^2=p^{2z}$, with $z=1$ or $z=2$. The value $z=1$ is easily excluded; for $z=2$ we obtain: $(2s+1)^2(p^2+3)-2p(p^2+1)(2s+1)+2p^2-2=0$, from which $2s+1=(p(p^2+1)\pm$ $\pm\sqrt{p^6-3p^2+6})/(p^3+3)$. Since $(p^3-1)^2<p^6-3p^2+6<(p^3)^2$, we get:

3.4. In $S_{3,q}$, $q=p^2$, $p\neq 2$, there are no k-sets of type $(0,m,n)$ with three exterior planes not belonging to a pencil.

(δ) Finally in this case the study of Δ' leads to the condition: $p^{4h-2a}+p^{3h-2a}-$ $-p^{3h}+p^{2h-2a}-p^{2h}+p^{2z+h}<0$, where $1\leq z\leq h$ and $n-m=p^a$. Thus:

3.5. If a k-set of type $(0,m,n)$ with three exterior planes not belonging to a pencil exists in $S_{3,q}$, $q=p^h$, $h\neq 2$, then $a<h/2$, $z>h-a$ $(n-m=p^a,\ \Delta'=p^{2z})$.

Since proposition 1.1. denies the existence of k-sets of the aforesaid type for $r>4$, we only have to consider $r=4$. In this case summarizing the previous results we obtain:

3.6. In $S_{4,q}$ a k-set of type $(0,m,n)$ with three hyperplanes not belonging to a pencil can not exist if: q is prime, $q=2^h$, $q=p^2$, q odd square and $m=(q-\sqrt{q})/2$, $n=(q+\sqrt{q})/2$. Furthermore if $q=p^h$, $h\neq 2$, a necessary condition for the existence of such a k-set is: $a<h/2$, $z>h-a$ $(n-m=p^a, \Delta'=p^{2z})$.

We can now consider the case in which $r=3$ and the three exterior planes belong to a pencil. In this case $k=|K|$ must satisfy:

$$k^2-k[1+(q^2+q+1)(n+m-1)]+mn(q^4+q^3-q^2-2q)=0 \quad (3.2)$$

Furthermore the general condition for the existence of such k-sets are the following:

3.7. In $S_{3,q}$, q prime, the only possible k-set of type $(0,m,n)$ with three exterior planes belonging to a pencil is $K=S_{3,q}\setminus\{\pi^1,\pi^2,\pi^3\}$. Moreover if a non trivial k-set of the aforesaid type exists in $S_{3,q}$, $q=p^h$, then $n\leq q-3$, $n-m$ divides

m,n,q and $n-m\neq 1,2$.

Since the proofs of 3.3., 3.4. and 3.5. are based on equation *(2.1)*, the analogous propositions are still true when the three planes pass through a line r. If q is an odd square $m=(q-\sqrt{q})/2$, $n=(q+\sqrt{q})/2$, from *(3.2)* we have $k=(q^3\pm \pm q\sqrt{2q^2+q-2})/2$ and $(q^3\pm q\sqrt{2q^2+q-2})/2=dq+(q-2)(q^2-q)/2$, where d is the number of planes through r intersecting K in a $(q^2+q)/2$-set (see [3]). Therefore: $7q^2-q(12d+13)+4d^2+8d+6=0$, from which $q=(12d+13\pm\sqrt{32d^2+88d+1})/14$. Moreover, if $r\not\subset\pi$, then $\pi\cap K$ is a $\hat{k}$-set of type $(0,(q-\sqrt{q})/2,(q+\sqrt{q})/2)$ with three exterior lines belonging to a pencil, thus $\hat{k}=(q^2+\sqrt{q}\sqrt{3q-2})/2$. From these observations follows:

3.8. If K is a k-set of type $(0,(q-\sqrt{q})/2,(q+\sqrt{q})/2)$ in $S_{3,q}$, q odd square, with three exterior planes passing through a line r, the following integers must be perfect squares: $2q\ +q-2$, $3q-2$, $32d\ +88d+1$, where d is the number of planes through r intersecting K in a $(q^2+q)/2$-set.

Finally, for $r>3$, proposition 1.2. can be extended as follows:

3.9. If K is a k-set of type $(0,(q-\sqrt{q})/2,(q+\sqrt{q})/2)$ with three exterior hyperplanes belonging to a pencil in $S_{r,q}$, $r>3$, q odd square, then $2q^{t-1}+q-2=a_t^2$, for any t such that $2\leq t\leq r$.

Proof. For every $S_{t,q}$ in $S_{r,q}$ non containing $\hat{S}_{r-2}$, $k_t=|K\cap S_{t,q}|$ must satisfy the equation: $k_t^2-k_tq^t+q^{t-1}/4(q^{t+1}-2q^{t-1}-q+2)=0$, whose discriminant must then be a perfect square.

REFERENCES

[1] Procesi, R. and Rota, R., Sui k-insiemi di tipo (O,m,n) di uno spazio di Galois $S_{r,q}$, in print.

[2] Tallini Scafati, M., Calotte di tipo (m,n) in uno spazio di Galois $S_{r,q}$, Rend.Acc.Naz.Lincei (8) 53 1972, pp. 71-81.

[3] Tallini Scafati, M., I k-insiemi di tipo (m,n) di uno spazio affine $A_{r,q}$, Rend.Mat. (1) 1981, vol. 1, Serie VII, pp. 63-79.

Annals of Discrete Mathematics 37 (1988) 385–390

AN ALGORITHM FOR L_s - COLOURATIONS

Luigia PUCCIO

Dipartimento di Matematica - Università di Messina
I-98100 Messina, Italy *

An algorithm for L_s - colourations of undirected graphs is described. This algorithm guarantees an optimal colouration and the correct value of the s-chromatic number for an arbitrary graph. Moreover, it can be used in case of directed graphs too.

1. INTRODUCTION

Let G=(V,E) be an undirected graph and K a mapping of V into a set C whose elements are called colours. K is an L_s - colouration of G if for any $x,y \in V$, $x \neq y$, $K(x) = K(y)$ implies $d(x,y) > s$, where $d(x,y)$ is the distance of y from x in G [10, 17], i.e. the minimum length of all chains from x to y. When $s = 1$ an L_1 - colouration is the usual colouration. If $K(x) \neq K(y)$ for all $x,y \in V$, then K is called an L_∞ - colouration.

Unless otherwise stated, by a graph we always mean a simple, i.e. without loops and multiple edges, connected and undirected graph.

The s-chromatic number is defined as the minimum number of colours, $\gamma_s(G)$, such that G has an L_s - colouration.

The determination of the s-chromatic number and some related parameters has been the subject of a good amount of research papers (see references).

It is well known that classical colourations have many applications. The same is true for L_s - colourations. Since their definition implies the idea of distance between vertices, a graph with an L_s - colouration can provide an interesting model to solve typical problems in Operations Research, e.g. the location of emergency centres, or medians, and transportation and assignment problems.

Let G=(V,E) be a graph; for any $s \in \mathbb{N}$, $s > 1$, and $x \in V$, define

$$\Gamma_s(x) = \{ y \in V : d(x,y) < s+1 \},$$
$$T_s(x) = \{ y \in V : d(x,y) > s \} = V \setminus \Gamma_s(x),$$
$$\mathcal{F}_s(x) = \{ X : X \subseteq V, \text{diam} X_G < s+1 \},$$

* Research supported by the Community of Mediterranean Universities.

where X_G is the subgraph of G whose vertices are in X;

$$d_s(G) = \max_{X \in \mathcal{F}_s} |X| ,$$

$d_s(G)$ is called the s-density of G;

$$\Delta_s(G) = \gamma_s(G) - d_s(G) ;$$

obviously, $\Delta_s(G) \geqq 0$.

The most important result is that, for every $s > 1$ and any $h \in \mathbb{N}$, there exists a planar graph such that $\Delta_s(G) = h$. Therefore,

$$\sup (\Delta_s(G) : s > 1, G \in \mathcal{P}) = \infty ,$$

where $\mathcal{P}$ is the set of all planar graphs. Fig. 1 shows such a planar graph; the black vertices yield the minimal configuration for $s = 2$ and $k = 2h - 1$.

In [8] the problem is posed of finding the values for the following parameters:

$$v_s(h) = \min \{ n \in \mathbb{N} : \exists G = (V,E) \ni' |V| = n \Rightarrow \Delta_s(G) = h \} ,$$

$$\delta_s(h) = \min \{ n \in \mathbb{N} : \exists G = (V,E) \ni' |V| = n + d_s(G) \Rightarrow \Delta_s(G) = h \} ,$$

$$m_s(h) = \min \{ n \in \mathbb{N} : \exists G = (V,E) \ni' |E| = n \Rightarrow \Delta_s(G) = h \} ,$$

where s, h are positive integers. The values for those parameters are known just in very few cases. In general, upper bounds are known [13] which might be improved.

The facts we just recalled seem to provide a good motivation for the research of an efficient algorithm for such L_s - colourations. Here we present an algorithm which guarantees both an optimal L_s - colouration and the correct value of the s-chromatic number for an arbitrary graph. Furthermore, this algorithm applies also to directed graphs for which L_s - colourations can be defined in a similar way [18].

2. DESCRIPTION OF THE ALGORITHM

There are many algorithms for the classical colouration. They are based on different procedure such as formulation as a zero-one programming [2] , sequential methods based on vertex ordering [14], backtrack methods [15] , optimal independent colourings [19] , formulation as a set covering problem [3]. Our algorithm is an appropriate combination of the last two ones. In fact, its procedure is a formulation as a set covering problem, where the sets of the coverings are maximal independent 1-chromatic sets, called maximal 1-subgraphs.

Recall the following definitions and theorem.

An independent set of a graph G=(V,E) is a subset of V such that no two of its vertices are adjacent in G. An independent set is maximal when there is no independent set containing it.

THEOREM [3] - If a graph is r-chromatic then it can be coloured with r (or

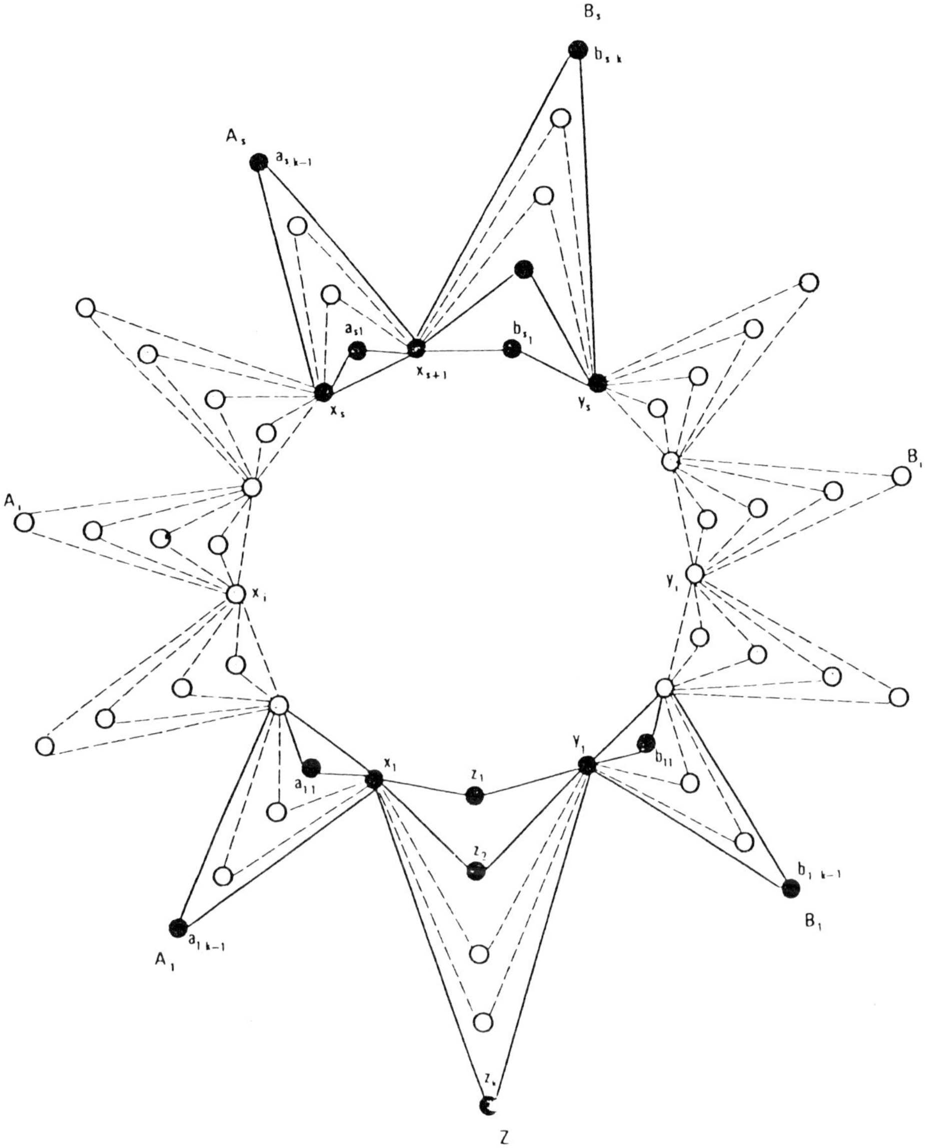

B_s
$b_{s\,k}$
A_s
$a_{s\,k-1}$
a_{s1}
b_{s1}
x_{s+1}
x_s
y_s
B_i
A_i
x_i
y_i
y_1
b_{11}
x_1
z_1
a_{11}
z_2
$b_{1\,k-1}$
B_1
$a_{1\,k-1}$
A_1
z_k
Z

Fig. 1

fewer) colours first colouring with one colour a maximal indepentent set $S_1[G]$, next colouring with another colour a set $S_1[<V \quad S_1[G]>]$ and so on, until all vertices are coloured.

This result can be used also for L_s- colourations as we next show. For a graph $G=(V,E)$ define the graph $G^s=(V',E')$ as follows. $V' = V$ and for any $x,y \in V$, $\{x,y\} \in E'$ if and only if $d_G(x,y) < s+1$. Thus in $G^s=(V',E')$ there is a link between x and every element in the set $\Gamma_s(x)$.

PROPOSITION - A classical colouration of $G^s=(V',E')$ is an L_s- colouration of $G=(V,E)$, and conversely.

Clearly, this result plays a fundamental role in the realization on our algorithm.. We try to find an optimal colouration of $G^s=(V',E')$ whit the help of the above mentioned theorem; consequentely, we get an optimal L_s-colouration for $G=(V,E)$ and its s-chromatic number.

2.1. Steps of the algorithm

We just outline the procedure. The program of its implementation is written in FORTRAN 77 and was tested on several graphs. Such directed and undirected graphs were constructed by means of a random generator program [4] . Details of the programs are available from the author.

STEP 1 - INPUT: n = number of vertices of $G=(V,E)$,
A = incidence matrix of G,
s = kind of colouration.

STEP 2: Calculate the incidence matrix B of $G^s=(V',E')$. By definition, G^s has no loop. Then , from the classical formula to calculate the reachability matrix it follows

$$B = A + A^2 + \dots + A^s .$$

STEP 3: Find all maximal 1-subgraph of G. Consider $T_s(x)$ for every $x \in V=V'$. Maximal independent sets are subsets of $T_s(x) \cup \{x\}$. They are found by depth-first search.

STEP 4: Store all maximal 1-subgraphs in an array Q in decreasing order of their numbers of vertices.

Such an ordering turns out to be useful to find the optimal colouration. All maximal 1-subgraphs yield a covering of the vertex set V=V'. We have to find a partition with the minimum number of elements since we want an optimal colouration. Thus, the element number of the partition is the s-chromatic number $\gamma_s(G)$.

STEP 5: Assign the first colour to the vertices of the first maximal 1-subgraph in Q, i.e. the 1-subgraph with the maximum vertex number.

STEP 6: Consider the sequential next maximal 1-subgraph in the array Q.

Three cases may occur:

(a) All its vertices are coloured. Return to step 6.

(b) No vertex is coloured. Assign to all vertices another colour. Return to step 6.

(c) Some vertices are not coloured. Put them in a list L with pointer. Return to step 6.

When all maximal 1-subgraphs are investigated, look at the list L.

STEP 7: a) If L is empty, go to step 8.

b) Colour vertices of list L which are distinct and not coloured. There is a different colour for every not vanished pointer.

STEP 8 - OUTPUT: List vertices of G with their own colours and the s-chromatic number $\gamma_s(G)$.

3. COMMENTS

We have the best implementation of the algorithm if the incidence matrices of $G=(V,E)$ and $G^s=(V',E')$, and the maximal 1-subgraphs storage array are logical. Obviously, the algorithm efficiency depends on the dimension of these matrices. However, if we use the algorithm to improve the first values of the parametres $v_s(h)$, $\delta_s(h)$, $m_s(h)$,then the incidence matrices of considered graphs are not too big. When the graph G is the model for a problem in operations research its incidence matrix can be very big but,usually, it is a sparse matrix. Therefore, we can use sparse matrix techniques. In the wrost cases, we can consider the complement $\bar{G}^s$ of G^s. Then an L_s - colouration of G is obtained assign the same colour to adjacent vertices of $\bar{G}^s$. If this is the case, we consider the set $\Gamma_s(G)$ instead of $T_s(G)$.

In [18] M_s- colourations are defined for directed graphs.

Let $G=(V,E)$ be a directed graph and K a mapping of V into a set C of colours. K is an M_s- colouration of G if for any $x,y \in V$, $x \neq y$, $K(x) = K(y)$ implies there is no path from x to y with length less than s+1. $\omega_s(G)$ is the s-chromatic number.

In case of directed graphs the incidence matrices are not symmetric. Nonetheless, our algorithm can be used, with no changes, also for M_s-colourations.

REFERENCES

[1] Antonucci, S., Generalizzazioni del concetto di cromatismo d'un grafo, Boll. Un. Mat. Ital., (5) 15-B (1978) 20-31.

[2] Berge, C. Graphes et hypergraphes, (Dunond, Paris, 1970).

[3] Christofides, N., Graph Theory, an algorithmic approach, (Academic Press, 1975).

[4] Gallo, G., Pallottino, S., Ruggeri, C. and Storchi, G., Metodi ed algoritmi per la determinazione di cammini minimi, Monografie di Software

Matematico N.29, I.A.C. (1984).

[5] Gionfriddo, M., Sulle colorazioni L_s d'un grafo finito, Boll. Un. Mat. Ital., (5) 15-A (1978) 444-454.

[6] Gionfriddo, M., Su un problema relativo alle colorazioni L_s d'un grafo planare e colorazioni L_s, Riv. Mat. Univ. Parma, (4) 6 (1980) 151-160.

[7] Gionfriddo, M., Alcuni risultati relativi alle colorazioni L_s d'un grafo, Riv. Mat. Univ. Parma, (4) 6 (1980) 125-133.

[8] Gionfriddo, M., Sul parametro Δ_s d'un grafo L_s- colorabile e problemi relativi, Riv. Mat. Univ. Parma, (4) 8 (1982) 1-7.

[9] Harary, F., Graph Theory, (Addison-Wesley, 1960).

[10] Kramer, F. and Kramer, H., Un problème de colrolation des sommets d'un graphe, C.R. Acad. Science Paris Ser. A 2 8 (1969) 46-48.

[11] Lizzio, A. and Milici, S., Determinazione di $v_3(1)$, $v_3(2)$, $v_4(1)$ in un grafo e condizioni sufficienti per $\Delta_s = 0$, Riv. Mat. Univ. Parma.

[12] Marino, M.C. and Puccio, L., Su alcuni parametri associati a colorazioni L_s in un grafo finito non orientato, Le Matematiche, vol. XXXV, fasc. I-II (1980) 301-310.

[13] Marino, M.C. and Puccio, L., Sul parametro $\Delta_s(G)$ d'un grafo planare, Riv. Mat. Univ. Parma, (4) 9 (1983) 9-13.

[14] Matula, D.W., Marble, G. and Isaacson, J.D., Graph colouring algirithms, from Graph Theory and Computing, (Ed. Read, R.C., Academic Press, 1972).

[15] Nijenhuis, A. and Wilf, E., Combinatorial Algorithms, (II Ed., Academic Press, 1978).

[16] Sedgewick, R., Algorithms, (Addison-Wesley, 1984).

[17] Speranza, F., Colorazioni di specie superiore d'un grafo, Boll. Un. Mat. Ital., (4) 12, supp. fasc. 3 (1975) 53-62.

[18] Speranza F., Sur les colorations des graphes orientés, Boll. Un. Mat. Ital., (5) 16-A (1979) 517-522.

[19] Wang, J., ACM 21, (1974), 385.

Annals of Discrete Mathematics 37 (1988) 391–394

A BLOCKING SET IN PG(3,q), $q \geq 5$

Sandro Rajola

Dipartimento di Matematica
Istituto "G. Castelnuovo"
Università degli Studi di Roma "La Sapienza"
I-00185 Rome, Italy

A blocking set is constructed in PG(3,q), $q \geq 5$, which turns out to be the first example of a blocking set when $q = 5$.

1. INTRODUCTION

A blocking set S in PG(3,q), the 3-dimensional projective space over the field GF(q), is a set of points meeting all lines and containing none.
In [2] it was proved that, for any prime power q, there exists an integer b(q) such that blocking sets exist in PG(r,q) provided that $r \leq b(q)$, whereas $r > b(q)$ rules out their existence. Furthermore, in [6] it was shown that $b(2) = 1$, $b(3) = 2$, and $b(q) \geq 3$ for any $q \geq 7$. The last result was achieved producing examples of blocking sets in PG(3,q), $q \geq 7$.
Therefore, the existence of blocking sets in PG(3,5) was an open problem. Here a blocking set is constructed in PG(3,q) for any $q \geq 5$. Consequently, $b(q) \geq 3$, $q \geq 5$.
In sect. 3 it is shown that a blocking set exhibited in [6] is reducible and an irreducible one is constructed starting from it.

2. THE CONSTRUCTION

Let O_1, O_2, O_3, and V be four independent points in PG(3,q), $q \geq 5$, i.e. the four vertices of a thetrahedron T. Take a point O_4 on the edge O_1V of T, other than O_1 and V, which is possible as $q \geq 5$. The plane through O_2O_3 and O_4 meets the faces O_1O_2V and O_1O_3V in the lines O_2O_4 and O_3O_4, respectively. Next, take any point A_1 on $O_1O_2 \setminus \{O_1, O_2\}$ and this is possible by the previous remark. Similarly, take any point B_1 on $O_2O_4 \setminus \{O_2, O_4\}$. Let $Z = A_1B_1 \cap O_2V$ and C_1 be a point on $O_2V \setminus \{Z, O_2, V\}$. Next, take A_2 on $O_1O_3 \setminus \{O_1, O_3\}$ and B_2 on $O_3O_4 \setminus \{O_3, O_4\}$. Let $T = A_2B_2 \cap O_3V$ and C_2 be a point on $O_3V \setminus \{T, O_3, V\}$. Finally, choose a point D on the line O_2O_3 other than O_2 and O_3 and non-collinear with any of the pairs A_1A_2, B_1B_2, and C_1C_2. Such a choice is possible by the following argument.

Denote by A', B', and C' the points on O_2O_3 collinear with the pairs A_1A_2, B_1B_2, and C_1C_2, respectively. Even if the five points O_2, O_3, A', B', C' are all distinct, at least one point remains on O_2O_3 to be taken as D. Denote by E_1, $E_2, \ldots, E_{q-2}$ the points on $O_1V \setminus \{O_1, O_4, V\}$, and $\pi(L,M,N)$ the plane spanned by the three independent points L, M, and N; whenever it is the case, the plane is considered only as the set of its points.

Define the set S by

$$S = (\pi(O_1,O_2,O_3) \cup \pi(O_1,O_2,V) \cup \pi(O_2,O_3,V) \cup \pi(O_3,O_1,V) \cup \pi(O_2,O_3,O_4)$$

$$\setminus (O_1O_2 \cup O_2O_3 \cup O_3O_1 \cup O_1V \cup O_2V \cup O_3V \cup O_2O_4 \cup O_3O_4))$$

$$\cup \{A_1,A_2,B_1,B_2,C_1,C_2,D,E_1,E_2,\ldots,E_{q-2}\}.$$

Notice some symmetries in S: the faces O_1O_2V and O_1O_3V are similarly constructed, as well as O_2O_3V, $O_2O_3O_4$, and $O_1O_2O_3$.

We claim that S is a blocking set.

Firstly, we prove that S contains no line. Since the points of S on the faces O_1O_2V and O_1O_3V are distributed in a similar way, it suffices to show that one of them contains no line. A line ℓ on $\pi(O_1,O_2,V)$, $\ell \neq O_1O_2$, O_1V, O_2V, O_2O_4, is contained in S iff it passes through A_1, B_1, C_1, a contradiction as those points are not collinear. Similarly, a line ℓ on $\pi(O_2,O_3,V)$, $\ell \neq O_2O_3$, O_2V, O_3V, belongs to S iff it is on C_1, C_2 and D which is impossible since such points are not collinear. The same argument proves that neither the face $O_2O_3O_4$ nor the face $O_2O_3O_1$ contains lines.

Next, any line ℓ in PG(3,q), not on a face of T, meets S at five points at most since S consists of points on five planes; thus, there is at least one point on $\ell \setminus S$.

Finally, we show that any line ℓ meets S. This is obviously true for the sides of the triangle O_1O_2V and for the lines on E_j, $j = 1,2,\ldots,q-2$, as well as for O_2O_4 on $\pi(O_1,O_2,V)$. On the other hand, by the construction of S, on any other line of $\pi(O_1,O_2,V)$ at most three points lie off S. The same argument applies to $\pi(O_1,O_3,V)$.

Next, we turn to $\pi(O_2,O_3,V)$. Of course, the lines O_2O_3, O_2V and O_3V meet S and the same occurs for the lines on C_1, C_2 and D, respectively. Keeping in mind the construction of S, on any other line of $\pi(O_2,O_3,V)$ three points at most lie off S. Similarly, all lines on $\pi(O_2,O_3,O_4)$ and $\pi(O_2,O_3,O_1)$ meet S.

Since S consists of points on five suitable planes, any line ℓ meeting those planes off the lines O_iO_j, O_sV meets S, $(i,j) = (1,2),(1,3),(2,3),(2,4),(3,4)$, $s = 1,2,3$. On the other hand, it is easy to check that all lines incident with both VO_i and O_jO_h meet S, (i,j,h) a permutation of $(1,2,3)$ or $(i,j,h)=(3,2,4)$,

(2,4,3).

Finally, if ℓ is a line incident with O_1O_2 and O_3O_4, then either ℓ is on one of the planes $\pi(O_2,O_3,O_4)$, $\pi(O_1,O_2,O_3)$, $\pi(O_1,O_2,V)$, and $\pi(O_1,O_3,V)$, so that ℓ meets S, or ℓ meets the plane $\pi(O_2,O_3,V)$ at a point on S. (Take into account that S is constructed by deleting some points from the edges of T.) The same argument shows that any line meets S which meets both O_1O_3 and O_2O_4. Obviously, any line on a vertex of T meets S. Finally, a line ℓ not on any vertex of T and incident with one edge of T meets one face of T at a point on S. Thus, S is a blocking set in PG(3,q) for any $q \geq 5$.

3. AN IRREDUCIBLE BLOCKING SET

First of all, observe that the blocking set constructed in sect. 2 may be reducible. In [6] a blocking set was exhibited in PG(3,q) for any even $q \geq 8$. Here we show that such a blocking set is reducible and produce an irreducible one sitting in it.

The construction in [6] is as follows. Take a plane π and a hyperoval on π, say C. Denote by O_j, $j = 1,2,3$, any three points on C and delete from π the points on $O_iO_j \setminus \{O_i,O_j\}$, $i,j = 1,2,3$; this yields a blocking set S on π. Let O_4 be any point off π and Γ the cone projecting C from O_4. Delete from Γ the points on $O_jO_4 \setminus \{O_j\}$; this gives a set Γ^*. Call π_j^* the plane $O_iO_hO_4$, (i,j,h) a permutation of (1,2,3), from which the above mentioned points have been removed. The set $K = S \cup \Gamma^* \cup (\pi_j^*: j = 1,2,3)$ is a blocking set in PG(3,q).

This blocking set is reducible; namely, we can delete from K the points on $C \setminus \{O_1,O_2,O_3\} = C^*$ and still have a blocking set, say K^*.

Obviously, $S \setminus C^* = S^*$ is a blocking set on π. Since K was a blocking set, it suffices to show that any line ℓ off π and Γ^* and meeting π on C^* actually meets K^*. Since ℓ meets at most one of O_4O_j, it meets π_i^* for some i. Hence, K^* is a blocking set.

Finally, we show that K^* is irreducible, i.e. that there is at least one tangent at each point. For any point P on S^*, PO_4 is a tangent at P. Next, take P on π_j^*; the plane $\omega = PO_4O_j$ meets Γ^* at a generator g. Let $G = g \cap C^*$. It is easy to check that GP is a tangent at P. Now suppose P is on Γ^*; take the plane PO_4O_j, any j; w.l.o.g., assume j = 1. This plane meets π in a line ℓ on O_1 and $A = \ell \cap O_2O_3 \neq O_2,O_3$; AP is a tangent at P.

REFERENCES

[1] Bruen, A.A., Blocking sets in finite projective planes, SIAM J. Appl. Math. 21 (1971), 380-392.

[2] Mazzocca, F., and Tallini, G., On the non-existence of blocking sets in PG(n,q) and AG(n,q) for all large enough n, Simon Stevin 59 (1985), 43-50.

[3] Rajola, S., Un esempio di blocking set in PG(3,q) per ogni $q \geq 5$, Quad. Sem. Geometrie Comb. n. 57, Nov. 1985, Dipart. Mat. Univ. Roma, "La Sapienza".

[4] Rajola, S., Blocking sets, k-insiemi e fibrazioni in PG(r,q), Tesi, Univ. Roma "La Sapienza", a.a. 1985-86.

[5] Tallini, G., Problemi e risultati sulle geometrie di Galois, Relaz. n. 30, Ist. Mat.,Univ. Napoli, 1973.

[6] Tallini, G., k-insiemi e blocking sets in PG(r,q) ed in AG(r,q), Ist. Mat., Univ. L'Aquila, 1982.

Annals of Discrete Mathematics 37 (1988) 395–398

A CHARACTERIZATION OF ALL ABELIAN GROUPS WHOSE LATTICE OF PRECOMPACT GROUP TOPOLOGIES REPRESENTS A PROJECTIVE GEOMETRY

Dieter Remus

c/o Institut für Mathematik, Lehrgebiet D
Universität Hannover, Welfengarten 1, D-3000 Hannover, F.R.G.

First, two structure theorems in the sense of the title are proved. Then all projective spaces are characterized for which the correspondent lattice of subspaces can be embedded in the lattice of precompact group topologies of some abelian group.

AMS Subject Classification. Primary 51 A05; Secondary 06 C10, 22 A99.

1. NOTATION AND CONVENTIONS

For lattices notation and definitions are used as in [4] . The order relation on a lattice L is denoted by $\leq$. If equality does not hold, one writes $<$. $\bigvee X$ means the supremum for every $X \subseteq L$. For $X = \{x_1, x_2\}$ one writes $x_1 \vee x_2$ instead of $\bigvee X$. Let $a, b \in L$. Then $a \prec b$ means that $a<b$, but there is no $c \in L$ with $a<c<b$. One says that **b covers a.**

In [3] one can find the notation and definitions from abelian group theory, in particular:

A group is defined to be

(a) a **p-group** iff the order of every element is a power of a fixed prime number p.

(b) **of bounded order** iff the set of orders of the elements of the group is bounded.

(c) **elementary** iff every element has a square-free order.

For algebraically isomorphic groups G and H one writes $G \cong H$.

The basic facts on topological groups can be found in [1] . $\mathbb{N}$ denotes the set of natural numbers and Z(p) the cyclic group of order p, where p is a prime number. The end of each proof is indicated by the symbol □ .

2. RESULTS

DEFINITION 1. A topological group is said to be **precompact** if it is precompact with respect to its left (right) uniformity.

REMARK. A Hausdorff precompact topological group is isomorphic to a dense subgroup of a compact group.

It is easy to show that the set of (not necessarily Hausdorff) precompact group topologies on a group G forms a complete lattice PK(G) with respect to inclusion [5] . In this paper some results of geometrical nature on the structure of the lattice PK(G) and an embedding theorem for the lattice of subspaces of Desarguesian projective spaces are stated.

From lattice theory one needs - see [4] -

DEFINITION 2. Let L be a lattice.
(a) $a \in L$ is called **compact** iff $a \leq \bigvee X$ for some $X \subseteq L$ implies $a \leq \bigvee X_1$ for some finite $X_1 \subseteq X$.
(b) L is called
(1) **semimodular** iff it satisfies the upper covering condition, that is, $a \prec b$ implies $a \vee c \prec b \vee c$ or $a \vee c = b \vee c$ for all a, b , $c \in L$.
(2) **algebraic** iff it is complete and every element of L is the join of compact elements.
(3) **geometric** iff it is semimodular, algebraic, and the compact elements of L are exactly the finite joins of atoms of L.

In [4] , Def. 8, p.184 a special closure space is defined to be a **geometry.** Using this concept, it is proved that there is a one-to-one correspondence between geometries and geometric lattices ([4] , Theorem 11, p.185).

The proof of Theorem 1 is carried out by using

LEMMA 1 ([5] , (2.37), p.52). Let G be an abelian group. The lattice PK(G) is complemented iff G is an elementary group of bounded order.

THEOREM 1. Let G be an abelian group. The lattice PK(G) is geometric iff G is an elementary group of bounded order.

PROOF. (a) Let PK(G) be geometric. By a theorem of G. Birkhoff and S. Mac Lane - [4] , Theorem 4, p.179 - every geometric lattice is complemented. Then apply Lemma 1 .

(b) Let G be an elementary group of bounded order. From Lemma 1 the lattice PK(G) is complemented. The modularity of PK(G) immediately follows from [5] , (2.1), p.36. Hence PK(G) is relatively complemented ([2] , 4.2, p.31). It is proved in [5] , (3.15), p.71 that PK(G) is continuous in the sense of [4] . Since G is of bounded order, it follows from [5] , (4.6), p.77 that PK(G) is atomic. Now apply a well-known result of G. Birkhoff and F. Maeda - [4] , Exercise 3, p.189 - stating that a lattice is geometric iff it is atomic, relatively complemented, continuous, and semimodular. □

To get a relation with projective spaces one needs from [4] , p. 204

DEFINITION 3. A geometry is called **projective** if the associated geometric lattice is modular.

Observe that the notion "projective space" in [4] , Def. 3, p. 202 includes the degenerated case "a line has exactly two points". Then one has a one-to-one correspondence between projective spaces, defined by points and lines, and projective geometries, defined by modular geometric lattices ([4] , Theorem 6, p. 204).

Since the lattice PK(G) is modular, one obtains from Theorem 1

COROLLARY. Let G be an abelian group. The lattice PK(G) represents a projective geometry iff G is an elementary group of bounded order.

Now it suggests itself to examine the existence of a class of groups G with the property that the lattice PK(G) represents a nondegenerated projective space for which a coordinatization by some vector space is possible. Theorem 2 gives an answer.

For the proof one uses

LEMMA 2 ([5] , (2.9)(b), p.40). For every abelian group G the lattice PK(G) is isomorphic to the lattice of subgroups of the (algebraic) dual group G^* of G.

THEOREM 2. Let G be an abelian group of cardinality m. The lattice PK(G) represents a Desarguesian nondegenerated projective space of dimension at least two iff G is an elementary p-group with $m \geq p^3$. Then the coordinatizing vector space equals $Z(p)^m$ if m is infinite, and $Z(p)^n$ if m is finite, where $p^n = m$ with some natural number $n \geq 3$.

PROOF. (a) If PK(G) represents a projective space, then PK(G) is geometric. By Theorem 1 G is an elementary group of bounded order.

Let L be a lattice which is isomorphic to the lattice of subspaces of a vector space. Then it holds ([2] , p.108):
For distinct atoms p_1, $p_2 \in L$ there is an atom $p_3 \in L$ being distinct from p_1, p_2 such that $p_1 \vee p_2 \geq p_3$.

If G is not a p-group, then it is easy to see by the aid of Lemma 2 that PK(G) does not fulfil the quoted property. Therefore, G is an elementary p-group if PK(G) possesses a representation described in the assertion.

(b) Let G be an elementary p-group. Then $G \cong \bigoplus_m Z(p)$ for $m \geq \aleph_0$ and $G = \bigoplus_n Z(p)$ for $m < \aleph_0$, where $p^n = m$ with some some $n \in \mathbb{N}$ ([3] , Theorem 8.5, p.43). For the dual group G^* of G one infers $G^* \cong Z(p)^m$ for $m \geq \aleph_0$ and $G^* \cong G$ for $m < \aleph_0$. [3] , Theorem 8.5, p.43 implies that every subgroup of G^* is also a subspace of the vector space G^* over the field Z(p). Combining this with Lemma 2 one concludes that PK(G) is isomorphic to the lattice of subspaces of the vector space G^*. Since only nondegenerated projective spaces of dimension at least 2 are considered, one gets the condition $m > p^3$. □

Theorem 2 shows that the lattice of subspaces of a Desarguesian nondegenerated space can be identified with the lattice of precompact group topologies of an abelian group only in a special case. In general, one has

THEOREM 3. Let P be a (not necessarily nondegenerated) projective space and L_p the lattice of subspaces of P. Then L_p can be embedded in the lattice of precompact group topologies of some abelian group iff P is Desarguesian.

PROOF. From [4] one needs the following results:

(1) Let L be a complemented modular lattice. Then L can be embedded in the lattice of subgroups of an abelian group iff L is arguesian (Theorem 20, p.213).

(2) Let L be a modular geometric lattice. Then L satisfies the arguesian identity iff Desargues' Theorem holds in the associated projective geometry (Theorem 8, p.205).

(a) If P is Desarguesian, then (1), (2) imply that L_p can be embedded in the lattice L(G) of subgroups of some abelian group G. By [5] , (2.52), p.59 the lattice L(G) is isomorphic to a sublattice of PK(H) of some abelian group H.

(b) Let L_p be isomorphic to a sublattice of PK(H) of some abelian group H. Using Lemma 2, L_p can be embedded in the lattice of subgroups of the dual group H*. Now apply (1), (2) to complete the proof. □

REFERENCES

[1] N. Bourbaki: General Topology, Part 1.
Addison-Wesley Publishing Company, Reading, 1966

[2] P. Crawley, R.P. Dilworth: Algebraic Theory of Lattices.
Prentice-Hall, Englewood Cliffs, N.Y., 1973

[3] L. Fuchs: Infinite Abelian Groups, Vol. I .
Academic Press, New York, 1970

[4] G. Grätzer: General Lattice Theory.
Birkhäuser-Verlag, Basel, 1978

[5] D. Remus: Zur Struktur des Verbandes der Gruppentopologien.
Dissertation, Universität Hannover, 1983.
English summary: Resultate Math. 6(1983), 151-152

Annals of Discrete Mathematics 37 (1988) 399–404

GROUPS OF HOMOLOGIES IN 4-DIMENSIONAL STABLE PLANES ARE CLASSICAL

Hans-Peter SEIDEL

Mathematisches Institut der Universität Tübingen
Auf der Morgenstelle 10
D-7400 Tübingen, W-Germany

It is shown that in a 4-dimensional stable plane $(M,\mathcal{L})$ every group of homologies $\Gamma_{[c,A]}$ with common center $c \in M$ and common axis $A \in \mathcal{L}$ is isomorphic to a closed subgroup of the multiplicative group $\mathbb{C}^\times$ of complex numbers.

1. DEFINITIONS AND STATEMENT OF RESULTS

1.1. Definition

A stable plane is a pair $(M,\mathcal{L})$ consisting of two locally compact Hausdorff topological spaces M and $\mathcal{L}$ of finite positive topological dimension, the point space M and the line space $\mathcal{L}$, respectively, such that the following axioms hold:

($\vee$) Any two distinct points $x,y \in M$ are on a unique line $x \vee y \in \mathcal{L}$, and the mapping

$$\vee: M \times M \setminus \{(x,x): x \in M\} \longrightarrow \mathcal{L}: (x,y) \longrightarrow x \vee y$$

is continuous.

($\wedge$) Any two distinct lines $K,L \in \mathcal{L}$ contain at most one common point $K \wedge L \in M$, the set $\mathfrak{D} := \{(K,L) \in \mathcal{L}\times\mathcal{L} : K \cap L \neq \emptyset\}$ of pairs of intersecting lines is nonempty and open in $\mathcal{L} \times \mathcal{L}$, and the mapping

$$\wedge: \mathfrak{D} \longrightarrow M: (K,L) \longrightarrow K \wedge L$$

is continuous.□

Classical examples of locally compact stable planes are the projective planes over $\mathbb{R}$, $\mathbb{C}$, $\mathbb{H}$, $\mathbb{O}$ and their open subsets like the classical affine or hyperbolic planes. But there also exists a variety of nonclassical stable planes and even planes that cannot be embedded in any topological projective plane at all. For further information see e.g. [14],[4] and references given there.

It is known in general that the point space of a stable plane can only have the classical dimensions 2, 4, 8, 16 and that the group of all continuous collineations of $(M,\mathcal{L})$, if endowed with the compact open topology, is a locally compact topological group, see [6],[4]. If dim $M \leq 4$ then M and all its lines $L \in \mathcal{L}$ are topological manifolds.

In this paper we shall study groups of homologies $\Gamma_{[c,A]}$ consisting of all collineations of $(M,\mathcal{L})$ having a comon center $c \in M$ and a common axis $A \in \mathcal{L}$ with $c \notin A$. It is easy to see and well known that in a locally compact stable plane with dim $M = 2$ every such homology group $\Gamma_{[c,A]}$ is isomorphic to a closed subgroup of the multiplicative group $\mathbb{R}^\times$ of the reals. In this paper we shall study the case dim $M = 4$ and prove the following theorem:

1.2. Theorem

Let $(M,\mathcal{L})$ be a stable plane with dim $M = 4$ and consider $c \in M$, $A \in \mathcal{L}$ with $c \notin A$. Then the homology group $\Gamma_{[c,A]}$ is isomorphic to a closed subgroup of the multiplicative group $\mathbb{C}^\times$ of complex numbers. More precisely, $\Gamma_{[c,a]}$ is isomorphic to one of the groups $\mathbb{Z}$, $\mathbb{Z}_k$, $\mathbb{Z} \times \mathbb{Z}_k$, $\mathbb{R}$, $SO_2\mathbb{R}$, $\mathbb{R} \times \mathbb{Z}_k$, $SO_2\mathbb{R} \times \mathbb{Z}$, $\mathbb{C}^\times$, where k is any integer.□

The following example shows that there are in fact nonprojective 4-dimensional stable planes $(M,\mathcal{L})$ admitting a nontrivial group of homologies $\Gamma_{[c,A]}$ with $c \in M$ and $A \in \mathcal{L}$:

1.3. Example

Let $P_2\mathbb{C}$ be the classical projective plane over the complex numbers and consider a point c and a line A with $c \notin A$. Let $\Psi := \mathrm{Aut}(P_2\mathbb{C})$ denote the full automorphism group of $P_2\mathbb{C}$ and let Σ be any nontrivial compact subgroup of the homology group $\Psi_{[c,A]} = \mathbb{C}^\times$. Consider the orbit K of a point $x \in P_2\mathbb{C}\backslash(A \cup \{c\})$ under Σ. Define $M := P_2\mathbb{C}\backslash K$ and let $\mathcal{L}$ denote the lineset that is induced on M by $P_2\mathbb{C}$. Then $(M,\mathcal{L})$ is a 4-dimensional nonprojective stable plane with $c \in M$ and $A \in \mathcal{L}$, and the group of homologies $\Gamma_{[c,A]} \leq \Gamma := \mathrm{Aut}(M,\mathcal{L})$ is isomorphic to the given group Σ.□

It is an open question to what extent the analogue to Theorem 1.2 remains true for 8- and 16-dimensional planes. There are however some partial results which indicate that this may be the case at least for topological translation planes [2].

2. DISCRETE SUBGROUPS OF $\Gamma_{[c,A]}$

2.1. Lemma

Let $(M,\mathcal{L})$ be a positive dimensional stable plane and let $\Sigma \leq \Gamma_{[c,A]}$ be a closed discrete subgroup of $\Gamma_{[c,A]}$. Furthermore let a be a point on A, and define

$L := c \vee a$. Then the following holds:

(1) Σ acts properly discontinuously on $X := L \setminus \{c,a\}$.

(2) The orbit space $Y := X/\Sigma$ is a locally compact Hausdorff space.

(3) The canonical projection $p: X \longrightarrow Y$ is a regular covering with group Σ.

Proof: We only have to prove (1) and (2). Since Σ operates freely on L, according to a criterion given in [10:p.166,Ex.8.5] it suffices to check that for every compact subset S of X the set $\Omega := \{\sigma \in \Sigma : \sigma(S) \cap S \neq \emptyset\}$ is finite. Otherwise Ω contains a sequence $\{\sigma_n\}$ such that $\{\sigma_n(x)\}$ converges to c for every $x \in X$, see [5:3.12]. Now consider points $x_n \in S$ with $\sigma_n(x_n) \in \sigma_n(S) \cap S \neq \emptyset$. Since S is compact we may assume $x_n \longrightarrow y$, $\sigma_n(x_n) \longrightarrow z$ with $y,z \in S$. But since Σ operates equicontinuously on X [5:3.13] this would imply that $\sigma_n(y)$ cannot converge to c, a contradiction.□

2.2. Proposition

Let $(M,\mathcal{L})$ be a 4-dimensional stable plane and let $\Sigma \leq \Gamma_{[c,A]}$ be a closed subgroup with $\dim \Sigma = 0$. Then Σ is either cyclic or isomorphic to the direct product $\mathbb{Z} \times \mathbb{Z}_k$ of the group of integers $\mathbb{Z}$ with some finite cyclic group $\mathbb{Z}_k$ of order k.

Proof: (1) Consider a and L as in Lemma 2.1 above. According to [4], L is homeomorphic to an open submanifold of the 2-sphere S_2, and since $\Gamma_{[c,A]}$ acts effectively on L, both $\Gamma_{[c,A]}$ and Σ are Lie groups, i.e. Σ is discrete. If Σ is finite, the assertion follows by a theorem of P. A. Smith [15] (Consider the action of Σ on the line pencil $\mathcal{L}_a$). Otherwise $X := L \setminus \{c,a\}$ is homeomorphic to $\mathbb{R}^2 \setminus \{0\}$, see [5:3.18], and Lemma 2.1 above implies that the fundamental group $\pi(Y)$ of the surface $Y := X/\Sigma$ contains an infinite cyclic normal subgroup $\langle\gamma\rangle = p_*(\pi(X)) < \pi(Y)$ such that Σ is isomorphic to the quotient group $\pi(Y)/\langle\gamma\rangle$.
(2) If Y is noncompact, its fundamental group $\pi(Y)$ is free [1]. On the other hand $\pi(Y)$ operates on the cyclic normal subgroup $\langle\gamma\rangle$ of $\pi(Y)$ and we get the relation $\alpha^{-2} \cdot \gamma \cdot \alpha^2 = \gamma$ for every $\alpha \in \pi(Y)$. Together this implies that $\pi(Y)$ must be infinite cyclic itself, and we obtain that $\Sigma = \pi(Y)/\langle\gamma\rangle$ is cyclic of finite order.
(3) We are left with the case that the surface Y is compact. Since according to [5:3.0] the action of Σ on L is orientation preserving Y is an orientable surface, of genus g, say. We claim that Y is homeomorphic to the torus T: The fundamental group $\pi(Y)$ is given by $\pi(Y) = \langle \alpha_1, \beta_1, \ldots, \alpha_g, \beta_g : r \rangle$ where r is the relation $r = \prod [\alpha_i, \beta_i]$. If Y has genus $g \geq 2$, then the Freiheitssatz [9:p.252] implies that the subgroup $\Xi := \langle \alpha_1, \beta_1 \rangle$ of $\pi(Y)$ is freely generated by α_1 and β_1 and the same argument as in (2) above shows that the generator γ of the infinite cyclic normal subgroup $\langle\gamma\rangle < \pi(Y)$ satisfies the relation

$\alpha_1^{-2} \cdot \gamma \cdot \alpha_1^2 = \gamma = \beta_1^{-2} \cdot \gamma \cdot \beta_1^2$. Thus γ cannot be contained in Ξ. But then according to a theorem of G. H. Bagherzadeh [8:p.203] the group $\Theta := \gamma \cdot \Xi \cdot \gamma^{-1} \cap \Xi$ is cyclic and contains both α_1^2 and β_1^2. Hence Ξ cannot be generated freely by α_1 and β_1, a contradiction. Therefore Y is in fact homeomorphic to the torus T and our group $\Sigma = \pi(Y)/\langle\gamma\rangle = (\mathbb{Z} \times \mathbb{Z})/\langle\gamma\rangle \lesssim \mathbb{Z} \times \mathbb{Z}_k$ is isomorphic to a subgroup of $\mathbb{Z} \times \mathbb{Z}_k$.□

3. PROOF OF THE MAIN THEOREM

We now complete the proof of Theorem 1.2:

Proof: (1) Throughout this section Σ will denote the full homology group $\Gamma_{[c,A]}$. Take a point $a \in A$ and define $L := c \cup a$. We consider the action of Σ on $X := L \setminus \{c,a\}$. According to section 2 and since Σ acts freely on X we are left with the case that Σ is a Lie group of dimension 1 or 2.

(2) Case 1: $\dim \Sigma = 2$.

It is easily seen that in this case Σ is noncompact. Thus according to [5:3.18] the set $X := L \setminus \{c,a\}$ must be homeomorphic to $\mathbb{R}^2 \setminus \{0\}$. Therefore Σ is isomorphic to $\mathbb{C}^\times$ and the assertion follows.

(3) Case 2: $\dim \Sigma = 1$.

Let Σ^1 denote the identity component of Σ. Then both Σ and Σ^1 act freely and equicontinuously on X and have closed orbits [5:3.8]. A criterion given in [3:p.140,1.6] thus implies that the quotient space $Y := X/\Sigma^1$ is a locally compact Hausdorff space and since the quotient group Σ/Σ^1 also operates freely on Y we obtain that Σ/Σ^1 is homeomorphic to each of its orbits $\Sigma(x)/\Sigma^1$.

(4) We first consider the case where Σ^1 is isomorphic to the circle group $SO_2\mathbb{R}$. Since the isotropy groups Σ^1_x are all trivial it follows from a result of Mostert [11] that Σ^1 has no singular orbit on X. Hence Y must be homeomorphic to the open unit interval (0,1). But then, according to a theorem of Hölder [12:p.8],[13], the group Σ/Σ^1 is isomorphic to some subgroup of the additive group $\mathbb{R}$ of the real numbers and, since Σ/Σ^1 has closed orbits, a result of R. Löwen [7:Th.2.6] implies that $\Sigma/\Sigma^1 = \langle\mu\rangle$ is infinite cyclic. It is now easy to check that $\Sigma = \Sigma^1 \rtimes \langle\mu\rangle$ is isomorphic to the semidirect product of its connected component Σ^1 with an infinite cyclic group $\langle\mu\rangle$. According to Proposition 2.2 μ commutes with every element $\delta \in \Sigma^1 = SO_2\mathbb{R}$ of finite order, hence this product must be even direct, and $\Sigma = SO_2\mathbb{R} \times \mathbb{Z}$ is isomorphic to a closed subgroup of $\mathbb{C}^\times$.

(5) We are left with the case that Σ^1 is isomorphic to the additive group $\mathbb{R}$ of the reals. Since Σ^1 is noncompact, it follows again from [5:3.12,3.18] that $X := L \setminus \{c.a\}$ is homeomorphic to $\mathbb{R}^2 \setminus \{0\}$ and that the connected orbits $\Sigma^1(x)$ $(x \in X)$ have nontrivial intersection with every neighbourhood of c or a. The

restriction of the canonical projection $p: X \longrightarrow X/\Sigma^1$ to the standard unit sphere S_1 of $\mathbb{R}^2\setminus\{0\}$ is therefore surjective and $Y := X/\Sigma^1$ is compact. Hence the discrete quotient group Σ/Σ^1 being homeomorphic to each of ist orbits is finite, and it follows e.g. from a theorem of Gaschütz [16:p.231] that $\Sigma = \Sigma^1 \rtimes F$ is a semidirect product of Σ^1 with some finite group F. According to section 2, F is cyclic and centralizes every $\delta \in \Sigma^1$. Therefore Σ is isomorphic to the direct product $\mathbb{R} \times \mathbb{Z}_k$ of the additive group of reals with some finite cyclic subgroup $\mathbb{Z}_k$ and Theorem 2.1 is proved.□

REFERENCES

[1] Ahlfors, L.V. and Sario, L., Riemann Surfaces (Princeton University Press, Princeton N.J., 1960)

[2] Buchanan, T. and Hähl, H., On the nuclei of 8-dimensional locally compact quasifields, Arch. Math. 24 (1977) 472-480

[3] Dugundji, J., Topology (Allyn and Bacon, Boston, 1970)

[4] Löwen, R., Vierdimensionale stabile Ebenen, Geom. Ded. 5 (1976) 239-294

[5] Löwen, R., Central collineations and the parallel axiom in stable planes, Geom. Ded. 10 (1981) 283-315

[6] Löwen, R., Topology and dimension of stable planes: On a conjecture of H. Freudenthal, Journ. reine u. angew. Math. 343 (1983) 108-122

[7] Löwen, R., Lacunary free actions on the real line, Topol. Appl. 20 (1985) 135-141

[8] Lyndon, R.C. and Schupp, P.E., Combinatorial Group Theory (Springer, Berlin, Heidelberg, New York, 1977)

[9] Magnus, W., Karrass, A. and Solitar, D., Combinatorial Group Theory (Interscience Publishers, New York, London, Sydney, 1966)

[10] Massey, W.S., Algebraic Topology: An Introduction (Springer, Berlin, Heidelberg, New York, 1977)

[11] Mostert, P.S., On a compact Lie group acting on a manifold, Ann. Math. 65 (1957) 447-455; Errata, Ann. Math. 66 (1957), 589

[12] Priess-Crampe, S., Angeordnete Strukturen (Springer, Berlin, Heidelberg, New York, 1983)

[13] Salzmann, H., Kompakte zweidimensionale projektive Ebenen, Arch. Math. 9 (1958) 447-454

[14] Salzmann, H., Topological planes, Adv. Math. 2 (1967) 1-60

[15] Smith, P.A., New results and old problems in finite transformation groups, Bull. Am. Math. Soc. 66 (1960) 401-415

[16] Suzuki, M., Group Theory I (Springer, Berlin, Heidelberg, New York, 1982)

Annals of Discrete Mathematics 37 (1988) 405–412

POLYNOMIAL SPECIES AND CONNECTIONS AMONG BASES OF THE SYMMETRIC POLYNOMIALS

Domenico SENATO Università di Napoli
Antonietta M. VENEZIA Univ. di Roma "La Sapienza"

1. INTRODUCTION

In two papers published in 1968 (see [8]) G.C. Rota, carrying to the limit the algebraic processes which Baxter and others introduced for the resolution of problems generated by theory of probability, shows that every identity in a Baxter algebra is equivalent to an identity between symmetric functions. In his second paper the Author proves,through combinatorial methods, classical identities between symmetric functions which translate identities in Baxter algebras of probability interest.

The concept of generating function of a function set, conveyed in these papers, is developed later by Doubilet, Rota and Stanley (see [4], [9]).These Authors introduce a process for the construction of algebras of generating functions, both classical and innovative, for the resolution of enumerative problems.

Using the concept of generating function of a function set and techniques involving the lattice of partition of a set, Doubilet (see [3]) derives many of the known results and new ones about symmetric functions. In many cases Doubilet utilizes the Möbius inversion formula, but he also succeds in giving bijective proofs of identities between symmetric functions. These proofs consist essentially in an interpretation of the functions that occur in the identities in terms of sets and in finding a bijections between them.

In this paper we use the theory of polynomial species (see [1]) which gives a systematic approach to this kind of proof. We prove with bijective arguments, some identities which occur among the classical bases of symmetric polynomials of degree n. The language is that of categories theory and this emphasizes the generality degree of the concept of species.

2. POLYNOMIAL SPECIES

Let Φ be the category of finite sets and bijections. A *finite species* (see [5]) is a functor M from Φ to Φ .

Let E, F $\in$ Ob(Φ), we shall denote by $M[E]$ the set of M-structures on E and by $M[u]$, $u \in \mathrm{Hom}(E,F)$, the bijection between

$M[E]$ and $M[F]$ obtained "transforming via u" every M-structure on E into an M-structure on F.

The concept of polynomial species is a generalization of that of finite species in the sense that the polynomial species functor, in addition to carrying the structures, defines a subset of functions from a finite set to a set of variables as specified in the following.

Let $X = \{x_i : i \in \mathcal{J}\}$ be a family of variables with indices in a non empty and totally ordered set $\mathcal{J}$. Let I be the functor from to the category **Ens** of sets and functions defined by:

$$I[E] = \bigcup_{A \subseteq E} \mathrm{Hom}(A,X)$$

and, for each $A \subseteq E$ and $f: A \longrightarrow X$,

$$I[u](f) = f o u^{-1} : u(A) \rightarrow X.$$

Let M be a finite species. We shall denote by $\mathrm{Pol}(M)$ the functor from Φ to **Ens** defined as follows:

$$\mathrm{Pol}(M)[E] = M[E] \times I[E]$$

$$\mathrm{Pol}(M)[u](s,f) = (M[u](s), I[u](f))$$

A *polynomial species* is any subfunctor P of $\mathrm{Pol}(M)$, i.e. for each $E \in \mathrm{Ob}(\Phi)$, $P[E]$ is a subset of $\mathrm{Pol}(M)[E]$ such that if $u \in \mathrm{Hom}(E,F)$ and $(s,f) \in P[E]$ then $(M[u](s), I[u](f)) \in P[F]$. If P is a subfunctor of $\mathrm{Pol}(M)$ we shall write $P \subseteq \mathrm{Pol}(M)$.

The category of polynomial species is defined as follows. Let M and N be finite species, and let φ be a natural transformation of M to N. We define a natural transformation $\bar{\varphi}$ of the polynomial species $\mathrm{Pol}(M)$ to the polynomial species $\mathrm{Pol}(N)$ as follows. The image of the set $\mathrm{Pol}(M)[E]$ is the set $(\varphi_E(s),f)$, as (s,f) ranges over $\mathrm{Pol}(M)[E]$. If P and Q are polynomial species, we define $\mathrm{Hom}(P,Q)$ to be the set of natural transformations of P to Q which are restrictions to P and Q of some natural transformation $\bar{\varphi} : \mathrm{Pol}(M) \longrightarrow \mathrm{Pol}(N)$, where $P \subseteq \mathrm{Pol}(M)$ and $Q \subseteq \mathrm{Pol}(N)$. In particular, we write $P = Q$, when P and Q are naturally equivalent in this category.

We associate to each polynomial species a generating function which considers both the structure and the subset of functions determined by the species.

Let C be a cofinite subset of X. We write, if $p \in Z[X]$, $p/_{C=0}$ to denote the polynomial obtained from p by setting to 0 all $x \in C$. If $p \in Z[X]$ we let $N(p,C)$ be the set of all $q \in Z[X]$ such that $p/_{C=0} = q/_{C=0}$. This defines a topology on the ring $Z[X]$. Let $Z[(X)]$ the completion of $Z[X]$ and $al(X)$ the algebra over

Z generated by **X**. A function $f: E \to X$ determines a function $\hat{f}: E \to al(X)$ defined by: $\hat{f}(e) = f(e)$ for each $e \in E$.

We shall call *generating monomial* of the function f the element of $Z[(X)]$ defined by:

$$gen(f) = \prod_{e \in E} \hat{f}(e) = \prod_i x_i^{|f^{-1}(x_i)|}$$

Let $P \subseteq Pol(M)$ be a polynomial species. We set:

$$gen(P[E]) = \sum_{(s,f) \in P[E]} gen(s,f)$$

where $gen(s,f) = gen(f)$. We note that $gen(P[E])$ depends on the cardinality of E alone and so we set: $gen(P[E]) = gen(P,n)$ for each set E such that $|E| = n$. We shall call $gen(P,n)$ nth-*coefficient polynomial of the species* **P**.

The *generating function* Gen(**P**,z) of the polynomial species **P** is the formal power series

$$Gen(P,z) = \sum_{n \geq 0} gen(P,n) \frac{z^n}{n!}$$

with coefficients in $Z[(X)]$.

THEOREM 2.1. *Polynomial species isomorphic have the same generating function.*

3. THE SPECIES OF THE ASSEMBLIES

We introduce now the notion of assembly of finite species. Let N be a finite species without constant term, i.e. $N[\emptyset] = \emptyset$. We shall call *assembly of structures of species* N on the finite set E a partition of E on every block of which a structure of species N is defined. Frmally an assembly is a pair (π, S_π) where $\pi = \{B: B \subseteq E\}$ is a partition of E and $S_\pi = \{s_B: s_B \in N[B]\}$.

The *species* $Exp_k(N)$ *of the assemblies of structures of species* N *of order* k is defined as follows:

$$Exp_k(N)[E] = \{\text{assemblies } (\pi, S_\pi) \text{ on } E \text{ with } |\pi| = k\}.$$

The *species* Exp(N) *of the assemblies of structures of species* N is

$$Exp(N)[E] = \{\text{assemblies on } E\}.$$

Let $\mathcal{P}(k)$ be the set of the partitions of E with k block and $P \subseteq Pol(M)$ a polynomial species without constant term (i.e. $P[\emptyset] = \emptyset$). An *assembly on* E *of order* k *of species* **P** is every pair (s,f) where:

i) $s = (\pi, S_\pi)$ is an assembly of species M with $\pi \in \mathcal{P}(k)$ and $S_\pi = \{s_B: B \in \pi, s_B \in M[B]\}$,

ii) there exist, for every $B \in \pi$, a function f_B such that $(s_B, f_B) \in P[B]$,

iii) if f_B is defined on a set $A \cap B$, then f is defined on $A = \bigcup_{B \in \pi} (A \cap B)$ and $f/_{A \cap B} = f_B$.

The *species of assemblies of P-structures of order* k is the polynomial species $Exp_k(P) \subseteq Pol(Exp_k(M))$ defined as follows: $Exp_k(P)[E] = \{$ assemblies of P-structures of order $k\}$.

THEOREM 3.1. *For any* $k \in \mathbf{N}$ *is:* $Gen(Exp_k(P), z) = \dfrac{Gen(P, z)^k}{k!}$

The *species of the assemblies of species* P is the polynomial species $Exp(P)$ defined by $Exp(P)[E] = \bigcup_{k \geq 0} Exp_k(P)[E]$.

THEOREM 3.2. $Gen(Exp(P), z) = e^{Gen(P, z)}$.

PROPOSITION 3.1. *If* P *and* Q *are isomorphic, then* $Exp_t(P) = Exp_t(Q)$.

Proof. Let ϱ an isomorphism between P and Q. An assembly $((\pi, S_\pi), f)$ of structures of species P is a partition π such that on each block $B \in \pi$ a structure $(s_B, f_B) \in P[B]$ is defined. The bijection $\overline{\varrho}_E : Exp_t(P)[E] \longrightarrow Exp_t(Q)[E]$ associates to $((\pi, S_\pi), f)$ the assembly of species Q related to partition π such that on each B the structure $\varrho_B(s_B, f_B)$ is defined. The bijection $\overline{\varrho}_E$ determines the requested natural isomorphism.

4. CONNECTIONS AMONG BASES OF SYMMETRIC POLYNOMIALS

Let n be an integer. A *partition* of n is any sequence $(\lambda) = (\lambda_1, \dots \lambda_q)$ of non negative integers in decreasing order $\lambda_1 \geqslant \dots \geqslant \lambda_q$ such that their sum is n. The non-zero λ_i are called the *parts* of (λ) and the number of the parts is the *length* of (λ). Sometime it is convenient to use a notation which indicates the numbers of time each integer occur as part: $(\lambda) = (1^{r_1} 2^{r_2} \dots)$ means that exactly r_i of the part of (λ) are equal to i.

We note that every partition π of a set E with $|E| = n$ determines the partition $(\pi) = (1^{r_1} 2^{r_2} \dots)$ of n, where r_i is the number of blocks of with i elements. We shall call (π) class of π.

The symmetric polynomials of degree n in the variables $x_1, \dots x_t$ with rational coefficients have four classical bases.

The elementary symmetric functions : a_λ

Let $p \in \mathbf{N}$ and $a_p = \sum x_{i_1} \dots x_{i_p}$ where the sum is over all the sequences $i_1, \dots i_p$ such that $i_1 < \dots < i_p \leq t$.

We set: $a_\lambda = a_{\lambda_1} a_{\lambda_2} \dots a_{\lambda_q}$

The monomial symmetric functions: k_λ

$$k_\lambda = \sum x_{i_1}^{\lambda_1} \cdots x_{i_q}^{\lambda_q}$$

where the sum is over all distinct monomials with distinct indices.

The homogeneus elementary functions: h_λ

Let $p \in \mathbf{N}$ and $h_p = \sum x_1^{i_1} \cdots x_t^{i_t}$ where the sum is over all the sequences $i_1, \ldots i_t$ such that $i_1 + \ldots + i_t = p$. We set:

$$h_\lambda = h_{\lambda_1} \cdots h_{\lambda_q}$$

The power sum functions: $\mathbf{s}_\lambda$

Let $p \in \mathbf{N}$ and $\mathbf{s}_p = \sum_{i=1}^{t} x_i^p$. We set:

$$\mathbf{s}_\lambda = \mathbf{s}_{\lambda_1} \cdots \mathbf{s}_{\lambda_q}$$

When (λ) ranges over all partitions of the integer n, the sets $\{a_\lambda\}$, $\{k_\lambda\}$, $\{h_\lambda\}$, $\{\mathbf{s}_\lambda\}$ are the classical bases of the symmetric polynomials of degree n.

Let $\bar{I}$ be the finite species defined by $\bar{I}[E] = \{E\}$. We denote with $\mathbf{S} \subseteq \mathrm{Pol}(\bar{I})$ the *power sum species* defined by $\mathbf{S}[E] = \{(E,f) : f : E \to \mathbf{X} \text{ constant}\}$. The nth-coefficient of the generating function of $\mathbf{S}$ is:

$$\mathrm{gen}(\mathbf{S}, n) = \mathbf{s}_n = \sum_{i \in \mathcal{I}} x_i^n, \text{ hence } \mathrm{Gen}(\mathbf{S}, z) = \sum_{n \geq 0} \mathbf{s}_n \frac{z^n}{n!}$$

Let $t \in \mathbf{N}$ and P_t the finite species defined by $P_t[E] = \{\text{partitions of } E \text{ with } t \text{ blocks}\}$. We denote with $\mathbf{K}_t$ the polynomial species $\mathbf{K}_t[E] = \{(\pi, f) : |\pi| = t,\ f : E \to \mathbf{X},\ \ker f \geq \pi\} \subseteq \mathrm{Pol}(P_t)[E]$. The nth-coefficient of the generating functions of $\mathbf{K}_t$ is:

$$\mathrm{gen}(\mathbf{K}_t, n) = \sum_{\pi \in P_t[E]} \ \sum_{\ker f \geq \pi} \mathrm{gen}(\pi, f)$$

The other hand: $\displaystyle\sum_{\ker f \geq \pi} \mathrm{gen}(\pi, f) = \sum_{\sigma \geq \pi} s_1! \ s_2! \cdots k_{(\sigma)}$

where $(\sigma) = (1^{s_1} 2^{s_2} \ldots)$. The sum on the rigth depends on $(\lambda) = (\pi)$ alone, hence setting $b_\lambda = \sum_{\sigma \geq \pi} s_1! \ s_2! \cdots k_{(\sigma)}$, we have

$$\mathrm{gen}(\mathbf{K}_t, n) = \sum_{\pi \in P_t[E]} b_\lambda = \sum_{\lambda_1 + \ldots + \lambda_t = n} \binom{n}{\lambda_1 \cdots \lambda_t} \frac{b_\lambda}{t!}$$

from which

$$\mathrm{Gen}(\mathbf{K}_t, z) = \sum_{n \geq 0} \ \sum_{\lambda_1 + \ldots + \lambda_t = n} \binom{n}{\lambda_1 \cdots \lambda_t} \frac{b_\lambda}{t!} \frac{z^n}{n!}$$

THEOREM 4.1. $\mathrm{Exp}_t(\mathbf{S}) = \mathbf{K}_t$

Proof. The bijection that to any pair $((\pi, S_\pi), f) \in \mathrm{Exp}_t(\mathbf{S})[E]$ associates the pair $(\pi, f) \in \mathbf{K}_t[E]$ determines the requested isomorphism.

COROLLARY 4.1. *If we set* $(\pi) = (\lambda)$, *we have:*

(i)
$$\mathbf{s}_\lambda = \sum_{\sigma \geq \pi} s_1!\, s_2! \ldots k_{(\sigma)}$$

Proof. From the theorems 3.1, 4.1 and 2.1 we have:

$$\mathrm{gen}(\mathrm{Exp}_t(\mathbf{S}),n) = \frac{1}{t!} \sum_{\lambda_1+\ldots+\lambda_t=n} \binom{n}{\lambda_1\ldots\lambda_t} \mathbf{s}_{\lambda_1}\ldots = \frac{1}{t!} \sum_{\lambda_1+\ldots+\lambda_t=n} \binom{n}{\lambda_1\ldots\lambda_t} \mathbf{s}_\lambda =$$

$$\mathrm{gen}(\mathbf{K}_t,n) = \frac{1}{t!} \sum_{\lambda_1+\ldots+\lambda_t=n} \binom{n}{\lambda_1\ldots\lambda_t} b_\lambda$$

The (i) gives the power sum functions in terms of the monomial symmetric functions.

Let S be the finite species defined by: $S[E]=\{\text{permutations on } E\}$. We shall denote with $\mathbf{H} \subseteq \mathrm{Pol}(\mathrm{Exp}(S))$ the *disposition species* defined by:
$\mathbf{H}[E] = \{(s,f):\ s = (\pi, S_\pi) \in \mathrm{Exp}(S)[E] \text{ and } f:E\to \mathbf{X} \text{ such that } \ker f = \pi\}$.
The nth-coefficient of the generating function of $\mathbf{H}$ is:

$$\mathrm{gen}(\mathbf{H},n) = n!\, h_n \qquad \text{(see [1])}$$

hence
$$\mathrm{Gen}(\mathbf{H},z) = \sum_{n\geq 0} n!\, h_n \frac{z^n}{n!}$$

The *cyclic species* $\mathbf{C} \subseteq \mathrm{Pol}(S)$ is defined by
$\mathbf{C}[E] = \{(\mu,f):\ \mu$ cyclic permutation of E, $f:E\to\mathbf{X}$ constant$\}$. The nth-coefficient of the generating function of species $\mathbf{C}$ is:

$$\mathrm{gen}(\mathbf{C},n) = (n-1)! \sum_{i\in \mathcal{I}} x_i^n = (n-1)!\, \mathbf{s}_n,$$

hence:
$$\mathrm{Gen}(\mathbf{C},z) = \sum_{n\geq 0} (n-1)!\, \mathbf{s}_n \frac{z^n}{n!}$$

We shall calculate now the nth-coefficient of the species $\mathrm{Exp}_t(\mathrm{Exp}(\mathbf{C}))$ that for brevity we shall denote by $\bar{\mathbf{C}}$. From theorem 3.1 we have:

$$\mathrm{gen}(\bar{\mathbf{C}},n) = \frac{1}{t!} \sum_{\lambda_1+\ldots+\lambda_t=n} \binom{n}{\lambda_1\ldots\lambda_t} \mathrm{gen}(\mathrm{Exp}(\mathbf{C}),\lambda_1)\ldots\mathrm{gen}(\mathrm{Exp}(\mathbf{C}),\lambda_t)$$

For any partition π of E $(|E|=n)$ such that $(\pi) = (\lambda_1,\ldots\lambda_t)$ we have:

$$\mathrm{gen}(\mathrm{Exp}(\mathbf{C}),\lambda_1)\ldots\mathrm{gen}(\mathrm{Exp}(\mathbf{C}),\lambda_t) = \sum_{((\pi,S_\pi),f)\in \bar{\mathbf{C}}[E]} \mathrm{gen}((\pi,S_\pi),f) =$$
$$\sum_{\sigma\leq\pi} (\nu_1-1)!\,(\nu_2-1)!\ldots \mathbf{s}_{(\sigma)}$$

where $(\sigma)=(\nu_1,\nu_2,\ldots)$. The sum on the right depends on (π) alone,

hence setting: $c_\lambda = \sum_{\sigma\leq\pi} (\nu_1-1)!\,(\nu_2-1)!\ldots \mathbf{s}_{(\sigma)}$, we have:

$$\mathrm{Gen}(\mathrm{Exp}_t(\mathrm{Exp}(\mathbf{C})),z) = \frac{1}{t!} \sum_{n\geq 0} \sum_{\lambda_1\ldots+\lambda_t=n} \binom{n}{\lambda_1\ldots\lambda_t} c_\lambda.$$

THEOREM 4.2. $\mathrm{Exp}(\mathbf{C}) = \mathbf{H}$ (see [1], [2]).

THEOREM 4.3. *If* $(\pi) = (\lambda_1,\ldots\lambda_t)$

(ii) $\lambda_1!\ \lambda_2!\ldots h_\lambda = \sum_{\sigma\le\pi} (\nu_1-1)!\ (\nu_2-1)!\ldots s_{(\sigma)}$

Proof. From the theorems 4.2 and 3.3, we have: $\mathrm{Exp}_t(\mathrm{Exp}(\mathbf{C}))=\mathrm{Exp}_t(\mathbf{H})$ thus, using the theorem 2.1, we obtain:

$$\mathrm{gen}(\mathrm{Exp}_t(\mathrm{Exp}(\mathbf{C})),n)\ \frac{1}{t!}\sum_{\lambda_1+\ldots+\lambda_t=n}\binom{n}{\lambda_1\ldots\lambda_t}\ c_\lambda=\mathrm{gen}(\mathrm{Exp}_t(\mathbf{H}),n) =$$

$$\frac{1}{t!}\sum_{\lambda_1+\ldots+\lambda_t=n}\binom{n}{\lambda_1\ldots\lambda_t}\ \lambda_1!h_{\lambda_1}\ \lambda_2!h_{\lambda_2}\ldots = \frac{1}{t!}\sum_{\lambda_1+\ldots+\lambda_t=n}\binom{n}{\lambda_1\ldots\lambda_t}\lambda_1!\ \lambda_2!\ldots h_\lambda$$

The (ii) gives the homogeneus elementary functions in terms of the power-sum functions.

We denote with $\mathbf{A}\subseteq \mathrm{Pol}(\bar{I})$ the *elementary symmetric species* defined by: $\mathbf{A}[E] = \{(E,f) : f: E\to X \text{ monomorphism}\}$. The coefficient of the generating function of the species $\mathbf{A}$ is: $\mathrm{gen}(\mathbf{A},n) = n!\ a_n$ hence:

$$\mathrm{Gen}(\mathbf{A},z) = \sum_{n\ge 0} n!\ a_n\ \frac{z^n}{n!}$$

Let $t\in\mathbf{N}$. We denote by $\mathbf{A_t}\subseteq \mathrm{Pol}(P_t)$ the species defined by: $\mathbf{A_t}[E] = \{(\pi,f) : |\pi| = t,\ f: E\to X\}$ and $\ker f\wedge\pi=\hat{0}$. The coefficient of $\mathbf{A_t}$ is:

$\mathrm{gen}(\mathbf{A_t},n) = \sum_{\pi\in P_t[E]}\ \sum_{\ker f\wedge\pi=\hat{0}} \mathrm{gen}(\pi,f)$. The other hand:
$\sum_{\ker f\wedge\pi=\hat{0}} \mathrm{gen}(\pi,f) = \sum_{\sigma\wedge\pi=\hat{0}} s_1!\ s_2!\ldots k_{(\sigma)}$ where $(\sigma) = (1^{s_1}2^{s_2}\ldots)$.

The sum on the right depends on (π) alone. Hence setting $d_\lambda = \sum_{\sigma\wedge\pi=\hat{0}} s_1!\ s_2!\ldots k_{(\sigma)}$ we have:

$$\mathrm{gen}(\mathbf{A_t},n) = \sum_{\pi\in P_t[E]} d_\lambda = \sum_{\lambda_1+\ldots+\lambda_t=n}\binom{n}{\lambda_1\cdots\lambda_t}\frac{d_\lambda}{t!}$$

from which

$$\mathrm{Gen}(\mathbf{A_t},z) = \sum_{n\geqslant 0}\ \sum_{\lambda_1+\ldots+\lambda_t=n}\binom{n}{\lambda_1\cdots\lambda_t}\frac{d_\lambda}{t!}\ \frac{z^n}{n!}$$

THEOREM 4.4. $\mathrm{Exp}_t(\mathbf{A}) = \mathbf{A_t}$

Proof. The bijection that to any pair $((\sigma,S_\sigma),f)\in \mathrm{Exp}_t(\mathbf{A})[E]$ associates the pair $(\sigma,f)\in\mathbf{A_t}[E]$ determinates the requested isomorphism.

COROLLARY 4.2. *If* $(\pi)=(\lambda)$, *we have:*

(iii) $\lambda_1!\ \lambda_2!\ldots a_\lambda = \sum_{\sigma\wedge\pi=\hat{0}} s_1!\ s_2!\ldots k_{(\sigma)}$

Poof. From the theorem 3.1, 4.4 and 2.1 follows:

$$\mathrm{gen}(\mathrm{Exp}_t(\mathbf{A}),n) = \frac{1}{t!}\sum_{\lambda_1+\ldots+\lambda_t=n}\binom{n}{\lambda_1\cdots\lambda_t}\ \lambda_1!\ a_{\lambda_1}\ \lambda_2!\ a_{\lambda_2}\ldots =$$

$$\frac{1}{t!}\sum_{\lambda_1+\ldots+\lambda_t=n}\binom{n}{\lambda_1\ldots\lambda_t}\ \lambda_1!\ \lambda_2!\ldots a_\lambda = \mathrm{gen}(\mathbf{A_t},n) = \frac{1}{t!}\sum_{\lambda_1+\ldots+\lambda_t=n}\binom{n}{\lambda_1\cdot\ \lambda_t}d_\lambda$$

The (iii) gives the elementary symmetric functions in terms of the monomial symmetric functions.

REFERENCES

[1] Bonetti, F., Rota, G.C., Senato, D., Venezia, A.M., On the foundation of Combinatorial Theory, X: a categorical setting for Symmetric Functions. To appear

[2] Bonetti, F., Rota, G.C., Senato, D., Venezia, A.M., Symmetric Functions and Symmetric Species, Annals of Discrete Math. 30 (1986) 107-114.

[3] Doubilet, P., On the foundation of Combinatorial Theory, VII: Symmetric Functions through the Theory of Distribution and Occupancy, Studies in App. Math. v. 51 (1972)

[4] Doubilet, P., Rota, G.C., Stanley, R., On the foundation of Combinatorial Theory, VI: The idea of Generating Function, Sixth Berkeley Symposium on Math. Statistic and Probability, v. 2, Berkeley Univ. Press (1972).

[5] Joyal, A., Une Théorie Combinatoire des séries formelles, Adv. in Math. v. 42 n.1 (1981)

[6] Macdonald, I.G., Symmetric Functions and Hall polynomials, (Clarendon press, Oxford, 1979)

[7] Mac Lane, S., Categories for the Working Mathematician, (Springer, New York, Heidelber, Berlin, 1971)

[8] Rota, G.C., Baxter Algebras and combinatorial identities I and II, Bull. Amer. Math. Soc. 75 (1969)

[9] Stanley, R., Theory and Application of Plane Partition, part 1 and 2, Studies in App. Math. v. L, n. 2 and 3 (1971)

Annals of Discrete Mathematics 37 (1988) 413–420

SET AND SEQUENCE CLOSURE FOR FINITE PERMUTATION GROUPS

Johannes Siemons*

1. INTRODUCTION

A permutation group G acting on a finite set Ω for each integer $k < |\Omega|$ has a natural faithful action on $\Omega^{\{k\}}$, the collection of all k-element subsets of Ω. We now define (see p.399 in [8]) the *{k}-closure* $G^{\{k\}}$ of G to be the largest subgroup in $Sym(\Omega)$ which has the same orbits on $\Omega^{\{k\}}$ as G. Thus G is a subgroup of $G^{\{k\}}$ and the map $G \longrightarrow G^{\{k\}}$ satisfies the usual properties of a closure operator as we shall see in Section 2. Set closure is relevant for automorphism groups of incidence structures: if (G,Ω) denotes the point action of some automorphism group G of an incidence structure $S = (\Omega,B)$ with block set B contained in $\Omega^{\{k\}}$ (i.e. all blocks are incident with precisely k points) then $G^{\{k\}}$ in particular preserves the blocks of S and hence consists of automorphisms. For this reason the full automorphism group of S coincides with its {k}-closure. This can be interpreted as a completeness property of full automorphism groups. A group for which $G = G^{\{k\}}$ will be called *{k}-closed*. Such groups are the subject of Section 3.

By analogy we may consider the action of a group G on the collection $\Omega^{(k)}$ of k-element sequences of distinct points of Ω. Here we define the *(k)-closure* $G^{(k)}$ of G to be the largest permutation group on Ω that has the same orbits on $\Omega^{(k)}$ as G. Then $G \longrightarrow G^{(k)}$ is also a closure operator in the usual sense. Sequence closure was first investigated by Wielandt in the Ohio Lecture Notes [11]. The definition of k-closure there, given in terms of the action on the cartesian product of k copies of Ω, does agree with the definition given above.

---

* Financial support from SERC through grant GR/D/30181 is acknowledged.

In [11] the completeness properties of $G^{(2)}$ formed an important step in the classification of uni-primitive groups of degree p^2.

In Section 2 we review the basic properties of closure groups such as inclusions in general, primitivity and permutation rank, both for set and for sequence closure.

In Section 3 we shall deal with conditions for a group to be {k}-closed for some k. A useful criterion (Result 4) depends on the existence of regular sets for closed over-groups. This problem has been investigated in collaboration with Jennifer Key, a survey [7] will be given at this conference.

2. GENERAL PROPERTIES OF CLOSURE

For the remainder of this note we shall use the following notation. G is a permutation group on a finite set Ω of n elements. The orbits of G are $\Omega_1,\ldots,\Omega_t$. The *orbitals* of G, denoted by $\Gamma_1,\ldots,\Gamma_r$, are the orbits of G acting on $\Omega \times \Omega$. When G is transitive, then there is just one diagonal orbital, say $\Gamma_1 = \{(\alpha,\alpha)|\alpha\in\Omega\}$ so that $\Gamma_2,\ldots,\Gamma_r$ are the orbits of G on $\Omega^{(2)}$. In this case $r(G) = r$ is the rank of G. When $\Gamma_j = \{(\alpha^g,\beta^g)|g \in G\}$ is an orbital then $\{\beta^g,\alpha^g)|g \in G\}$ is the orbital *paired* to Γ_j, denoted by $\Gamma_{j'}$. If $\Gamma_j = \Gamma_{j'}$ then Γ_j is said to be *self-paired*. In the transitive case the number of self-paired orbitals is given by the *Frobenius-Schur indicator* $\eta_G = |G|^{-1}\Sigma_{g\in G}\,\pi(g^2)$ where π denotes the permutation character of G. As a reference to these facts see for instance Chapter 1 in Cameron's article [3]. A non-diagonal orbital Γ_j may be viewed as a directed graph with vertex set Ω and (α,β) an edge if and only if (α,β) belongs to Γ_j. Its full automorphism group is denoted by $\mathrm{Aut}(\Gamma_j)$. The undirected graph underlying Γ_j will be denoted by $(\Gamma_j+\Gamma_{j'})$ with automorphism group $\mathrm{Aut}(\Gamma_j+\Gamma_{j'})$. Thus Γ_j and $(\Gamma_j+\Gamma_{j'})$ represent the same graph if and only if Γ_j is self-paired.

Let F be a field and H some subgroup of $\mathrm{Sym}(\Omega)$. Then the *centralizer algebra* $C_\Omega(H)$ over F consists of all n×n matrices A (with entries in F) for which $AM_h = M_hA$ for all h in H where M is the permutation matrix representing h. As a vector space the centralizer algebra is spanned by the adjacency matrices associated to the orbitals of H , including diagonal orbitals. In particular, if H is transitive, then the dimension of $C_\Omega(H)$ is r(H). For a reference see for instance Chapter V in Wielandt's book [10]. All the remaining notation is standard.

Comparing the actions of G on $\Omega^{(k)}$ and Ω^k, the cartesian product of k copies of Ω , it is clear that G-orbits on $\Omega^{(k)}$ determine G-orbits on Ω^k and vice versa. This shows that our definition of (k)-closure agrees with the definition of k-closure given by Wielandt in [11]. The first result to be stated now shows that both set- and sequence-closure satisfy the usual properties of a closure operator,

RESULT 1:

(1) If $H \subseteq G$ then $H^{\{k\}} \subseteq G^{\{k\}}$, $H^{(k)} \subseteq G^{(k)}$ and $G^{(k)} \subseteq G^{\{k\}} = G^{\{n-k\}}$ for $1 \leqslant k \leqslant n$.
(2) If $k \leqslant \ell$ then $(G^{(\ell)})^{(k)} = (G^{(k)})^{(\ell)} = G^{(k)}$.
(3) If $k \leqslant \ell \leqslant n-k$, then $(G^{\{\ell\}})^{\{k\}} = (G^{\{k\}})^{\{\ell\}} = G^{\{k\}}$.

PROOF: (1) follows directly from the definitions. The second statement is theorem 5.10, page 16 in [11], while (3) follows from theorem 5.1 in [8].□

The next theorem is concerned with the lattice of closure groups in general.

RESULT 2: Let $G \subseteq Sym(\Omega)$, let n^* be the integer with $n-1 \leqslant 2n^* \leqslant n$ and suppose that $2 \leqslant k \leqslant \ell \leqslant n-k$. Then

(1) $G = G^{(n)} \subseteq \ldots \subseteq G^{(\ell)} \subseteq \ldots \subseteq G^{(k)}$,
(2) $G \subseteq G^{\{n^*\}} \subseteq \ldots \subseteq G^{\{\ell\}} \subseteq \ldots \subseteq G^{\{k\}}$,
(3) $G^{(2)} = \bigcap_{1,,r} Aut(\Gamma_j) \subseteq G^{(1)} = \bigotimes_{1,,t} Sym(\Omega_i)$,
(4) $G^{\{2\}} = \bigcap_{1,,r} Aut(\Gamma_j + \Gamma_{j'}) \subseteq G^{\{1\}} = G^{(1)}$.

PROOF: The inclusions in (1) and (2) are theorem 5.8, page 15 in [11] and theorem 5.1 in [8]. The equations in (3) and (4) can be established directly from the definitions.□

REMARKS: **A.** It is worthwhile to comment on the inclusions in Result 2.(1) and 2.(2). While 2.(1) is straightforward to establish due to the inductive nature of sequence closure the inclusions in 2.(2) are due to rather different arguments: no obvious inductive arguments apply. The proof given in [8] relies upon the fact that the incidence matrix of k-element subsets versus ℓ-element subsets of Ω has maximal linear rank. Note also that 2.(2) remains true for infinite permutation groups. This result, relying on a weak form of Ramsey's theorem, is due to Bercov and Hobby [1].

B. No information in general seems to be known about the index of $G^{(k)}$ in $G^{\{k\}}$. Its range can be seen from the following two examples. The cyclic group $G = C_n$, generated by an n-cycle, has $G^{(2)} = G \lhd G^{\{2\}} = D_{2n}$, the dihedral group of order $2n$, while $G = PSL(2,q)$, acting on the $q+1$ points of the projective line, leads to $G = G^{(3)} \subset G^{\{3\}} = Sym(q+1)$ if $q \equiv 3 \mod 4$.

C. As a consequence of Result 2.(2) the group $G^{\{n^*\}}$ has the same orbits as G when acting on the collection of subsets of *any* size and it is the largest group with this property. In [8], page 401, I have conjectured that, under suitable general hypotheses, G should coincide with $G^{\{n^*\}}$. In this direction theorem A in [9] shows that a primitive group G for which some prime divides the order of $G^{\{n^*\}}$ but not that of G must be one of seven groups of degree at most nine. Result 4 in Section 3 will give more information about this question.

We now turn to the questions of primitivity and permutation rank of the closure groups.

RESULT 3: Let G be a transitive group on Ω and suppose that $2 \leqslant k \leqslant n-2$.
(1) The groups G, $G^{\{k\}}$ and $G^{(k)}$ all share the same block systems of imprimitivity. Thus G is primitive if and only if any of its closure groups $(1 < k < n-1)$ is primitive.
(2) For the centralizer algebras we have $C_\Omega(G) = C_\Omega(G^{(k)})$ and in particular $r(G) = r(G^{(k)})$.
(3) When η_G denotes the Frobenius-Schur indicator of G then $r(G) \geqslant r(G^{\{k\}}) \geqslant 2^{-1}(r(G) + \eta_G)$.

PROOF: (1). As $G \subseteq G^{(k)} \subseteq G^{\{k\}} \subseteq G^{\{2\}}$ by Result 1.(1) and 2.(2), this follows from theorem 5.2 in [8] where it is shown that G and $G^{\{2\}}$ have the same block systems of imprimitivity. See also theorem 4.10, page 10 in [11].
(2). As $G \subseteq H$ implies $C_\Omega(H) \subseteq C_\Omega(G)$ in general, by Result 2(1) we only have to show that $C_\Omega(G) = C_\Omega(G^{(2)})$. By definition G and $G^{(2)}$

have the same orbitals, and hence, by the remark in the introduction of this section, their centralizer algebras coincide.
(3). As $G \subseteq G^{\{k\}} \subseteq G^{\{2\}}$, the rank of these groups satisfies $r(G) \geq r(G^{\{k\}}) \geq r(G^{\{2\}})$ so that only the lower bound for $r(G^{\{2\}})$ needs to be established. Hence compare orbitals of G with orbitals of $G^{\{2\}}$. An orbital of $G^{\{2\}}$ can be the union of at most two G-orbitals and this happens only if they are paired orbitals for G. Thus $r(G^{\{2\}})$ is at least the number of self paired G-orbitals plus the number of orbital pairs of G. This number is $2^{-1}(r(G)+\eta_G)$ by the remarks at the beginning of this Section.□

3. CLOSED GROUPS

A group G on Ω is $\{k\}$*-closed* or (k)*-closed* if $G = G^{\{k\}}$ or $G = G^{(k)}$ respectively. As a consequence of Results 1.(1),2.(1) and 2.(2) a $\{k\}$-closed group is also (k)-, $\{\ell\}$- and (ℓ)-closed for $k \leq \ell \leq n - k$ and of course $G^{\{k\}}$ is always $\{k\}$-closed. In this section we are concerned with conditions for a group to be $\{k\}$-closed for some k. As already mentioned in the introduction, examples of $\{k\}$-closed groups are provided by full automorphism groups of structures on Ω in which blocks have constant size k. Our main criterion in general is

RESULT 4: Let $G \subseteq Sym(\Omega)$ be given and suppose that some permutation group H on Ω containing G satisfies

I : H is $\{k\}$-closed for some k, $2 \leq k \leq n - 2$, and

II: for some subset $\Delta \subset \Omega$ of size ℓ, $k \leq \ell \leq n - k$, the set stabilizer of Δ in H is the identity on Ω.

Then G is $\{\ell\}$-closed and a group on Ω having the same orbits on $\Omega^{\{\ell\}}$ as G coincides with G.

PROOF: Let $K = G^{\{\ell\}}$. Then $G \subseteq K \subseteq G^{\{k\}} \subseteq H^{\{k\}} = H$ by Results 1.(1), 2.(1) and hypothesis I. Therefore $K_{\{\Delta\}} = 1 = G_{\{\Delta\}}$ by hypothesis II and as $\Delta^G = \Delta^L$ we have $G = K$. Suppose now that some group L has the same orbits on $\Omega^{\{\ell\}}$ as G. Thus, by definition, $G^{\{\ell\}} = L^{\{\ell\}}$ so that L is a subgroup of H. But then, as above, $L = L^{\{\ell\}}$ so that $L = G$.□

Hypothesis I in Result 4 in itself can be satisfied quite easily by choosing H to be the $\{k\}$-closure of any group containing G. The relevant requirement now is the existence of a subset Δ as in hypothesis II. Such a set is said to be a *regular* set for H. In the survey [7] by J.Key the existence of regular sets for several infinite families of groups will be discussed, including $A\Gamma L(d,q)$, $P\Gamma L(d,q)$, the Ree groups $R(q)$ and $P\Gamma U(3,q^2)$ in their natural representations. These groups generally are {3}- or {4}-closed and are shown to have regular sets of sufficiently small size. For details see also [6] and [5]. These results are independent of the classification theorem of finite simple groups. A theorem in [4] which however does depend on the classification shows that all primitive groups of sufficiently large degree not containing the alternating group of the same degree have regular sets. Note finally that a group G as in Result 4 is uniquely identified through its orbits on the power set over Ω.

A strengthening of the notion of {k}-closure is that of a k-*geometric* action, due to Betten [2]. Here a permutation group G on Ω is said to be k-*geometric* if there is a system B of k-element subsets of Ω such that G is the *full* automorphism group of the structure (Ω,B). Clearly a k-geometric group is {k}-closed but the converse does not hold in general.

RESULT 5: Let G be a subgroup of $A\Gamma L(d,q)$ or $P\Gamma L(d+1,q)$ where $d \geq 2$ if $q \geq 3$ and $d \geq 6$ if $q = 2$. Then G acts k-geometrically for some k on the points of the corresponding affine or projective space.

The values for k are given in [6] and [5] but are too cumbersome to reproduce here. We give an indication of the

PROOF: There is a regular set Δ for $A\Gamma L(d,q)$ or $P\Gamma L(d+1,q)$ respectively, say of size k. Now let $B_1 = \Delta^G$ and let B_2 be the collection of all k-element subsets of points in the corresponding affine (projective) space that are contained in a d_0-dimensional coset (subspace), where $d_0 < d$ is taken to be minimal. Setting $B = B_1 \cup B_2$ and $A = Aut(\Omega,B)$ it can be shown that ${B_2}^A = B_2$ so that A is a group of collineations. As $A_{\{\Delta\}} = 1 = G_{\{\Delta\}}$ and $\Delta^A = \Delta^G$ it follows that $G = A$. For the details of this proof see theorems 3.1 and 3.2 in [5] and theorem 5.1 in [6].□

BIBLIOGRAPHY

[1] R.D.Bercov and C.R.Hobby, 'Permutation groups on unordered sets', Math. Z. 115, 165-168, (1970).

[2] D. Betten, 'Geometrische Permutationsgruppen', Mitt. Math. Gesellschaft Hamburg, 10, 317-324, (1977).

[3] P.J.Cameron, 'Suborbits in transitive permutation groups', "Combinatorics", Part 3, Mathematical Centre Tracts, Amsterdam, 1975.

[4] P.J. Cameron, P.M. Neumann and J.Saxl, 'On groups with no regular orbits on the set of subsets', Arch. Math. 43, 295-296, (1984).

[5] F. Dalla Volta, 'Regular sets for the affine and projective groups over the field of two elements', to appear in Journal of Geometry.

[6] J.D. Key and I.J. Siemons, 'Regular sets and geometric groups', Resultate der Mathematik, 11, 97-116 (1987).

[7] J.D.Key, 'Regular sets in Geometries', Proceedings of this conference.

[8] I.J.Siemons,'On partitions and permutation groups on unordered sets', Arch. Math. 38 391-403, (1982).

[9] I.J.Siemons and A.Wagner, 'On finite permutation groups with the same orbits on unordered sets', Arch. Math. 45, 492-500, (1985).

[10] H.Wielandt, "Finite permutation groups", New York 1984.

[11] H.Wielandt, "Permutation groups through invariant relations and invariant functions", Lecture Notes, Ohio State University, Ohio 1969.

Department of Mathematics,
University of Birmingham,
Birmingham
B15 2TT
United Kingdom

School of Mathematics and Physics,
University of East Anglia,
Norwich NR4 7TJ
United Kingdom

Annals of Discrete Mathematics 37 (1988) 421–426

P-CYCLIC HYPERGROUPS WITH THREE CHARACTERISTIC ELEMENTS

Thomas N.VOUGIOUKLIS and Stefanos H.SPARTALIS

Democritus University of Thrace, 67100 Xanthi, Greece.

Let $(G,\cdot)$ be a group and P a non-empty subset of G. The set G with hyperoperation $P^\star$ given by the relation $xP^\star y = xPy$ becomes a hypergroup. These hypergroups have been introduced by the first author and have been called P-hypergroups. In this paper we study the simplest case of 3-dimensional P-hypergroups.

§ 1. Let $(G,\cdot)$ be a group and P a non-empty subset of G. In G we can define hyperoperation

$$P^\star : G\mathrm{x}G \longrightarrow \mathscr{P}(G) : (x,y) \longmapsto xP^\star y = xPy.$$

Then G becomes a hypergroup in the sense of Marty (1934), i.e. $P^\star$ is associative and the reproduction axiom $xP^\star G = GP^\star x = G$, $\forall x \in G$, is valid. The above hypergroup denoted $\langle G,P^\star\rangle$ introduced in [6] and generalized in [7] was called *P-hypergroup*. The P-hyperoperations $P^\star$ are $|P|$-dimensional, see [1], where $|P| = \text{card } P$.

In this paper we focus our attention on P-hypergroups when G is a finite cyclic group, say $(H_n,\cdot)$. The hypergroup $\langle H_n,P^\star\rangle$ is a cyclic hypergroup [8]; i.e. denoted by $h^{[v]}$ the v-th power of the element h with respect to the hyperoperation $P^\star$, there exists a generator h of H_n such that $H_n = h^{[1]} \cup \ldots \cup h^{[v]}$. The minimal such number v is called the *period* of h and if all generators have the same period then we call the $\langle H_n,P^\star\rangle$ *cyclic with period.* Finally, $\langle H_n,P^\star\rangle$ is called *single-power cyclic* if we can write $H_n = h^{[v]}$; see [6 and 2].

If P is a singleton then $P^\star$ becomes an operation. The simplest case of P-hypergroups with two elements is $P = \{e,x\}$, $e \neq x$ and this case for cyclic hypergroups was studied in [6]. In this paper we study the simplest case of 3-dimensional P-cyclic hypergroups $\langle H_n,P^\star\rangle$, i.e. $P = \{e,x,x^{-1}\}$, $e \neq x \neq x^{-1}$. In this case in order to simplify our notation we shall write the hyperoperation $x_\star$ instead of $P^\star$.

§ 2. Henceforth we denote a generator of $(H_n,\cdot)$ by a; therefore the elements of H_n are a^s, $s \in \{0,1,\ldots,n-1\}$. We observe that $\langle H_n,x_\star\rangle \equiv \langle H_n,x_\star^{-1}\rangle$; so we study the hypergroups $\langle H_n,a_\star^\varkappa\rangle$ with $\varkappa \in \{1,\ldots,[\frac{n}{2}]\}$ where $[\frac{n}{2}]$ is the integer part of $\frac{n}{2}$. Obviously if n is an even, we take $\varkappa \in \{1,\ldots,[\frac{n}{2}]-1\}$.

The powers of the element a^s with respect to the hyperoperation $a_\star^\varkappa$ are

$a^{s[1]} = \{a^s\}$, $a^{s[2]} = \{a^{2s}, a^{2s+\varkappa}, a^{2s-\varkappa}\},\ldots,$

$a^{s[t]} = \{a^{ts}, a^{ts+\varkappa},\ldots,a^{ts+(t-1)\varkappa}, a^{ts-\varkappa},\ldots,a^{ts-(t-1)\varkappa}\},\ldots;$

so $a^{s[t]} = \{a^{ts+v\varkappa} \mid -t<v<t\}$.

The set of unit elements (two-sided) of $\langle H_n, a^{\varkappa}_{\star}\rangle$ is $P = \{e, a^{\varkappa}, a^{-\varkappa}\}$, i.e. if $x \in P$ we have

$x a^{\varkappa}_{\star} a^s = a^s a^{\varkappa}_{\star} x \ni a^s, \quad \forall a^s \in H_n.$

The set of inverse elements (two-sided) of $a^s \in H_n$ with respect to the unit $x \in P$ is $a^{-s}x^{-1}P$ and the set of inverse elements of a^s with respect to some unit is $a^{-s}P^2$. Finally, the only total inverse element of a^s is a^{-s}, see [7]. We note that every P-hypergroup is left and right inversible in itself; see [3 and 7].

A subset H of H_n is called a P-subhypergroup of $\langle H_n, a^{\varkappa}_{\star}\rangle$ if $\{e, a^{\varkappa}, a^{-\varkappa}\} \subset H \subset H_n$ and $\langle H, a^{\varkappa}_{\star}\rangle$ is a hypergroup. It is obvious that if $\langle H, a^{\varkappa}_{\star}\rangle$ is a P-subhypergroup of H_n then H is a subgroup of $(H_n, \cdot)$. Finally we note that every P-subhypergroup of a hypergroup is ultraclosed; see [7 and 5].

PROPOSITION 1
The union of the powers, with respect to the P-hyperoperation, of an element a^s of H_n is a P-subhypergroup of $\langle H_n, a^{\varkappa}_{\star}\rangle$.

Proof
The union of the powers of a^s is

$S = a^{s[1]} \cup a^{s[2]} \cup \ldots = \{a^{ts+v\varkappa} \mid t \in \mathbb{N}_o,\ v \in \mathbb{Z}\}.$

For $t = n$ and $v \in \{-1,0,1\}$ we obtain $\{e, a^{\varkappa}, a^{-\varkappa}\} \subset S$.

Now let $a^{t_o s+v_o\varkappa} \in S$; then

$$a^{t_o s+v_o\varkappa}\, a^{\varkappa}_{\star}\, S = \bigcup_{a^{ts+v\varkappa} \in S} a^{t_o s+v_o\varkappa}\, a^{\varkappa}_{\star}\, a^{ts+v\varkappa} =$$

$$= \bigcup_{a^{ts+v\varkappa} \in S} a^{(t_o+t)s+(v_o+v)\varkappa} \{e, a^{\varkappa}, a^{-\varkappa}\} \subset S.$$

Moreover when we take $0 \leqslant r_o < n$ such that $t_o \equiv r_o \bmod n$, then

(i) if $r_o = 0$ obviously $a^{t_o s+v_o\varkappa}\, a^{\varkappa}_{\star} S = S$, and

(ii) if $r_o > 0$, for $t \in \{n-r_o+1,\ n-r_o+2,\ldots\}$ we have $a^{(t_o+t)s+(v_o+v)\varkappa} \equiv a^{s+(v_o+v)\varkappa}, a^{2s+(v_o+v)\varkappa},\ldots;$ so again $a^{t_o s+v_o\varkappa}\, a^{\varkappa}_{\star} S = S$.
Therefore from the commutativity of $a^{\varkappa}_{\star}$ we conclude that the reproduction axiom is valid.

PROPOSITION 2
In $\langle H_n, a^{\varkappa}_{\star}\rangle$ if there exists an element a^s and a number $t \geqslant 2$ such

that $a^{s[t]} \subset a^{s[1]} \cup \ldots \cup a^{s[t-1]}$, then $a^{s[t+1]} \subset a^{s[1]} \cup \ldots \cup a^{s[t-1]}$.

Proof
Let $y \in a^{s[t+1]}$; then $y = a^{(t+1)s+v\varkappa}$ where $-t \leqslant v \leqslant t$.
We have two cases.

(i) $y = a^{ts+s+v\varkappa}$ with $0 \leqslant v \leqslant t$. From the hypothesis for the element $a^{ts+(v-1)\varkappa} \in a^{s[t]}$ there exists an r where $1 \leqslant r \leqslant t-1$ such that $a^{ts+(v-1)\varkappa} = a^{rs+u\varkappa}$ with $-r < u < r$. Consequently
$y = a^{ts+(v-1)\varkappa+s+\varkappa} = a^{(r+1)s+(u+1)\varkappa} \in a^{s[1]} \cup \ldots \cup a^{s[t-1]}$.

(ii) $y = a^{ts+s+v\varkappa}$ with $-t \leqslant v \leqslant 0$. Then in a similar way we have

$a^{ts+(v+1)\varkappa} \in a^{s[t]}$; so $a^{ts+(v+1)\varkappa} = a^{rs+u\varkappa}$, $1 \leqslant r \leqslant t-1$, $-r < u < r$.
Therefore $y = a^{ts+(u+1)\varkappa-\varkappa+s} = a^{(r+1)s+(u-1)\varkappa} \in a^{s[1]} \cup \ldots \cup a^{s[t-1]}$.

§ 3. THEOREM 1
The P-cyclic hypergroup $\langle H_n, a_\star^\varkappa \rangle$ is single-power cyclic iff $(k,n) = 1$ and in this case every element of H_n is a generator with period $[\frac{n}{2}]+1$.

Proof
We observe that for an element a^s to be a generator of the single-power cyclic hypergroup H_n of period t, it must satisfy the relation
$|a^{s[t]}| = n$. But, since $a^{s[t]} = \{a^{ts+v\varkappa} \mid -t < v < t\}$,
we have $|a^{s[t]}| \leqslant 2t-1$ so $t \geqslant [\frac{n}{2}]+1$.

For $t = [\frac{n}{2}]+1$, we have

$$a^{s\left[[\frac{n}{2}]+1\right]} = a^{s([\frac{n}{2}]+1)}\{a^{v\varkappa} \mid -[\tfrac{n}{2}] \leqslant v \leqslant [\tfrac{n}{2}]\} = a^{[\frac{n}{2}](s-\varkappa)+s}\{e, a^\varkappa, \ldots, a^{2[\frac{n}{2}]\varkappa}\}.$$

Now, if $n = 2w$ we have $a^{2[\frac{n}{2}]\varkappa} = e$, and if $n = 2w+1$ we have
$a^{2[\frac{n}{2}]\varkappa} = a^{(n-1)\varkappa}$; therefore $a^{s\left[[\frac{n}{2}]+1\right]} = a^{[\frac{n}{2}](s-\varkappa)+s}\{e, a^\varkappa, \ldots, a^{(n-1)\varkappa}\}$.

Also, for every $r \in \mathbb{N}_o$, we have

$$a^{s\left[[\frac{n}{2}]+1+r\right]} = a^{s([\frac{n}{2}]+1+r)}\{a^{v\varkappa} \mid -[\tfrac{n}{2}]-r \leqslant v \leqslant [\tfrac{n}{2}]+r\} =$$

$$a^{([\frac{n}{2}]+r)(s-\varkappa)+s}\{e, a^\varkappa, \ldots, a^{2([\frac{n}{2}]+r)\varkappa}\} = a^{([\frac{n}{2}]+r)(s-\varkappa)+s}\{e, a^\varkappa, \ldots, a^{(n-1)\varkappa}\}.$$

So $\left|a^{s\left[[\frac{n}{2}]+1\right]}\right| = \left|a^{s\left[[\frac{n}{2}]+1+r\right]}\right|$.

This means that the element a^s is a generator of the single-power cyclic hypergroup H_n

iff $\left|a^{s\left[[\frac{n}{2}]+1\right]}\right| = n$, i.e. iff $(k,n) = 1$.

We conclude that every element a^s of H_n is a generator with period $[\frac{n}{2}] + 1$.

§ 4. THEOREM 2
In the P-cyclic hypergroup $<H_n, a^{\varkappa}_{\star}>$ the element a^s is a generator iff $(s,\varkappa,n) = 1$.

Proof
We observe that if $a \in a^{s[t]}$, $t \in \mathbb{N}_o$, then for every $v \in \mathbb{N}_o$ we have $a^v \in a^{[vt]}$. Therefore a^s is a generator of H_n iff a belongs to a power of a^s. That means that a^s is a generator iff we can find $t \in \mathbb{N}_o$, $v \in \mathbb{Z}$ such that $a^{ts+v\varkappa} = a$ iff $ts + v\varkappa \equiv 1 \bmod n$, $t \in \mathbb{N}_o$, iff $(s,\varkappa,n) = 1$.

THEOREM 3
If $(\varkappa,n) = 1$ then the generators $a^{\varkappa}, a^{-\varkappa}$ of the P-cyclic hypergroup $<H_n, a^{\varkappa}_{\star}>$ have period $[\frac{n}{2}] + 1$.

Proof
We have $a^{\varkappa[t]} = \{a^{r\varkappa} \mid r = 1,\dots,2t-1\}$; so

$a^{\varkappa[t]} = a^{\varkappa[t-1]} \cup \{a^{(2t-2)\varkappa}, a^{(2t-1)\varkappa}\}$ and

$a^{\varkappa[t-1]} \subsetneqq a^{\varkappa[t]}$ for $t \in \{2,\dots,[\frac{n}{2}] + 1\}$.

Moreover, since $(\varkappa,n) = 1$, we deduce that $a^{\varkappa\left[[\frac{n}{2}]+1\right]} = H_n$.
In a similar way we deduce that $a^{-\varkappa}$ has period $[\frac{n}{2}] + 1$.

THEOREM 4
In the P-cyclic hypergroup $<H_n, a^{1}_{\star}>$, $n > 5$, the generator $a^{[\frac{n}{2}]}$ has period at most $[\frac{n}{2}]$.

Proof
We have $a^{[\frac{n}{2}]\left[[\frac{n}{2}]\right]} = a^{[\frac{n}{2}]^2 - [\frac{n}{2}] + 1}\{e,a,\dots,a^{n-2}\}$.

(i) If $n = 2w$, i.e. $[\frac{n}{2}] = w$, $w > 2$, then
$H_n = a^{w^2-w+1}\{e,a,\dots,a^{n-1}\} = a^{w[w]} \cup \{a^{w^2-w+1+2w-1}\} = a^{w[w]} \cup \{a^{w(w-1)}\}$.
But $a^{w(w-1)} \in a^{w[w-1]}$; so $H_n = a^{w[w]} \cup \dots \cup a^{w[1]}$.

(ii) If $n = 2w+1$, i.e. $[\frac{n}{2}] = w$, $w > 2$, then
$H_n = a^{w^2-w+1}\{e,a,\dots,a^{n-1}\} = a^{w[w]} \cup \{a^{w^2-w+1+2w-1}, a^{w^2-w+1+2w}\} =$
$= a^{w[w]} \cup \{a^{w(w-1)-1}, a^{w(w-1)}\}$.

But $a^{w(w-1)-1}, a^{w(w-1)} \in a^{w[w-1]}$; so $H_n = a^{w[w]} \cup \dots \cup a^{w[1]}$.
Therefore this theorem is valid for $n > 5$.
Notice that this theorem is also valid for $n = 4$. So the only cases for which it is not valid are $n \in \{3,5\}$.

THEOREM 5
If $n>5$, $\varkappa>1$, then the period of the generator a^1 of the P-cyclic hypergroup $\langle H_n, a_\star^\varkappa\rangle$ is at most $[\frac{n}{2}]$.

Proof
We have the following four cases.

(i) $n>9$ and $1<\varkappa<[\frac{n}{2}]-1$. We obtain

$$a^{[1]} \cup a^{[2]} \cup \ldots \cup a^{\left[[\frac{n}{2}]\right]} = \{a^{t+v\varkappa} \mid t=1,\ldots,[\tfrac{n}{2}] \text{ and } -t<v<t\}.$$

So $a^r \in a^{[r]}$ for $r\in\{1,\ldots,[\frac{n}{2}]\}$. Moreover, with the above assumptions for $r\in\{[\frac{n}{2}]+1,\ldots,n\}$, we observe that

if $r-[\frac{n}{2}]=\varkappa d$, $d\in\mathbb{N}_0$ then $a^r = a^{[\frac{n}{2}]+\varkappa d}$ implies $a^r\in a^{\left[[\frac{n}{2}]\right]}$ and

if $r-[\frac{n}{2}]=\varkappa d+u$, $0<u<\varkappa$, then

$$a^r = a^{[\frac{n}{2}]+\varkappa d+u} = a^{[\frac{n}{2}]-\varkappa+u+(d+1)\varkappa} \quad \text{implies} \quad a^r\in a^{\left[[\frac{n}{2}]-\varkappa+u\right]}.$$

(ii) $n>9$ and $\varkappa=[\frac{n}{2}]-1$.

If $r\in\{1,\ldots,[\frac{n}{2}]\}$, then $a^r\in a^{[r]}$;

if $r\in\{[\frac{n}{2}]+1,\ldots,2[\frac{n}{2}]-1\}$, then $a^r\in a^{\left[r-[\frac{n}{2}]+1\right]}$.

Also, for n even, we have $a^n\in a^{[\frac{n}{2}-1]}$ and for n odd, we have $a^{n-1}\in a^{\left[[\frac{n}{2}]-2\right]}$, $a^n\in a^{\left[[\frac{n}{2}]-1\right]}$.

(iii) $n>9$ and $\varkappa=[\frac{n}{2}]$. We examine this case only for odd n, since for even n we have a degenerate case for P.

If $r\in\{1,\ldots,[\frac{n}{2}]\}$, then $a^r\in a^{[r]}$;

if $r\in\{[\frac{n}{2}]+2,\ldots,2[\frac{n}{2}]\}$, then $a^r\in a^{\left[r-[\frac{n}{2}]\right]}$;

if $r\in\{[\frac{n}{2}]+1, 2[\frac{n}{2}]+1\}$, then $a^r\in a^{\left[[\frac{n}{2}]\right]}$.

In the above three cases we can see that the period of a^1 is at most $[\frac{n}{2}]$.

(iv) We can see by inspection that the theorem is valid for $n\in\{6,7,8,9\}$.

§ 5. THEOREM 6
If $(\varkappa,n)=1$ then the P-cyclic hypergroup $\langle H_n, a_\star^\varkappa\rangle$, $n>5$, is not cyclic with period (c.f. [4]).

Proof
From theorem 3 we derive that, since $(\varkappa,n)=1$, the generator $a^\varkappa$ has

period $[\frac{n}{2}]+1$. In theorem 4 it was shown that, for $\varkappa=1$, the generator $a^{[\frac{n}{2}]}$ has period at most $[\frac{n}{2}]$; for $\varkappa>1$, from theorem 5 the generator a also has period at most $[\frac{n}{2}]$.
Therefore $<H_n, a_{\star}^{\varkappa}>$ is not cyclic with period.

Remark
If $(\varkappa,n)=1$, then the only hypergroups $<H_n, a_{\star}^{\varkappa}>$ cyclic with period are for $n\in\{3,5\}$.

REFERENCES

[1] Corsini,P., Sur les homomorphismes d'hypergroupes, Rend. Sem. Mat. Univ. Padova, Vol. 52, p.p. 117-140 (1974).
[2] De Salvo, M.-Freni,D., Semi-ipergruppi ciclici, Atti Sem. Mat. Fis. Univ. Modena, XXX, p.p. 44-59 (1981).
[3] Dresher,M.-Ore,O., Theory of Multigroups, Amer. J. Math. V. 60, p.p. 705-733 (1938).
[4] Konguetsof,L.-Vougiouklis,T.-Kessoglides,M.-Spartalis,S., On cyclic hypergroups with period, to appear in Acta Un. Carolinae - Math. et Physica.
[5] Sureau,Y., Sous-hypergroupe engendré par deux sous-hypergroupes et sous-hypergroupe ultra-clos d'un hypergroupe, C.R.Acad. Sc. Paris, t.284 (2 Mai 1977) Série A. p.p. 983-984.
[6] Vougiouklis,T., Cyclicity in a Special Class of Hypergroups, Acta Un. Carolinae-Math. et Physica, Vol. 22, No 1, p.p. 3-6 (1981).
[7] Vougiouklis,T., Generalization of P-hypergroups, to appear in Rendiconti del Circolo Matematico di Palermo.
[8] Wall,H.S., Hypergroups, American Journal of Mathematics, Vol. 59, p.p.77-98 (1937).

Annals of Discrete Mathematics 37 (1988) 427–432

ORDER AND UNIFORM STRUCTURE IN PROJECTIVE GEOMETRY

Horst SZAMBIEN

Kaiser Wilhelms Gymnasium
Seelhorststraße 52
D-3000 Hannover 1
Fed. Rep. of Germany

Riassunto. In questa nota studieremo uno spazio proiettivo ordinato: Esiste una struttura uniforme compatibile. Il complementamento di questo spazio é strutturabile a sopra-spazio proiettivo ordinato. AMS-Subject-Classification: Primary 51H05, secondary 54E15, 06B30.

1. INTRODUCTION

Uniform structures have been studied in planar geometry by Guggenheimer [Gu 59], Breitsprecher [Br 67] and Szambien [Sz 81], [Sz 86] mainly in the context of projective planes. In this paper the notion of uniform projective geometry is defined for the case of arbitrary dimension. It shall be a projective geometry (i. e. subspace lattice of some vector space) provided with a uniformity, such that the following condition and its dual hold: Joining of skew subspaces is uniformly continuous outside an arbitrary uniform neighbourhood of the closed set of non-skew pairs of subspaces. Any uniform projective geometry becomes a topological projective geometry when it is provided with the uniform topology, [Sz 86]. Proposition 1 shows that every finite-dimensional projective geometry over a topological field is a topological projective geometry. So, in this class the converse problem arises: For which topological projective geometries does there exist a compatible uniformity, which turns it into a uniform projective geometry? The problem will be solved for ordered geometries:

In a projective space (i. e. the set of 1-dimensional subspaces of some vector space) one can consider an order structure for instance in terms of a 4-ary separation relation or some equivalent concept; cf. [Le 65] . The so-called special separation relations are then equivalent to orderings of the coordinatizing division ring. In this paper finite-dimensional pappian projective geometries, whose correspondent spaces carry an order structure, are considered. They are of course topological projective geometries w. r. to their order topology, [Mi 68;S.51,Satz11]. It is shown that they can be provided with a uniformity derived from the order structure to become uniform projective geometries. The construction of this uniformity is carried out using the well known Grassmann-mapping. This mapping embeds the set of subspaces of a given vector space into the point set of some projective space of higher dimension. It then suffices to define a suitable uniformity on this projective space. This is done in proposition 2. In proposition 3 the existence of the Weil-completion is proved for any ordered pappian projective geometry; as a corollary to this one recognizes in proposition 4 the Dedekind-completion and the Weil-completion as isomorphic objects.

2. RESULTS

[Bo 66], [Le 65], and [Bi 79] are taken as general references for notions of general topology, algebra, geometry, and lattice theory, resp. . Throughout this paper "$\restriction$" will denote the restriction of mappings, topologies, and uniformities, M^n the n-fold cartesian product of the set M; furthermore L is a lattice, where s+t and s·t denote the join, resp. meet of lattice elements s and t. If L is an irreducible, meet-continuous, atomistic, modular lattice, then it will be termed projective geometry. The length of this lattice minus 1 will as usual be called the dimension of the geometry. If L has bottom 0 and top 1, then $S:=\{(s,t)\in L^2: s\cdot t=0\}$ is the set of skew pairs and dually $S^*:=\{(s,t)\in L^2: s+t=1\}$.Let σ be a topology on L. (L,σ) is called a topological projective geometry (TPG) iff L is a projective geometry, and $+\restriction S$ and $\cdot\restriction S^*$ are continuous mappings with open domains. Examples of TPGs are constructed as follows:

Let (K,Ω) be a commutative,topological field with topology Ω , Γ the exterior algebra on K^n, the n-dimensional vector space over K, n a natural number ≥ 3, Γ_k the $\binom{n}{k}$-dimensional K-vector space of k-vectors; thus Γ is the direct sum of $\{\Gamma_k: 0\leq k\leq n\}$. Let $\Theta_k \subset \Gamma_k$ be the set of decomposable k-vectors for $1\leq k\leq n$, $\Theta_o:= K\setminus\{0\}$, $\Theta:=\cup\{\Theta_k:0\leq k\leq n\}$. Now take the lattice $L:=L(K)$ of subspaces of K , $L_k:= \{s\in L : \dim s = k\}$, P the set of 1-dimensional subspaces of the vector space $\Gamma;\Gamma^\times:= \Gamma\setminus\{0\}$, $K^\times:=K\setminus\{0\}$, $\Gamma^\times/K^\times:= \{K^\times g : g \in \Gamma\}$, $D_k:= \Theta_k^\times/K^\times$, $D:=\Theta^\times/K^\times$. P will be identified with $\Gamma^\times/K^\times$. Let Γ carry the product topology of Ω. Provide P with the topology π, the finest topology which renders the canonical projection $\Gamma \longrightarrow \Gamma^\times/K^\times$ continuous. Since there exists a bijection between L and D, one gets an embedding $\Phi: L \longrightarrow P$; provide L with σ, the initial topology w. r. to Φ. σ is called the coordinate topology.

PROPOSITION 1. (L,σ) is a TPG.

Proof. Let $\wedge$ denote the exterior multiplication on , and $\Sigma:= \{(s,t) \in \Theta^2 : s\wedge t \neq 0\}$; then $\wedge$ maps Σ into Θ ; furthermore it is continuous with Σ open, because the coordinates of $s\wedge t$ are calculated by rational operations from the coordinates of s and t. Hence it induces a continuous mapping $S \longrightarrow L$, which coincides with the lattice-join +, making the following diagram commutative:

$$\begin{array}{ccc} \Sigma & \longrightarrow & \Theta \\ {\scriptstyle \alpha\times\alpha}\downarrow & & \downarrow{\scriptstyle \alpha} \\ S & \xrightarrow[+]{} & L \end{array}$$

Therein α is the map, which associates with each $a \in \Theta$ the subspace of K determined by it. + becomes continuous, because α, and also $\alpha\times\alpha \restriction \Sigma$ are quotient mappings.

The continuity of · remains to be shown. To this end consider Γ', the exterior algebra on the dual vector space of K ; likewise introduce $\wedge'$, L', S', Σ', Θ' as before. Now $\delta : \Gamma\longrightarrow \Gamma'$, mapping a multivector to its dual, is a homeomorphism and induces another homeomorphism $L \longrightarrow L'$, mapping a subspace to its annullator. Define $\Sigma^*:= (\delta\times\delta)^{-1}[\Sigma']$. Now the following diagram commutes - making the lower row mapping $S^* \longrightarrow L$ continuous. This in turn coincides with $\cdot\restriction S^*$.

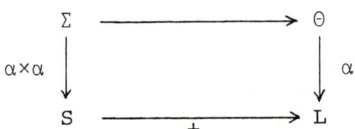

$$\begin{array}{ccccccc}
\Sigma^* & \longrightarrow & \Sigma' & \xrightarrow{\ \wedge'\ } & \Theta' & \longrightarrow & \Theta \\
\downarrow{\scriptstyle \alpha\times\alpha} & & \downarrow & & \downarrow & & \downarrow{\scriptstyle \alpha} \\
S^* & \longrightarrow & S' & \xrightarrow[+']{} & L' & \longrightarrow & L
\end{array}$$

q. e. d.

Let μ be a uniformity on L, $N \in \mu$ an entourage; $S_N := L^2\setminus N\circ(L^2\setminus S)\circ N$, $S_N^* := L^2\setminus N\circ(L^2\setminus S^*)\circ N$. (L,μ) is called a uniform projective geometry (UPG) iff L is a projective geometry such that for every $N \in \mu$ the mappings $+ \mid S_N$ and $\cdot \mid S_N^*$ are uniformly continuous, and furthermore S and S^* are open w. r. to the uniform topology $T(\mu)$. A class of examples is furnished by

PROPOSITION 2. Let L be an ordered pappian projective geometry of finite dimension at least two. Then there can be constructed a uniform structure μ on L inducing the coordinate topology, such that (L,μ) is a UPG. μ is called the coordinate uniformity on L.

Proof. With the notations of proposition 1 one considers the totally ordered coordinate field $(K,<)$ of L which of course is a topological field w. r. to its order topology . Take a natural number $n \geq 3$, $m := 2^n$, $g := (g_1,\dots,g_m) \in K^m$;
$R_i := \{g : -1 \leq g_j \leq 1 \text{ for } 1\leq j\leq i-1,\ g_i = 1,\ -1 < g_j < 1 \text{ for } i+1\leq j\leq m\}$ and $R := R(K) := \cup\{R_i : i=1,\dots,m\}$. There is a bijection $r : R \longrightarrow P$ whose inverse assigns to each $K^\times g \in P$ the vector

(*) $\operatorname{sgn}(g)\max(|g_1|,\dots,|g_m|)^{-1}g \in R$, where $\operatorname{sgn}(g)$ denotes the sign of the component g_j of highest index j such that the equation $|g_j| = \max(|g_1|,\dots,|g_m|)$ holds. $r \mid R\cap\Theta$ therefore maps onto L. On the topological group K^m there is given the left(or right) uniformity w. r. to addition; restrict this uniformity to $R\cap\Theta$ and take on L the final uniformity w. r. to $r \mid R\cap\Theta$; call this μ, the coordinate uniformity on L. Since the coordinates are taken only from a bounded subset of K, the exterior product $s\wedge t$ of s and t depends on $(s,t) \in (R\cap\Theta)^2$ in a uniformly continuous manner. Take an entourage N in $(R\cap\Theta)^2$, $\Sigma_N := \{(s,t)\in(R\cap\Theta)^2 : N(s)\wedge N(t) \neq 0\}$, $S_N := (r\times r)[\Sigma_N]$; then certainly there exists a neighbourhood V of zero in Γ , such that Σ_N is mapped by $\wedge$ into $\Theta\setminus V$; any image multivector g can be normed as indicated by (*) above; the combined mapping $\Sigma_N \longrightarrow R\cap\Theta$ is uniformly continuous and induces a uniformly continuous mapping $S_N \longrightarrow L$, which coincides with + :

$$\begin{array}{ccc}
\Sigma_N & \longrightarrow & R\cap\Theta \\
\downarrow{\scriptstyle r\times r} & & \downarrow{\scriptstyle r} \\
S_N & \xrightarrow[+]{} & L
\end{array}$$

One now proceeds along the lines of the proof in proposition 1 to show the uniform continuity of the meet: Define Σ_N', R', S_N' as above, and $\Sigma_N := (\delta\times\delta)^{-1}[\Sigma_N']$. Since $\delta[R] \subset R'$, we have the following commutative diagram:

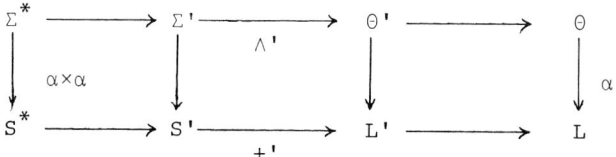

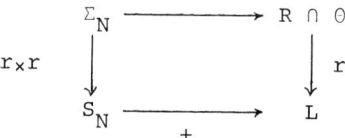

$$\begin{array}{ccccccc} \Sigma_N & \xrightarrow{\delta\times\delta} & \Sigma_N' & \longrightarrow & R' \cap \Theta' & \xrightarrow{\delta} & R \cap \Theta \\ {\scriptstyle r\times r}\downarrow & & \downarrow & & \downarrow & & \downarrow{\scriptstyle r} \\ S_N & \longrightarrow & S_N' & \xrightarrow[+']{} & L' & \longrightarrow & L \end{array}$$

By the same arguments as above it contains a uniformly continuous upper row mapping. Now r is uniformly continuous and r×r is a quotient mapping; hence the lower row mapping which coincides with · is continuous, too. Finally, μ induces the coordinate topology, because scalar multiplication is continuous. q. e. d.

Any uniform space can be densely embedded into a unique complete uniform space, called its Weil-completion; continuation of the geometric structure is possible for UPGs, as is shown by

PROPOSITION 3. The Weil-completion (w. r. to its coordinate uniformity) of an ordered pappian projective geometry of finite dimension at least two is again a UPG.

Proof. Let (K,<) be the coordinatizing field, K^ its Weil-completion, which again carries the structure of an ordered field by [Ni 65;p.14,Thm.2]. L^ will denote the lattice of subspaces of $(K^\wedge)^n$, $n \geq 3$, $m := 2^n$; $L_{\hat{k}}$, $R^\wedge := R(K^\wedge)$, and P^ are used analogously to the notations in the proof of proposition 2. $(K^\wedge)^m$ is certainly complete, R^ as a closed subset as well. Thus P^carries a complete uniform structure. Since the Grassmann-manifolds $L_{\hat{k}}$ are Zariski-closed projective varieties in P, they are closed in the projective coordinate topology as well, [Le 65;p.337,Satz 1.2]; thus they are complete. Hence L^ is complete, and it is a UPG by proposition 2. Since it contains L densely, L^ is uniformly isomorphic to the Weil-completion of L. q. e. d.

It is shown by Massaza that the Dedekind-completion by cuts of an ordered field is an extension field K˜ of K. Furthermore K˜ is isomorphic to the Weil-completion K^ of K; [Ma 69;p.331,333] or [Pri 83;p.72,Satz 9]. Considering ordered projective geometries L and L˜ of finite dimension over K resp. K˜, one terms L˜ the Dedekind-completion of L. Then propositiom 3 yields an analogue of Massaza's theorem:

PROPOSITION 4. The Dedekind-completion of an ordered pappian projective geometry of finite dimension not less than two is isomorphic to its Weil-completion.

Proof. $L^\wedge = L(K^\wedge) = L(K^\sim) = L^\sim$. q. e. d.

From Sz 81;(3.11) one derives that the coordinate field of any pappian UPG is of type V (called locally retrobounded in [Bo 66; TG, Ch.III.6, Exerc.]). Following [Pre 78;(5.4)] a field of type V is orderable iff the set of all sums of squares is disjoint with some neighbourhood of -1. This condition singles out the orderable UPGs among the pappian UPGs, and it can of course be translated into geometric language.

REFERENCES

[Bi 79] Birkhoff, G.: Lattice theory, Amer. Math. Society Colloq. Publications Vol. 25, Providence RI 1960

[Bo 66] Bourbaki, N.: Algèbre, Topologie Générale; Paris 1966

[Br 67] Breitsprecher, S.: Uniforme projektive Ebenen; Mathematische Zeitschrift 95 (1967) 139-168

[Gu 59] Guggenheimer, H.: The topology of elementary geometry; Mathematica Japonicae 5 (1959) 1-26

[Le 65] Lenz, H.: Vorlesungen über projektive Geometrie; Leipzig 1965

[Ma 69] Massaza, C.: Sul completamento dei campi ordinati; Rendiconti del Seminario Matematico, Universitá e Politecnico di Torino 29 (1969/1970) 329-348

[Mi 68] Misfeld, J.: Topologische projektive Räume, Dissertation Hamburg 1968

[Ni 65] Nieminen, T.: The completions of orderable topological groups and fields; Annales Academiae Scientiarum Fennicae Series A 365 (1965) 1-15

[Pre 78] Prestel, A., Ziegler, M.: Model theoretic methods in the theory of topological fields; Journal für Reine und Angewandte Mathematik 299/300 (1978) 318-341

[Pri 83] Prieß-Crampe, S.: Angeordnete Strukturen; Berlin 1983

[Sz 81] Szambien, H.: Desarguessche uniforme Ebenen; Dissertation, Hannover 1981; summary (German) in Resultate der Mathematik 5 (1982) 96-98

[Sz 86] Szambien, H.: Uniform structures in projective geometry; Proceedings of the Sixth Prague Topological Symposium 1986, Berlin ca. 1987

Annals of Discrete Mathematics 37 (1988) 433–450

ON BLOCKING SETS IN FINITE PROJECTIVE AND AFFINE SPACES

Giuseppe TALLINI

Dipartimento di Matematica, Universita` di Roma, "La Sapienza" (Italia)

This paper is a new version of the first "Quaderno" [9] (written in italian) of the "Seminario di Geometria Combinatoria" published by the University of L'Aquila in 1982. In this Quaderno we proved several new results for blocking sets (with respect to lines) in PG(r,q) and AG(r,q). Since this paper has found great interest (see references [11],...,[42]), it seems reasonable to make it available to a wider audience. (Note that in [24] a result of [9] was proved in a different way).

1. k-SETS IN PROJECTIVE GALOIS SPACES

Let S_r = PG(r,q) be a projective Galois space of dimension $r \geq 2$ and order $q=p^h$ (p prime). If S_d denotes a d-dimensional subspace of S_r, we have:

$$(1.1)\qquad |S_d| = \sum_{i=0}^{d} q^i := \vartheta_d .$$

It is well-known that the number of d-dimensional subspaces S_d in S_r is given by

$$(1.2)\qquad \gamma_{r,d} = \prod_{i=0}^{d} \vartheta_{r-i} / \vartheta_{d-i}$$

Denote by K a k-set of PG(r,q) with $K \neq \emptyset$ and $K \neq$ PG(r,q). Denote by m and n respectively the smallest and the greatest positive number of points in which K is intersected by a d-dimensional subspace S_d $(1 \leq d \leq r)$ of S_r. In other words we set:

$$(1.3)\qquad m=\min\left\{|K\cap S_d| \ : K \cap S_d \neq \emptyset\right\}$$

$$(1.4)\qquad n=\max\left\{|K\cap S_d| \ : K \cap S_d \neq \emptyset\right\}$$

We have, obviously

$$(1.5)\qquad 1 \leq m \leq n \leq |S_d| = \vartheta_d .$$

If $s=0,1,\ldots,\vartheta_d$, by $t_s=t_s^d$ we denote the number of d-dimensional subspaces of S_r, each of which contains exactly s points of a fixed k-set K. The numbers t_s are called the characters of index s of set K, with respect to dimension d. Fix h+1 integers $m_0,m_1,\ldots,m_h$ with $0 \leq m_0 < m_1 < \ldots < m_h$. A k-set K is said to be of class $[m_0, m_1,\ldots, m_h]_d$, with respect to the dimension d, if $t_s=0$ for any s different from $m_0,m_1,\ldots,m_h$. Moreover, a k-set K of class $[m_0,m_1,\ldots,m_h]_d$, with the additional condition $t_s \neq 0$ for $s=m_0,m_1,\ldots m_h$, is said to be of type $(m_0,m_1,\ldots, m_h)_d$, with respect to dimension d, cf. [7].

Note that, if m=n, each S_d of S_r intersects K in zero or m points. So if r=2, either we have $K=\emptyset$ or $K=S_2$ or K is a set of type $(0,m)_1$ with $m=p^t$ $(0 \leq t \leq h)$ (cf. [7]). If $r \geq 3$, we have two possibilities:

(i) d-dimensional subspaces external to K do not exist. (In this case we know that either $K=\emptyset$ or $K=S_r$).

(ii) Suppose that there is an S_d without points in common with K. (Then either K is one point or the complement of a hyperplane (cf. [7], Prop. I, XIV)). So, we suppose:

$$(1.6) \qquad 1 \leq m < n \leq \vartheta_d \quad .$$

Let K be a k-set of S_r . Denote by m and n, respectively the smallest and the greatest cardinality of the set $S_d \cap K$, if $S_d \cap K \neq \emptyset$.Then K is called a $(k;\ 0,m,n)_d$-set. The characters of a k-set K of S_r satisfy the following conditions (cf. [7]):

$$(1.7) \qquad \left| \begin{aligned} t_o + \sum_{s=m}^{n} t_s &= \gamma_{r,d} \\ \sum_{s=m}^{n} s t_s &= k\gamma_{r-1,d-1} \\ \sum_{s=m}^{n} s(s-1)t_s &= k(k-1)\gamma_{r-2,d-2} \end{aligned} \right. .$$

$(1.7)_{II}$ and $(1.7)_{III}$ imply:

$$(1.8) \qquad \sum_{s=m}^{n} s^2 t_s = k\left[(k-1)\gamma_{r-2,d-2} + \gamma_{r-1,d-1}\right] .$$

Fix two integers M,N. By (1.7) and (1.8) we have:

$$(1.9) \qquad \sum_{s=m}^{n} (N-s)(s-M)t_s = -F$$

$$(1.10) \quad F=F(k,M,N,t_o,r\ ,\ d,q) = \gamma_{r-2,d-2}k^2 - k\left[(M+N-1)\gamma_{r-1,d-1} + \gamma_{r-2,d-2}\right] + MN(\gamma_{r,d} - t_o).$$

Suppose $N \geq n$, $M \leq m$, then

$$(1.11) \qquad F \leq 0 \quad .$$

Moreover, we claim that

$$(1.12) \qquad F=0 \iff N=n,\ M=m,\ t_{m+1}=\ldots=t_{n-1}=0.$$

In fact $N>n$ implies $F \neq 0$ (since $t_n \neq 0$) and $M<m$ implies $F\neq 0$ too. So, when F=0 it follows N=n and M=m. Consequently, we have $t_{m+1}=\ldots=t_{n-1}=0$. The converse is obvious. By (1.12) we have

$$(1.13) \qquad F=0 \iff N=n,\ M=m,\ K \text{ is of class } [0,m,n]_d \ .$$

Since $F\leq 0$, if $N\geq n$ and $M\leq m$ it follows

$$(1.14) \qquad \Delta = \left[(M+N-1)\gamma_{r-1,d-1} + \gamma_{r-2,d-2}\right]^2 - 4MN\gamma_{r-2,d-2}(\gamma_{r,d} - t_o) \geq 0.$$

Then, if $N\geq n$ and $M\leq m$, we have:

$$(1.15) \qquad t_o \geq \gamma_{r,d} - \left[(M+N-1)\gamma_{r-1,d-1} + \gamma_{r-2,d-2}\right]^2 / 4MN\gamma_{r-2,d-2} \quad .$$

In (1.15) the equality holds if, and only if, $\Delta=0$. Since $F\leq 0$, the equality in (1.15) holds if, and only if, F=0 and

$$k=k_o := \left[(m+n-1)\gamma_{r-1,d-1} + \gamma_{r-2,d-2}\right] / 2\gamma_{r-2,d-2}$$

that is if, and only if, (1.13) holds with $k=k_o$.

Condition (1.15) does not depend on k. When its right-hand side is positive, (1.15) gives us a lower bound for t_o. When this lower bound is significant, we can exclude the existence of some k-sets.

If M=m and N=n, it follows $F \leq 0$. So we have a special case in which (1.14) holds and consequently (1.15). Then we can deduce the bound for k:

(1.16) $$k_1 \leq k \leq k_2$$

where k_1 and k_2 are the solutions of the equation F=0 (in which we put M=m, N=n). In (1.16) $k=k_1$ (or $k=k_2$) holds if, and only if, F=0. That is if, and only if, (1.13) holds.

Moreover, we note that if N=n, M=m, (1.15) becomes:

(1.17) $$t_o \geq \gamma_{r-2,d-2}\left\{[4mn\,\gamma_{r,d}/\gamma_{r-2,d-2}] - [1+(m+n-1)\gamma_{r-1,d-1}/\gamma_{r-2,d-2}]^2\right\}/4mn.$$

We set:

(1.18) $$D := 4mn-(m+n-1)^2 .$$

By (1. 2), we have:

$$\gamma_{r-1,d-1}/\gamma_{r-2,d-2} = \vartheta_{r-1}/\vartheta_{d-1} \quad , \quad \gamma_{r,d}/\gamma_{r-2,d-2} = \vartheta_r\vartheta_{r-1}/\vartheta_d\vartheta_{d-1} .$$

So (1.17) can be written in the form:

(1.19) $$t_o \geq \gamma_{r-2,d-2}\,[Dq^{2r+d-2}+P(q)]\,/4mn\,\vartheta_d\vartheta_{d-1}^2$$

where

(1.20) $$P(q):=D[\vartheta_{r-1}^2\vartheta_{d-1} + q^d\vartheta_{r-2}(\vartheta_{r-2}+2q^{r-1})] - 4mnq^d\vartheta_{r-1}\vartheta_{r-d-1} - \vartheta_d\vartheta_{d-1}[2(m+n-1)\vartheta_{r-1}+\vartheta_{d-1}]$$

is a polynomial function whose degree is less than 2r+d −2. Now we set

(1.21) $$A(q):=D\,q^{2r+d-2}+P(q).$$

So (1.19) becomes:

(1.22) $$t_o \geq \gamma_{r-2,d-2}\,A(q)/4mn\,\vartheta_d\,\vartheta_{d-1}^2 .$$

Suppose $D=4mn-(m+n-1)^2>0$. Since $A(0)=-(m-n)^2<0$, there exists in the set of real positive numbers R^+, at least one solution of the equation A(q)=0. Denote by q_o the greatest solution of the equation A(q)=0 in R^+. We have $A(q)>0$ for each $q>q_o$, and

$$D=4mn-(m+n-1)^2 = -\left[n-(\sqrt{m}-1)^2\right]\cdot\left[n-(\sqrt{m}+1)^2\right] .$$

Since $n-(\sqrt{m}-1)^2>0$, we can write:

(1.23) $$n<(\sqrt{m}+1)^2 \iff D>0 \implies t_o>0, \quad \forall \quad q>q_o .$$

Condition (1.23) implies the following theorem:

I. If $q>q_o$ in PG(r,q) there is no $(k;0,m,n)_d$-set with $t_o=0$ and $n<(\sqrt{m}+1)^2$. (We should remember that q_o is the greatest real positive solution of the equation A(q)=0, with A(q) defined in (1.21)).

Suppose r=2, d=1. By (1.22), we have

$$(1.24)\qquad t_o \geq A(q)/\vartheta_1 = \left\{4mn\,\vartheta_2 - \left[(m+n-1)\,\vartheta_1 + 1\right]^2\right\}/4mn.$$

The importance of (1.22), (1.23) and (1.24) can be shown in several special cases. The following are examples of such cases. Let K be a k-set in PG(2,q), with m=1 and n=3. Then condition (1.24) holds and q_o=4,309... . So the following is a Corollary of Theorem I.

II. In PG(2,q) if $q > 4$, a k-set with m=1, n=3, admits at least t_o external lines, with $t_o \geqslant [3q^2-12q-4]/12$.
(Note that an irreducible cubic is such k-set. When $q \leqslant 4$, the existence of irreducible cubics without external lines is well known. One of these is the unital of equation $x^3+y^3+z^3=0$).
It follows that in PG(r,q), $q>4$, each k-set of class $[0,1,2,3]_1$, with respect to lines, always admits external lines.

Let K be a k-set in PG(2,q) with m=2 and $n \leqslant 5$. Condition (1.23) holds and we have q_o=11,20.... Then

III. A k-set of PG(2,q) of class $[0,2,3,4,5]_1$ admits at least t_o external lines if $q > 11$, where $t_o > [4q^2-44q-9]/40$. Then in PG(r,q), $q > 11$, a k-set of class $[0,2,3,4,5]_1$ always admits external lines.

Let us now consider a particular case of Prop.III. Suppose $q > 11$ and q even in PG(2,q); let K_1 and K_2 denote two distinct (q+2)-arcs in such a plane. Suppose that an algebric curve $\mathscr{C}$ of degree 5 in PG(2,q), $q > 11$, satisfies the following conditions:

(i) $\mathscr{C}$ contains no lines,
(ii) $\mathscr{C}$ has no external lines.

Then $\mathscr{C}$ admits tangent lines and $K_1 \cup K_2$ admits external lines.

2. THE k-SETS OF AG(r,q)

Let A_r=AG(r,q) be an affine Galois space of dimension $r \geqslant 2$ and order q. Let K be a k-set of A_r with $K \neq \emptyset$ and $K \neq A_r$. Denote by m and n respectively the smallest and the greatest positive number of points common to K and a d-subspace A_d of A_r, with $A_d \cap K = \emptyset$. Let us put therefore

$$(2.1)\qquad m=\min\left\{|K \cap A_d| \quad : \quad A_d \cap K \neq \emptyset\right\}$$

$$(2.2)\qquad n=\max\left\{|K \cap A_d| \quad : \quad A_d \cap K \neq \emptyset\right\}.$$

Obviously, we have

$$1 \leqslant m \leqslant n \leqslant |A_d| = q^d.$$

Denote by $t_s=t_s^d$ $(s=0,1,\ldots,q^d)$ the number of d-dimensional subspaces of A_r, each of which contains exactly s points of set K. The numbers t_s are called the characters. As with the projective case, the k-set K is said to be of class $[m_o, m_1,\ldots,m_\ell]_d$, with respect to the dimension d, if $t_s=0$ for every $s \neq m_0, m_1, \ldots, m_\ell$. Suppose K is of class $[m_0,m_1,\ldots,m_\ell]_d$. If in addition $t_s \neq 0$ when $s=m_0, m_1, \ldots, m_\ell$, then K is said to be of type $(m_0, m_1,\ldots, m_\ell)_d$.

If m=n, we have either $t_0=0$ or $t_0 > 0$. In the case $t_0=0$, K is of type $(m)_d$. Consequently, neither $K=\emptyset$ nor K= AG(r,q), a contradiction. Suppose now $t_0 > 0$. In this case K is of type $(0, m)_d$. So, if we have the additional condition $r \geqslant 3$, set K is one point (cf.[5]).

In the sequel let us suppose

$$(2.3)\qquad 1 \leq m < n \leqslant q^d.$$

The characters of a k-set K satisfy the following equations, (the proof is similar to the projective case).

(2.4)
$$\begin{cases} t_o + \sum_{s=m}^{n} t_s = \gamma_{r,d} - \gamma_{r-1,d} = q^{r-d}\gamma_{r-1,d-1} \\ \sum_{s=m}^{n} st_s = k\,\gamma_{r-1,d-1} \\ \sum_{s=m}^{n} s(s-1)t_s = k(k-1)\,\gamma_{r-2,d-2}\,. \end{cases}$$

Conditions $(2.4)_{II}$ and $(2.4)_{III}$ imply:

(2.5)
$$\sum_{s=m}^{n} s^2 t_s = k\left[(k-1)\,\gamma_{r-2,d-2} + \gamma_{r-1,d-1}\right].$$

Now we fix two integers M and N. We have:

(2.6)
$$\sum_{s=m}^{n} (N-s)(s-M)t_s = -\gamma_{r-2,d-2}F$$

where

(2.7)
$$F=F(k,M,N,q,r,d,t_o) := k^2-k\left[1+(M+N-1)\,\vartheta_{r-1}/\vartheta_{d-1}\right] + $$
$$+ MN(q^{r-d}\vartheta_{r-1}/\vartheta_{d-1} - t_o/\gamma_{r-2,d-2}).$$

In the sequel, we suppose $N \geq n$ and $M \leq m$. These conditions imply

$$0 \leq \sum_{s=m}^{n} (N-s)(s-M)t_s = -\gamma_{r-2,d-2}F\,.$$

So, it follows

(2.8)
$$F \leq 0.$$

We have (as with (1.12)):

(2.9)
$$F=0 \iff N=n, \quad M=m, \quad t_{m+1}=\ldots=t_{n-1}=0$$

that is,

(2.10) $F=0 \iff$ N=n, M=m, and K is either of type $(0,m,n)_d$ or of type $(m,n)_d$ according to $t_o > 0$, or $t_o = 0$.

As $F \leq 0$, we have

(2.11)
$$\Delta = \left[1+(M+N-1)\,\vartheta_{r-1}/\vartheta_{d-1}\right]^2 -$$
$$- 4MN\left[q^{r-d}\vartheta_{r-1}/\vartheta_{d-1} - t_o/\gamma_{r-2,d-2}\right] \geq 0$$

So

(2.12)
$$t_o \geq \gamma_{r-2,d-2}\left\{4MN\,q^{r-d}\,\vartheta_{r-1}/\vartheta_{d-1} -\right.$$
$$\left. - \left[1+(M+N-1)\,\vartheta_{r-1}/\vartheta_{d-1}\right]^2\right\}/4MN.$$

Conditions (2.11) and (2.12) hold in the case M=m and N=n, too. In (2.12) equality holds if, and only if, $\Delta=0$. As $F \leq 0$, we have:

In (2.12) equality holds if, and only if, F=0 with

$$k=k_0 := \left[1+(m+n-1)\,\vartheta_{r-1}/\vartheta_{d-1}\right]/2;$$

that is if, and only if, (2.10) holds with $k=k_0$.

If the right-hand side of (2.12) is positive, we get a lower bound for t_o. In this case we can exclude the existence of particular k-sets.

If M=m and N=n, it follows $F \leq 0$ and either condition (2.11) or (2.12) holds. So, we can deduce the bounds for k

$$(2.13) \qquad k_1 \leq k \leq k_2$$

where k_1 and k_2 are the solutions of (the equation) F=0 (in which M=m and N=n).

In (2.13) we have $k=k_1$ or $k=k_2$ if, and only if, F=0; that is if, and only if, (2.10) holds.

We observe again that, if N=n and M=m, (2.12) becomes

$$(2.14) \qquad t_o \geqslant \gamma_{r-2,d-2}\{4mnq^{r-d}\vartheta_{r-1}/\vartheta_{d-1} - [(m+n-1)\vartheta_{r-1}/\vartheta_{d-1}+1]^2\}/4mn.$$

By (1.18) and (2.14) we have:

$$(2.15) \qquad t_o \geqslant \gamma_{r-2,d-2}[Dq^{2r-2}+P'(q)]/4mn\vartheta^2_{d-1},$$

where P'(q) is the following polynomial of q with a degree less than 2r-2:

$$(2.16) \qquad P'(q):=D\vartheta_{r-2}(q^{r-1}+\vartheta_{r-1})- -2\vartheta_{r-1}[2mn\vartheta_{r-d-1}+(m+n-1)\vartheta_{d-1}] - \vartheta^2_{d-1}.$$

Put

$$(2.17) \qquad B(q) := Dq^{2r-2}+P'(q) \quad , \qquad D:=4mn-(m+n-1)^2 .$$

So (2.15) becomes

$$(2.18) \qquad t_o \geqslant \gamma_{r-2,d-2}B(q)/4mn\vartheta^2_{d-1} .$$

If r=2 and d=1, (2.18) becomes

$$(2.19) \qquad t_o \geqslant B(q)/4mn= \left\{ Dq\vartheta_1 - \vartheta_1\left[(m+n)^2-1\right]-1\right\}/4mn.$$

Since $B(0)=-(n-m)^2<0$, if $D>0$, the equation B(q)=0 has real positive solutions. Denote by q'_o the greatest positive solution of B(q)=0. Then $B(q)>0$ for each $q>q'_o$. We have:

$$D=4mn-(m+n-1)^2=-\left[n-(\sqrt{m}-1)^2\right]\cdot\left[n-(\sqrt{m}+1)^2\right].$$

Since $n-(\sqrt{m}-1)^2>0$, by previous equality it follows:

$$(2.20) \qquad n<(\sqrt{m}+1)^2 \iff D>0 \implies t_o>0, \quad \forall\, q>q'_o .$$

By (2.20) we obtain

IV. If $q>q'_o$, then in AG(r,q) there is no k-set with $t_o=0$ and $n<(\sqrt{m}+1)^2$.

3. SPREADS WITH LINES IN PG(r,q)

Let S_r=PG(r,q) be a projective Galois space of dimension r and order $q=p^h$ (p prime). A family $\mathscr{S}(\neq\emptyset)$ of lines of PG(r,q) mutually skew is called a spread in PG(r,q). If the union S of the lines of $\mathscr{S}$ coincides with PG(r,q), the spread $\mathscr{S}$ is said to be total or a spread of PG(r,q); if it is not, $\mathscr{S}$ is said to be partial. A spread $\mathscr{S}$ in PG(r,q) is said to be maximal if in PG(r,q) there exists no spread $\mathscr{S}'$ such that $\mathscr{S} \subset \mathscr{S}'$.

Let $\mathscr{S}$ be a spread in PG(r,q), $r \geq 4$. Consider the subspaces S_3 of PG(r,q) that contain at least one line of $\mathscr{S}$. Denote by m and n respectively the smallest and the greatest number of lines of $\mathscr{S}$ contained in a S_3. That is, if $\mathscr{S} \cap S_3$ denotes the set of lines of $\mathscr{S}$ contained in S_3, then:

$$m=\min\left\{|\mathscr{S}\cap S_3| \;:\; S_3\cap\mathscr{S}\neq\emptyset\right\} \tag{3.1}$$

$$n=\max\left\{|\mathscr{S}\cap S_3| \;:\; S_3\cap\mathscr{S}\neq\emptyset\right\}. \tag{3.2}$$

Obviously, it holds

$$1 \leq m \leq n \leq 1+q^2 . \tag{3.3}$$

We define character t_s $(s=0,1,\dots,1+q^2)$ of a spread $\mathscr{S}$ in PG(r,q) with respect to dimension 3, the number of subspaces S_3 of PG(r,q) containing exactly s lines of $\mathscr{S}$ (cf. [8]).

A spread $\mathscr{S}$ is said to be of class $[m_0,m_1,\dots,m_\ell]$ where $0\leq m_0<m_1<\dots<m_\ell\leq q^2+1$, with respect to the dimension 3, if $t_s=0$ for each $s\neq m_0,m_1,\dots,m_\ell$. If $\mathscr{S}$ is of class $[m_0,m_1,\dots,m_\ell]$ and moreover $t_s\neq 0$ $\forall\, s=m_0,m_1,\dots,m_\ell$, then $\mathscr{S}$ is said to be of type $(m_0,m_1,\dots,m_\ell)$ with respect to dimension 3.

If m=n in (3.3), necessarily $\mathscr{S}$ is of class [0,m] with respect to dimension 3. In PG(r,q) there are no spreads with a single character different from zero. Moreover, spreads of type (0,m) exist only if r=4 (cf. [8]). Consequently, if $r\geq 5$, then $m<n$ in (3.3).
The characters of a spread $\mathscr{S}$ satisfy the following equations, where $|\mathscr{S}|=k$ (cf. [8]):

$$\begin{cases} t_0 + \sum_{s=m}^{n} t_s = \gamma_{r,3} = \vartheta_r\,\vartheta_{r-1}\vartheta_{r-2}\vartheta_{r-3}/\vartheta_3\,\vartheta_2\,\vartheta_1 \\ \sum_{s=m}^{n} s t_s = k\,\gamma_{r-2,1} \\ \sum_{s=m}^{n} s(s-1)t_s = k(k-1) \end{cases} \tag{3.4}$$

Conditions $(3.4)_{III}$ and $(3.4)_{II}$ imply

$$\sum_{s=m}^{n} s^2\, t_s = k\left[k-1+\gamma_{r-2,1}\right] . \tag{3.5}$$

However we fix two integers M and N, it holds

$$\sum_{s=m}^{n} (N-s)(s-M)\, t_s = -\varphi \tag{3.6}$$

where

$$\varphi = \varphi(k,M,N,q,r,t_0) = k^2-k\left[1+(M+N-1)\gamma_{r-2,1}\right] + MN(\gamma_{r,3}-t_0) . \tag{3.7}$$

In the sequel, we suppose $N\geq n$, $M\leq m$. The conditions imply

$$\sum_{s=m}^{n} (N-s)\,(s-M)\, t_s \geq 0.$$

So, it follows

$$\varphi \leq 0. \tag{3.8}$$

Suppose $m<n$. As with (1.12) and (2.9), we have

$$\varphi=0 \quad\Leftrightarrow\quad N=n,\; M=m,\; t_{m+1}=\dots=t_{n-1}=0 \tag{3.9}$$

that is

$$(3.10)\qquad \varphi=0 \iff \begin{array}{l} N=n,\ M=m,\ \mathcal{S} \text{ of type } (0,m,n) \\ \text{if } t_o>0, \text{or } \mathcal{S} \text{ of type } (m,n) \\ \text{if } t_o=0, \text{ being } r=4,5 \text{ necessarily, [8].} \end{array}$$

If N=n and M=m, (3.6) and (3.8) imply:

$$(3.11)\qquad t_o \geqslant \{ k^2-k\,[1+(m+n-1)\gamma_{r-2,1}] + mn\gamma_{r,3}\} /mn,$$

where the equality holds if, and only if, (3.10) holds.

Since $\varphi\leq 0$, we have

$$(3.12)\qquad \Delta = [(M+N-1)\gamma_{r-2,1}+1]^2 -4MN(\gamma_{r,3} -t_o)\geq 0,$$

which implies

$$(3.13)\qquad t_o\geq \{4MN\gamma_{r,3} - [1+(M+N-1)\gamma_{r-2,1}]^2\}/4MN.$$

In particular (3.12) and (3.13) hold if M=m and N=n.

In (3.13) the equality holds if, and only if, $\Delta=0$. Since $\varphi\leq 0$, we have that:
equality holds in (3.13) if, and only if, $\varphi=0$ with

$$(3.14)\qquad k= [1+(m+n-1)\gamma_{r-2,1}]/2 \ ,$$

that is if, and only if, (3.10) and (3.14) hold.

If, in particular, we put N=n, M=m in (3.7), since $\varphi\leq 0$ and (3.12) or (3.13) are satisfied, we obtain the bounds

$$(3.15)\qquad k_1 \leq k \leq k_2 ,$$

where k_1 and k_2 are the solutions of $\varphi=0$. It results $k=k_1$ or $k=k_2$ in (3.15) if, and only if, (3.10) holds.

Moreover, we note that, if N=n, M=m, m<n, $r\geq 4$, (3.13) becomes

$$(3.16)\qquad t_o\geq \{4mn\gamma_{r,3} - [1+(m+n-1)\gamma_{r-2,1}]^2\} /4mn.$$

Condition (3.16) is significant, if its right-hand side is positive. It can be written (cf. (1.18)):

$$[D\,q^{4r-6}+\bar{P}(q)]\,/4mn\,\vartheta_1\vartheta_2\vartheta_3$$

where $\bar{P}(q)$ is a polynomial in q with a degree less than 4r-6 (which can be determined easily). If we put

$$(3.17)\qquad C(q)=Dq^{4r-6}+\bar{P}(q),$$

we obtain

$$(3.18)\qquad t_o \geq C(q)/4mn\vartheta_1\vartheta_2\vartheta_3 \ .$$

Since

$$C(q)= \vartheta_1\vartheta_2\vartheta_3\ \{4mn\gamma_{r,3} - [1+ [m+n-1]\,\gamma_{r-2,1}]^2\} ,$$

then $C(0)=-(n-m)^2< 0$. So, if $D=4mn-(m+n-1)^2 > 0$, the equation C(q)=0 has real positive solutions. Denote by $\bar{q}_o$ the greatest real positive solution of C(q)=0. Then C(q)>0 for each $q>\bar{q}_o$. We have $D=4mn-(m+n-1)^2 = [n-(\sqrt{m}-1)^2]\cdot$ $\cdot[n-(\sqrt{m}+1)^2]$. By this, since $n-(\sqrt{m}-1)^2 > 0$, we obtain

(3.19) $n < (\sqrt{m}+1)^2 \iff D > 0 \implies t_o < 0, \quad \forall \quad q > \bar{q}_o$.

By (3.19) it follows:

V. In PG(r,q) with $r \geqslant 4$, for each $q > \bar{q}_o$, there are no spreads with $t_o=0$ and $n < (\sqrt{m}+1)^2$.

4. BLOCKING SETS IN PG(r,q)

A subset S of PG(r,q) is called a blocking set with respect to lines, if it satisfies the following conditions.
(i) Each line of PG(r,q) intersects S.
(ii) S contains no lines of PG(r,q).
It follows immediately that:

VI. The complement of a blocking set is a blocking set.

VII. Let H be any subspace of PG(r,q). The set $H \cap S$ is a blocking set of H with respect to the lines of H.

A blocking set S is said to be irreducible if for any point $P \in S$ the set $S-\{P\}$ is not a blocking set, i.e. if at least one tangent passes through each point P of S. Clearly, each blocking set S contains some irreducible blocking set.

VIII. In PG(2,q) the complement of an irreducible blocking set is a reducible blocking set.

PROOF. Let S be an irreducible blocking set of PG(2,q), such that its complement S' is irreducible too. If $P \in S$, there exists at least one tangent t of S through P. Denote by Q_1 , Q_2 two points with $Q_1 \neq Q_2$ and Q_1 , $Q_2 \in t-\{P\}$. A tangent r_i to S' necessarily passes through Q_i (i=1,2), i.e. a q-secant of S. Moreover, $r_1 \neq r_2$ and $r_i \neq t$. Since S is irreducible, necessarily $S=(r_1-\{Q_1\}) \cup \cup (r_2-\{Q_2\}) \cup \{P\}$. (If we add a point, we have a contradiction). Then $q > 2$ and S'=PG(2,q)-S. Hence, S' is reducible, a contradiction.

Another proof of VIII is the following.
Let S and S'=PG(2,q)-S be irreducible blocking sets of PG(2,q). Through each point P of S there is at least one tangent to S and through each point P' of S' there is at least a q-secant S . Moreover, the lines passing through distinct points of these lines are distinct. Denote by t_s the character of index s of S (cf.n.1). Then $t_1+t_q \geqslant \vartheta_2$. Since $t_1+\dots+t_q=\vartheta_2$, we have $t_1+t_q=\vartheta_2$, $t_2=\dots=t_{q-1}=0$, that is S is of type (1,q). This is a contradiction, since sets of type (1,q) do not exist (cf.[7]).

Problem. In PG(r,q), $q \geqslant 3$, are there irreducible blocking sets whose complements are irreducible?

An example of irreducible blocking set in PG(2,q), with q square, is the Baer subplane. In fact, $PG(2,\sqrt{q})$ is a $(q+\sqrt{q}+1)$-set of type $(1,1+\sqrt{q})$. Other examples can be found in the proof of XII.

The following theorem of Bruen ([2],[3]) concerns the blocking sets of PG(2,q).

IX. Let S be a blocking set of PG(2,q). Then

(4.1) $$q+\sqrt{q}+1 \leqslant |S| \leqslant q^2-\sqrt{q} .$$

The equality holds on the left-hand side if, and only if, S is a Baer subplane (that on the right-hand side holds if, and only if, S is a complement of a

Baer subplane).

No blocking set exists in PG(2,2). It can be proved directly or by (4.1). From VII it follows

X. No blocking set exists in PG(r,2), $r \geq 2$.

Now we prove that

XI. In PG(2,3) there are exactly two blocking sets, up to isomorphisms. One of these has 6 points, the other has 7 points.

PROOF. Each blocking set of PG(2,3) has 6 or 7 points by (4.1). Since PG(2,3) has 13 points, if S is a blocking set with $|S|=6$, the complement S' of S is a blocking set with 7 points, and vice versa. The following example proves the existence of a blocking set S with $|S|=6$ (see also [1]). Consider the points (using homogeneous coordinates): $O_1(1,0,0)$, $O_2(0,1,0)$, $O_3(0,0,1)$, $U_1(0,1,1)$, $U_2(1,0,1)$, $U_3(1,1,0)$, $V(0,1,-1)$, $W(1,0,-1)$. It is very easy to verify that the set $\{O_3, U_1, U_2, U_3, V, W\}$ is an irreducible blocking set with 6 points.

Now we prove that any other blocking set with 6 points is projectively equivalent to the previous blocking set. It is sufficient to prove that each blocking set S with 6 points necessarily has two 3-secant lines, which meet at a point of S. Since S has the smallest cardinality, S is irreducible. Then through each point O_3 there is at least one tangent to S. Two tangents to S cannot pass through O_3, because the other 5 points of S should lie on the other two lines through O_3. This is a contradiction, because a line should be contained in S. Hence, every line through O_3, different from the tangent, contains at least a further point of S and at least one point of the complement of S. Then the remaining two points of S lie necessarily on two distinct lines through O_3, which are 3-secant of S. Let O_3 be (0,0,1). Assume the two points of the 3-secant and outside S as $O_1=(1,0,0)$ and $O_2=(0,1,0)$. Moreover, let $U_3=(1,1,0)$ be the point common to S and to a line through O_3, which is not a 3-secant of S. So we obtain exactly the blocking set of the previous example.

A different proof of this theorem can be found in Hirschfeld [5].

We recall the following:

THEOREM (Mazzocca-Tallini [6]). For any prime power q there exists an integer $b=b_p(q)$, which depends only on q such that in PG(r,q) no blocking set exists if, and only if, r is greater than b .

By X and XII it is:

$$b_p(2)=1 \quad , \quad b_p(3) \geq 2. \tag{4.2}$$

Now we prove the following

XII. In any PG(2,q) with $q \geq 3$, there are blocking sets.

PROOF. We give several examples of blocking sets for every $q \geq 3$.
a) Denote by O_1, O_2, O_3 three independent points of PG(2,q).
Define the set

$$S=\left\{ PG(2,q)- \{ O_1O_2 \cup O_2O_3 \cup O_3O_1 \} \right\} \cup \{ O_1, O_2, O_3 \}.$$

It is easy to verify that S is a blocking set with $|S|=q^2-2q+4$. The complement of S is an irreducible blocking set.
b) Let C be a conic, T a point of C and t the tangent on T to C. The set $S=C \cup t - \{T\}$, if q is odd, is an irreducible blocking set with $|S|=2q$. Set S is of type (1,2,3,q) with respect to the lines.

c) Suppose q odd. Let C be a conic of PG(2,q), denote by S the set of external points of C. C is of type ((q-1)/2, (q+1)/2,q). Then S is a blocking set with $|S| = q(q+1)/2$.
d) Suppose q square. In PG(2,q) a hermitian form H and a Baer subplane S are blocking sets with $|H| = q\sqrt{q}+1$ and $|S| = q+\sqrt{q}+1$. They are irreducible.
e) Let T be the point-set of the sides of a triangle in PG(2,q). Let us consider $T-\{V_1, V_2\}$, where V_1, V_2 are two of the three vertices of T, this is a blocking set with 3q-2 points.

From Prop. XII it follows that

$$(4.3) \qquad q \geq 3 \implies b_p(q) \geq 2.$$

Now we prove that

XIII. If $r \geq 3$, there is no blocking set in PG(r,3). Hence, we have

$$(4.4) \qquad b_p(3) = 2 \quad .$$

PROOF. By VII, it is sufficient to prove that no blocking set exists in PG(3,3). Suppose that S is a blocking set in PG(3,3). By VII, if π is a plane of PG(3,3), $\pi \cap S$ is a blocking set of π. Then S is of type (6,7) with respect to the planes, as a consequence of XI. Denote by t_6, t_7 the characters of S with respect to the planes. We have

$$(4.5) \qquad \begin{cases} t_6 + t_7 = \vartheta_3 = 40 \\ 6t_6 + 7t_7 = |S|\,\vartheta_2 = 13|S| \\ 30t_6 + 42t_7 = 4\,|S|(|S|-1) \end{cases}$$

By $(4.5)_{I}$ and $(4.5)_{II}$ we obtain

$$(4.6) \qquad t_6 = 280 - 13|S| \quad , \quad t_7 = 13|S| - 240 \quad .$$

If we substitute these in $(4.5)_{III}$, we have

$$(4.7) \qquad |S|^2 - 40|S| + 420 = 0 \quad .$$

So we obtain a contradiction, since (4.7) has no real solution. Consequently the assertion is proved.

We note that each blocking set S of PG(2,4) is such that $7 \leq |S| \leq 14$. Moreover the proof of XII implies the existence of a blocking set for each possible cardinality.

XIV. In PG(3,q), with q even and $q > 4$, there are blocking sets. Hence,

$$(4.8) \qquad q > 4, \quad q \text{ even} \implies b_p(q) \geq 3.$$

PROOF. The following is an example of a blocking set in PG(3,q) with q even and $q > 4$. Let $\mathscr{C}$ be a (q+2)-arc in a plane π and O_1, O_2, O_3 three of its points. Let O_4 be a point outside π. Denote by Γ the cone projecting $\mathscr{C}$ from O_4. Consider the subset K given by $\Gamma - (O_1O_4 \cup O_2O_4 \cup O_3O_4)$ and
a) the points of π outside lines O_1O_2, O_2O_3, O_3O_1 and in addition the points O_1, O_2, O_3;
b) the points of planes $O_1O_2O_4$, $O_2O_3O_4$, $O_3O_1O_4$ deprived of the points of the edges of the tetrahedron $O_1O_2O_3O_4$, excluded above. If π_i is the face of tetrahedron $O_1O_2O_3O_4$ opposite to O_i, K can be written as

$K:=[\Gamma\cup\pi_1\cup\pi_2\cup\pi_3\cup\pi_4-(O_1O_2\cup O_2O_3\cup O_3O_1\cup O_4O_1\cup O_4O_2\cup O_4O_3)]\cup\{O_1,O_2,O_3\}$.

Now we prove that K is a blocking set if $q>4$. We begin by observing that no line is contained in K. This is trivial for the lines of the faces of the tetrahedron $O_1O_2O_3O_4$ and for the lines of the cone. The other lines have at most six point in common with K. So no line is contained in K, since $q\geqslant 8$. Moreover, each line intersects K. This is obvious for the lines of the faces of the tetrahedron, for those of the cone and also for the lines which intersect the faces of the tetrahedron outside the edges. Now suppose that ℓ is a line through a point of the edge O_1O_2 and a point of the edge O_3O_4 (different from O_4). The plane spanned by O_3O_4 and ℓ contains a line ℓ' of Γ. Then $\ell'-\{O_4\}$ is contained in K, and so $\ell\cap\ell'\in K$. The same arguments hold for the pair of edges O_2O_3, O_1O_4 and O_3O_1, O_2O_4. So, the assertion follows.

XV. In PG(3,q), with q odd and $q\geqslant 7$, there are blocking sets. Hence,

$$q\geq 7\ ,\ q\ \text{odd}\implies b_p(q)\geq 3. \tag{4.9}$$

PROOF. We give an example of a blocking set in PG(3,q) with q odd and $q\geq 7$. Let π be a plane of PG(3,q), $\mathscr{C}$ a conic of π, O_1, O_2, O_3 three points on $\mathscr{C}$. Denote by t_1, t_2, t_3 the tangents to $\mathscr{C}$ on O_1, O_2, O_3 respectively. Put $T_1=t_2\cap t_3$, $T_2=t_1\cap t_3$, $T_3=t_1\cap t_2$. Let O_4 be a point outside π. Consider the set K given by:
a) the points of π outside the lines O_1O_2, O_2O_3, O_3O_1 but with the points O_1, O_2, O_3;
b) the points different from O_4 of the cone Γ projecting from O_4 the set $(\mathscr{C}\cup\{T_1, T_2, T_3\})-\{O_1, O_2, O_3\}$.
c) The points of the planes $O_1O_2O_4$, $O_1O_3O_4$, $O_2O_3O_4$ which are not on the edges of the tetrahedron $O_1O_2O_3O_4$, but with the points O_1, O_2, O_3.

The proof that K is a blocking set is similar to the proof of XIV.

5. BLOCKING SETS IN AG(r,q)

Let AG(r,q) be an affine Galois space of dimension r and order q. A subset S of AG(r,q) is called a blocking set with respect to the lines of AG(r,q) if it satisfies the following conditions:
(i) Each line of AG(r,q) intersects S.
(ii) No line of AG(r,q) is contained in S.

The Mazzocca-Tallini Theorem also holds in the affine case: for any q, there exists a positive integer $b_a(q)$ such that in AG(r,q) no blocking set exists if, and only if, r is greater than $b_a(q)$. (cf. [6]).

It is easy to prove that

XVI. Let S_{r-1} be a hyperplane of PG(r,q). If K is a blocking set of $AG(r,q)=PG(r,q)-S_{r-1}$ and K_o is a blocking set of S_{r-1}, then $K\cup K_o$ is a blocking set of PG(r,q).

Now we prove that

XVII. Let $b_p(q)$ be the greatest value of r for which there is a blocking set in PG(r,q) (cf. n. 4). Then

$$b_a(q)\leq b_p(q)\ . \tag{5.1}$$

PROOF. Suppose $b_a(q)>b_p(q)$. Consider a hyperplane $S(b_p(q))=PG(b_p(q),q)$ of $PG(b_p(q)+1,q)$. Let K' be a blocking set of $S(b_p(q))$. In $AG(b_p(q)+1,q)=PG(b_p(q)+1,q)-S(b_p(q))$ there is a blocking set K, since $b_p(q)+1\leq b_a(q)$. So, by XVI, $K\cup K'$ is a blocking set of $PG(b_p(q)+1,q)$. This is a contradiction, since

$b_p(q)+1 > b_p(q)$. Then the assertion is proved.

It is easy to prove that

XVIII. No blocking set exists in AG(2,q) with q=2,3. Hence

(5.2) $$b_a(2) = 1, \qquad b_a(3)=1.$$

Now we prove that

XIX. In AG(2,q), with $q \geq 5$, there are blocking sets.

PROOF. We prove the existence with the following example. Denote by (x,y) the affine coordinates in AG(2,q). Consider the set K whose points are on the lines of equations: x=0, y=0, x+y=1 respectively, except the points (0,0),(0,1), (1,0) and in addition the points (1,1), (-1,1), (1,-1). If q is odd, K is a (3q-3)-set, while if q is even, K is a (3q-5)-set. It is easy to prove that if $q>4$, K is a blocking set. If q=4, K is a 7-set, which is not a blocking set since, for example, it contains the line y=ix+i+1 (where i is such that $i^2+i+1=0$).

XX. In AG(2,q), with $q \geq 4$, there are blocking sets.

PROOF. We prove the existence with the following example. Denote by a,b two parallel lines. Let A,B be two points with $A \in a$, $B \in b$. Set

$$K := (a-\{A\}) \cup (b-\{B\}) \cup (AB-\{A,B\}).$$

K is a blocking set with 3q-4 points.

Propositions XIX and XX imply that

(5.3) $$b_a(q) \geq 2 \qquad \text{if } q \geq 4 .$$

XXI. Each blocking set of AG(2,4) has eight points.

PROOF. Let K be a blocking set of AG(2,4), then K is of class [1,2,3]. Necessarily $t_3 \neq 0$, otherwise K would be an arc, which always has external lines. Moreover, $t_1 \neq 0$ since the complement of K has also 3-secant lines (it is a blocking set too).

Put $|K| = k$. Then by (2.8) and (2.10) we have

$$k^2-16k+60 \leq 0 \quad \Longrightarrow \quad 6 \leq k \leq 10$$

and

$$k = 6 \text{ or } k = 10 \quad \Longleftrightarrow \quad K \text{ is of type } (1,3).$$

Moreover, we have (K' is the complement of K):

$$k = 6 \iff |K'| = 10 \quad (K \text{ of type } (1,3))$$

$$k = 7 \iff |K'| = 9 \quad (K \text{ of type } (1,2,3))$$

$$k = 8 \iff |K'| = 8 \quad (K \text{ of type } (1,2,3)) .$$

Hence, we can suppose k = 6,7,8; in fact, we obtain the other possible cardinalities by considering the complements. We have

(5.4) $$\begin{cases} t_1 + t_2 + t_3 = 20 \\ t_1 + 2t_2 + 3t_3 = 5k \\ t_2 + 3t_3 = k(k-1)/2. \end{cases}$$

If k=6, from (5.4) we have $t_3=5$, $t_2=0$, $t_1=15$. Let a,b be two lines which are 3-secant of K. Necessarily they meet at a point of K, otherwise the lines through their common point would be external to K. Then the points of K are those on two 3-secant a,b and another point P, which necessarily lies on line AB, where $\{A\} = a \setminus K$, $\{B\} = b \setminus K$. Set $C:=a \cap b$, then CP is a 2-secant of K, a contradiction (K is of type (1,3)). So K cannot have 6 points.

If k=7, from (5.4) we have $t_3=6$, $t_2=3$, $t_1=11$. Let a,b be two lines which are 3-secant of K. They meet necessarily at a point O of K. Set $\{A\} := a \setminus K$, $\{B\} := b \setminus K$, denote by a' the line through B and parallel to a, and by b' the line through A parallel to b. Finally put $C:=a' \cap b'$. Five points of K lie on $a \cup b$. Necessarily one of the remaining two points of K, say D, lies on AB, and the other lies on a' and b' namely it is point C. CD is different from the lines a', b' and from OC too.
(It is possible to choose O=(0,0), A=(1,0), B=(0,1) and C=(1,1), where (x,y) are affine coordinates in AG(2,4). It follows that AB and OC have equations x+y=1 and x-y=x+y=0, respectively). Then CD intersects a and b at two distinct points, which are different from A and B, namely at two points of K. So CD is contained in K, a contradiction. So, K cannot have 7 points.

Finally, if k=8, the existence of blocking sets is proved by the example given in the proof of XX.

From proposition XVI, it follows that

XXII. No blocking set exists in AG(r,4), with $r \geq 3$. Hence,

(5.5) $$b_a(4)=2$$

PROOF. Suppose that K is a blocking set of AG(r,4) with $r\geq 3$. Each plane α intersects K in a blocking set. Hence $|\alpha \cap K| = 8$, by XXI. Then K would have a unique character different from 0 with respect to the planes. Consequently, we have $K=\emptyset$ or K=AG(r,4) (cf. [7]), a contradiction.

Now we prove:

XXIII. Suppose that in AG(2,q) there exists a blocking set K such that every line has at least two points in K and two points outside K. Then a blocking set exists in AG(r+1,q). Consequently, each blocking set of $AG(b_a(q),q)$ is such that either $t_1 \neq 0$, or $t_{q-1} \neq 0$.

PROOF. Let π=AG(r,q) be a hyperplane of AG(r+1,q), and K a blocking set in π, having $t_1=0$ and $t_{q-1}=0$. Fix a direction δ, non-parallel to π. For each point $P \in \pi - K$ consider the line r_P through P, parallel to δ. Define the set

$$K' = K \cup \Big(\bigcup_{P\in\pi-K} (r_P - \{P\}) \Big).$$

We prove that K' is a blocking set.

Each line parallel to δ intersects either K' or its complement in at least one point. Let S be any line of AG(r+1,q) non-parallel to δ. Denote by α the plane through S parallel to δ. Obviously α is not parallel to π and intersects π in a line S'. Denote by $P_1, P_2, \ldots, P_t$ the points common to S' and $\bar{K}$ ($\bar{K}$ is the complement of K), denote by $Q_{t+1}, Q_{t+2}, \ldots, Q_q$ the points common to S' and K, with $2 \leq t \leq q-2$.

Then we have

$$K' \cap \alpha = \{Q_{t+1}, Q_{t+2}, \ldots, Q_q\} \cup \Big(\bigcup_{i=1}^{t} r_{P_i} \Big)$$

Since $2 \leq t \leq q-2$, $K' \cap \alpha$ is a blocking set of α, and so S ($\subset \alpha$) intersects either K' or its complement; namely K' is a blocking set of $AG(r+1,q)$. So, the assertion follows.

Now we construct an example of blocking set in $AG(2,q)$, where $q \geq 7$, with $t_1=0$, $t_{q-1}=0$.

Fix in $AG(2,q)$, four distinct lines a,b,c,d with $a \parallel c$, $b \parallel d$ and a non-parallel line to b. Set $A:=a\cap d$, $B:=a\cap b$, $C:=c\cap b$, $D:=c\cap d$, $E:=DB\cap AC$. Fix a line $b' \parallel b$ with $E \notin b'$, $b' \neq b$, $b' \neq d$. Finally set $F:=BD\cap b'$, $G:=AC\cap b'$. It is easy to verify that

$$K = a\cup b\cup c\cup d \cup \{E,F,G\} - \{A,B,C,D\}$$

is a blocking set with the required property.

By the previous proposition it follows that

$$(5.6) \qquad b_a(q) \geq 3 \qquad \text{if} \quad q \geq 7$$

Conditions (5.6) and (5.1) again imply (4.8), (4.9).

6. A CHARACTERIZATION OF HERMITIAN ARCS

In this section we characterize hermitian arcs. So, we generalize a result of Bruen and Thas which is about irreducible blocking sets (cf. [4]).

Let π_q be a projective plane of order q, desarguesian or not. Denote by S a s-set of π_q with $t_0=0$ and $t_1 \geq |S|$. (Each irreducible blocking set of π_q is an example of such a set). Consider the characters t_i of set S. Denote by n the maximum of $|r \cap S|$, where r is a line of π_q. Then the characters satisfy (1.7). If we multiply $(1.7)_I$ by n and subtract $(1.7)_{II}$ from it, we obtain

$$(6.1) \qquad \sum_{i=2}^{n-1} (n-i)t_i = n(q^2+q+1)-s(q+1)-(n-1)t_1 ,$$

where $|S| = s$.

The right-hand side of (6.1) is never less than 0; but it is 0 if, and only if, $t_2=\ldots=t_{n-2}=0$, namely if, and only if, S is of type (1,n). It follows that

$$(6.2) \qquad t_1 \leq [n(q^2+q+1)-s(q+1)] / (n-1),$$

where equality holds if, and only if, S is of type (1,n). By our assumption it is $t_1 \geq s$ and by (6.2) we have

$$(6.3) \qquad s \leq n(q^2+q+1)/(q+n),$$

where equality holds if, and only if, $s=t_1$ and equality holds in (6.2), then S is of type (1,n); in this case, from (1.7) we have that

$$t_n = q^2+q+1-s = sq/n = s(s-1)/n(n-1),$$

by which we obtain $s=q\sqrt{q}+1$, $n=\sqrt{q}+1$, namely S is a $(q\sqrt{q}+1)$-set of type $(1,1+\sqrt{q})$, i.e. S is a hermitian arc. Then we proved

XXIV. Let S a s-set of π_q without external lines and such that the number t_1 of 1-secant linesof S is greater or equal to the cardinality s of S. Denote by n the greatest number of points common to S and a line. Then it follows that

$$(6.4) \qquad s \leq n(q^2+q+1)/(q+n).$$

Equality holds if,and only if, S is a hermitian arc.

It is easy to prove that

(6.5) $n(q^2+q+1)/(q+n) \leq q\sqrt{q}+1 \iff n \leq \sqrt{q}+1$

From XXIV we obtain

XXV. Let S be a s-set of PG(2,q) without external lines and such that $s \leq t_1$ and $n \leq \sqrt{q}+1$. Then it is:

(6.6) $s \leq q\sqrt{q}+1$,

where equality holds if, and only if, S is a hermitian arc.

We conclude this paper with the following tables about the existence of blocking sets (y=yes).

PG(r,q)

r \ q	2	3	4	5	7	8	9,...,13	16
2	NO	y	y	y	y	y	y	y
3	NO	NO	?	y	y	y	y	y
4	NO	NO	?	?	?	?	?	y
5	NO	NO	?	?	?	?	?	?

AG(r,q)

r \ q	2	3	4	5	7	8	9,...,13	16
2	NO	NO	y	y	y	y	y	y
3	NO	NO	NO	?	y	y	y	y
4	NO	NO	NO	?	?	?	?	y
5	NO	NO	NO	?	?	?	?	?

REFERENCES

[1] A. Beutelspacher, Blocking sets and partial spreads in finite projective spaces, Math. Z., 145 (1975), 211-229.

[2] A.A.Bruen, Baer subplane and blocking set, Bull. Amer. Math. Soc., 76 (1970), 342-344.

[3] A.A.Bruen, Blocking sets in projective planes, Siam. J. Appl. Math. 3 (1971), 380-392.

[4] A.A.Bruen and J.A. Thas, Blocking sets, Geometriae Dedicata, 6 (1977), 193-203.

[5] J.W.P. Hirschfeld, Projective Geometries over Finite Fields, Clarendon Press, Oxford, (1979).

[6] F.Mazzocca and G.Tallini,On the non existence of blocking sets in PG(n,q) and AG(n,q) for all large enough n, Simon Stevin 1 (1985), 43-50.

[7] G.Tallini, Problemi e risultati sulle Geometrie di Galois, Relazione n. 30, Ist. Mat. Univ. Napoli 1973.

[8] G. Tallini, Fibrazioni in rette di PG(r,q), "Le Matematiche" Catania, 37 (1982), 8 - 27.

[9] G.Tallini, k-insiemi e blocking sets in PG(r,q) e in AG(r,q), Quaderno n.1, Sem. Geom. Comb. Ist. Mat. Applicata, Fac.Ingegneria L'Aquila (1982), 1-36.

[10] M.Tallini Scafati, Sui {k;n}-archi di un piano grafico finito, con particolare riguardo a quelli con due caratteri, nota I e II Rend. Acc. Naz. Lincei, 8,40 (1966) 812-818 and 1020-1025.

REFERENCE II

[11] L.M. Batten, Embedding the complement of a minimal blocking set in a projective plane, to appear in Discrete Math.

[12] L.Berardi, Irreducible blocking sets in a symmetric design, Boll. Un. Mat. Ital. 1-A (1987), 39-47.

[13] L.Berardi and A. Beutelspacher, Blocking sets of the known biplanes, to appear.

[14] L. Berardi, A.Beutelspacher and F. Eugeni, On (s,t;h)-blocking sets in Finite Projective and Affine Spaces, Atti Sem. Mat. Fis. Univ. Modena, 32 (1983), 130-157.

[15] L.Berardi and F.Eugeni, On the cardinality of blocking sets in PG(2,q), J. Geom. 22 (1984), 5-14.

[16] L. Berardi and F. Eugeni, On Blocking sets in affine planes, J.Geom. 22 (1984), 167-177.

[17] L. Berardi, F.Eugeni and O. Ferri, Sui blocking sets nei sistemi di Steiner, Boll. Un. Mat. Ital. Algebra e Geometria 1 (1984), 141-164.

[18] L.Berardi and F. Eugeni, Blocking sets in projective plane of order four, In these Proceedings.

[19] L. Berardi, Blocking sets in the large Mathieu design, I: the case S(3,6,22), In these Proceedings.

[20] L. Berardi, Blocking sets in the large Mathieu design, II: the case S(4,7,23), Preprint 1986.

[21] L. Berardi and F. Eugeni, Blocking sets in the large Mathieu design, III: the case S(5,8,24), Preprint 1986.

[22] A.Beutelspacher and F.Eugeni, On the type of partial t-spreads in finite projective spaces, Discrete Math. 54 (1985), 241-257.

[23] A. Beutelspacher and F. Eugeni, Sui blocking sets di dato indice con particolare riguardo all'indice tre, Boll. Un. Mat. Ital. 4-A (1985), 441-450.

[24] A.Beutelspacher and F. Eugeni, A lower bound for partial t-spreads of level t-1 in PG(2t+1,q), Atti Sem. Mat. Fis. Univ. Modena, 33 (1984),1-8.

[25] A. Beutelspacher and F. Eugeni, On n-fold blocking sets, Proceedings "Combinatorics 84" Bari, in Annals of Discrete Math. 30 (1986), 31-38.

[26] A. Beutelspacher and F. Eugeni, On blocking sets in projective and affine spaces of large order,Communicated in Oberwolfach, October 1985, Rend. Mat. (Roma), to appear.

[27] A. Beutelspacher and F. Mazzocca, Blocking sets in infinite projective and affine spaces, J. of Geometry, 28 (1987), 111-116.

[28] T.C. Brown, Monocromatic affine lines in finite vector space, J. Combinatorial Theory (A), 39 (1985), 35-41.

[29] P. Cameron and F. Mazzocca, Bijections which preserve blocking sets, to appear on Geometriae Dedicata.

[30] M.J. de Resmini, On blocking sets in symmetric BIBD's with $\lambda \geq 2$, J. Geometry 18 (1982), 194-198.

[31] M.J. de Resmini, On 2-blocking sets in projective planes, Ars Combinatoria 20-B (1985), 59-69.

[32] M.J. de Resmini, On 3-blocking sets in projective planes, Preprint 1987.

[33] D.A. Drake, Blocking sets in block designs, J. Combin. Theory A, 2 (1985), 459-462.

[34] F. Eugeni, Sulla esistenza di t-fibrazioni in PG(r,q) di fissato tipo, Boll. Un. Mat. Ital.Algebra e Geometria, 1 (1984), 19-43.

[35] F. Eugeni and E. Mayer, On blocking sets of index two, In these Proceedings.

[36] F. Eugeni and S. Innamorati, On fixed parity sets, In these Proceedings.

[37] F. Eugeni and S. Innamorati, Arcs and blocking sets in symmetric designs, Preprint 1987.

[38] M.Huber, A characterization of Baer cones in finite projective spaces, Geometriae Dedicata. To appear.

[39] F. Mazzocca, Some results on blocking sets, Announcements at Conference "Combinatorics 84" Bari (1984) and "Finite Geometries" Oberwolfach (1985).

[40] C. O'Keefe, A. Venezia, Blocking sets in AG(r,5), Sem Geom. Comb. Univ. Roma "La Sapienza" n. 56 (1983).

[41] S. Rajola, Un esempio di blocking set in PG(3,q) per ogni $q \geq 5$, In these Proceedings.

[42] G. Tallini, Blocking sets nei sistemi di Steiner e d-blocking sets in PG(r,q), Quaderno n. 3, Sem. Geom. Comb. Ist. Mat. Appl. Univ. L'Aquila (1983).

Annals of Discrete Mathematics 37 (1988) 451–458

SYMMETRIC DESIGNS WITHOUT OVALS AND EXTREMAL SELF-DUAL CODES

Vladimir D. Tonchev*

Institute of Mathematics, Sofia 1090, P. O. Box 373, Bulgaria

ABSTRACT. Constructions of doubly-even self-dual codes from symmetric designs are discussed. It is shown that the absence of ovals in a design is usually necessary, and sometimes even sufficient condition for the corresponding code to be extremal.

1. INTRODUCTION

The terminology and notations from design and coding theory used in this paper are in accordance with those from [3], [5], and [8] respectively.

A binary (n,k) code C is a k-dimensional subspace of the n-dimensional vector space V_n over $GF(2)$. Given an (n,k) code C, the $(n,n-k)$ code $C^{\perp} = \{x \in V_n : yx = 0 \text{ for each } y \in C\}$ is called the orthogonal, or dual of C. A matrix with the property that the linear span of its rows generates the code C, is a generator matrix of C. The generator matrices of the dual code $C^{\perp}$ are called parity check matrices of C. We shall often reffer to the elements of a code as codewords, or words only. The weight of a codeword is the number of its nonzero positions, and the minimum weight of a code is the weight of a lightest nonzero codeword. An (n,k,d) code is an (n,k) code with minimum weight d.

A code C is self-orthogonal (resp. self-dual) if $C \subseteq C^{\perp}$ (resp. $C = C^{\perp}$). The weights of all words in a self-orthogonal code are even. If in addition all weights are divisible by 4, the code is called doubly-even. A doubly-even self-dual $(n,n/2)$ code exists if and only if $n \equiv 0 \pmod 8$, and the minimum weight d of such a code is bounded by

$$d \leqq 4[n/24] + 4, \tag{1}$$

(cf. [8]). A code satisfying the equality in (1) is called extremal. The codewords of minimum weight in an extremal doubly-even self-dual code yield 1-, 3-, or 5-design provided that $n \equiv 16, 8, \text{ or } 0 \pmod{24}$.

The number of extremal codes is finite (but unknown), the smallest open case being for $n = 72$.

The following theorem describes a construction of doubly-even self-dual codes from symmetric designs.

THEOREM 1.1. Let A be an incidence matrix of a symmetric 2-(v,k,λ) design of an odd order $k-\lambda$. Then

(i) If $k \equiv 3 \pmod 4$, the code generated by the matrix

*Research partially supported by the CSCMB under contract 37/1987.

$$(I,A) \tag{2}$$

is a doubly-even self-dual (2v,v) code.

(ii) If $k \equiv 2 \pmod 4$, then the code generated by the matrix

$$\left|\begin{array}{ccc} & & 1\ldots10 \\ & & 1 \\ I & A & \vdots \\ & & 1 \end{array}\right| \tag{3}$$

is a doubly-even self-dual (2v+2, v+1) code.

This statement is a variant of a more general construction (cf. [2],[4]).

In this paper we are interested in necessary and sufficient conditions for a code constructed in this way to be extremal. In other words, what combinatorial properties should a design possess in order to produce an extremal code?

A notion playing an important role in the study of symmetric designs, especially projective planes and biplanes, is that of "arc" or "oval". An arc in a design is a subset S of points such that each block intersects S in at most 2 points. The cardinality of an arc S in a nontrivial symmetric 2-(v,k,λ) design of an odd order $k-\lambda$ is bounded by

$$|S| \leqq (k+\lambda-1)/\lambda. \tag{4}$$

An arc with maximum number of points in the sense of bound (4) is called an oval [2].

It will be seen by the next exposition that the absence of ovals in a symmetric design is usually necessary, and sometimes even sufficient condition for the corresponding self-dual code to be extremal.

2. ARCS AND CODES

Let D be a symmetric 2-(v,k,λ) design with an arc S of size s, i.e. $|S| = s$. Denoting by n_i $(i = 0,1,2)$ the number of blocks having exactly i common points with S, we have:

$$\begin{aligned} n_0 + n_1 + n_2 &= v, \\ n_1 + 2n_2 &= sk, \\ n_2 &= \binom{s}{2}\lambda. \end{aligned}$$

Consider the code with generator matrix (2) defined by D. Clearly, the weight of the sum (over GF(2)) of s rows of the generator matrix (2) indexed by the points of an arc of size s is $s+n_1 = s(k+1)-(s-1)\lambda)$. Similarly, the sum of s rows of the matrix (3), indexed by the points of an arc of size s has weight

$$s+n_1+(1+(-1)^{n+1})/2 = s(k+1-(s-1)\lambda)+(1+(-1)^{s+1})/2.$$

Thus we have the following

THEOREM 2.1. Let D be a symmetric 2-(v,k,λ) design satisfying the assumption of Theorem 1.1, and admitting an arc S of size s. Then the minimum weight of the code defined by D as in Theorem 1.1 is bounded as follows:

$$d \leq \begin{cases} s(k+1-(s-1)\lambda & \text{in case (i);} \\ s(k+1-(s-1)\lambda)+(1+(-1)^{s+1})/2 & \text{in case (ii).} \end{cases} \tag{6}$$

It turns out that if the arc S in an oval, the inequalities (6) are often sharper than (1). Thus a code arising from a design with ovals is usually not extremal. We shall illustrate this by some examples.

In Table 1 the parameters of symmetric 2-(v,k,λ) designs with $k-\lambda \leq 9$, yielding doubly-even codes by the construction of Theorem 1.1. are listed.

TABLE 1

No.	k-λ	v	k	λ	Codes
1	1	3	2	1	Extended Hamming (8,4,4) code; extremal.
2	3	11	6	3	Extended Golay (24,12,8) code; extremal.
3	5	19	10	5	Three extremal (40,20,8) codes.
4	7	31	10	3	Extremal (64,32,12) code from any design without ovals (Theorem 2.5).
5	7	27	14	7	An extremal (56,28,12) code from the quadratic residue design.
6	9	35	18	9	Is there any extremal (72,36,16) code?
7	9	36	15	6	
8	9	40	27	18	A (80,40,d ≥ 12) code from any design without ovals.

Parameters No. 1,2,3,5,6 in Table 1 are of Hadamard type, i.e. of the form 2-(4t-1,2t,t). The size of an oval in a Hadamard 2-(4t-1,2t,t) (or 2-(4t-1, 2t-1,t-1)) design is 3, thus the bound from Theorem 2.1 for the minimum weigth of a code of type (ii) is 4. However, a Hadamard 2-(4t-1,2t,t) design with odd $t > 1$ does not admit any ovals; this follows from a result of Morgan [9], and can be also easily seen directly. For, if a 2-(4t-1,2t,t) design possesses an arc of size 3, then the sum of three rows of its incidence matrix A indexed by the points of the arc, as well as the sum of all remaining rows over GF(2) is the zero vector. Consequently, the rank of the incidence matrix over GF(2) must be less or equal to 4t-3. On the other hand, since t is odd, the matrix (I,J-A), where J denotes the all-one matrix, generates a self-dual code. Thus the rank of J-A over GF(2) is 4t-1, whence the rank of A is 4t-2, a contradiction.

Estimating the weight of a sum of at most 4 rows of a matrix of the form (3), where A is an incidence matrix of a Hadamard 2-(4t-1,2t,t) design with odd t, say $t = 2m+1$, the following can be proved:

THEOREM 2.2. [13]. The minimum weight of a (16m+8,8m+4) code with generator matrix (3) defined by a Hadamard 2-(8m+3,4m+2,2m+1) design is equal to 4 if $m = 0$, and is at least 8 if $m > 0$.

In fact, the codes derived from Hadamard 2-(8m+3,4m+2,2m+1) designs for $m \leqq 2$ are all extremal. It is worth noting that Hadamard 2-(4t-1,2t,t) designs which are extendable into isomorphic Hadamard 3-(4t,2t,t-1) designs, yield equivalent codes. For instance, there exist exactly 6 nonisomorphic 2-(19,10,5) designs and 3 nonisomorphic 3-(20,10,4) designs, producing 3 inequivalent extremal (40,20,8) codes [13].

The absence of ovals in a 2-(8m+3,4m+2,2m+1) design with $m > 2$ is not a sufficient condition for the extremality of the related code. In this case some additional conditions are to be fulfiled, including the absence of ovals in the complementary 2-(8m+3,4m+1,2m) design as well.

THEOREM 2.3. [13]. A necessary condition for an (16+8,8m+4) code defined by a Hadamard 2-(8m+3,4m+2,2m+1) design D to have minimum weight $d \geqq 12$ is any triple of points of D to be contained in at least 2 blocks, and at most 2m-1 blocks of D.

Formulated for the complementary 2-(8m+3,4m+2,2m) design $\overline{D}$, this condition states that each triple of points must occur in at most 2m-2 blocks, and at least one block of $\overline{D}$. In particular, $\overline{D}$ cannot have any ovals.

The enumeration of all 2-(27,14,7) designs up to isomorphism is not yet completed. However, the only primes which can divide the order of the automorphism group of a Hadamard matrix of order 28 are 13, 7, 3 and 2, and the matrices possessing automorphisms of order 13 or 7 are already known [14], [15]. Among the designs arising from Hadamard matrices of order 28 with automorphisms of order 13 or 7, only those related to the Hadamard matrix of quadratic residue type yield an extremal (56,28,12) code.

Any 2-(35,18,9) design defines a doubly-even (72,36) code with minimum weight $d \geqq 8$. Although the existence of an extremal (72,36,16) code is still in doubt, it is known that such a code cannot be obtained from a 2-(35,18,9) design with an automorphism of order 17 [16]. More generally, such a code cannot possess automorphisms of order 17 or any larger prime order [6], [11], [12].

Another approach for construction of doubly-even (72,36) codes is by means of symmetric 2-(36,15,6) designs (see Theorem 1.1, (i)). In this case Theorem 2.1 gives the following result:

THEOREM 2.4. A necessary condition for a 2-(36,15,6) design to yield a $(72,36,d \geqq 12)$ code is the absence of any arcs of size 3.

The next theorem is an example of a better-behaved situation.

THEOREM 2.5. A doubly-even self-dual (64,32) code defined by a 2-(31,10,3) design D is extremal (i.e. has minimum weight 12) if and only if D does not possess any ovals.

Proof. The first row of the generator matrix (3), where A is now an incidence matrix of a symmetric 2-(31,10,3) design D, has weight 32, while all remaining rows are of weight 12. The weight of a sum of two rows of (3) is 24 provided that one of the rows is the first row of (3), or 16 if both rows are other than the first row of (3). The sum of 3 rows of (3) including the first row is 20. Consider now the sum of a triple of rows not including the first row. Let n_i denote the number of blocks containing exactly i points of the triple of points of D corresponding to the choosen triple of rows. We have:

$$\begin{aligned} n_0 + n_1 + n_2 + n_3 &= 31, \\ n_1 + 2n_2 + 3n_3 &= 3.10, \\ n_2 + 3n_3 &= 3.3. \end{aligned}$$

This system has the following solutions:

n_0	n_1	n_2	n_3
10	12	9	0
9	15	6	1
8	18	3	2
7	21	0	3

Evidently, the weight of the sum of the considered triple of rows is $4+n_1+n_3 \geqq 16$.

Furthermore, the weight of a sum of 4 rows of (3) including the first row is $4+n_0+n_2+1 \geqq 12$.

Now we estimate the weight of the sum of a quadruple of rows of (3) not including the first row. Denoting by n_i $(0 \leqq i \leqq 4)$ the number of blocks of D containing exactly i points from the quadruple of points corresponding to the choosen quadruple of rows, one has:

$$\begin{aligned} n_0 + n_1 + n_2 + n_3 + n_4 &= 31, \\ n_1 + 2n_2 + 3n_3 + 4n_4 &= 4.10, \\ n_2 + 3n_3 + 6n_4 &= 6.3, \end{aligned}$$

whence $n_1 = 3n_3 + 8n_4 + 4$.

The weight of the considered sum is $4+n_1+n_3 = 8+4n_3+8n_4$. By our assumption D does not possess any ovals, hence $n_3+n_4 > 0$, and consequently the weight of the sum of four rows is at least 12.

Thus the weight of any linear combination of at most 4 rows of the generator matrix (3) is at least 12.

A symmetric design of an odd order $k-\lambda$ contains an oval if and only if the dual design contains an oval [2]. The following matrix

$$\left|\begin{array}{ll} \begin{matrix}1\\ \vdots \\ 1\end{matrix}\; A^t & I \\ 01\ldots1 & \end{array}\right| \qquad (7)$$

is a parity check matrix of the code generated by (3), and since the code is self-dual, the matrix (7) is also a generator matrix of the same code. Since A^t is an incidence matrix of the dual design of D, we can apply the same arguments for the weight of the sum of at most 4 rows of (7). Since a codeword of weight 8 or less must be sum of at most 4 rows of one of the matrices (3) or (7), this completes the proof.

The 2-(31,10,3) design listed in Hall book [7] possesses ovals. Since there are no cyclic difference sets with parameters (31,10,3), it can be easily seen by use a result of Aschbacher [1] that the greatest prime which can be an order of an automorphism of a 2-(31,10,3) design is 7. We enumerated all such designs finding exactly 4 nonisomorphic solutions, one of them being without ovals [17]. Assume that an automorphism ß of order 7 acts on the points and blocks as follows:

ß = (1,2,...,7)(8,9,...,14)(15,16,...,21)(22,23,...,28)(29)(30)(31).

Then a 2-(31,10,3) design without ovals is defined by the following base blocks:

B_1 = (1,8,13,14,15,18,21,22,25,27), B_{22} = (1,4,6,10,14,19,20,22,28,31),
B_8 = (1,6,7,8,11,16,18,24,28,29),
B_{15} = (1,4,7,9,11,15,20,26,27,30),
B_{29} = (8,9,10,11,12,13,14,29,30,31),
B_{30} = (15,16,17,18,19,20,21,29,30,31),
B_{31} = (22,23,24,25,26,27,28,29,30,31).

The full automorphism group of this design is of order 42. We do not know whether the code obtained from the above design is equivalent to the extremal (64, 32) code constructed by Pasquier [10].

Finally, let us consider the parameters 2-(40,27,18). By arguments similar to those from the proof of the preceding theorem the following proposition can be proved.

THEOREM 2.6. The minimum weight d of a doubly-even (80,40) code obtained from a 2-(40,27,18) design D is 8 if and only if the complementary 2-(40,13,4) design D possesses an oval. Otherwise, $d \geq 12$.

Let us remark that for extremality one needs $d = 16$. As a first candidate, we checked the complement of the 2-(40,13,4) design formed by the hyperplanes in PG(3,3). However, it turns out that the resulting code has minimum weight $d = 12$.

REFERENCES

[1] M. Aschbacher, On collineation groups of symmetric block designs, J. Combin. Theory, A 11 (1971), 272-281.
[2] E.F. Assmus, Jr., and J.H. van Lint, Ovals in projective designs, J. Combin. Theory, A 27 (1979), 307-324.
[3] Th. Beth, D. Jungnickel, H. Lenz, "Design Theory", B.I. Wissenschaftsverlag,

Zürich 1985.

[4] V.K. Bhargava, J.M. Stein, (v,k,λ) configurations and self-dual codes, Information and Control, 28 (1975), 352-355.

[5] P.J. Cameron and J.H. van Lint, "Graphs, Codes and Designs", London Math. Soc. Lecture Note Ser. 43, Cambridge University Press, Cambridge 1980.

[6] J.H. Conway, V. Pless, On primes dividing the group order of a doubly-even (72,36,16) code and the group order of a quaternary (24,12,10) code, Discrete Math. 38 (1982), 143-156.

[7] M. Hall, Jr., "Combinatorial Theory", Ginn (blaisdell), Boston 1967.

[8] F.J. MacWilliams and N.J.A. Sloane, "The Theory of Error-Correcting Codes", Noth-Holland, Amsterdam 1977.

[9] E.J. Morgan, Arcs in block designs, Ars Combinatoria 4 (1977), 3-16.

[10] G. Pasquier, A binary extremal doubly even self-dual code (64,32,12) obtained from an extended Reed-Solomon code over F_{16}, IEEE Trans, Inform. Theory, 27 (1981), 807-808.

[11] V. Pless, 23 does not divide the order of the group of a (72,36,16) doubly even code, IEEE Trans. Inform. Theory, 28 (1982), 113-117.

[12] V. Pless, J.G. Thompson, 17 does not divide the order of the group of a (72,36,16) doubly even code, IEEE Trans, Inform. Theory, 28 (1982), 537-541.

[13] V.D. Tonchev, Block designs of Hadamard type and self-dual codes, Problemi peredatchi informatsii, 19 (1983), No. 4, 25-30.

[14] V.D. Tonchev, Hadamard matrices of order 28 with automorphisms of order 13, J. Combin. Theory, A 35 (1983), 43-57.

[15] V.D. Tonchev, Hadamard matrices of order 28 with automorphisms of order 7, J. Combin. Theory, A 40 (1985), 62-81.

[16] V.D. Tonchev, R.V. Raev, Cyclic 2-(17,8,7) designs and related doubly-even codes, Compt. rend. Acad. bulg. Sci., 35 (1982), 1367-1370.

[17] V.D. Tonchev, Symmetric 2-(31,10,3) designs with automorphisms of order 7, Annals of Discrete Math. (to appear).

Annals of Discrete Mathematics 37 (1988) 459–468

GROUPS IN HYPERGROUPS

Thomas VOUGIOUKLIS

Democritus University of Thrace, 67100 Xanthi, Greece.

We study the fundamental relation introduced by Koskas as the transitive closure of a basic relation in a given hypergroup. We use the quotient set, which is a group, in order to define a semi-direct hyperproduct of two hypergroups. We obtain an extension of hypergroups by hypergroups. Moreover we reveal some hypergroups in a given hypergroup.

1. INTRODUCTION

The class of hypergroups in the sense of Marty [8] is the largest class of multivalued systems that satisfies group-like axioms: $<H,\cdot>$ is a hypergroup if $\cdot:H\times H \longrightarrow \mathcal{P}(H)$ is a hyperoperation such that the associative law and the reproduction axiom are satisfied; i.e. $x(yz)=(xy)z$ and $xH=Hx=H$, $\forall x,y,z \in H$. Applications of hypergroups have mainly appeared in special subclasses. For example, canonical hypergroups [10] are used as an organizing device for characters of finite groups; cogroups and polygroups, especially the chromatic ones, are used to study color algebras [1], [2]; join spaces [11] are used to study classical geometries. Generalizations of certain above subclasses, as the feebly canonical hypergroups [4], or closely related subclasses, as the reversible hypergroups [9], have been also introduced and investigated.

The purpose of this paper is, first to study the fundamental group of a hypergroup, second to introduce the fundamental hypergroups closely related to the fundamental group, and third to define a semidirect product of hypergroups in order to have an extension of a hypergroup by a hypergroup.

2. THE FUNDAMENTAL GROUP

DEFINITION 1

Let $<H,\cdot>$ be a hypergroup. In H the *fundamental equivalcnce relation* $\beta^\star_H$ is defined as follows:

$a\beta^\star_H b$ if we can find $x_1,x_2,\ldots,x_{\varkappa-1}, h_{i_1}, h_{i_2},\ldots,h_{i_\varkappa} \in H$

and $I_1,\ldots,I_\varkappa$ finite sets of indices, such that

$$\{a,x_1\} \subset \prod_{i_1\in I_1} h_{i_1},\quad \{x_1,x_2\} \subset \prod_{i_2\in I_2} h_{i_2},\ldots,\{x_{\varkappa-1},b\} \subset \prod_{i_\varkappa\in I_\varkappa} h_{i_\varkappa}.$$

The fundamental relation introduced by Koskas in [7,p.167] was defined as the transitive closure of the relation β_H defined by setting

$x\beta_H y \Longleftrightarrow \exists x_1,\ldots,x_n \in H$:

$$\{x,y\} \subset \prod_{i=1}^{n} x_i .$$

The fundamental relation β^*_H, which is a two-sided strong regular relation, has been studied in several papers such as [3], [5], [6] and many interesting results have been obtained. The most useful result remains that H/β^*_H is a group; that is why β^*_H is called the fundamental relation. We shall call H/β^*_H the *fundamental group* of H.

Let us denote by $F_H(x)$, $x \in H$ the elements of H/β^*_H. The product of classes has the simple form $F_H(a)\cdot F_H(b) = F_H(ab) = F_H(x)$, $\forall x \in ab$, (see [3]), instead of the usual form:

$$F_H(a)\cdot F_H(b) = \{F_H(x) \mid x \in a'b', \ a'\beta^*_H a, \ b'\beta^*_H b\}$$

it is called the *fundamental property*.

An element x of H will be called a *fundamental element* iff $F_H(x) = \{x\}$; i.e for any $a,b \in H$ we have $ab = \{x\}$ or $x \notin ab$. If H is a group then all elements are fundamental; therefore in this case the fundamental group coincides with H. The fundamental elements are the unique images of the same pairs for every fundamental map needed for the definition of a hypergroup algebra; see [13].

All unit elements of H, if they exist , belong to the identity of H/β^*_H because if, for example, e is a left unit in H then $x \in ex$, $\forall x \in H$, and so $F_H(e)\cdot F_H(x) = F_H(ex) = F_H(x)$. The kernel of the canonical map $\varphi : H \longrightarrow H/\beta^*_H$ is called the *nucleus* of H and denoted by w_H.

Remark 2.1 If h is a subhypergroup of H then

$$F_h(x) \subset F_H(x), \quad \forall x \in h.$$

Let h be a subhypergroup of H; then we do not necessarily have the relation $|h/\beta^*_h| \leqslant |H/\beta^*_H|$. Example: In [4] a generalization of the canonical hypergroups, the feebly canonical hypergroups, have been introduced and proved to be join spaces [11]. An example of those is the following one [4]: Let G be an abelian group and S a set. One supposes that $|G| > 1 < |S|$ and $G \cap S = \emptyset$. In $G \cup S$ a hyperproduct $(\cdot)$ was defined as follows:

$\forall (g,s) \in G\text{x}S$, $g\cdot s = s\cdot g = S$; $\forall (g_1,g_2) \in G^2$, $g_1 \cdot g_2 = g_1 g_2$;

$\forall (s_1,s_2) \in S^2$, $s_1 \cdot s_2 = G$. Then $\langle G \cup S, \cdot \rangle$ becomes a feebly canonical hypergroup. For $|G| > 2$ we obtain

$$|G/\beta^*_G| > |(G \cup S)/\beta^*_{G\cup S}| = 2 .$$

PROPOSITION 1
Let $\langle H, \cdot \rangle$ be a hypergroup and $h_1,\dots,h_s,k_1,\dots,k_s \in H$ such that $h_j \beta^*_H k_j$, $j \in \{1,\dots,s\}$. Then, for every $c_1 \in h_1 \dots h_s$ and $c_2 \in k_1 \dots k_s$, we have $c_1 \beta^*_H c_2$.

Proof
It is enough to prove this proposition for $h_j = k_j, j \in \{2,\dots,s\}$.

$$h_1 \beta^*_H k_1 \Longleftrightarrow \{h_1, z_1\} \subset \prod_{i_1 \in I_1} \mu_{i_1}, \dots, \{z_{\lambda-1}, k_1\} \subset \prod_{i_\lambda \in I_\lambda} \mu_{i_\lambda}.$$

Therefore

$$(h_1h_2\dots h_s)\cup(z_1h_2\dots h_s)\subset \prod_{i_1\in I_1}\mu_{i_1}\,h_2\dots h_s,\dots,$$

$$(z_{\lambda-1}h_2\dots h_s)\cup(k_1h_2\dots h_s)\subset \prod_{i_\lambda\in I_\lambda}\mu_{i_\lambda}\,h_2\dots h_s$$

so if we choose $y_r\in z_rh_2\dots h_s$, $r\in\{1,\dots,\lambda-1\}$ we obtain that for every $c_1\in h_1h_2\dots h_s$ and $c_2\in k_1h_2\dots h_s$ we have $c_1\beta^\star_H c_2$.

Let us now consider the P-hypergroups introduced in [12] and generalized in [14], [15]. They are defined as follows. Given a group $(G,\cdot)$ and a non-empty subset P of G then G becomes a hypergroup with hyperoperation $P^\star$ such that $xP^\star y=xPy$; we denote this by $\langle G,P^\star\rangle$.

THEOREM 1
Let Z be the center of the group G ; then

$$\langle G,Z^\star\rangle/\beta^\star_G\cong G/Z.$$

Proof
Let us denote by $\star\Pi$ the hyperproduct in $\langle G,Z^\star\rangle$ and $\cdot\Pi$ the product in $(G,\cdot)$. Then

$a\beta^\star_G b \iff$ there exist x_i,h_{i_j} such that

$$\{a,x_1\}\subset \star\prod_{i_1\in I_1}h_{i_1},\dots,\{x_{\varkappa-1},b\}\subset \star\prod_{i_\varkappa\in I_\varkappa}h_{i_\varkappa}\iff$$

$$\{a,x_1\}\subset \cdot\prod_{i_1\in I_1}h_{i_1}Z\quad,\dots,\{x_{\varkappa-1},b\}\subset \cdot\prod_{i_\varkappa\in I_\varkappa}h_{i_\varkappa}Z\iff$$

$$a=h_{a,x_1}z_a,\ x_1=h_{a,x_1}z_{x_1},\dots,x_{\varkappa-1}=h_{x_{\varkappa-1},b}z_{x_{\varkappa-1},b}=h_{x_{\varkappa-1},b}z_b\iff$$

$$h_{a,x_1}=az_a^{-1}=x_1z_{x_1}^{-1},\dots,h_{x_{\varkappa-1},b}=x_{\varkappa-1}z_{x_{\varkappa-1}}^{-1}=bz_b^{-1}\iff$$

$$a=x_1z_{a,x_1},\dots,x_{\varkappa-1}=bz_{x_{\varkappa-1},b}\iff a=bz_{x_{\varkappa-1},b}\dots z_{a,x_1}\iff a=bz'\iff$$

$a\equiv b \bmod Z$.

A polygroup [2] is a system $\mathcal{M}=\langle M,\cdot,e,{}^{-1}\rangle$ where $e\in M$, ${}^{-1}$ is a unitary operation on M, $\cdot$ maps M^2 into nonempty subsets of M, and the following axioms hold for all x,y,z in M:

$$(x\cdot y)\cdot z=x\cdot(y\cdot z)\ ,\qquad e\cdot x=x=x\cdot e\ ,\qquad \text{and}$$

$$x\in y\cdot z\quad\text{implies}\quad y\in x\cdot z^{-1}\quad\text{and}\quad z\in y^{-1}x.$$

In [1] an extension of polygroups by polygroups have been introduced in the following way:
Suppose $\mathcal{U}$ and $\mathcal{B}$ are polygroups whose elements have been renamed so that $A\cap B=\{e\}$ where e is the identity of both $\mathcal{U}$ and $\mathcal{B}$. A new system

$\mathcal{U}[\mathcal{B}] = \langle M,\star,e,^{I}\rangle$, called the extension of $\mathcal{U}$ by $\mathcal{B}$, is formed in the following way: Set $M = \{x \in A : x \neq e\} \cup \{x \in B : x \neq e\} \cup \{e\}$ and let $e^{I} = e$, $x^{I} = x^{-1}$, $e\star x = x\star e = x$ for all $x \in M$, and for all $x,y \in \{x \in M : x \neq e\}$,

$$x\star y = \begin{cases} x\cdot y & \text{if} \quad x,y \in A \\ x & \text{if} \quad x \in B,\ y \in A \\ y & \text{if} \quad x \in A,\ y \in B \\ x\cdot y & \text{if} \quad x,y \in B,\ \ y \neq x^{-1} \\ x\cdot y \cup A & \text{if} \quad x,y \in B \ \text{ and } \ y = x^{-1} \end{cases}$$

The extension $\mathcal{U}[\mathcal{B}]$ is a polygroup which preserves being chromatic.

THEOREM 2

$$\mathcal{U}[\mathcal{B}]/\beta^{\star}_{\mathcal{U}[\mathcal{B}]} \cong \mathcal{B}/\beta^{\star}$$

Proof
We have by definition: $x\star x^{-1} = x\cdot x^{-1} \cup A$ therefore all the elements of A belong to the fundamental class $F_{\mathcal{U}[\mathcal{B}]}(e)$. Moreover for the elements of B we have $b\beta^{\star}_{\mathcal{U}[\mathcal{B}]}b_1$ iff $b\beta^{\star}_{\mathcal{B}}b_1$;

that is because two elements of B can be contained only in the hyperproducts of elements of B.
Therefore the extension $\mathcal{U}[\mathcal{B}]$ has in the unit fundamental class all the elements of A and the other classes have exactly the elements of $\mathcal{B}/\beta^{\star}_{\mathcal{B}}$.

3. THE FUNDAMENTAL HYPERGROUPS

DEFINITION 2
Let $\langle H,\cdot\rangle$ be a hypergroup. We define the L^{x}_{H} - *fundamental relation* as follows:

$aL^{x}_{H}b$ if we can find $z_1,z_2,\ldots,z_{\varkappa-1},h_1,h_2,\ldots,h_{\varkappa} \in H$ such that

$\{a,z_1\} \subset xh_1$, $\{z_1,z_2\} \subset xh_2$, ... , $\{z_{\varkappa-1},b\} \subset xh_{\varkappa}$.

Then L^{x}_{H} is an equivalence relation for every x of H. One can also define the R^{x}_{H} - *fundamental relation* using right multiplication by x. Let us denote by $L^{x}_{H}(a)$ the equivalence class of a, then the hyperproduct in the quotient set H/L^{x}_{H} is given in the same way as in any equivalence relation [10]; i.e.

$$L^{x}_{H}(a)\cdot L^{x}_{H}(b) = \{L^{x}_{H}(c) \mid c \in a'b' \quad \text{where} \quad a'L^{x}_{H}a, \quad b'L^{x}_{H}b\}.$$

The quotient H/L^{x}_{H} becomes a hypergroup which we shall call the *fundamental hypergroup* of H corresponding to the element x.

Remark 3.1

$$L^{x}_{H}(h) \subset F_{H}(h), \qquad \forall x,h \in H.$$

Remark 3.2
Given a reproductive generalized permutation

$$f:H \longrightarrow \mathscr{P}(H), \quad \text{with} \quad \bigcup_{h\in H} f(h) = H, \quad \text{see} \quad [13],$$

we obtain a similar definition as for the fundamental hypergroups by setting $a f b$ if we can find $z_1,\ldots,z_{\varkappa-1},\ h_1,\ldots,h_\varkappa \in H$ such that $\{a,z_1\} \subset f(h_1),\ldots,\{z_{\varkappa-1},b\} \subset f(h_\varkappa)$.

In this sense the L_H^x - fundamental relations correspond to the inner generalized permutations of H.

Remark 3.3
For every left scalar element x of H we have $H/L_H^x \cong H$.

THEOREM 3
For every hypergroup H and every x of H we have

$$H/\beta_H^\star \cong (H/L_H^x)/\beta^\star_{H/L_H^x}\ .$$

Proof
We consider the mapping

$$s:H/\beta_H^\star \longrightarrow (H/L_H^x)/\beta^\star_{H/L_H^x} \ :\ F_H(h) \longmapsto F_{H/L_H^x}\,(L_H^x(h))$$

We have for $h_1,h_2 \in H$ and for every $z \in h_1h_2$

$$s(F_H(h_1)\cdot F_H(h_2)) = s(F_H(h_1h_2)) = s(F_H(z)) = F_{H/L_H^x}\,(L_H^x(z)).$$

On the other hand we have

$$F_{H/L_H^x}\,(L_H^x(h_1))\cdot F_{H/L_H^x}\,(L_H^x(h_2)) = F_{H/L_H^x}\,(L_H^x(h_1)\cdot L_H^x(h_2)) = F_{H/L_H^x}\,(Z),$$

$\forall Z \in L_H^x\,(h_1)\cdot L_H^x\,(h_2)$.

But the above Z's have the form $Z = L_H^x\,(c)$ where $c \in h_1'h_2'$ with $h_1' L_H^x\, h_1$, $h_2' L_H^x\, h_2$ and some c is the z in h_1h_2 chosen above. Therefore s is a homomorphism.

Now let us take $F_H(h_1) \neq F_H(h_2)$ and suppose $F_{H/L_H^x}\,(L_H^x(h_1)) = F_{H/L_H^x}\,(L_H^x(h_2))$; then

$$\{L_H^x(h_1),\ L_H^x(z_1)\} \subset \prod_{i_1\in I_1} L_H^x\,(k_{i_1}),\ldots,\{L_H^x\,(z_{\varkappa-1}),\ L_H^x\,(h_2)\} \subset \prod_{i_\varkappa\in I_\varkappa} L_H^x\,(k_{i_k}).$$

From the first inclusion we obtain

$$\{L_H^x\,(h_1),\ L_H^x\,(z_1)\} \subset \{L_H^x\,(c)\ \mid\ c \in \prod_{i_1\in I_1} k'_{i_1} \quad \text{where} \quad k'_{i_1}\ L_H^x\ k_{i_1},\ \forall i_1 \in I_1\};$$

but by remark 3.1 $\quad k'_{i_1}\ L_H^x\ k_{i_1} \Longrightarrow k'_{i_1}\ \beta_H^\star\ k_{i_1}$.

So we can take c_1,c_2 such that $L_H^x\,(h_1) = L_H^x\,(c_1) \subset F_H(c_1)$ and $L_H^x\,(z_1) = L_H^x\,(c_2) \subset F_H(c_2)$. From the construction of c_1 and c_2 and proposition 1, we obtain $c_1 b_H^\star c_2$ and $F_H(c_1) = F_H(c_2)$. Therefore $F_H(h_1) = F_H(z_1)$ and similarly $F_H(z_{\varkappa-1}) = F_H(h_2)$; so $F_H(h_1)=F_H(h_2)$ which is a contradiction.

COROLLARY

For every element h of a hypergroup H we have

$$F_H(h) = \bigcup_{L_H^x(w) \in F_{H/L_H^x}(L_H^x(h))} L_H^x(w)$$

4. THE SEMIDIRECT HYPERPRODUCT

DEFINITION 3

Let A,B be hypergroups. We consider the group Aut A (see WALL [16]) and the fundamental group B/β_B^*.

Let

$$\hat{}: B/\beta_B^* \longrightarrow \mathrm{Aut}A : F_B(b) \longmapsto \widehat{F_B(b)} \overset{\text{denote}}{=\!=} \hat{b}$$

be a homomorphism. Then in $A\times B$ we can define a hyperproduct as follows:

$$(a,b)(a_1,b_1) = \{(x,y) \mid x \in a\hat{b}(a_1),\ y \in bb_1\} = (a\hat{b}(a_1), bb_1)$$

and we shall call this the *semidirect hyperproduct* of A and B. From the fundamental property we have for every $x,y \in B$ that $\widehat{xy} = \hat{x}\circ\hat{y} = \hat{z}$, $\forall z \in xy$. The semidirect hyperproduct is associative, since

$$((a,b)(a_1,b_1))(a_2,b_2) = \bigcup_{\substack{x \in a\hat{b}(a_1)\\ y \in bb_1}} (x,y)(a_2,b_2) =$$

$$= \bigcup_{\substack{x \in a\hat{b}(a_1)\\ y \in bb_1}} \{(z,w) \mid z \in x\hat{y}(a_2),\ w \in yb_2\} = \{(z,w) \mid z \in a\hat{b}(a_1)\,\widehat{bb_1}(a_2),\ w \in bb_1b_2\}$$

and $$(a,b)((a_1,b_1)(a_2,b_2)) = \bigcup_{\substack{x \in a_1\hat{b}_1(a_2)\\ y \in b_1b_2}} (a,b)(x,y) =$$

$$= \bigcup_{\substack{x \in a_1\hat{b}_1(a_2)\\ y \in b_1b_2}} \{(z,w) \mid z \in a\hat{b}(x),\ w \in by\} =$$

$$= \{(z,w) \mid z \in a\hat{b}(a_1\hat{b}_1(a_2)),\ w \in bb_1b_2\} = \{(z,w) \mid z \in a\hat{b}(a_1)\ \hat{b}(\hat{b}_1(a_2)),\ w \in bb_1b_2\} =$$

$$= \{(z,w) \mid z \in a\hat{b}(a_1)\,\hat{b}\circ\hat{b}_1(a_2),\ w \in bb_1b_2\} = \{(z,w) \mid z \in a\hat{b}(a_1)\ \widehat{bb_1}(a_2),\ w \in bb_1b_2\}.$$

It is also easy to see that the reproduction axiom is valid in $A\times B$; therefore $A\times B$ equipped with the semidirect hyperproduct becomes a hypergroup which we denote $A\hat{\times}B$.

Remark 4.1

Using the fundamental group one can define the wreath product as well.

Remark 4.2

In case that in A there exists an absolute unit element e_a then for every b of B which $\hat{b}(e_a) = e_a$ we obtain $\{e_a\}\hat{\times} B \cong B$. Similarly if e_b is any unit element of B such that $e_b e_b = e_b$, then $A\hat{\times}\{e_b\} \cong A$. This remark can be applied, for example, for reversible hypergroups [9], canonical hypergroups [10], and polygroups [1,2].

Now let us consider in $A\times B$ the equivalence relation $\mathscr{A}$ such that $(a,b)\,\mathscr{A}\,(a_1,b_1) \Longleftrightarrow b = b_1$. Let us denote the equivalence class of $b\in B$ by $(A,b) = \{(a,b) \mid a\in A\}$. Then we obtain that the map $s:(A,b) \longrightarrow b$ is a hyperisomorphism between the hypergroups $(A\hat{\times}B)/\mathscr{A}$ and B. Therefore we have the following.

THEOREM 4
For every hypergroup A and B, $(A\hat{\times}B)/\mathscr{A} \cong B$.

Remark 4.3
The above theorem allows us to say that $A\hat{\times}B$ is an extension of the hypergroup A by the hypergroup B.

LEMMA 4.1
Let A,B be hypergroups and $a\in A$, $b\in B$; then $\hat{b}(F_A(a)) = F_A(\hat{b}(a))$.

Proof
Let $x\in\hat{b}(F_A(a))$; then there exists c in $F_A(a)$ such that $x = \hat{b}(c)$. So

$$\{c,z_1\} \subset \prod_{i_1\in I_1} h_{i_1},\ldots,\{z_{\lambda-1},a\} \subset \prod_{i_\lambda\in I_\lambda} h_{i_\lambda} \Longrightarrow$$

$$\{\hat{b}(c),\ \hat{b}(z_1)\} \subset \prod_{i_1\in I_1} \hat{b}(h_{i_1}),\ldots,\{\hat{b}(z_{\lambda-1}),\ \hat{b}(a)\} \subset \prod_{i_\lambda\in I_\lambda} \hat{b}(h_{i_\lambda})$$

i.e. $\hat{b}(c)\,\beta_A^*\,\hat{b}(a)$. Therefore $x\in F_A(\hat{b}(a))$.

Conversely, let $x\in F_A(\hat{b}(a))$ or $x\beta_A^*\hat{b}(a)$. Then

$$\{x,z_1\} \subset \prod_{i_1\in I_1} h_{i_1},\ldots,\{z_{\lambda-1},\hat{b}(a)\} \subset \prod_{i_\lambda\in I_\lambda} h_{i_\lambda}.$$

If we now apply the inverse automorphism $\hat{b}^{-1}$ of $\hat{b}$, we obtain

$$\{\hat{b}^{-1}(x),\hat{b}^{-1}(z_1)\} \subset \prod_{i_1\in I_1} \hat{b}^{-1}(h_{i_1}),\ldots,\{\hat{b}^{-1}(z_{\lambda-1}),a\} \subset \prod_{i_\lambda\in I_\lambda} \hat{b}^{-1}(h_{i_\lambda})$$

so $\hat{b}^{-1}(x)\beta_A^*\,a$ or $\hat{b}^{-1}(x)\in F_A(a)$, and therefore $x\in\hat{b}(F_A(a))$.

LEMMA 4.2
Let A,B be hypergroups and $a,a_1\in A$, $b,b_1\in B$. Then $\{(a,b),(a_1,b_1)\} \subset (h_1,k_1)\ldots(h_\nu,k_\nu)$ for some $h_i\in A$, $k_i\in B$ $\Longleftrightarrow$ $\{a,a_1\}\subset h_1'\ldots h_\nu'$ and $\{b,b_1\} \subset k_1\ldots k_\nu$ for some $h_i'\in A$, $k_i\in B$.

Proof
We have

$$(h_1,k_1)\dots(h_\nu,k_\nu)=(h_1\cdot\hat{k}_1(h_2)\dots\widehat{k_1k_2\dots k_{\nu-1}}(h_\nu),k_1\dots k_\nu).$$

Therefore " $\Longrightarrow$ " is clear.

" $\Longleftarrow$ " is also clear because for given h_i' such that

$$\{a,a_1\}\subset h_1'\dots h_\nu'$$

we can take the elements

$$h_1=h_1',\quad h_2=\hat{k}_1^{-1}(h_2'),\dots,h_\nu=\widehat{k_1\dots k_{\nu-1}}^{-1}(h_\nu').$$

THEOREM 5

Let A,B be hypergroups, then

$$(A\hat{\times}B)/\beta^\star_{A\hat{\times}B}\cong A/\beta^\star_A\ \hat{\times}\ B/\beta^\star_B,$$

where the homomorphism $\wedge$ used or the groups $A/\beta^\star_A$, $B/\beta^\star_B$ is the one induced by the homomorphism $\wedge$ for the hypergroups A,B.

Proof

The mapping

$$\varphi:F_{A\hat{\times}B}(a,b)\longmapsto(F_A(a),F_B(b))$$

is a homomorphism because, using lemma 4.1, we have

$$\varphi(F_{A\hat{\times}B}(a,b)\cdot F_{A\hat{\times}B}(a_1,b_1))=\varphi(F_{A\hat{\times}B}((a,b)(a_1,b_1)))=\varphi(F_{A\hat{\times}B}(a\hat{b}(a_1),bb_1))=$$

$$=(F_A(a\hat{b}(a_1)),F_B(bb_1))=(F_A(a)\cdot F_A(\hat{b}(a_1)),F_B(b)\cdot F_B(b_1))=$$

$$=(F_A(a)\cdot\hat{b}(F_A(a_1)),F_B(b)F_B(b_1))=(F_A(a),F_B(b))(F_A(a_1),F_A(b_1))=$$

$$=\varphi(F_{A\hat{\times}B}(a,b))\cdot\varphi(F_{A\hat{\times}B}(a_1,b_1)).$$

Finally using lemma 4.2 we have

$$(a,b)\beta^\star_{A\hat{\times}B}(a_1,b_1)\qquad\text{iff}$$

$$\{(a,b),(x_1,y_1)\}\subset\prod_{i_1\in I_1}(h_{i_1},k_{i_1}),\dots,\{(x_{\lambda-1},y_{\lambda-1}),(a_1,b_1)\}\subset\prod_{i_\lambda\in I_\lambda}(h_{i_\lambda},k_{i_\lambda})$$

iff $a\beta^\star_A a_1$ and $b\beta^\star_B b_1$.

REFERENCES

[1] Comer,S., Extension of polygroups by polygroups and their representations using color schemes, Lecture notes in Math., no 1004, Universal Algebra and Lattice Theory, p.p. 91-103 (1984).

[2] Comer,S., Polygroups Derived from Cogroups, Journal of Algebra, V.89, No 4, p.p. 397-405 (1984).

[3] Corsini,P., Contributo alla teoria degli ipergruppi, Atti Soc. Pelor. Sc. Mat. Fis. Nat., Messina, (1980).

[4] Corsini,P., Feebly Canonical and 1-Hypergroups, Acta Universitatis Carolinae - Mathematica et Physica, V. 24(2), p.p. 49-56 (1983).

[5] De Salvo, M., Sugli ipergruppi completi finiti, Riv. Mat. Univ. Parma (4)8, p.p. 269-280 (1982).

[6] De Salvo,M.-Freni,D., Ipergruppi finitamente generati, Bolletino U.M.I. (1984).

[7] Koskas,M., Groupoides, demi-hypergroupes et hypergroupes, J. Math. pures et appl., V. 49, p.p. 155-192 (1970).

[8] Marty,F., Sur une généralisation de la notion de groupe, Huitième congrès de mathématiciens Scandinaves, Stockholm, p.p. 45-49 (1934).

[9] McMullen,J.R.-Price,J.F., Reversible hypergroups, Rendiconti del Seminario Matematico e Fisico di Milano, V.XLVII (1977), p.p. 67-85 (1979).

[10] Mittas,J., Hypergroupes canoniques, Mathematica Balkanica, V.2, p.p. 165-179 (1972).

[11] Prenowitz,W.-Jantosciak,J., Join Geometries, U.T.M., Springer - Verlag (1979).

[12] Vougiouklis,T., Cyclicity in a special class of hypergroups, Acta Univ. Carolinae - Mathematica et Physica, V. 22(1), p.p. 3-6 (1981).

[13] Vougiouklis,T., Representations of hypergroups. Hypergroup algebra, Proceedings in: Convegno su: Ipergruppi, altre strutture multivoche e loro appl. Udine (1985) p.p. 59-73.

[14] Vougiouklis,T., Generalization of P-Hypergroups, to appear in Rend. Circolo Mat. di Palermo.

[15] Vougiouklis,T.-Konguetsof,L., P-Hypergroupes, to appear in Acta Univ. Carolinae - Math. et Physica.

[16] Wall,H., Hypergroups, American Journal of Mathematics, V. 59, p.p. 77-98 (1937).

Annals of Discrete Mathematics 37 (1988) 469–478

THE PERRON-FROBENIUS PROJECTION IN THE THEORY OF GRAPHS, DIGRAPHS, DESIGNS AND STOCHASTIC PROCESSES

Karl Erich WOLFF

Fachhochschule Darmstadt, Fachbereich Mathematik und Naturwissenschaften, Schöfferstr. 3, D-6100 Darmstadt, FRG.

Problem 5 in "Spectra of graphs" by CVETKOVIC, DOOB, SACHS [2,p.266] says: "Find the relation between the theory of Markov chains and the theory of graph spectra."
We show that the investigation of spectral properties of graphs, digraphs, designs and stochastic matrices are ruled by the PERRON-FROBENIUS projection and three matrix representations of it, which yield a common generalization of the HOFFMAN theorems for graphs [4], directed graphs [5] and point stable designs [9] as well as the ergodic theorem [3] for stochastic matrices and the ranking procedure of WEI [8] and KENDALL [6] for tournaments.

1. THE PERRON-FROBENIUS PROJECTION

In the HOFFMAN theorems [4,5,9], in the ergodic theorem and in the ranking procedure of WEI [8] and KENDALL [6] there appears a non-negative $n\times n$-matrix M, its PERRON-FROBENIUS eigenvalue α and a projection matrix P such that

$$MP = \alpha P = PM \ .$$

For a more general description of this situation we need the following notations: Let $(F)_n$ denote the set of $n\times n$-matrices over a field F , E the identity matrix in $(F)_n$. For $M \in (F)_n$ let *specM* be the set of eigenvalues of M. For $\alpha \in \mathrm{spec}M$ we denote by $a(\alpha)$ the algebraic, by $g(\alpha)$ the geometric multiplicity of α. For $\vec{x} \in F^n$ let $\langle\vec{x}\rangle := \{ t\vec{x} \mid t \in F \}$.

Definition:

Let $M \in (F)_n$, $\alpha \in F$. M is called *α-quasi-orthogonal* ,
if $\ker(M - \alpha E) \cap \mathrm{Im}(M - \alpha E) = \{ \vec{o} \}$
(hence $\ker(M - \alpha E) \oplus \mathrm{Im}(M - \alpha E) = F^n$).

Lemma 1:

Let $M \in (F)_n$, $\alpha \in \mathrm{spec}M$. Then for any $P \in (F)_n$ the following conditions (0) and (1) are equivalent:

(0) M is α-quasi-orthogonal and P is the matrix of the projection onto $\ker(M - \alpha E)$ along $\mathrm{Im}(M - \alpha E)$,

(1) $P^2 = P$, $MP = \alpha P = PM$, $\mathrm{rank}(P) \geq g(\alpha)$.

The proof is obvious.

If (1) holds, we say: *P is the (matrix of the) (M,α)-projection.*

Lemma 2:

For $M \in (F)_n$, $\alpha \in \operatorname{spec}M$ the following properties are equivalent:

a) M is α-quasi-orthogonal ,

b) $a(\alpha) = g(\alpha)$,

c) $\ker(M - \alpha E) = \ker(M - \alpha E)^2$,

d) there exists a polynomial $S(X) \in F[X]$ such that $(X - \alpha)S(X)$ is the minimal polynomial of M and $S(\alpha) \neq 0$.

This uniquely determined polynomial $S(X)$ is called the *(M,α)-polynomial* .

Proof:

Let $M \in (F)_n$, $\alpha \in \operatorname{spec}M$, $f(\vec{x}) := M\vec{x}$ $(\vec{x} \in F^n)$.

a) ==> b): Since M is α-quasi-orthogonal, the characteristic polynomial $\chi(X)$ of M satisfies $\chi(X) = (X - \alpha)^{g(\alpha)}\chi_2(X)$, where $\chi_2(X)$ is the characteristic polynomial of the restriction $f|\operatorname{Im}(M - \alpha E)$. Then $\chi_2(\alpha) \neq 0$, since otherwise there would exist a vector $\vec{x} \neq \vec{o}$ such that $\vec{x} \in \ker(M - \alpha E) \cap \operatorname{Im}(M - \alpha E)$, contradicting a). Hence $a(\alpha) = g(\alpha)$.

b) ==> c): Clearly the characteristic polynomial $\chi_1(X)$ of the restriction $f|E_1$, where E_1 is the generalized eigenspace of α , satisfies $\chi_1(X) = (X - \alpha)^{g(\alpha)}$, hence $\chi_1(X)$ decomposes into linear factors. Therefore the matrix of $f|E_1$ is similar to a Jordan matrix, and therefore similar to a diagonal matrix, since $a(\alpha) = g(\alpha)$. Hence $E_1 = \ker(M - \alpha E)$, which is equivalent to c). Clearly c) is equivalent to d).

c) ==> a): $\ker(M - \alpha E) \cap \operatorname{Im}(M - \alpha E) \neq \{\vec{o}\}$ is equivalent to

$$\{\vec{o}\} < \ker(M - \alpha E) < \ker(M - \alpha E)^2 .$$

The main examples of α-quasi-orthogonal real matrices are:

Ex.1: Any symmetric matrix M is α-quasi-orthogonal for any eigenvalue α of M.

Ex.2: Any stochastic matrix is 1-quasi-orthogonal (cf.[2],2.6.d).

Ex.3: Any irreducible non-negative matrix is α-quasi-orthogonal for its PERRON-FROBENIUS eigenvalue α (since $a(\alpha) = 1 = g(\alpha)$, cf.[7]).

Definition:

Let M be α-quasi-orthogonal. Then the (M,α)-projection is called the *PERRON-FROBENIUS projection of M* , if M is non-negative and α is the PERRON-FROBENIUS eigenvalue of M.

2. THE POLYNOMIAL REPRESENTATION OF THE (M,α)-PROJECTION

THEOREM 1:

Let $M \in (F)_n$, $\alpha \in \operatorname{spec}M$, M α-quasi-orthogonal and $S(X)$ the (M,α)-polynomial.

Then $$P := \frac{S(M)}{S(\alpha)}$$ is the (M,α)-projection.

Main case:

If $S(X) = \prod_{\substack{\lambda \neq \alpha \\ \lambda \in \mathrm{spec}M}} (X - \lambda)^{s_\lambda}$ (e.g. for $F = \mathbb{C}$), then

$$(2) \qquad P = \prod_{\substack{\lambda \neq \alpha \\ \lambda \in \mathrm{spec}M}} \left(\frac{M - \lambda E}{\alpha - \lambda}\right)^{s_\lambda} .$$

Proof:

Let $P = S(M)/S(\alpha)$. From $(M - \alpha E)S(M) = 0$ we get $MP = \alpha P = PM$. Clearly

(3) $P\vec{x} = \vec{x}$ for all $\vec{x} \in \ker(M - \alpha E)$,

hence $\ker(M - \alpha E) \leq \mathrm{Im}(P)$. On the other side we obtain from $(M - \alpha E)S(M) = 0$ $\mathrm{Im}(P) = \mathrm{Im}(S(M)) \leq \ker(M - \alpha E)$, hence

(4) $\mathrm{Im}(P) = \ker(M - \alpha E)$.

From (4) and (3) we get $P^2 = P$ and therefore (1) because of (4). Hence P is the (M,α)-projection according to Lemma 1.

3. THE DYADIC REPRESENTATION OF THE (M,α)-PROJECTION

Let $M \in (F)_n$, $\alpha \in \mathrm{spec}M$,

$R(\alpha) := \{ \vec{x} \in F^n \mid M\vec{x} = \alpha\vec{x} \}$ the righteigenspace of α ,

$L(\alpha) := \{ \vec{x} \in F^n \mid M^T\vec{x} = \alpha\vec{x} \}$ the lefteigenspace of α .

<u>THEOREM 2</u>:

Let $M \in (F)_n$, $\alpha \in \mathrm{spec}M$ and $a(\alpha) = 1$.

Then there exist $\vec{r} = (r_1,\dots,r_n)^T$, $\vec{\ell} = (\ell_1,\dots,\ell_n)^T \in F^n$, such that $R(\alpha) = \langle\vec{r}\rangle$, $L(\alpha) = \langle\vec{\ell}\rangle$ and for any such $\vec{r},\vec{\ell}$

$$(5) \quad P := \left(\sum_{k=1}^{n} r_k\ell_k\right)^{-1} \cdot \begin{pmatrix} r_1\ell_1 & \cdots & r_1\ell_n \\ \vdots & & \vdots \\ r_n\ell_1 & \cdots & r_n\ell_n \end{pmatrix} = \frac{\vec{r}\,\vec{\ell}^T}{\vec{\ell}^T\vec{r}}$$

is the (M,α)-projection.

Proof:

Since $a(\alpha) = 1$, we have $a(\alpha) = g(\alpha)$, hence M is α-quasi-orthogonal. Let P denote the matrix of the (M,α)-projection. From (1) we have $MP = \alpha P$, hence the columns of P are elements of $R(\alpha) = \langle\vec{r}\rangle$, since $g(\alpha) = 1$. Hence there exists a vector $\vec{c} = (c_1,\dots,c_n)^T \in F^n$ such that $P = (r_ic_j) = \vec{r}\,\vec{c}^T$. Since $PM = \alpha P$, the rows of P are elements of $L(\alpha) = \langle\vec{\ell}\rangle$, since $\dim L(\alpha) = \dim R(\alpha) = 1$. Hence from $P^2 = P \neq 0$ (by (1) and $g(\alpha) = 1$) we obtain $\vec{o} \neq \vec{\ell}^T = \vec{\ell}^TP$, consequently there exists a j such that $0 \neq \ell_j = \vec{\ell}^Tc_j\vec{r}$, hence $\vec{\ell}^T\vec{r} \neq 0$ and $c_k = \ell_k/(\vec{\ell}^T\vec{r})$ for all $k \in \{1,\dots,n\}$, therefore (5) holds.

4. WHEN IS P A MULTIPLE OF J ?

Let $J_{n,1} := (1,\dots,1)^T \in F^n$, $J := J_{n,1}J_{n,1}^T$ (the all-one-matrix).
If in the dyadic representation (5) $\vec{r} = \vec{\ell} = J_{n,1}$, then $P = \frac{1}{n} J$.
The following simple, but useful theorem answers the question : " When is P a multiple of J ?" , i.e. $P \in \langle J\rangle := \{tJ \mid t \in F\}$.

THEOREM 3:
Let $M \in (F)_n$, $\alpha \in \text{spec}M$, M α-quasi-orthogonal, P the matrix of the (M,α)-projection. Then

a) $P \in \langle J_{n,1}\cdot\vec{d}^T\rangle$ for some $\vec{d} \in F^n$ iff $R(\alpha) = \langle J_{n,1}\rangle$,
b) $P \in \langle\vec{c} \cdot J_{n,1}^T\rangle$ for some $\vec{c} \in F^n$ iff $L(\alpha) = \langle J_{n,1}\rangle$,
c) $P \in \langle J\rangle$ iff $R(\alpha) = \langle J_{n,1}\rangle = L(\alpha)$.

Proof:
Since P is the (M,α)-projection, $\text{rank}(P) = g(\alpha) \geq 1$. If $P \in \langle J_{n,1}\vec{d}^T\rangle$ for some $\vec{d} \in F^n$, then $\text{rank}(P) \leq 1$, hence $\text{rank}(P) = 1$ and $R(\alpha) = \langle J_{n,1}\rangle$, since the columns of P generate $R(\alpha)$. For the converse take $\vec{d} \in L(\alpha)$, $\vec{d} \neq \vec{o}$ and use Theorem 2. This proves a).
b) The rows of P generate $L(\alpha)$, since $PM = \alpha P$ and $\text{rank}(P) = g(\alpha) = \dim L(\alpha)$. Hence the "left version" of the proof of a) shows b).
c) follows from a) and b).

5. APPLICATION OF THE POLYNOMIAL AND DYADIC REPRESENTATION

In this section we introduce the HOFFMAN equation for connected designs and show how three well-known theorems of HOFFMAN, HOFFMAN-McANDREW and the author can be derived easily from our previous results.

5.1. The HOFFMAN equation of a connected design

Let D denote a connected design (= incidence structure), A its v×b-incidence matrix, $N := AA^T \in (\mathbb{R})_v$ its connection matrix and α the PERRON-FROBENIUS eigenvalue of N. Then the righteigenspace $R(\alpha)$ is one-dimensional (since N is irreducible, because D is connected). From $N = N^T$ we have $L(\alpha) = R(\alpha) = \langle\vec{r}\rangle$ for some $\vec{r} \in \mathbb{R}^v$. Hence we obtain from Theorem 1 and Theorem 2 the following HOFFMAN equation for connected designs

$$(6) \qquad \prod_{\substack{\lambda<\alpha \\ \lambda\in\text{spec}N}} \frac{N - \lambda E}{\alpha - \lambda} = \frac{\vec{r}\,\vec{r}^T}{\vec{r}^T\vec{r}} ,$$

since both sides of this equation represent the PERRON-FROBENIUS projection of $N = N^T$.

5.2. POINT STABLE DESIGNS

From Theorem 3c) we obtain

COROLLARY 1:

Let N denote the connection matrix of a design D and α its PERRON-FROBENIUS eigenvalue. Then N is α-quasi-orthogonal and the PERRON-FROBENIUS projection P satisfies

$$P \in \langle J \rangle \quad \text{iff} \quad NJ = \alpha J \text{ and } D \text{ is connected.}$$

Proof:

Because of the symmetry of N, N is α-quasi-orthogonal and the right side of Theorem 3c) is equivalent to "$NJ=\alpha J$ and D is connected", since D is connected iff $a(\alpha) = 1$.

The following definition was introduced by the author in [9].

Definition

A design with connection matrix N is called *point stable*, if $NJ = \alpha J$ for some $\alpha \in \mathbb{R}$.

The most important examples of point stable designs are the (r,λ)-designs, the 1-designs and therefore also the regular graphs.

The direct generalization of the HOFFMAN theorem for connected regular graphs to connected point stable designs is the following (cf.[9])

THEOREM 4:

a) A design D with connection matrix N is connected and point stable iff there exists a polynomial $f(X) \in \mathbb{R}[X]$ such that $f(N) = J$.

b) If D is connected and point stable, then there exists exactly one polynomial $h(X)$ of minimal degree such that $h(N) = J$, namely the HOFFMAN polynomial

$$h(X) = v \prod_{\substack{\lambda<\alpha \\ \lambda\in \mathrm{spec}N}} \frac{X - \lambda}{\alpha - \lambda} .$$

Remark:

The HOFFMAN equation $h(N) = J$ of a connected point stable design is just the special case of (6), where $\vec{r} = J_{v,1}$. Note that Corollary 1 gives a characterization of the connected point stable designs in terms of the PERRON-FROBENIUS projection $P = v^{-1}h(N)$. Therefore we call Corollary 1 the "projection version" of Theorem 4.

5.3. GRAPHS AND DIGRAPHS

Now we give the projection versions of the HOFFMAN theorem for connected regular graphs [4] and of the HOFFMAN-McANDREW theorem for strongly connected regular digraphs [5].

COROLLARY 2: (projection version of the HOFFMAN theorem)
Let A denote the adjacency matrix of a graph G and d its PERRON-FROBENIUS eigenvalue. Then A is d-quasi-orthogonal and the PERRON-FROBENIUS projection P satisfies

$P \in \langle J \rangle$ iff $AJ = dJ$ and G is connected (i.e. G is regular and connected).

The proof is the "same" as the proof of Corollary 1.

COROLLARY 3: (projection version of the HOFFMAN-McANDREW theorem)
Let $A \in (R)_n$ denote the adjacency matrix of a digraph G, d its PERRON-FROBENIUS eigenvalue. Then the following conditions a) and b) are equivalent:

a) A is d-quasi-orthogonal and the matrix P of the PERRON-FROBENIUS projection satisfies $P \in \langle J \rangle$,

b) G is strongly connected and regular.

Proof:

a) ==> b): By Theorem 3c) we obtain $AJ = dJ = A^T J$, hence G is regular. G is strongly connected, since $a(d) = g(d) = \text{rank}(P) = 1$ and $AJ = dJ = A^T J$ (cf.[2],p.19,Th.0.4).

b) ==> a): Since G is strongly connected, A is irreducible, hence $a(d) = 1 = g(d)$ by the PERRON-FROBENIUS theorem [7], hence A is d-quasi-orthogonal. From the regularity we obtain $AJ = \bar{d}J = A^T J$ for some $\bar{d}$, hence $\bar{d} = d$ and $R(d) = \langle J_{n,1} \rangle = L(d)$, since $g(d) = 1$, hence $P \in \langle J \rangle$ by Theorem 3c).

If G is a strongly connected and d-regular digraph with adjacency matrix $A \in (\mathbb{R})_n$, then we obtain from Theorem 1 and Theorem 2 the HOFFMAN-McANDREW equation

$$(7) \qquad \frac{S(A)}{S(d)} = \frac{1}{n} J \,,$$

where $S(X)$ is the (A,d)-polynomial.

6. THE LIMIT REPRESENTATION OF THE (M,α)-PROJECTION

The ergodic theorem (cf.[3]) mainly says that the limit $\lim_{k\to\infty} S^k$ of a stochastic matrix S is a special dyadic matrix of the form (5) with $\vec{r} = J_{n,1}$. More generally we consider the limit $\lim_{k\to\infty} M^k$ for $M \in (\mathbb{C})_n$ and use the following well-known

Lemma 3:

For any matrix $M \in (\mathbb{C})_n$ the limit $\lim_{k\to\infty} M^k$ exists iff $|\lambda| < 1$ for all $\lambda \in \text{spec}M \setminus \{1\}$ and M is 1-quasi-orthogonal.

THEOREM 5:

Let $M \in (\mathbb{C})_n$. If $P := \lim_{k\to\infty} M^k$ exists, then P is the (M,α)-projection (with $\alpha = 1$) or $P = 0$.

Proof:

Let $P := \lim_{k\to\infty} M^k \neq 0$. Then $P^2 = P$, $MP = P = PM$, and therefore $1 \in \operatorname{spec}M$. Exactly as in [3, p.17,37] one can prove that $\operatorname{rank}(P) = g(1)$. Hence from Lemma 1 we obtain that P is the $(M,1)$-projection.

From Theorem 2 and Theorem 5 we get

THEOREM 6:

Let $M \in (\mathbb{C})_n$. If $\lim_{k\to\infty} M^k \neq 0$ exists and the geometric multiplicity of the eigenvalue 1 of M satisfies $g(1) = 1$, then

$$\lim_{k\to\infty} M^k = \frac{\vec{r}\,\vec{\ell}^{\,T}}{\vec{\ell}^{\,T}\vec{r}} \quad , \tag{8}$$

where $R(1) = \langle\vec{r}\rangle$, $L(1) = \langle\vec{\ell}\rangle$.

Remarks:

(i) Since any stochastic matrix S is 1-quasi-orthogonal, we obtain from the ergodic theorem and Lemma 1 that the PERRON-FROBENIUS projection P of S satisfies

$$P = \lim_{k\to\infty} \frac{1}{k} \sum_{i=0}^{k-1} S^i \quad .$$

(ii) For stochastic matrices we obtain from (8) with $\vec{r} = J_{n,1}$ the main statement of the ergodic theorem. This central part of the ergodic theorem is generalized by the following corollary.

COROLLARY 4:

Let $M \in (\mathbb{C})_n$, $1 \in \operatorname{spec}M$, M 1-quasi-orthogonal and $|\lambda| < 1$ for all $\lambda \in \operatorname{spec}M\setminus\{1\}$. Then

$\lim_{k\to\infty} M^k = J_{n,1}\vec{d}^{\,T}$ for some $\vec{d} \in \mathbb{C}^n$ iff $MJ = J$ and $a(1) = 1$.

The proof results from Lemma 3 and Theorem 3a).

Finally we mention as a consequence of Theorem 1 and Theorem 5 another algebraic possibility to calculate the limit $\lim_{k\to\infty} M^k$.

THEOREM 7:

Let $M \in (\mathbb{C})_n$. If $\lim_{k\to\infty} M^k \neq 0$ exists, then

$$\lim_{k\to\infty} M^k = \frac{S(M)}{S(1)} = \prod_{\substack{|\lambda|<1 \\ \lambda\in\operatorname{spec}M}} \left(\frac{M - \lambda E}{1 - \lambda}\right)^{s_\lambda} \qquad \text{(notations as in (2)).} \tag{9}$$

7. THE PERRON-FROBENIUS PROJECTION IN THE RANKING PROCEDURE OF WEI AND KENDALL

For the calculation of a "meaningful" ranking of the participants of a tournament first determine (cf.[1,6,8]) the linear order of the strong components of the tournament digraph T. In each strong component of T rank the m ($\neq 2$) participants in the following way:

If $m = 3$, all three participants get the same rank.
If $m > 3$, then the adjacency matrix A of the given strong component C is primitiv (cf.[1]). If α is the PERRON-FROBENIUS eigenvalue of A, then for

$$A_1 := \frac{1}{\alpha} A$$

the limit $P := \lim_{k\to\infty} A_1^k \neq 0$ exists according to Lemma 3, since $|\lambda| < 1$ for all $\lambda \in \mathrm{spec}A_1 \setminus \{1\}$ and $a(1) = 1 = g(1)$ by the PERRON-FROBENIUS theorem for primitive matrices (cf.[7]). Further, from Theorem 5 follows that P is the PERRON-FROBENIUS projection onto the one-dimensional kernel $\ker(A_1 - E) = \ker(A - \alpha E)$. Hence there exists exactly one positive vector $\vec{r}$ such that

$$\langle\vec{r}\rangle = \ker(A - \alpha E) \quad \text{and} \quad \sum_{i=1}^{m} r_i = 1 .$$

This vector $\vec{r}$ is taken as "ranking vector":
Participant i is ranked higher than participant j, if $r_i > r_j$.
The combinatorial meaning of this ranking procedure of WEI and KENDALL can be explained easily using Theorem 2:
Let $s_{ki} := (A^k J_{m,1})_i$ be the i-coordinate of $A^k J_{m,1}$, hence s_{ki} is the number of paths of length k starting at vertex i (in the given strong component C). Then

$$\text{(10)} \qquad \lim_{k\to\infty} \frac{s_{ki}}{s_{kj}} = \frac{r_i}{r_j} \qquad (i,j = 1,\dots,m).$$

Proof:
Clearly $\lim_{k\to\infty} \frac{s_{ki}}{\alpha^k} = (\lim_{k\to\infty} A_1^k J_{m,1})_i = (PJ_{m,1})_i$. According to Theorem 2 we have $P = \vec{r}\,\vec{\ell}^T/(\vec{\ell}^T\vec{r})$, hence $PJ_{m,1} = c\vec{r}$, where $c := \vec{\ell}^T J_{m,1}/(\vec{\ell}^T\vec{r}) > 0$. Therefore $\lim_{k\to\infty} \frac{s_{ki}}{\alpha^k} = (PJ_{m,1})_i = cr_i$, hence (10) holds.

Remark: The corresponding result of WEI [8] for graphs (cf. [2], p.104) can be proved with Theorem 2 in the same way.

ACKNOWLEDGEMENT
I wish to thank A.W.M. DRESS (Bielefeld) for his valuable comments on quasi-orthogonal matrices.

REFERENCES

[1] Bondy, J.A. and Murty, U.R.S., Graph theory with applications (Amer. Elsevier Publ. Co. , Inc., New York, 1976)
[2] Cvetković, D.M. and Doob, M. and Sachs, H., Spectra of graphs (Academic Press, New York, 1979)
[3] Fritz, F.J. and Huppert, B. and Willems, W., Stochastische Matrizen (Springer Verlag, Berlin, 1979)
[4] Hoffman, A.J., On the polynomial of a graph, Amer. Math. Monthley 70 (1963) 30-36.

[5] Hoffman, A.J. and McAndrew, M.H., The polynomial of a directed graph, Proc. Amer. Math. Soc. 16 (1965) 303-309.
[6] Kendall, M.G., Further contributions to the theory of paired comparisons, Biometrics 11 (1955) 43-62.
[7] Seneta, E., Non-negative matrices (George Allen and Unwin Ltd., London, 1973)
[8] Wei, T.H., The algebraic foundations of ranking theory (Thesis, Cambridge, 1952)
[9] Wolff, K.E., Punkt-stabile und semi-partial-geometrische Inzidenzstrukturen Mitt. math. Sem. Gießen 121 (1976).
[10] Wolff, K.E., Punkt-stabile Inzidenzstrukturen und stochastische Matrizen in: Grabmeier, J. and Kerber, A., (eds.), Séminaire lotharingien de combinatoire (Publ. IRMA, Strasbourg, 1985) pp. 124-126.
[11] Wolff, K.E., Zur kombinatorischen Bedeutung der PERRON-FROBENIUS-Projektion in: Strehl, V., (ed.), Séminaire lotharingien de combinatoire (Publ. IRMA, Strasbourg, 1986) pp. 153-162.

Annals of Discrete Mathematics 37 (1988) 479–484
Elsevier Science Publishers B.V. (North-Holland)

ON THE NON-EXISTENCE OF CERTAIN DIFFERENCE SETS

N. Zagaglia Salvi

Dipartimento di Matematica
Politecnico di Milano
P.za L. da Vinci 32, Milano, Italy*

SUMMARY- In this paper we determine some properties concerning a (v,k,λ)-difference set when v is even.
In particular it is proved in certain cases the non-existence of circulant Hadamard matrices.

INTRODUCTION

Recall that a subset $\mathcal{D}$ of Z_v is a (v,k,λ)-difference set if $|\mathcal{D}| = k$ and the difference $d_i - d_j$ ($d_i, d_j \in \mathcal{D}$, $d_i \neq d_j$) takes each non-zero value in Z_v λ times.
A matrix is circulant if each row differing from the first is derived from its previous row by shifting it cyclically one position to the right.
An Hadamard matrix has every entry ± 1 and its rows are orthogonal to each other.
It is proved that a circulant Hadamard matrix has order $4N^2$ and corresponds to a difference set of parameters $(4N^2, 2N^2 - N, N^2 - N)$.
It has been conjectured the non-existence of circulant Hadamard matrices of order $v > 4$.
Up to now Turyn [6] proved the conjecture for N even; moreover it is proved for every odd $N < 55$.
Brualdi [3] was able to prove the non-existence of symmetric circulant Hadamard matrix. In its paper it was implicitely proved the conjecture for reflective circulant, where we call a circulant (0,1)-matrix reflective if, arranged the elements of the first row regularly on a circle, there exists a diameter of the circle with respect to which 1's are symmetric.
In this paper we determine some properties concerning a (v,k,λ)-difference set, when v is even.
In particular we determine in certain cases the non-existence of a $(4N^2, 2N^2-N, N^2-N)$-difference set.
Finally we give a construction of a subset of Z_v satisfying the property of a difference set only for odd values.

1. First we determine the following property of a difference set.

PROPOSITION 1.1 - *Let* $\mathcal{D} = \{d_1, d_2, \ldots, d_k\}$ *be a* (v,k,λ)- *difference set where* v *is even and* $k-\lambda = N^2$. *Then the number of the odd elements of* $\mathcal{D}$ *is* $\frac{1}{2}(k + N)$ *or* $\frac{1}{2}(k-N)$.

* This research was supported by the Ministero della Pubblica Istruzione.

Proof. Let r and s be the numbers of the even and odd elements of $\mathcal{D}$.
As v is even, the difference $\alpha = d_i - d_j \bmod v$, where $d_i, d_j \in \mathcal{D}$, is odd if and only if d_i is even and d_j odd or conversely.
Moreover if α is odd, also $-\alpha$ is odd.
So the number of odd differences is 2rs and it has to coincide with $\frac{v}{2}\lambda$.
Then the parameters r,s satisfy the relations r+s = k and r s = v/4.
So we obtain that $r = \frac{1}{2}(k + N)$ and $s = \frac{1}{2}(k - N)$ or conversely. <>

We note that in the case of a $(4N^2, 2N^2 - N, N^2 - N)$ -difference set we have that the numbers of odd and even elements are N^2 and $N^2 - N$ or conversely.

2. DEFINITION 2.1 - Let $\underline{\alpha} = [c_1\ c_2\ \ldots\ c_v]$ a sequence of v numbers, with $v = hk$.
We said $\underline{\alpha}$ has <u>period</u> h if the first h elements of $\underline{\alpha}$ are repeated consecutively k times.
If h is the minimum integer that satisfies this relation, we call h <u>real</u> period.

PROPOSITION 2.2 -*If* $\underline{r}_i$, $1 \leq i \leq v$, *are the rows of a circulant matrix* C *of order* v= hk, *then the vector* $\underline{r}_1 + \underline{r}_{h+1} + \ldots + \underline{r}_{(k-1)h+1}$ *has period* h.

Proof. Denoted $\underline{r}_1 = [c_1 c_2 \ldots c_v]$, we have $\underline{r}_{ih+1} = [c_{v-ih+1}\ c_{v-ih+2} \ldots c_{v-ih+v}]$, $1 \leq i \leq k-1$.
If $\underline{\alpha} = [a_1\ a_2\ \ldots\ a_v]$ is the vector $\underline{r}_1 + \underline{r}_{h+1} + \ldots + \underline{r}_{(k-1)h+1}$, we have
$a_i = c_i + c_{v-h+i} + \ldots + c_{v-(k-1)h+i}$, $1 \leq i \leq v$.
So we obtain $a_{1+rh} = c_{1+rh} + c_{v-h+1+rh} + \ldots + c_{v-(k-1)h+1+rh}$, for $0 \leq r \leq k-1$.
It follows $a_1 = a_{1+h} = \ldots = a_{1+(k-1)h}$ and, analogously, $a_j = a_{j+h} = \ldots = a_{j+(k-1)h}$, $2 \leq j \leq h$.
Then the sequence $a_1\ a_2\ \ldots\ a_h$ is repeated k times and $\underline{\alpha}$ has period h. <>

3. Denoted by H a circulant Hadamard matrix, we consider the (0,1)-circulant matrix $K = \frac{1}{2}(H + J)$, where J is all one matrix.
It is well known that the positions of the elements 1 on the first row of K determine a difference set of parameters $(4N^2, 2N^2 \underline{+} N, N^2 \underline{+} N)$.
Since the different values of k and λ correspond to complementary difference sets, there is not loss of generality in assuming that $(v,k,\lambda) = (4N^2, 2N^2 - N, N^2 - N)$.

We call <u>odd</u> or <u>even</u> an element of a sequence if it is in odd or even position.
Let C be the linear code over Z_2 generated by the rows of K.
Let $\underline{r}_i$, $1 \leq i \leq v$ and $\underline{r}_1 = [c_1\ c_2\ \ldots\ c_v]$, be the rows of K.
Consider the word $\underline{e}_1 = \underline{r}_1 + \underline{r}_3 + \ldots + \underline{r}_{v-1}$.
By Prop. 2.1 $\underline{e}_1$ has period 2. The first [second] element coincides with the sum of elements of $\underline{r}_1$ in odd [even] position.
As k is odd, one number is odd and the other even. So $\underline{e}_1$ is obtained by repeating [10] or [01]. Suppose $\underline{e}_1 = [1010\ldots10]$.
Then the word $\underline{e}_2 = \underline{r}_2 + \underline{r}_4 + \ldots + \underline{r}_v$ coincides with [0101...01].

As, by Prop. 2.1, $\underline{r}_1 + \underline{r}_{1+v/2} = \underline{a}$ has period $2N^2$, consider the possibility that $\underline{a}$ has real period 2, that is $\underline{a} = \underline{e}_1$ or $\underline{a} = \underline{e}_2$.

Without loss of generality, we can suppose $\underline{a} = \underline{e}_1$.

REMARK 3.1 - *The condition* $\underline{r}_1 + \underline{r}_{1+v/2} = \underline{e}_1$ *contradicts the existence of* K.

Proof. As $\det K = k(k-\lambda)^{(v-1)/2} \equiv 1 \bmod 2$, the rows of K are linearly independent in Z_2.
The condition $\underline{r}_1 + \underline{r}_{1+v/2} = \underline{e}_1$ implies $\underline{r}_i + \underline{r}_{i+v/2}$ is $\underline{e}_1$ for odd i and $\underline{e}_2$ for even i.
So the rows $\underline{r}_{j+v/2}$, $1 \leq j \leq v/2$, depent on $\underline{r}_j$, $1 \leq j \leq v/2$, and $\underline{e}_1$, $\underline{e}_2$.
Then the number of linearly independent rows of K is $\leq v/2 + 2$, a contradiction. <>

The condition $\underline{a} = \underline{e}_1$ implies $c_i + c_{i+v/2} = 1$ for odd i and $c_i + c_{i+v/2} = 0$ for even i.
It follows odd [even] elements at distance v/2 are distinct [coincident].
So half of the odd elements are equal to 1, that is N^2 is the number of $c_i=1$ for odd i.
On the contrary, if i is even and $c_i=1$, also $c_{i+v/2} = 1$. So this element determines a difference equal to v/2.
As these differences are $\lambda = N^2 - N$, it follows the number of the even elements equal to 1 is $N^2 - N$. These numbers agree with those of Prop. 1.1.
However the following Theorem 3.5 proves this situation is impossible.

LEMMA 3.2 - *Let* $\underline{a} = [c_2\ c_4\ c_6 \dots c_{4q}]$ *be, where* $c_i \in Z_2$, $c_i + c_{i+2q} \equiv 0$ *mod* 2, *for* $2 \leq i \leq 4q$. *Moreover, let* $\mathcal{D} = \{d_1, d_2, \dots, d_k\}$ *be the set of the positions of* 1*'s in* $\underline{a}$.
Then every difference $\alpha = d_i - d_j$ *mod* 4q, *where* $d_i, d_j \in \mathcal{D}$, *occurs an even number of times.*

Proof. From the condition $c_i + c_{i+2q} \equiv 0 \bmod 2$, we obtain that, if $c_{d_i} = c_{d_j} = 1$, also the elements corresponding to $d_i' = d_i + 2q$ and $d_j' = d_j + 2q$ coincide to 1.
Denoted $\alpha = d_i - d_j$, also the difference $d_i' - d_j'$ coincides with α mod 4q.
So every difference occurs an even number of times. <>

LEMMA 3.3 - *Let* $\underline{b} = [c_1 c_3 c_5 \dots c_{4q-1}]$ *be, where* q *is odd*, $c_i \in Z_2$ *and* $c_i + c_{i+2q} \equiv 1$ *mod* 2, $1 \leq i \leq 4q-1$.
Moreover let $\mathcal{B} = \{d_1\ d_2 \dots d_k\}$ *be the set of the positions of* 1*'s in* $\underline{b}$. *Then the number* 4 *occurs an odd number of times as difference* $d_i - d_j$ *where* $d_i, d_j \in \mathcal{B}$.

Proof. Let $v=4q$ be, $\underline{h}_1 = [c_1\ c_5\ c_9 \dots c_{v-3}]$ and $\underline{h}_2 = [c_3\ c_7\ c_{11} \dots c_{v-1}]$.
We note that $d_i - d_j \equiv 4 \bmod v$, where $d_i, d_j \in \mathcal{B}$, if and only if $d_i \equiv d_j + 4 \bmod v$, that is $c_{d_i} = c_{d_j} = 1$ are two adjacent elements in $\underline{h}_1$ or $\underline{h}_2$.
As it is $c_i + c_{i+v/2} \equiv 1 \bmod 2$ and q is odd, it follows that, if $c_i=1$ belongs to $\underline{h}_1$, then $c_{i+v/2} = 0$ belongs to $\underline{h}_2$ and conversely.

Let r,s respectively the numbers of ordered couples $[1,1]$, $[0,0]$ and m the number of ordered couples $[1,0]$ or $[0,1]$ in $\underline{h}_1$.

It is clear that whenever there is an ordered couple $[1,0]$, there is also an ordered couple $[0,1]$.

So m is even.

Denote by r', s', m' the analogous values for $\underline{h}_2$.

We note that for every couple $[c_{i+4}, c_i] = [1,1]$ of $\underline{h}_1$ $[\underline{h}_2]$ there is the couple $[c_{i+4+v/2}, c_{i+v/2}] = [0,0]$ of $\underline{h}_2$ $[\underline{h}_1]$.

So r'=s and s'=r, while clearly it is m'=m.

We note that the number of ordered couples $[1,1]$, $[0,0]$, $[1,0]$ and $[0,1]$ in $\underline{h}_1$ coincides with the number of elements of $\underline{h}_1$, that is r+s+m=q.

Moreover the number of couples $[c_{i+4}, c_i] = [1,1]$ of $\underline{h}_1$ and $\underline{h}_2$ is r+r' = r+s and this number is odd because q is odd and m even. <>

THEOREM 3.4 -*Let* $\underline{r}_1$ *be the first row of a circulant* (0,1)-*matrix of order* $v = 4N^2$ *with odd* N. *Moreover, let* $\mathcal{D} = \{d_1, d_2, \ldots, d_k\}$ *be the set of positions of* 1'*s in* $\underline{r}_1$.
Then, if $\underline{r}_1$ *satisfies* $\underline{r}_1 + \underline{r}_{1+v/2} = \underline{e}_1$, *where the sum is mod* 2, $\mathcal{D}$ *is not a difference set.*

Proof. If $\underline{r}_1 + \underline{r}_{1+v/2} = \underline{e}_1$, then the odd elements of $\underline{r}_1$ at distance v/2 are different, while the even elements coincide.

Suppose $\mathcal{D}$ is a difference set with the parameters (v,k,λ).

As $k(k-1) = (v-1)\lambda$, λ is even.

Denoted by D and P the sets of odd and even elements, we note that a difference $\alpha = d_i - d_j$, where $d_i, d_j \in \mathcal{D}$, is even if and only if both d_i, d_j belong to D or to P.

By Lemma 3.2, if $d_i, d_j \in P$, α occurs an even number of times.

By Lemma 3.3, if $d_i, d_j \in D$, we can note that the number 4 occurs an odd number of times.

So 4 occurs an odd number of times, while λ is even. <>

4. If $\underline{a}$ is a (0,1)-sequence , we denote by $\underline{a}_h$ the sequence obtained by shifting cyclically every element of h positions to the right.

THEOREM 4.1 -*Let* $\underline{a}$ *be a* (0,1)-*sequence of length* v=4q *and* $\mathcal{D} = \{d_1, d_2, \ldots, d_k\}$ *the set of the positions of* 1'*s in* $\underline{a}$.
If $\underline{a}$ *satisfies* $\underline{a} + \underline{a}_{v/2} = \underline{e}_1$, *where the sum is mod* 2, *then every odd value in* Z_v *occurs* $\lambda = k - q$ *times as difference* $d_i - d_j$, *where* $d_i, d_j \in \mathcal{D}$.

Proof. Let D and P be the sets of odd and even elements of $\mathcal{D}$.

As v is even, the difference $d_i - d_j$, where $d_i, d_j \in \mathcal{D}$, is odd if and only if an integer belongs to D and the other to P.

Let $\underline{a} = [c_1 c_2 \ldots c_v]$ be .

If $\underline{a} + \underline{a}_{v/2} = \underline{e}_1$, the odd [even] elements of $\underline{a}$ at distance $v/2$ are different [coincident].

This implies the odd elements are $v/4 = q$ and hence $k \geq q$.

Let c_i be, where i is odd, an element of $\underline{a}$.

We have two possibilities:

1) $c_i = 1$

2) $c_i = 0$.

In the first case, let d_j be an element of P; so also $d_i + v/2 \in P$.

Denoted $\alpha = i - d_j$, we have also $\alpha' = i - (d_j+v/2) = \alpha + v/2$.

In the second case, as $c_{i+v/2} = 1$, we obtain the differences $\beta = i+v/2 - d_j =$ $= \alpha + v/2$ and $\beta' = i+v/2 - (d_j + v/2) = \alpha$.

Hence, for every $d_j \in P$, we have the same differences in both of the cases 1 and 2.

Then, in order to calculate all the odd differences, we can suppose $c_i = 1$ for every odd $i \in [1, v/2]$, that is $D = \{1, 3, \ldots, \frac{v}{2} - 1\}$.

Now, consider an element $d_j \in P$.

We determine the differences

$$\alpha_i = d_j - i \tag{1}$$

where $i \in D$ and also

$$\alpha_i + v/2 = (d_j + v/2) - i. \tag{2}$$

So for every $d_j \in P$, the differences (1) and (2) give every odd number in $[1, v]$ exactly once.

As there are also the differences $-\alpha_i$, $-(\alpha_i+v/2)$, that determine again every odd number of $[1, v]$ exactly once, we note that for every couple d_j, $d_j + v/2$ belonging to P, we obtain every odd number exactly twice.

As the couples d_j, $d_j+v/2$ of P are $\frac{1}{2}(k - q)$, we have proved that every odd number in Z_v occurs exactly $k - q$ times as difference $d_i - d_j$, where $d_i, d_j \in \mathcal{D}$.<>

REFERENCES

[1] Baumert, L.D., Cyclic Difference Sets, Lecture Notes in Mathematics n. 182 (Springer Verlag, 1971).

[2] Biggs, N., Discrete Mathematics (Clarendon Press, 1985).

[3] Brualdi, R.A., A note on multipliers of difference sets, J. of Res. Nat. B. of Standards, vol. 69 B (1965), pp. 87-89.

[4] Davis, P.J., Circulant matrices (A Wiley-Interscience Publication, 1979).

[5] Van Lint, J.H., Coding Theory, Lecture Notes in Mathematics n.201 (Springer Verlag, 1971).

[6] Turyn, R., Character Sums and Difference Sets, Pacific J. Math. 15 (1965) pp. 319-346.

[7] Zagaglia Salvi, N., Combinatorial structures corresponding to reflective circulant (0,1)-matrices, Annals of Discrete Mathematics 30 (1986), pp. 363-372.

Annals of Discrete Mathematics 37 (1988) 485–492

ON COMPLETE 12-ARCS IN PROJECTIVE PLANES OF ORDER 12

Corrado Zanella

Dipartimento di Matematica, Università di Roma "La Sapienza",
I-00185 Roma, Italy.

Let K be a complete 12-arc in a projective plane π_{12} of order 12. Extending a result in [6] we show that π_{12} has no point of index 10. Moreover, there are at most two points of index 8. This enables us to investigate all possible configurations of the tangents to K; there may occur exactly 17 cases.

1. ON q-ARCS IN π_q, q EVEN

In a projective plane π_q of even order q, let K be a q-arc, i.e. a set of q points no three of which are collinear. On any point of K there are exactly two tangents. Thus K has exactly 2q tangents and q(q-1)/2 secants. Furthermore, through any point of π_q there is an even number of tangents. A point $P \in \pi_q$ is said to be of index j, briefly a *j-point*, if there are exactly j tangents to K through it. Let t_j ($j = 0,1,\dots,q$) be the number of j-points in π_q. Then $t_{2h+1} = 0$ for every h. Moreover [6],

$$\sum_{j=0}^{q/2} t_{2j} = \theta_2, \quad \sum_{j=0}^{q/2} j t_{2j} = q\theta_1, \quad \sum_{j=0}^{q/2} j(2j-1) t_{2j} = q(2q-1), \quad (\theta_h = 1+q+q^2+\dots+q^h). \qquad (1.1)$$

Obviously,

$$t_q = 0 \textit{ iff } K \textit{ is complete.} \qquad (1.2)$$

If π_q is Desarguesian, then K is never complete [1,4]. In what follows, K is always supposed to be complete; therefore, π_q is not Desarguesian. Whenever K is incomplete, it is contained in a (q+2)-arc. Hence, if π_q contains no (q+2)-arc (as it happens for q = 10, [2]) then any q-arc is complete. Therefore

$$q \geq 10. \qquad (1.3)$$

Let ℓ be a non-tangent line and denote by u_j the number of j-points on it. Then

$$\sum_{j=0}^{q/2} u_{2j} = q+1, \quad \sum_{j=0}^{q/2} j u_{2j} = q. \qquad (1.4)$$

If t is a tangent to K, and v_j is the number of j-points on t, then

$$\sum_{j=1}^{q/2} v_{2j} = q+1, \quad \sum_{j=1}^{q/2} (2j-1) v_{2j} = 2q-1. \qquad (1.5)$$

In [6] the following results are proved.

I. *Any complete* q*-arc admits at most one* (q-2)*-point.*

II. *If* t *is a tangent to* K *through a* (q-2)*-point* P *then there exists a* 4*-point* Q (≠P) *on* t *and any point on* t, *other than* P *and* Q, *is of index* 2. *Consequently*

$t_{q-2} = 1,\ t_4 \geq q-2,\ t_2 \geq (q-1)(q-2).$

III. *If* t *is a tangent to* K *through a* (q-4)*-point* P, *then either a* 6*-point* Q ($\neq$P) *exists on* t *and any other point on* t *is of index* 2, *or there are two distinct* 4*-points on* t *and any other point on* t *is of index* 2.

IV. *Assume* K *is a complete* q*-arc admitting a* (q-2)*-point* P. *Then,*

$$t_{q-2} = 1,\quad t_{2j} = 0 \quad \text{for} \quad j = 3,4,\ldots,(q-4)/2,$$
$$t_4 = (q-2)(q+4)/8,\quad t_2 = 3(q^2+4)/4,\quad t_0 = (q-2)(q+8)/8. \tag{1.6}$$

Moreover, on any tangent through P *a* 4*-point lies, the remaining points being of index* 2; *any tangent not on* P *has* (q-2)/2 4*-points and* (q+4)/2 2*-points. On any non-tangent not on* P *there lie* $v_4 \leq 2$ 4*-points and* $v_0 = 1+v_4$ 0*-points. The remaining points have index* 2.

V. *If a complete* q*-arc admits a* (q-2)*-point, then* q = 12.

Therefore, the case q = 12 seems to be interesting. The aim of this note is to deal with it.

2. ON COMPLETE 12-ARCS IN π_{12}

VI. *Let* K *be a complete* 12*-arc in* π_{12}. *Then there is no* 10*-point.*

Proof. Suppose that P is a 10-point. Then, by IV,

$$t_{10} = 1,\quad t_6 = t_8 = 0,\quad t_0 = 25,\quad t_2 = 111,\quad t_4 = 20.$$

Again by IV, on any of the ten tangents on P there is exactly one 4-point, the remaining ones being 2-points. Let r be the unique secant through P. From (1.4), since $u_{10} = 1$,

$$u_4 = u_0 - 5 \tag{2.1}$$

follows. Since $u_2 \geq 2$, we have $u_0 + u_4 \leq 10$. This, together with (2.1), gives $u_4 \leq 2$. On the other hand, $t_4 = 20$ and on any tangent through P there is exactly one 4-point. Therefore, on the remaining (external) lines through P there lie at least eight 4-points. Thus, there is a line ℓ on P which is external to K and contains at least four 4-points, a contradiction (use (1.4)). □

From V and VI it follows that

VII. *No complete* q*-arc in* π_q, *any even* q, *admits a* (q-2)*-point.* □

VIII. *A complete* 12*-arc* K *in* π_{12} *admits at most two* 8*-points.*

Proof. Suppose A, B, C are three distinct 8-points. Firstly, assume that A, B, and C are non-collinear. By III, the lines AB, AC, BC are non-tangents. Thus the 8 tangents through C meet the line AB in 8 points other than A and B. Hence, each of them is of positive index. Since $u_8 \geq 2$, $(1.4)_2$ implies that there are at most four points distinct from A and B, which are of positive index, a contradiction.

Next, assume that A, B, C lie on the line ℓ. By III, ℓ is non-tangent. Again (1.4) hold; hence,

$$u_8 = 3,\quad u_0 = 10,\quad u_j = 0 \quad \text{for} \quad j \neq 0,8. \tag{2.2}$$

Consider a point Q $\notin$ AB of index at least 4. Then the tangents through Q meet

AB in four distinct points of positive index. This contradicts (2.2). □

Let K be a complete 12-arc, and suppose $t_8 = 2$. The line joining A with B is non-tangent (by III). From (1.1) we get

$$t_0 = 25-t_6 , \quad t_2 = 112+3t_6 , \quad t_4 = 18-3t_6, \quad t_8 = 2. \tag{2.3}$$

IX. *If* $t_8 = 2$, *then* $t_0 \geq 23$.

Proof. Let A and B be the two 8-points, and r one of the four non-tangents through A and not through B. Then $u_8 = 1$, and, by (1.4),

$$u_0 = 4+u_4+2u_6 \Rightarrow u_0 \geq 4.$$

The eight tangents through B meet r in eight points ($\neq$A) of positive index, hence

$$u_0 = 4. \tag{2.4}$$

Let n be the number of 0-points on AB. By (1.4), $n \geq 7$. Since no tangent on A has a 0-point, by (2.4),

$$t_0 = 16+n, \quad n \geq 7. \ \square \tag{2.5}$$

X. *If* $t_8 = 2$, *then either* $t_0 = 25$, *and the line joining the two* 8*-points is external to* K, *or* $t_0 = 23$.

Proof. By (2.3) and IX, $23 \leq t_0 \leq 25$. If $t_0 = 24$, then (2.5) implies n = 8 (n is defined in the proof to IX), and $(2.3)_1$ implies $t_6 = 1$. Call P the unique 6-point. If P ∈ AB, then n = 9. If P ∉ AB, then the tangents on P meet AB in six distinct points of positive index; so, $n \leq 7$. In both cases we get a contradiction. Therefore, $t_0 \neq 24$.

If $t_0 = 25$, then from (2.3) $t_6 = 0$ follows. By (2.5), n = 9, and (1.4) give $u_2 = 0$ for the line AB. Since the points on K have index 2, AB is external to K. □

XI. *If* $t_8 = 2$ *and* $t_0 = 23$, *then, denoting by* A *and* B *the* 8*-points, any non-tangent through* A, *but not through* B, *has eight* 2*-points and four* 0*-points (this is also true if* $t_0 = 25$*). The line* AB *has seven* 0*-points and four* 2*-points. Through* A *there are eight tangents, two of them have one* 6*-point each, whereas the remaining ones are of index* 2. *The remaining six tangents have two* 4*-points each, all other points being of index* 2.

Proof. The first assertion immediately follows from the proof to IX and (1.4). If $t_0 = 23$, then n = 7 and $t_6 = 2$. The remaining statements can be obtained with the help of equations (1.4) and (1.5). □

XII. *If* $t_8 = 2$ *and* $t_0 = 23$, *then the line through the two* 6*-points is non-tangent to* K.

Proof. Let A and B be the 8-points, and C and D the 6-points. By XI, P = AB ∩ CD is distinct from A, B, C, D. If CD is a tangent, then P, again by XI, is of index 2. By (1.5),

$$v_2 = 10, \qquad v_4 = 1, \qquad v_6 = 2.$$

Besides P, on the line AB there are five other points of positive index. Through C there are five tangents besides CD; each of them intersects AB in a point of positive index. Therefore, the lines joining each of these points with C are tangent. The same argument applies to D. Consequently, the tangents through the

2-points other than P on AB, meet CD in C and D. CD has one 4-point Q. The three tangents through Q, other than CD, must intersect AB in three points of positive index different from 2, by the previous argument. Since just two such points exist, we get a contradiction. □

Next, a more detailed investigation is carried out in case $t_0 = 23$. Let A, B be the 8-points, and C, D the 6-points. Let S be the set consisting of the four 2-points on AB.

W.l.o.g. we may suppose that $A \neq P = AB \cap CD$. Let u_j be the number of j-points on CD. By XI, CD contains neither A nor B; thus, $u_8 = 0$. The eight tangents on A meet CD in points of positive index. Thus,

$$u_2 + u_4 \geq 6. \tag{2.6}$$

With the help of (1.4) we obtain

$$u_0 = 5, \quad u_2 = 6, \quad u_4 = 0, \quad u_6 = 2, \quad u_8 = 0. \tag{2.7}$$

Notice that CD has eight points of positive index, namely the intersections with CD of the tangents through A. Hence,

$$\underline{P = AB \cap CD \text{ is of index zero.}} \tag{2.8}$$

The six tangents through C intersect AB in points of positive index. Since on AB there lie exactly six points of positive index, the line joining each of them with C is a tangent. The same argument applies to D. In particular,

<u>The tangents through any 2-point on AB are</u> (2.9)

<u>the lines joining it with C and D.</u>

Let t be any of such tangents. By (1.5),

$$v_2 = 9, \quad v_4 = 3, \quad v_6 = 1. \tag{2.10}$$

Consequently, all the twelve 4-points lie on the union of the tangents on C. Since the same argument applies also to D, we conclude that any 4-point is the intersection of two tangents, CS_i and DS_j, $S = \{S_1, S_2, S_3, S_4\}$, $i \neq j$. On the other hand, the eight lines CS_j, DS_j, meet off $S \cup \{C, D\}$ in precisely twelve points. Therefore,

XIII. *Suppose* $t_0 = 23$; *let* A *and* B *be the two* 8-*points,* C *and* D *the two* 6-*points,* S *the set of the four* 2-*points on* AB. *Then there are precisely twelve* 4-*points, namely the points in the set*

$$X = \{CQ \cap DR \mid Q, R \in S,\ Q \neq R\}. \ \square$$

3. THE CASE $t_8 = 1$

Throughout this section we denote by K a complete 12-arc in π_{12}, such that $t_8 = 1$. By (1.1),

$$t_0 = 28 - t_6, \quad t_2 = 104 + 3t_6, \quad t_4 = 24 - 3t_6. \tag{3.1}$$

XIV. *The* 6-*points form a* t_6-*arc* H.

Proof. Let ℓ be a non-tangent; by $(1.4)_2$, $u_6 \leq 4$. Moreover, if $u_6 = 4$, then $u_j = 0$ for $j \neq 0,6$. Denote by P the unique 8-point. Then $P \notin \ell$ and the eight tangents on P meet ℓ in points of positive index, a contradiction. If $u_6 = 3$, then again $P \notin \ell$. Indeed, by eq. $(1.4)_2$, $4u_8 \leq 12 - 3u_6$. Therefore ℓ has at most six points of positive index, a contradiction. Thus, ℓ has at most two 6-points.

Next, assume that t is a tangent. By (1.5), $v_6 \leq 2$. □

XV. *If* $t_8 = 1$, *then* $t_6 \leq 6$. (*Proposition* XIX *yields a better result.*)

Proof. Suppose Q is any 6-point. Then rhere are at least five tangents through Q which are not on the 8-point P. By (1.5), according to the possible values for v_6 ($v_6 = 1$ or $v_6 = 2$), we have $v_4 = 3$ or $v_4 = 1$. In any case, a tangent t, on Q and not on P, contains a 4-point; therefore, $t_4 \geq 5$. The statement follows from (3.1). □

XVI. *Let* P *be the unique* 8-*point and* H *the set of all* 6-*points. If* $t_6 \neq 2$, *then* $H \cup \{P\}$ *is a* (t_6+1)-*arc.*

Proof. In case $t_6 \geq 3$, we prove that on any line on P one 6-point lies at most. Let A be a 6-point. If PA is a tangent, then III implies that A is the unique 6-point on PA. If PA is non-tangent, by (1.4), $u_6 \leq 2$. Suppose $u_6 = 2$, then (1.4) imply

$$u_0 \geq 8. \tag{3.2}$$

Let B be a 6-point not on PA. The six tangents through B must meet PA in points of positive index. This contradicts (3.2). □

We were not able to prove XVI when $t_6 = 2$. Thus, we investigate the case in which P and the two 6-points A, B lie on a line, say ℓ. Firstly, ℓ is non-tangent (by III). Next, from (1.4) either

$$u_0 = 9,\quad u_2 = 0,\quad u_4 = 1,\quad u_6 = 2,\quad u_8 = 1, \text{ or} \tag{3.3}$$

$$u_0 = 8,\quad u_2 = 2,\quad u_4 = 0,\quad u_6 = 2,\quad u_8 = 1 \tag{3.4}$$

follow.

By (1.5)

A tangent t without 6-points and 8-points has exactly five 4-points, the remaining ones being of index 2. (3.5)

If (3.3) holds, then ℓ has four points of positive index. Denote by C the 4-point on ℓ. By (3.5), the union of the tangents through C contains 17 4-points. From (3.1) $t_4 = 18$ follows; thus another 4-point exists, say Q. The tangents on Q meet ℓ in four points of positive index other than C, a contradiction. Therefore, (3.4) must hold. Denote by C and D the two 2-points on ℓ, and by S the set of the 18 4-points. If $Q \in S$, then the tangents through Q intersect ℓ in points of positive index. Consequently, at least one tangent through Q contains either C or D. By (3.5), on the union of the tangents on C (or D) there are ten 4-points; hence, there are exactly two points $U,V \in S$ such that UC,UD,VC,VD are tangents. We claim that

The tangents on U (or V) meet ℓ in A,B,C,D. If $Q \in S-\{U,V\}$, then PQ is a tangent. (3.6)

If t is a tangent through A (or B), then by (1.5) t has three 4-points. Thus S is contained in the union of the six tangents through A (or B) and the first statement follows. Next, by III, on any of the eight tangents on P there lie two 4-points. Therefore, the remaining two 4-points must coincide with U and V.

Next, we investigate, in more detail, the (t_6+1)-arc

$$M = \{P \in \pi_q \mid P \text{ of index at least } 6\}.$$

Unless otherwise stated, $t_6 \geq 3$ up to the end of this section.

XVII. *Let* A *be a* 6-*point. If* $t_6 > 3$, *then there is no* 4-*point on* PA.

Proof. Suppose that PA has a 4-point B. Then PA is non-tangent and (1.4) hold. Since the tangents on any 6-point C ($\neq$ A) intersect PA in points of positive index, eqs. (1.4) have the unique solution $u_0 = 7$, $u_2 = 3$, $u_4 = 1$, $u_6 = 1$, $u_8 = 1$. There are exactly six points of positive index on PA. Therefore,

$$\text{If C is any 6-point, then AC is a tangent.} \tag{3.7}$$

Notice that, by (1.5),

If t is a tangent through two 6-points, then t has exactly one 4-point. If on a tangent s there is just one 6-point and no 8-point, then s has exactly three 4-points. (3.8)

On A there are t_6-1 tangents, each of them joining it with a 6-point. None of the remaining $7-t_6$ tangents contains a point ($\neq$ A) of index 8 or 6. Since B is a 4-point that does not lie on a tangent through A, (3.8) implies $t_4 \geq (t_6-1)+3(7-t_6)+1 = 21-2t_6$. Eq. $(3.1)_3$ gives $t_6 \leq 3$, a contradiction. □

If t is a tangent on P without 6-points, then by III t has two 4-points. If any tangent on P has no 6-point, then $t_4 \geq 16$ and $(3.1)_3$ yields $t_6 \leq 2$. Since, by assumption, $t_6 \geq 3$,

A 6-point A exists such that PA is a tangent. (3.9)

Let A and B be two 6-points. Then, either AB is a tangent, in which case AB contains exactly one 4-point, or it is a non-tangent and contains no 4-point. (3.10)

The first statement is a consequence of (3.8). If AB is non-tangent, then $P \notin AB$ (by XVI) and the tangents on P meet AB in points of positive index. Hence, eqs. (1.4) have the unique solution $u_0 = 5$, $u_2 = 6$, $u_4 = 0$, $u_6 = 2$, $u_8 = 0$.

Now assume $t_6 > 3$, and consider a 6-point A. If t is a tangent on A with neither 6-points (other than A) nor 8-points, then, by (3.8), t has three 4-points. If any tangent on A has no other 6-point, then $t_4 \geq 15$ as at least five of such tangents are not on P. Thus eq. $(3.1)_3$ gives $t_6 \leq 3$. Hence,

If $t_6 > 3$ and A is a 6-point, then a 6-point B exists such that AB is a tangent. (3.11)

XVIII. *If* $t_6 = 6$, *then each secant to* M *is a tangent to* K.

Proof. By (3.1), $t_4 = 6$. Fix any 6-point A, and denote by x the number of tangents to K through A each of which contains another 6-point. Thus, $x \leq 5$. By (3.8), on the union of these tangents there are x 4-points. On A there are at least 5-x tangents to K having no point of index 8 or 6 other than A. By (3.8),

$$6 = t_4 \geq x+3(5-x) = 15-2x \Rightarrow x = 5.$$

Therefore, all lines which join A with any other 6-point are tangent to K. Now suppose PA is non-tangent. Thus through A there exists a tangent t to K, on which A is the unique point of index ≥ 4. By (3.8), t has three 4-points. Since on the union of the above considered tangents to K on A there are five 4-points, we get $t_4 \geq 8$, a contradiction. □

XIX. *If* $t_8 = 1$, *then* $t_6 \leq 5$.

Proof. Suppose $t_6 = 6$. As a consequence of the proof to XVIII,

A tangent joining a 4-point with a 6-point contains another 6-point, the remaining ones being of index 2. (3.12)

Indeed, if t is a tangent through a 6-point A, then either t contains P, in which case it has no 4-point, or t has another 6-point and precisely one 4-point. Now fix a 6-point A. There are five tangents through it and not through P, each of them has one 4-point. There are exactly two tangents on P containing 4-points; namely, those tangent to M too. Each of such tangents has two 4-points (by XVIII and III). Therefore, a 4-point Q exists, such that AQ is tangent and PQ is not. Suppose that a tangent t exists through Q, such that on t there is no point of index > 4. By (3.5), t has five 4-points. There are two tangents on P containing 4-points. Each of them has at least one 4-point not on t. Hence, $t_4 \geq 7$, a contradiction. Consequently, since (3.12) holds, any tangent through Q has exactly two 6-points. Therefore, on the union of the tangents on Q there are eight 6-points. This contradiction and XV prove the statement. □

4. THE CASE $t_8 = 0$

In this section K denotes a complete 12-arc in π_{12} with $t_8 = 0$. By (1.1),

$$t_0 = 31-t_6, \quad t_2 = 96+3t_6, \quad t_4 = 30-3t_6. \tag{4.1}$$

Under these assumptions, by (1.5),

Suppose that t is a tangent through a 6-point A. Then either t has exactly two 6-points and one 4-point, the remaining ones being of index 2, or t has one 6-point (A) and three 4-points, the remaining points being of index 2. (4.2)

XX. *If* $t_8 = 0$, *then* $t_6 \leq 8$.

Proof. By (4.2), on any tangent through a 6-point A there is at least one 4-point. Therefore, $t_4 \geq 6$. The statement follows (take into account $(4.1)_3$). □

XXI. *Denote by* H *the set of all* 6-*points. If* $t_6 \geq 7$, *then* H *is an incomplete* t_6-*arc.*

Proof. Assume that A, B and C are three collinear 6-points. Let x be the number of tangents through A containing no further 6-point. Since AB is non-tangent and the 6-points, distinct from A, B, C, number t_6-3, we have $x \geq 6-(t_6-3) = 9-t_6$. Thus, on the tangents through A there are $3x+(6-x) = 2x+6 \geq 24-2t_6$ 4-points. By $(4.1)_3$, $24-2t_6 \leq 30-3t_6 \Rightarrow t_6 \leq 6$. To prove that H is incomplete, we distinguish two cases, according to the value of t_6.

$t_6 = 8$. Let A be a 6-point. By (4.2), on the tangents through A there are $3x+(6-x) = 6+2x$ 4-points. By (4.1), $t_4 = 6$; hence, $x = 0$, i.e. any tangent on A has another 6-point.

Notice that the number of tangents to K having two 6-points equals the half of the cardinality of the set

$$\{(P,Q) \mid P,Q \in H \text{ and } PQ \text{ is tangent to } K\}.$$

Indeed, let t be the tangent to K through $P,Q \in H$. With t the two ordered pairs (P,Q) and (Q,P) are associated. Since there are eight possible choices for P and just six for Q, the tangents to K number $8 \cdot 6/2 = 24$ which are also secant to H. Consequently, they exhaust all tangents to K. Hence,

Any tangent to K is secant to H. (4.3)

As a consequence of (4.3), there are precisely four secants to H which are non-tangent to K.

Let ℓ be one of the six tangents to H through an arbitrary 6-point, say A. By (4.3), ℓ is non-tangent to K. Hence, by (1.4), $u_0 = 3$, $u_2 = 9$, $u_6 = 1$. Denote by S the set of those 0-points which lie on the tangents to H through A; thus,

$|S| = 18$. Let r be a secant to H which is non-tangent to K. There are four such lines. By (1.4) with $u_4 = 0$, we get $u_0 = 5$, $u_2 = 6$, $u_6 = 2$. If $A \in r$, then $r \cap S = \emptyset$. On the other hand, if $A \notin r$, then $|r \cap S| \leq 5$. Since there are precisely three lines which are secant to H, non-tangent to K and not on A, a point $P \in S$ exists, such that there is no secant to H through it. Hence, H is incomplete.

$t_6 = 7$. First of all, we prove that

There exist at most three lines which are secant to H and non-tangent to K. (4.4)

Fix a point $A \in H$, and denote by x the number of lines through A and tangent to both H and K. By (4.2), on the union of the tangents to K on A there are 6+2x 4-points. Eqs. (4.1) imply $t_4 = 9$; thus, $x \leq 1$. Obviously, x is also the number of those lines on A which are secant to H and non-tangent to K. Therefore, through any point of H at most one line passes secant to H and non-tangent to K. Hence, (4.4) follows. As a consequence of (4.4), a point Q exists in H, such that any secant to H on Q is a tangent to K, and vice-versa. Let r be a tangent to H at Q. There are seven such tangents. Since r is non-tangent to K, by (1.4), $u_2+2u_4 = 9 \Rightarrow u_2+u_4 \leq u_2+2u_4 = 9$ which, together with eq. $(1.4)_1$, implies $u_0 \geq 3$.

Next, denote by S' the set of those 0-points lying on the tangents to H at Q. Thus, $|S'| \geq 21$. Assume a secant, say p, exists to H, which is non-tangent to K. Take $X \in p \cap H$, then $X \neq Q$. Five of the six tangents to K on X are secant to H, whereas the remaining one is tangent to H. Hence on the union of the tangents to K through X there are eight 4-points. Since $t_4 = 9$, there is at most one 4-point on p. Therefore, by (1.4), either $u_0 = 5$ or $u_0 = 6$, which implies $|p \cap S'| \leq 6$. Thus, the same argument as in case $t_6 = 8$ yields the existence of a point $P \in S$ such that $H \cup \{P\}$ is an arc. □

5. CONCLUSION

XXII. *Equations* (1.1) *have at most* 17 *solutions with a geometrical meaning* (i.e. *associated with a complete* 12-*arc in* π_{12}).

Proof. The statement follows from props. VI, VIII, X, XIX, and XX. □

REFERENCES

[1] Hirschfeld, J.W.P., Projective geometry over finite fields, Clarendon Press, Oxford 1979.

[2] Lam, C.W.H., Thiel, L., Swiercz, S., Mc Kay, J., The non-existence of ovals in a projective plane of order 10, Discrete Math. 45 (1983), 319-321.

[3] Martin, G.E., On arcs in a finite projective plane, Canad. J. Math. 19 (1967), 376-393.

[4] Tallini, G., Sui q-archi di un piano lineare finito di caratteristica p=2, Rend. Accad. Lincei (8) 23 (1957), 242-245.

[5] Tallini, G., Lezioni di Geometria III, anno acc. 1983-84, Dip. Mat. Univ. Roma "La Sapienza", 1984.

[6] Tallini, G., Sui q-archi completi di un piano proiettivo non desarguesiano di ordine q pari, Quaderni Sem. Geom. Comb. 54, Marzo 1985, Dip. Mat. Univ. Roma "La Sapienza".

[7] Zanella, C., Sui 12-archi completi di un piano proiettivo di ordine 12, Quaderni Sem. Geom. Comb. 58, Dicembre 1985, Dip. Mat. Univ. Roma "La Sapienza".

Annals of Discrete Mathematics 37 (1988) 493–496

BLOCK DESIGNS ADMITTING FLAG TRANSITIVE GROUPS OF AUTOMORPHISMS

Paul-Hermann Zieschang

Mathematisches Seminar der Universität Kiel
Olshausenstraße 40
D-2300 Kiel 1
West Germany

In recent years, a large number of papers (e. g. [2], [3], [5], [9], [10]) has been investigating finite block designs with $\lambda = 1$ admitting flag transitive groups of automorphisms. Most of these papers consider additional assumptions which restrict either the structure of the design or that of the acting group. In the following we will discuss finite block designs with $\lambda = 1$ admitting a flag transitive group G of automorphisms such that

(*) each two point stabilizer G_{xy} acts trivially on the unique block determined by x and y.

The following proposition is the main result of [9].

PROPOSITION 1. Let D be a block design with $\lambda = 1$ and point set X and let G be a flag transitive group of automorphisms of D that satisfies (*).

Let $x \in X$ and let A be a non trivial normal p-subgroup of G_x.

Then each one point stabilizer G_r contains exactly one conjugate A^r of A and the family of fixed point sets $\{Fix(A^r \cap G_s) : r,s \in X, r \neq s\}$ forms a block design with $\lambda = 1$ on X on which G acts as a group of automorphisms.

Proposition 1 is needed to prove the following result of [9].

THEOREM 2. Let D be a block design with $\lambda = 1$ and point set X and let G be a flag transitive group of automorphisms of D that satisfies (*).

Let $x \in X$ and assume that G_x has an abelian normal subgroup which is not semiregular on $X - \{x\}$.

Then, for some integer $n \geq 3$ and some prime power q, we have $PSL(n,q) \leq G \leq P\Gamma L(n,q)$.

Theorem 2 generalizes Theorem A of O'Nan's paper [6]. It is crucial in the proof of the following theorem [10].

THEOREM 3. Let D be a block design with $\lambda = 1$ and point set X and let G be a flag transitive group of automorphisms of D that satisfies (*).

Let $x \in X$ and assume that G_x has a normal subgroup which is a T.I. set in G and not semiregular on $X - \{x\}$.

Then the conclusion of Theorem 2 holds.

Theorem 3 is a generalization of Theorem A of [7].

The following result [8] is similar to Bender's theorem [1]. Its proof uses complex character theory and a lot of results due to Hiller [4].

THEOREM 4. Let D be a block design with $\lambda = 1$ and point set X and let G be a flag transitive group of automorphisms of D that satisfies (*).

Assume that the one point stabilizers of G have odd order and that the two point stabilizers are cyclic.

Then either G is solvable or G acts doubly transitively on X.

REFERENCES

[1] H. Bender, Endliche zweifach transitive Permutationsgruppen, deren Involutionen keine Fixpunkte haben, Math. Z. 104 (1968), 175 - 204.

[2] F. Buekenhout, A. Delandtsheer, J. Doyen, Finite linear spaces with flag-transitive and locally primitive groups. To appear.

[3] A. R. Camina, Groups acting flag-transitively on designs, Arch. Math. 32 (1979), 424 - 430.

[4] W. Hiller, 7/4-transitive Permutationsgruppen, Dissertation, Math. Sem. der Univ. Kiel, 1980.

[5] W. M. Kantor, Primitive permutation groups of odd degree, and an application to finite projective planes, J. Algebra 106 (1987), 15 - 45.

[6] M. O'Nan, A characterization of $L_n(q)$ as a permutation group, Math. Z. 127 (1972), 426 - 439.

[7] M. O'Nan, Normal structure of the one-point stabilizer of a doubly-transitive permutation group. I., Trans. Amer. Math. Soc. 214 (1975), 1 - 42.

[8] P.-H. Zieschang, Über eine Klasse von Permutationsgruppen, Dissertation, Math. Sem. der Univ. Kiel, 1983.

[9] P.-H. Zieschang, Fahnentransitive Automorphismengruppen von Blockplänen, Geom. Dedicata 18 (1985), 173 - 180.

[10] P.-H. Zieschang, A theorem of O'Nan for finite linear spaces. To appear.

Annals of Discrete Mathematics 37 (1988) 497–502

AN INDEPENDENCE THEOREM ON THE CONDITIONS FOR INCIDENCE LOOPS

Elena ZIZIOLI

Dipartimento di Matematica, Università Cattolica
via Trieste 17
25121 Brescia, Italia *

1. INTRODUCTION

A *fibered loop* $(P,\mathscr{F},\cdot)$ is a loop $(P,\cdot)$ together with a set $\mathscr{F}$ $(|\mathscr{F}|>1)$ of subloops of $(P,\cdot)$ such that for any $a\in P\setminus\{1\}$ there is exactly an element $F\in\mathscr{F}$ with $a\in F$.

In the recent paper [5] ,we have stated algebraic conditions in order to provide a fibered loop $(P,\mathscr{F},\cdot)$ with a geometric structure either of an incidence loop or of an incidence space with parallelism or of a kinematic loop.

In this note we prove (Theorem 1) that the conditions stated there are orderly independent.

2. PRELIMINARY NOTIONS

First of all we recall some definitions and properties.

An *incidence space* is a pair $(P,\mathscr{R})$ where:

i) $\mathscr{R}$ is a set of subset of P

ii) $\forall\, a,b\in P$, $a\neq b$ $\exists_1 R\in\mathscr{R}$ such that $a,b\in R$

iii) $\forall\, R\in\mathscr{R}$ $|R|\geq 2$.

In an incidence space $(P,\mathscr{R})$ let us assume the set P to be a loop $(P,\cdot)$ and let us consider the following properties:

(L) $\forall\, a\in P$ the map $a_l : P\to P;\ x\to ax$ is an element of $Aut(P,\mathscr{R})$;

(R) $\forall\, a\in P$ the map $a_r : P\to P;\ x\to xa$ is an element of $Aut(P,\mathscr{R})$;

(K) $\forall\, X\in\mathscr{R}$ with $1\in X$ $(X,\cdot)$ is a subloop of $(P,\cdot)$;

then we shall say (cf. [2] §7) that $(P,\mathscr{R},\cdot)$ is:

an *incidence loop* if (L) holds ;

a *2-sided incidence loop* if (L) and (R) hold ;

a *kinematic loop* if (L),(R) and (K) hold .

A *fibered loop* $(P,\mathscr{F},\cdot)$ is a loop $(P,\cdot)$ together with a subset $\mathscr{F}$ of the powerset of P such that:

i) $\forall\, F\in\mathscr{F}$ $(F,\cdot)$ is a subloop of $(P,\cdot)$ and $|\mathscr{F}|>1$,

ii) $\forall\, a\in P\setminus\{1\}$ $\exists_1 F\in\mathscr{F}$ such that $a\in F$;

the set $\mathscr{F}$ will be called a *fibration* of $(P,\cdot)$.

By definition a kinematic loop $(P,\mathscr{R},\cdot)$ is always a fibered loop with respect to the fibration $\mathscr{F}:=\{X\in\mathscr{R}\mid 1\in X\}$.

An incidence space $(P,\mathscr{R})$ is called an *incidence space with parallelism*

* Research supported by the Italian Ministry of Education (M.P.I. 40%) .

$(P, \mathcal{R}, \|)$ if there is a relation "$\|$" on $\mathcal{R}$ such that:
(E) "$\|$" is an equivalence relation
(P) $\forall x\varepsilon P$, $\forall Y\varepsilon \mathcal{R}$ $\exists_1 Z\varepsilon \mathcal{R}$ with $x\varepsilon Z \| Y$.

In an incidence space with parallelism $(P, \mathcal{R}, \|)$ we denote by $\{x\|Y\}$ the unique element $Z\varepsilon\mathcal{R}$ such that $x\varepsilon Z$ and $Z\|Y$.

From now on let $(P, \mathcal{F}, \cdot)$ be a fibered loop and let $\mathcal{R}:=\{aX \mid a\varepsilon P , X\varepsilon \mathcal{F}\}$.
Moreover we define the relation "$\|$" on $\mathcal{R}$ in the following way:
$aX \| bY \;:\Leftrightarrow \exists a',b'\varepsilon P , \exists Z\varepsilon \mathcal{R}$ such that $aX = a'Z$, $bY = b'Z$.
In [5] we introduced the following conditions for a fibered loop $(P, \mathcal{F}, \cdot)$:
(F1) $\{aX \mid a\varepsilon P , X\varepsilon \mathcal{F}\} = \{Yb \mid b\varepsilon P , Y\varepsilon \mathcal{F}\}$;
(F2) $\forall A\varepsilon \mathcal{F}$ $\forall x,y\varepsilon P$ $\exists B\varepsilon \mathcal{F}$: $x(yA) = (xy)B$;
(F2') $\forall A\varepsilon \mathcal{F}$ $\forall x,y\varepsilon P$ $\exists C\varepsilon \mathcal{F}$: $(Ax)y = C(xy)$;
(F3) $\forall A\varepsilon \mathcal{F}$, $\forall a\varepsilon A$, $\forall x\varepsilon P$ $(xa)A = xA$;
and we proved (see [5] (9)c),(16),(10)):
(1) $(P, \mathcal{R}, \cdot)$ is an incidence loop if and only if (F2) holds.
(2) $(P, \mathcal{R}, \cdot)$ is a kinematic loop if and only if (F1,2,2') hold.
(3) $(P, \mathcal{R}, \|)$ is an incidence space with parallelism if and only if (F3) holds.

We now consider for the loop $(P, \mathcal{F}, \cdot)$ the following condition as well:
(F3') $\forall A\varepsilon \mathcal{F}$, $\forall a\varepsilon A$, $\forall x\varepsilon P$ $A(ax) = Ax$;
and we define on $\mathcal{R}$ the following relation (right-parallelism):
$aX \|_r bY \;:\Leftrightarrow \exists a',b'\varepsilon P, \exists Z\varepsilon \mathcal{F}$ such that $aX = Za'$, $bY = Zb'$,
then we have:
(4) $(P, \mathcal{R}, \|, \|_r)$ is an incidence space with parallelism if for the fibered loop $(P, \mathcal{F}, \cdot)$ the conditions (F1) and (F3') hold.
Proof.
By (F1) $\mathcal{R} = \{Yb \mid Y\varepsilon \mathcal{F} , b\varepsilon P\}$, hence the proof follows from prop. (10) of [5] .

Therefore from a fibered loop $(P, \mathcal{F}, \cdot)$ verifying (F1,3,3') we can obtain an incidence space with two parallelisms $(P, \mathcal{R}, \|, \|_r)$.
A *double space* (cf. [3]) is an incidence space $(P, \mathcal{R})$ with two parallelisms $\|$, $\|_r$ such that the following condition holds:
(D) let $A,B\varepsilon\mathcal{R}$, let $a\varepsilon A$, $b\varepsilon B$ and let $B':=\{a\|B\}$, $A':=\{b\|_r A\}$ then
$A'\cap B' \neq \emptyset$ if $A\cap B \neq \emptyset$.
But our fibered loop $(P, \mathcal{F}, \cdot)$ fulfilling (F1,3,3') is not a double space because the condition (D) is only proved if $(P, \cdot)$ is a fibered group such that for any $a\varepsilon P$ $a\mathcal{F}a^{-1} \subset \mathcal{F}$ (cf. [3]). Otherwise, for a proper fibered loop (i.e. a loop which is not a group), we have only the specialization of (D) where $1 \varepsilon A\cap B$.

Let $\{p_1,p_2,\ldots,p_n\}$ be a set of conditions in a given structure. We recall the following definitions:
i) the conditions $p_1,p_2,\ldots p_n$ are *absolutely independent* if for any $i\varepsilon\{1,2,\ldots,n\}$ the condition p_i cannot be proved by means of the remaining $p_1,\ldots,p_{i-1},p_{i+1},\ldots,p_n$,
ii) the conditions $p_1,p_2,\ldots p_n$ are *orderly independent* if for any $i\varepsilon\{1,2,\ldots,n\}$ the condition p_i cannot be proved by means of the previous $p_1,\ldots,p_{i-1}$ ones.
The definition i) implies ii). Viceversa B.Levi proved [1]) that if $\{p_1,p_2,\ldots$

$.,p_n\}$ are orderly independent it is always possible to find a set of conditions $\{p'_1,p'_2,\dots,p'_n\}$ absolutely independent and equivalent to the set $\{p_1,p_2,\dots,p_n\}$.

In §4 we shall prove the following main Theorem:

Theorem 1 . There exist proper fibered loops $(P,\mathscr{F},\cdot)$ fulfilling the conditions (F1,2,3,3');moreover these conditions are orderly independent.

3. FIBERED LOOPS AND TACTICAL CONFIGURATIONS

In order to prove Theorem 1 we recall the following notions (cf. [1]):

A *tactical configuration* is a pair $(P,\mathscr{B})$ where:

i) $\mathscr{B}$ is a subset of the powerset of P,

ii) $\forall X,Y\in\mathscr{B}\quad \exists k\in\mathbb{N} : |X| = |Y| = k$,

iii) $\exists\, r\in\mathbb{N}$ such that for all $p\in P \quad r = |\{X\in\mathscr{B} \mid p\in X\}|$.

We now assume that the fibered loop $(P,\mathscr{F},\cdot)$ has all the fibers of the same cardinality and we introduce the following numerical constants.Let $v:=|P|$, $n:=|\mathscr{F}|$, $k:=|X|$ for $X\in\mathscr{F}$. Then $v = n(k-1)+1$ and $n\geq 3$.
If we set $\mathscr{L}:=\{a_l \mid a\in P\}$ and $\mathscr{B}:=\langle\mathscr{L}\rangle\,\mathscr{F} = \{\gamma X \mid \gamma\in\langle\mathscr{L}\rangle,\ X\in\mathscr{F}\}$ then $\mathscr{R}\subset\mathscr{B}$ and $(P,\mathscr{B})$ is a tactical configuration in which any two points $a,b\in P$ are contained in at least one block $B\in\mathscr{B}$.
Here the group $\langle\mathscr{L}\rangle$ generated by all the permutations $a_l\in\mathscr{L}$ turns out to be a subgroup of the automorphism group $Aut(P,\mathscr{B})$ of the tactical configuration.
Further if $b:=|\mathscr{B}|$ then $v\cdot r = b\cdot k$.
If the fibered loop fulfils (F3) then by (3) $(P,\mathscr{R},\|)$ is an incidence space with parallelism and in this case $v/k\in\mathbb{N}$ and it is the cardinal number of any class of parallel lines.

4. PROOF OF THEOREM 1

In this section we shall prove Theorem 1 in four steps by showing at each step that there exist fibered loops fulfilling the first n but not the last (4-n) conditions of Theorem 1 (with n = 1,2,3,4) .

I Case n = 1 .

We consider the set P = { 1,a,b,c,ab,bc,ca,(ab)c,(bc)a,(ca)b} with the composition law defined by the following multiplication table:

1	a	b	c	ab	bc	ca	(ab)c	(bc)a	(ca)b
a	1	ab	ca	b	(bc)a	c	(ca)b	bc	(ab)c
b	ab	1	bc	a	c	(ca)b	(bc)a	(ab)c	ca
c	ca	bc	1	(ab)c	b	a	ab	(ca)b	(bc)a
ab	b	a	(ab)c	1	(ca)b	(bc)a	c	ca	bc
bc	(bc)a	c	b	(ca)b	1	(ab)c	ca	a	ab
ca	c	(ca)b	a	(bc)a	(ab)c	1	bc	ab	b
(ab)c	(ca)b	(bc)a	ab	c	ca	bc	1	b	a
(bc)a	bc	(ab)c	(ca)b	ca	a	ab	b	1	c
(ca)b	(ab)c	ca	(bc)a	bc	ab	b	a	c	1

(tab.1)

It is easy to verify that (P,·) is a commutative loop of order 10 with the fibration $\mathscr{F}$:= {{1,a,b,ab} , {1,c,(bc)a,(ca)b} , {1,bc,ca,(ab)c}} and k = 4. Then this loop fulfils (F1) but since 10/4 and (10·3)/4 are not integers (P,$\mathscr{F}$,·) fulfils neither (F2) nor (F3) nor (F3').
In a subsequent note we shall introduce a constructive method to obtain finite fibered loops (in general non commutative) with all the fibers of the same cardinality such that only the condition (F1) holds for k>2.

II Case n = 2 .

To obtain fibered loops verifying (F1) and (F2) but not the remaining conditions, it is enough to choose k = 2 and v odd.The smallest example is the set P = {1,a,b,c,d} where the multiplication table is given by:

1	a	b	c	d
a	1	c	d	b
b	d	1	a	c
c	b	d	1	a
d	c	a	b	1

(tab. 2)

Moreover we have also other examples with k>2 :e.g. the remark 6 of theorem (18) in [5] gives us a class of examples of kinematic loops (P,$\mathscr{F}$,·) where the fibration $\mathscr{F}$ violates the laws (F3) and (F3') and where (P,$\mathscr{R}$) is an affine space of dimension 5.

III Case n = 3 .

Now we show that there are examples of fibered loops which are not groups but where (F1),(F2) and (F3) are valid.
Since v/k has to be integer and since any loop with four elements is a group the smallest order for such a possible loop is at least 6 .The following multiplication table gives us such an example:

1	a	b	c	d	e
a	1	d	e	c	b
b	c	1	d	e	a
c	b	e	1	a	d
d	e	a	b	1	c
e	d	c	a	b	1

(tab. 3)

Here $\mathscr{F}$:= {{1,a} ,{1,b} ,{1,c} , {1,d} ,{1,e}} and one has (xy)y = x for all x,y$\in$P.Therefore (xy)Y = {xy,x} = xY if Y = {1,y} .
But {1,a}·(ab) = {d,c} $\neq$ {b,d} = {1,a}·b , hence (F3') is not valid.

IV Case n = 4 .

A large class of examples of proper fibered loops with (F1,2,3,3') can be obtained in the following way.By remark 3 of §6 in [5] (see also [2]) we know that there are proper fibered loops (P,$\mathscr{F}$,·) for any $k = p^n$, $p \neq 2$ a prime, $n \in \mathbb{N}$ and $|P| = k^7$ such that all the four conditions of theorem 1 hold and (P,$\mathscr{R}$) is an affine space of dimension 7.

This achieves the proof of Theorem 1.

We may investigate the case of the smallest cardinality for the loop $(P,\mathcal{F},\cdot)$ fulfilling the conditions (F1,2,3,3').Hence let $k = 2$ (i.e. $x^2 = 1$ for all $x\in P$).Then we have:

(5) Let $(P,\mathcal{F},\cdot)$ be a fibered loop with $k = 2$.Then

a) (F3) is equivalent to : $\forall\, a,b\in P \quad (ab)b = a$;

b) (F3') is equivalent to : $\forall\, a,b\in P \quad a(ab) = b$;

c) the conditions (F3) and (F3') hold if and only if for all $a,b\in P\setminus\{1\}$,$a \neq b$ the subloop $\langle a,b\rangle$ is a Klein 4-group;

d) the conditions (F3) and (F3') imply that $(P,\cdot)$ is a commutative loop.

Proof.

c) Let us assume (F3) and (F3') and let $c:=ab$ .Then $ac = a(ab) = b$ and $bc = (ac)c = a$, $ba = b(bc) = c$ and so $ab = c = ba$.
Conversely for any $a,b\in P\setminus\{1\}$, $a \neq b$, we have $\langle a,b\rangle = \{1,a,b,ab\}$ where $a(ab) = b$ and $(ab)b = a$ that is,by a) and b) , (F3') and (F3) .

With these observations one recognizes that the smallest proper loop with (F3) and (F3') and $k = 2$ (hence with (F1) and (F2)) is the set $P = \{1,a,b,c,ab,bc,ca,(ab)c,(bc)a,(ca)b\}$ with the fibration $\mathcal{F}^*:=\{\ \{1,x\}\ |\ x\in P\setminus\{1\}\ \}$ and the multiplication table of (tab.1). Hence $|P| = 10$.

If we set now $k = 3$,the smallest cardinality for a fibered loop $(P,\mathcal{F},\cdot)$ with (F1,2,3,3') can be only 9.But for $v = 9$ the corresponding incidence loop with two parallelisms $(P,\mathcal{R},\ \|_\ell,\|_r)$ can be only the affine plane (cf. [4]) and the loop $(P,\cdot)$ is the abelian group $\mathbb{Z}_3 \times \mathbb{Z}_3$.
Therefore we have the smallest proper fibered loop $(P,\mathcal{F},\cdot)$ fulfilling (F1,2,3,3') for $k = 2$ and $|P| = 10$.

Remark.

In this note we did not consider the independence of the condition (F2') .
Actually we don't possess examples of loops verifying the conditions (F1,2) but not (F2') .
In this context we can only prove the absolute independence of (F3) and (F3') from (F1),(F2),(F2'),as the following examples show.
Let $P:= \{1,a,a^2,b,b^2,c,c^2,d,d^2\}$.We consider the composition law "$\cdot$" defined on P by the following multiplication table:

1	a	a^2	b	b^2	c	c^2	d	d^2
a	a^2	1	c	d	b	d^2	b^2	c^2
a^2	1	a	c^2	d^2	b^2	d	b	c
b	c^2	d	b^2	1	d^2	a	c	a^2
b^2	c	d^2	1	b	d	a^2	c^2	a
c	d^2	b^2	d	a	c^2	1	a^2	b
c^2	d	b	d^2	a^2	1	c	a	b^2
d	b	c^2	a	c	a^2	b^2	d^2	1
d^2	b^2	c	a^2	c^2	a	b	1	d

(tab. 4)

Then $(P,\cdot)$ is a proper loop with a fibration $\mathcal{F}:=\{\ \{1,a,a^2\}\ ,\ \{1,b,b^2\}\ ,\ \{1,c,c^2\}\ ,\ \{1,d,d^2\}\}$ of 4 subgroup of order 3.It is easy to verify that in this fibered loop (F1),(F2),(F2') and (F3') are not valid but (F3) is fulfilled.
If we consider now the new operation "$\circ$" defined on the same set P as follows:
for all $x,y\in P \qquad x\circ y:= yx$

the multiplication table of $(P,\circ)$ is given by transposing the matrix of (tab.4); hence $(P,\mathcal{F},\circ)$ is a fibered loop and in this case only the condition (F3') is fulfilled.

NOTES AND REFERENCES

[1]) Atti dell'Accademia delle Scienze - Torino. (2) 54 (1904) p.284

[1] DEMBOWSKI,H.P. : Finite geometries . Berlin-Heidelberg-New York 1968

[2] KARZEL,H. and KIST,G.P. : Kinematic algebras and their geometries. Rings and geometry (Kaya et al. eds.) NATO ASI series - C 160 (1985), 437-509

[3] ______,KROLL,H.J. and SÖRENSEN,K. : Invariante Gruppenpartitionen und Doppelräume. J. reine angew. Math. 262/263 (1973) 153-157

[4] KARZEL,H. and PIEPER,I. : Bericht über geschlitzte Inzidenzgruppen. Jahresber. Deutsch. Math. Verein 72 (1970) 70-114

[5] ZIZIOLI,E. : Fibered incidence loops and kinematic loops. J.of Geometry (to appear)